Quantitative Particle Physics
Cargèse 1992

NATO ASI Series

Advanced Science Institutes Series

A series presenting the results of activities sponsored by the NATO Science Committee, which aims at the dissemination of advanced scientific and technological knowledge, with a view to strengthening links between scientific communities.

The series is published by an international board of publishers in conjunction with the NATO Scientific Affairs Division

A Life Sciences	Plenum Publishing Corporation
B Physics	New York and London
C Mathematical and Physical Sciences	Kluwer Academic Publishers
D Behavioral and Social Sciences	Dordrecht, Boston, and London
E Applied Sciences	
F Computer and Systems Sciences	Springer-Verlag
G Ecological Sciences	Berlin, Heidelberg, New York, London,
H Cell Biology	Paris, Tokyo, Hong Kong, and Barcelona
I Global Environmental Change	

Recent Volumes in this Series

Volume 306 — Ionization of Solids by Heavy Particles
    edited by Raúl A. Baragiola

Volume 307 — Negative Differential Resistance and Instabilities in 2-D
    Semiconductors
    edited by N. Balkan, B. K. Ridley, and A. J. Vickers

Volume 308 — Photonic Band Gaps and Localization
    edited by C. M. Soukoulis

Volume 309 — Magnetism and Structure in Systems of Reduced Dimension
    edited by Robin F. C. Farrow, Bernard Dieny, Markus Donath,
    Albert Fert, and B. D. Hermsmeier

Volume 310 — Integrable Quantum Field Theories
    edited by L. Bonora, G. Mussardo, A. Schwimmer, L. Girardello,
    and M. Martellini

Volume 311 — Quantitative Particle Physics: *Cargèse 1992*
    edited by Maurice Lévy, Jean-Louis Basdevant, Maurice Jacob,
    Jean Iliopoulos, Raymond Gastmans, and Jean-Marc Gérard

Volume 312 — Future Directions of Nonlinear Dynamics in Physical and Biological
    Systems
    edited by P. L. Christiansen, J. C. Eilbeck, and R. D. Parmentier

Series B: Physics

Quantitative Particle Physics

Cargèse 1992

Edited by

Maurice Lévy and Jean-Louis Basdevant

Université Pierre et Marie Curie
Paris, France

Maurice Jacob

CERN
Geneva, Switzerland

Jean Iliopoulos

Ecole Normale Supérieure
Paris, France

Raymond Gastmans

Katholieke Universiteit Leuven
Leuven, Belgium

and

Jean-Marc Gérard

Université Catholique de Louvain
Louvain-la-Neuve, Belgium

Springer Science+Business Media, LLC

Proceedings of a NATO Advanced Study Institute on
Quantitative Particle Physics,
held July 20–August 1, 1992,
in Cargèse, France

NATO-PCO-DATA BASE

The electronic index to the NATO ASI Series provides full bibliographical references (with keywords and/or abstracts) to more than 30,000 contributions from international scientists published in all sections of the NATO ASI Series. Access to the NATO-PCO-DATA BASE is possible in two ways:

—via online FILE 128 (NATO-PCO-DATA BASE) hosted by ESRIN, Via Galileo Galilei, I-00044 Frascati, Italy

---

Library of Congress Cataloging in Publication Data

Lévy, Maurice, 1922–
  Quantitative particle physics: Cargèse 1992 / edited by Maurice Lévy . . . [et al.]
     p.   cm.—(NATO ASI series. Series B, Physics; vol. 311)
  Published in cooperation with NATO Scientific Affairs Division.
  "Proceedings of a NATO Advanced Study Institute on Quantitative Particle Physics, held July 20–August 1, 1992, in Cargèse, France"—T.p. verso.
  Includes bibliographical references and index.
  ISBN 978-1-4613-6279-1   ISBN 978-1-4615-2944-6 (eBook)
  DOI 10.1007/978-1-4615-2944-6
  1. Electroweak interactions—Congresses. 2. Standard model (Nuclear physics)—Congresses. 3. Mathematical physics—Congresses. I. North Atlantic Treaty Organization. Scientific Affairs Division. II. NATO Advanced Study Institute on Quantitative Particle Physics (1992: Cargèse, France) III. Series: NATO ASI series. Series B, Physics; v. 331.
QC794.8.E44Q35   1993   93-22901
539.7'544—dc20   CIP

ISBN 978-1-4613-6279-1

© 1993 Springer Springer Science+Business Media New York
Originally published by Plenum Press, New York in 1993
Softcover reprint of the hardcover 1st edition 1993

PREVIOUS CARGÈSE SYMPOSIA PUBLISHED IN THE
NATO ASI SERIES B: PHYSICS

Volume 261   Z° PHYSICS: *Cargèse 1990*
              edited by Maurice Lévy, Jean-Louis Basdevant, Maurice Jacob, David Speiser,
              Jacques Weyers, and Raymond Gastmans

Volume 223   PARTICLE PHYSICS: *Cargèse 1989*
              edited by Maurice Lévy, Jean-Louis Basdevant, Maurice Jacob, David Speiser,
              Jacques Weyers, and Raymond Gastmans

Volume 173   PARTICLE PHYSICS: *Cargèse 1987*
              edited by Maurice Lévy, Jean-Louis Basdevant, Maurice Jacob, David Speiser,
              Jacques Weyers, and Raymond Gastmans

Volume 156   GRAVITATION IN ASTROPHYSICS: *Cargèse 1986*
              edited by B. Carter and J. B. Hartle

Volume 150   PARTICLE PHYSICS: *Cargèse 1985*
              edited by Maurice Lévy, Jean-Louis Basdevant, Maurice Jacob, David Speiser,
              Jacques Weyers, and Raymond Gastmans

Volume 130   HEAVY ION COLLISIONS: *Cargèse 1984*
              edited by P. Bonche, Maurice Lévy, Phillippe Quentin, and Dominique Vautherin

Volume 126   PERSPECTIVES IN PARTICLES AND FIELDS: *Cargèse 1983*
              edited by Maurice Lévy, Jean-Louis Basdevant, David Speiser, Jacques Weyers,
              Maurice Jacob, and Raymond Gastmans

Volume 85   FUNDAMENTAL INTERACTIONS: *Cargèse 1981*
              edited by Maurice Lévy, Jean-Louis Basdevant, David Speiser, Jacques Weyers,
              Maurice Jacob, and Raymond Gastmans

Volume 72   PHASE TRANSITIONS: *Cargèse 1980*
              edited by Maurice Lévy, Jean-Claude Le Guillou, and Jean Zinn-Justin

Volume 61   QUARKS AND LEPTONS: *Cargèse 1979*
              edited by Maurice Lévy, Jean-Louis Basdevant, David Speiser, Jacques Weyers,
              Raymond Gastmans, and Maurice Jacob

Volume 44   RECENT DEVELOPMENTS IN GRAVITATION: *Cargèse 1978*
              edited by Maurice Lévy and S. Deser

Volume 39   HADRON STRUCTURE AND LEPTON–HADRON INTERACTIONS: *Cargèse 1977*
              edited by Maurice Lévy, Jean-Louis Basdevant, David Speiser, Jacques Weyers,
              Raymond Gastmans, and Jean Zinn-Justin

Volume 26   NEW DEVELOPMENTS IN QUANTUM FIELD THEORY AND STATISTICAL
              MECHANICS: *Cargèse 1976*
              edited by Maurice Lévy and Pronob Mitter

Volume 13   WEAK AND ELECTROMAGNETIC INTERACTIONS AT HIGH ENERGIES:
              Cargèse 1975 (Parts A and B)
              edited by Maurice Lévy, Jean-Louis Basdevant, David Speiser, and Raymond Gastmans

PREFACE

The 1992 Cargèse Summer Institute on Quantitative Particle Physics was organized by the Université Pierre et Marie Curie, Paris (M. Lévy and J.-L. Basdevant), CERN (M. Jacob), the Ecole Normale Supérieure, Paris (J. Iliopoulos), the Katholieke Universiteit te Leuven (R. Gastmans) and the Université Catholique de Louvain (J-M. Gérard), which, since 1975, have joined their efforts and worked in common. It was the tenth Summer Institute on High Energy Physics organized jointly at Cargèse by these three universities.

The 1992 School centered on quantitative tests of the Standard Model for electroweak and strong interactions. First, Professor T.D. Lee reviewed the fascinating history of weak interactions. Professor R. Barbieri then discussed the implications of the latest experimental results of LEP presented by Professor Foà. Professor G. Ecker described in detail the interplay between electroweak and strong interactions at low energy. Professors K. Berkelman and J-M. Gérard stressed the necessity to study the effects of CP-violation in both B- and K- physics. The first results of the HERA machine were presented by Professor G. Wolf, while Professor M. Shochet reviewed heavy flavor physics in hadron collider experiments. Recent non-accelerator experiments in neutrino physics were presented by Professor B. Barish. Finally, Professor M. Turner reviewed Cosmology after COBE.

We owe many thanks to all those who have made this Summer Institute possible!

Special thanks are due to the Scientific Committee of NATO and its President for a generous grant. We are also very grateful for the financial contribution given by the Centre National de la Recherche Scientifique and the Institut National de Physique Nucléaire et de Physique des Particules (IN^2P^3).

We also want to thank Ms. M.-F. Hanseler for her efficient organizational help, Mr. and Ms. Ariano and Ms. Cassegrain for their kind assistance in all material matters of the school, and, last but not least, the people from Cargèse for their hospitality.

Mostly, however, we would like to thank all the lecturers and participants: their commitment to the school was the real basis for its success.

M. Lévy	J. Iliopoulos
J.-L. Basdevant	R. Gastmans
M. Jacob	J-M. Gérard

CONTENTS

History of Weak Interactions 1
    T.D. Lee

Physics at LEP 29
    L. Foà

Electroweak Precision Tests : What do we Learn? 71
    R. Barbieri

Chiral Perturbation Theory 101
    G. Ecker

CP- and T- Violations in the Standard Model 149
    J-M. Gérard

Heavy Flavor Physics 173
    K. Berkelman

Physics at HERA 211
    G. Wolf

Physics with Hadron Colliders 259
    M.J. Shochet

Neutrino Physics 301
    B.C. Barish

Inflation After COBE: Lectures on Inflationary Cosmology 341
    M.S. Turner

Oblique Electroweak Parameters and Additional Fermion Generations 399
    G. Bhattacharyya

Electroweak Symmetry Breaking from the Top 407
    N. Evans

Higgs Mass Limits from Electroweak Baryogenesis 413
    S. Myint

Carbon 60 425
    T.D. Lee

Index 433

HISTORY OF WEAK INTERACTIONS

T. D. Lee

Columbia University, New York, N.Y. 10027

We may separate the history of weak interactions into three periods:

1. Classical Period, 1898-1949
2. Transition Period, 1949-1956
3. Modern Period, 1956-

1. CLASSICAL PERIOD (β DECAY)

In 1898 Lord Rutherford[1] discovered that the so-called Becquerel ray actually consisted of two distinct components: one that is readily absorbed, which he called alpha radiation, and another of a more penetrating character, which he called beta radiation. With that began the history of the weak interaction. Then, in 1900[2], the Curies measured the electric charge of the β particle and found it to be negative.

Sometimes when we think of physics in those old days, we have the impression that life was more leisurely and physicists worked under less pressure. Actually, from the very start the road of discovery was tortuous and the competition intense. A letter written in 1902 by Rutherford (then 32) to his mother expressed the spirit of research at that time[3,4]:

"I have to keep going, as there are always people on my track. I have to publish my present work as rapidly as possible in order to keep in the race. The best sprinters in this road of investigation are Becquerel and the Curies... ."

Most of the people in this room can appreciate these words. Rutherford's predicament is still very much shared by us to this day. Soon many fast runners came: Hahn, Meitner, Wilson, Von Baeyer, Chadwick, Ellis, Bohr, Pauli, Fermi and many others.

In preparing this lecture, I was reminded once more of how relatively recent these early developments are. We know that to reach where we are today took more than 90 years and a large cast of illustrious physicists. I recall that when Lise Meitner came to

Quantitative Particle Physics, Edited by
M. Lévy *et al.*, Plenum Press, New York, 1993

New York in the mid '60s, I had lunch with her at a restaurant near Columbia. Later K.K. Darrow joined us. Meitner said, "It's wonderful to see young people." To appreciate this comment, you must realize that Darrow was one of the earliest members of the American Physical Society and at that lunch he was over 70. But Lise Meitner was near 90.

I was quite surprised when Meitner told me that she started her first postdoc job in theory with Ludwig Boltzmann. Now, Boltzmann was a contemporary of Maxwell. That shows us how recent even the "ancient" period of our profession is.

After Boltzmann's unfortunate death in 1906, Meitner had to find another job. She said she was grateful that Planck invited her to Berlin. However, upon arrival she found that because she was a woman she could only work at Planck's institute in the basement, and only through the servant's entrance. At that time, Otto Hahn had just set up his laboratory in an old carpenter shop nearby. Lise Meitner decided to join him and to become an experimentalist. For the next thirty years, their joint work shaped the course of modern physics.

In 1906, Hahn and Meitner published a paper[5] stating that the β ray carries a unique energy. Their evidence was that the absorption curve of a β ray shows an exponential decrease along its path when passing through matter, like the α ray.

Then W. Wilson[6], in 1909, said "no", the β ray does not have a unique energy. By observing the absorption curve through matter of an electron of unique energy, Wilson found electrons to exhibit totally different behavior from the α particle; the absorption curve of a unique energy electron is not exponential. Consequently, Wilson deduced that the apparent exponential behavior of the absorption curve of β decay implies that the β does not have a unique energy, the same experimental observation on β but with a totally opposite conclusion.

In 1910, Von Baeyer and Hahn[7] applied a magnetic field to the β ray; they found the β to have several discrete energies. In this way, they also reconciled the conclusion of W. Wilson. Then, in 1914, Chadwick[8] said "no". The β energy spans a continuous spectrum, instead of discrete values. The discrete energy observed by Von Baeyer and Hahn was due to the secondary electron from a nuclear γ transition, with the γ energy absorbed by the atomic electron. In this process, the discrete energy refers to the nuclear γ emission.

Then came World War I and scientific progress was arrested. In 1922, Lise Meitner[9] again argued that the β energy should be discrete, like α and γ. The apparent continuum manifestation is due to the subsequent electrostatic interaction between β and the nucleus. From 1922 to 1927, through a series of careful measurements, Ellis[10] again said "no" to Meitner's hypothesis. The β energy is indeed continuous. Furthermore, Ellis proved that the maximum β energy equals the difference of the initial and final nuclear energy.

There would then appear a missing energy. This was incorporated by Niels Bohr[11], who proposed the hypothesis of non-conservation of energy.

Very soon, Pauli said "no" to Bohr's proposition. Pauli[12] suggested that in the β decay energy is conserved, but accompanying the β particle there is always emission of a neutral particle of extremely small mass and with almost no interaction with matter. Since such a weakly interacting neutral particle is not detected, there appears to be an apparent nonconservation of energy.

Fermi[13] then followed with his celebrated theory of β decay. This in turn stimulated further investigation of the spectrum shape of the β decay, which did not agree with Fermi's theoretical prediction. This led Konopinski and Uhlenbeck[14] to introduce the derivative coupling. The confusion was only cleared up completely after World War II, in 1949, by Wu and Albert[15], signalling the end of one era and the beginning of a new one.

2. CLASSICAL PERIOD (OTHER WEAK INTERACTIONS)

When I began my graduate study of physics at the University of Chicago, in 1946, the pion was not known. Fermi and Teller[16] had just completed their theoretical analysis of the important experiment of Conversi, Pancini and Piccioni[17]. I attended a seminar by Fermi on this work. He cut right through the complex slowing-down process of the mesotron, the capture rate versus the decay rate, and arrived at the conclusion that the mesotron could not possibly be the carrier of strong forces hypothesized by Yukawa. Fermi's lectures were always superb, but that one to me, a young man not yet twenty and fresh from China, was absolutely electrifying. I left the lecture with the impression that, instead of Yukawa's idea, perhaps one should accept Heisenberg's suggestion[18] that the origin of strong forces could be due to higher-order processes of β interaction. As was known, these were highly singular.

At that time, the β interaction was thought to be reasonably well understood. Fermi's original vector-coupling form,

$$G\left(\psi_n^\dagger\,\gamma_4\gamma_\lambda\,\psi_p\right)\left(\psi_4^\dagger\,\gamma_4\gamma_\lambda\gamma_5\,\psi_\nu\right)$$

was, after all, too simple; to conform to reality, it should be extended to include a Gamow-Teller term. Fermi told me that his interaction was modelled after the electromagnetic forces between charged particles, and his coupling G was inspired by Newton's constant. His paper was, however, rejected by *Nature* for being unrealistic. It was published later in Italy, and then in *Zeitschrift für Physik*[13]. Fermi wrote his γ matrices explicitly in terms of their matrix elements. His lepton current differs from his hadron current by a γ_5 factor; of course the presence of this γ_5 factor has no physical significance. Nevertheless, it is curious why Fermi should choose this particular expression, which resembles the V-A interaction, but with parity conservation. Unfortunately, by 1956, when I noticed this, it was too late to ask Fermi.

A year later, the discovery of the pion through its decay sequence $\pi \to \mu \to e$ by Lattes, Muirhead, Occhialini and Powell[19] dramatically confirmed the original idea of Yukawa. The fact that the higher-order β interaction is singular is not a good argument that it should simply become the strong force.

In January 1949 my fellow student, Jack Steinberger, submitted a paper[20] to *The Physical Review* in which he established that the μ meson disintegrates into three light particles, one electron and two neutrinos. This made it look very much like any other β decay, and stimulated Rosenbluth, Yang and myself to launch a systematic investigation. Are there other interactions, besides β decay, that could be described by Fermi's theory?

We found that if μ decay and μ capture were described by a four-fermion interaction similar to β decay, all their coupling constants appeared to be of the same magnitude. This was the beginning of the *Universal Fermi Interaction.* We then went on to speculate that, in analogy with electromagnetic forces, the basic weak interaction could be carried by a universal coupling through an *intermediate heavy boson*[21], which I later called $W^{\pm}$ for weak. Naturally I went to my thesis adviser, Enrico Fermi, and told him of our discoveries. Fermi was extremely encouraging. With his usual deep insight, he immediately recognized the further implications beyond our results. He put forward the problem that if this is to be the universal interaction, then there must be reasons why some pairs of fermions should have such interactions, and some pairs should not. For example, why does

$$p \not\to e^+ + \gamma,$$
$$p \not\to e^+ + 2\nu ?$$

A few days later, he told us that he had found the answer; he then proceeded to assign various sets of numbers, $+ 1$, $- 1$ and 0, to each of these particles. This was the first time to my knowledge that both the laws of baryon-number conservation and of lepton-number conservation were formulated together to give selection rules. However, at that time (1948), my own reaction to such a scheme was to be quite unimpressed: surely, I thought, it is not necessary to explain why $p \not\to e + \gamma$, since everyone knows that the identity of a particle is never changed through the emission and absorption of a photon; as for the weak interaction, why should one bother to introduce a long list of mysterious numbers, when all one needs is to say that only three combinations $(\bar{n}p)$, $(\bar{e}\nu)$ and $(\bar{\mu}\nu)$ can have interactions with the intermediate boson. (Little did I expect that soon there would be many other pairs joining these three.)

Most discoveries in physics are made because the time is ripe. If one person does not make it, then almost inevitably another person will do it at about the same time. In looking back, what we did in establishing the Universal Fermi Interaction was a discovery of exactly this nature. This is clear, since the same universal Fermi coupling observations were made independently by at least three other groups, Klein[22], Puppi[23], and Tiomno and Wheeler[24], all at about the same time. Yet Fermi's thinking was of a more profound nature. Unfortunately for physics, his proposal was never published. The full significance of these conservation laws was not realized until years later. While this might be the first time that I failed to recognize a great idea in physics when it was presented to me, unfortunately it did not turn out to be the last.

Thus, in 1949, there existed a simple theoretical frame based on the Fermi theory, describing the three weak interaction processes:

$$n \to p + e + \nu,$$
$$\mu + p \to n + \nu$$

and

$$\mu \to e + 2\nu.$$

4

So, at the end of the classical period, we moved from the observation of β decay to the discovery of the Universal Fermi Interaction. '

3. TRANSITION PERIOD (1949-1956)

Beginning in 1949, extensive work was done on the shape of the electron spectrum from μ decay. From the analysis of L. Michel[25], it was found that this distribution is given by

$$N(x) = x^2 \left\{ \left(2 - \frac{4}{3}\rho \right) - \left(2 - \frac{16}{9}\rho \right) x \right\}$$

where

$$x = (\text{momentum of } e)/(\text{maximum } e \quad \text{momentum}),$$

and ρ is the well-known Michel parameter, which can be any real number between 0 and 1, and measures the height of the end point at $x = 1$, as shown in Figure 1.

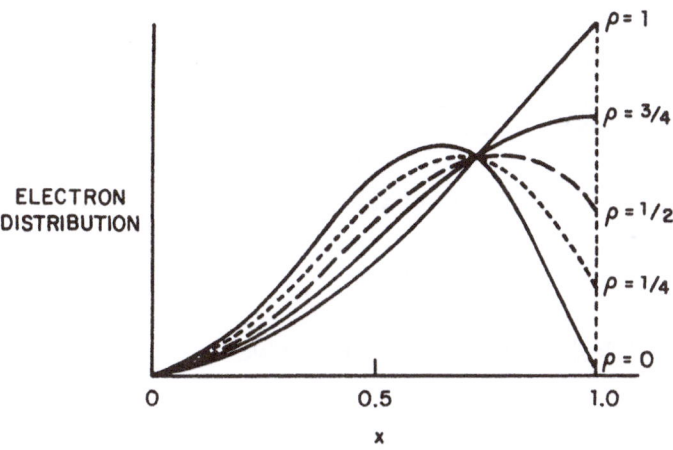

Figure 1. The ρ parameter in μ decay.

It is instructive to plot the experimental value of ρ against the year when the measurement was made. As shown in Figure 2, historically it began with $\rho = 0$ in 1949, at the beginning of the transition period. Then it slowly drifted upwards; only after the end of the transition period with the theoretical prediction in 1957 did it gradually become $\rho = \frac{3}{4}$. Yet, it is remarkable that at no time did the 'new' experimental value lie outside the error bars of the preceding one.

5

In the same period (1949-56), a large amount of effort was also made on β decay experiments. By then, the Konopinski-Uhlenbeck interaction was definitely ruled out. The absence of the Fierz interference term[26] in the spectrum shows that the β interaction must be either V, A or S, T. These two possibilities were further resolved by a series of $\beta - \nu$ angular correlation experiments. In an allowed transition, the distribution for the angle θ between β and ν is given by (neglecting the Fierz term)

$$[1 + \lambda(P/E)_e \cos\theta] \, d\cos\theta,$$

where the subscript e refers to the momentum P and energy E of the electron. For a $\Delta J = 1$ transition,

$$\lambda = \begin{cases} +\frac{1}{3}, & \text{for } T \\ -\frac{1}{3}, & \text{for } A. \end{cases}$$

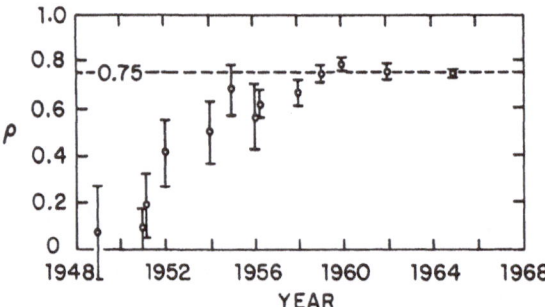

Figure 2. Variation of the ρ parameter over time.

The experiment on ^{6}He decay by Rustad and Ruby gave[27]

$$\lambda = +0.34 \pm 0.09,$$

which seemed to establish unquestionably that the β decay interaction should be S, T with perhaps some unknown admixture of an additional pseudoscalar interaction.

I was quite depressed at that time because, with this new result, the theoretical idea of the intermediate boson seemed to be definitely ruled out. It is bad enough to assume the possiblity of two kinds of intermediate bosons of different spin-parity, one for the Fermi coupling and the other for the Gamow-Teller coupling. However, a tensor interaction with no derivative coupling simply cannot be transmitted by a spin-2 boson, since the former is described by an antisymmetric tensor and the latter by a symmetric one.

4. NEW HORIZON IN THE TRANSITION PERIOD

We now come to the $\theta - \tau$ puzzle.

During a recent physics graduate qualifying examination in a well-known American university, one of the questions was on the $\theta - \tau$ problem. Most of the students were puzzled over what θ was; of course they all knew that τ is the heavy lepton, the charged member of the third generation. So much for the history of physics.

In the early 1950s, θ referred to the meson which decays into 2π, whereas τ referred to the one decaying into 3π:

$$\theta \rightarrow 2\pi$$

and

$$\tau \rightarrow 3\pi.$$

The spin-parity of θ is clearly $0^+, 1^-, 2^+$, etc. As early as 1953, Dalitz[28] had already pointed out that the spin-parity of τ can be analyzed through his Dalitz plot and, by 1954, the then-existing data were more consistent with the assignment 0^- than 1^-. Although both mesons were known to have comparable masses (within ~ 20 MeV), there was, at that time, nothing too extraordinary about this situation. The masses of θ and τ are very near three times the pion mass, the phase space available for the θ decay is much bigger than that for τ decay; therefore one expects the θ decay rate to be much faster. However, when accurate lifetime measurements were made in 1955, it turned out that θ and τ have the same lifetime (within a few percent, which was the experimental accuracy). This, together with a statistically much more significant Dalitz plot of τ decay, presented a very puzzling picture indeed. The spin-parity of τ was determined to be 0^-; therefore it appeared to be definitely a different particle from θ. Yet, these two particles seemed to have the same lifetime, and also the same mass. This was the $\theta - \tau$ puzzle.

My first efforts were all on the wrong track. In the summer of 1955, Jay Orear and I proposed[29] a scheme to explain the $\theta - \tau$ puzzle within the bounds of conventional theory. We suggested a cascade mechanism, which turned out to be incorrect.

The idea that parity is perhaps not conserved in the decay of $\theta - \tau$ flickered through my mind. After all, strange particles are by definition strange, so why should they respect parity? The problem was that, after you say parity is not conserved in $\theta - \tau$ decay, then what do you do? Because if parity nonconservation exists only in $\theta - \tau$, *then we already have all the observable facts*, namely the same particle can decay into either 2π or 3π with different parity. I discussed this possibility with Yang, but we were not able to make any progress[30]. So we instead wrote papers on parity doublets, which was another wrong try[31].

5. THE BREAKTHROUGH (1956)

The Rochester meeting on high energy physics was held from April 3 to 7, 1956.

At that time, Steinberger and others were conducting extensive experiments on the production and decay of the hyperons Λ^0 and Σ^- :

$$\pi^- + p \rightarrow \begin{cases} \Lambda^0 + K^0 \\ \Sigma^- + K^+ \end{cases} \tag{1}$$

and

$$\left.\begin{array}{c} \Lambda^0 \\ \Sigma^- \end{array}\right\} \rightarrow \pi + N . \tag{2}$$

The dihedral angle ϕ between the production plane and the decay plane (which will be defined below) is of importance for the determination of the hyperon spin.

Let $\vec{\pi}$, $\vec{\Lambda}$ and $\vec{N}$ be the momenta of π, Λ in process (1) and N in (2), all, say, in the respective center-of-mass systems of the reactions. The normal to the production plane is parallel to $\vec{\pi} \times \vec{\Lambda}$, and that to the decay plane to $\vec{\Lambda} \times \vec{N}$. Hence the dihedral angle ϕ is defined through its cosine:

$$\cos\phi \propto (\vec{\pi} \times \vec{\Lambda}) \cdot (\vec{\Lambda} \times \vec{N}) . \tag{3}$$

Its distribution is

$$D(\phi) = \begin{cases} 1 & \text{if the hyperon-spin is } \frac{1}{2}, \\ 1 + \alpha\cos^2\phi & \text{if the hyperon-spin is } \frac{3}{2}, \end{cases} \tag{4}$$

etc. By this definition, ϕ varies from 0 to π. Furthermore, $D(\phi)$ is identical to $D(\pi - \phi)$. At the Rochester Conference, Jack Steinberger gave a talk and plotted his data on $D(\phi)$ with ϕ varying from 0 to π. However other physicists, W.D. Walker and R.P. Shutt, plotted $D(\phi) + D(\pi - \phi)$; in this way ϕ can only vary from 0 to $\frac{\pi}{2}$. After the conference, Jack came to my office to discuss a letter which he had just received from R. Karplus. In this letter Karplus questioned why Jack did not join the others, since the total number of events was (at that time) quite limited, and a folding of $D(\pi - \phi)$ onto $D(\phi)$ would increase the experimental sensitivity of the spin determination. Jack wanted to know how certain was the relationship that $D(\phi)$ is an even function of $\cos\phi$.

The dihedral angle, as defined by expression (3), has nothing to do with parity, since it is a scalar. In the course of explaining to Jack the $\cos^2\phi$ dependence of $D(\phi)$, I suddenly realized that if one changes the definition of ϕ to be the angle of rotation around the Λ momentum-vector, which is the intersection of these two planes, then the range of ϕ can be extended from 0 to 2π; that is, in place of (3), one defines ϕ through the pseudoscalar

$$\sin\phi \propto (\vec{\pi}_\perp \times \vec{N}_\perp) \cdot \vec{\Lambda}, \tag{5}$$

where $\vec{\pi}_\perp$ and $\vec{N}_\perp$ refer to the components of $\vec{\pi}$ and $\vec{N}$ perpendicular to $\vec{\Lambda}$, as shown in Figure 3. In this case, ϕ can vary from 0 to 2π.

If parity is not conserved in strange particle decays, there could be an asymmetry between events with ϕ from 0 to π and those with ϕ from π to 2π. *This is the missing key!* I was quite excited, and urged Jack to re-analyse his data immediately and test the idea experimentally. This led to the very first experiment on parity nonconservation. Very soon, within a week, Jack and his collaborators (Budde, Chrétien, Leitner, Samios and Schwartz) had their results, and the data were published[32] even before the theoretical paper[33] on parity nonconservation. The odds turned out to be 13 to 3 in Σ^- decay and 7 to 15 in Λ^0 decay (see Figure 4).

Of course, because of the limited statistics, no definitive conclusion on parity violation could be drawn. Nevertheless, had the statistics been ten times more, then with the same kind of ratio one could have made a decisive statement on parity conservation. This showed clearly that parity violation could be tested experimentally provided one measured a pseudoscalar, such as (5).

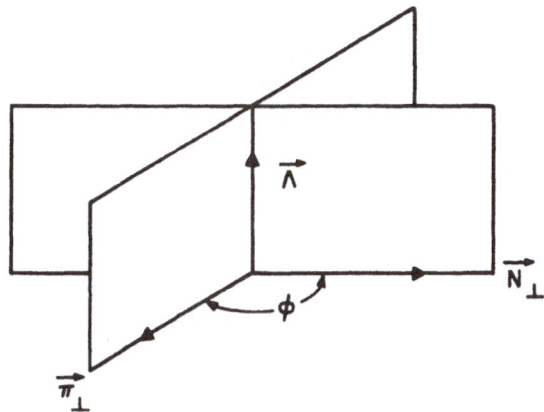

Figure 3. The dihedral angle ϕ between the production plane and the decay plane.

However, on the theoretical side there was still the question of parity conservation in ordinary β decay. In this connection, at the beginning of May, C.N. Yang came to see me and wished to join me in the examination of β decay. This led to our discovery that, in spite of the extensive use of parity in nuclear physics and β decay, there existed no evidence at all of parity conservation in any weak interaction.

Several months later followed the decisive experiments by Wu, Ambler, Hayward, Hoppes and Hudson, at the end of 1956, on β decay[34], and by Garwin, Lederman and Weinrich[35] and by Friedman and Telegdi[36] on $\pi - \mu$ decay. Table 1 lists the important papers on symmetry violation in weak interactions.

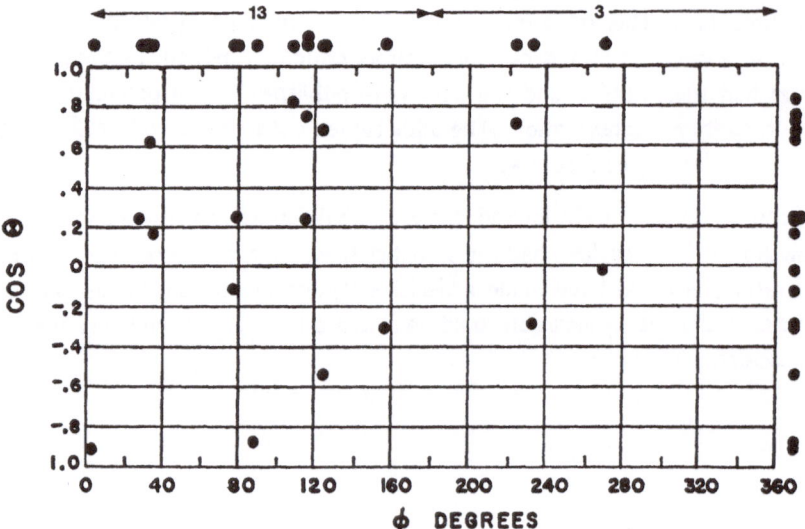

Figure 4a. Angular correlation plot of the cosine of the polar angle against the azimuthal angle ϕ for the reaction $\pi^- + p \to \Sigma^- + K^+$, $\Sigma^- \to \pi^- + n$. The point histograms on the edges of the figure represent the data after integration over the other coordinate. (Taken from Ref. 32)

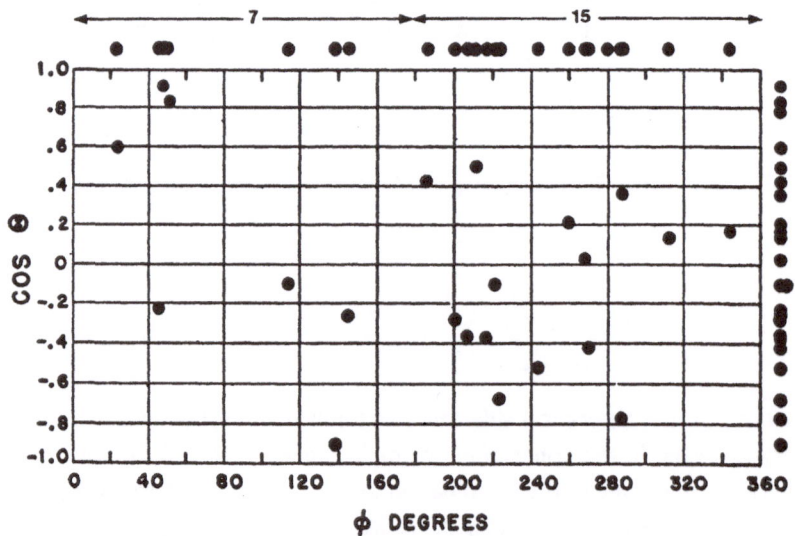

Figure 4b. Angular correlation plot for the reaction $\pi^- + p \to \Lambda^0 + \theta^0$, $\Lambda^0 \to \pi^- + p$. (Taken from Ref. 32)

Table 1

T.D.L. and C.N. Yang,
  "Question of Parity Nonconservation in Weak Interactions,"
    Phys.Rev. **104**, 254 (1956).
  "A Two-component Theory of the Neutrino,"
    Phys.Rev. **105**, 1671 (1957).
T.D.L., R. Oehme and C.N. Yang,
  "Possible Noninvariance of T, C and CP,"
    Phys.Rev. **106**, 340 (1957).
C.S. Wu, E. Ambler, R.W. Hayward, D. Hoppes and R.P. Hudson,
  "Experimental Test of Parity Nonconservation in Beta Decay,"
    Phys.Rev. **105**, 1413 (1957).
R.L. Garwin, L.M. Lederman and M. Weinrich,
  "Nonconservation of P and C in Meson Decays,"
    Phys.Rev. **105**,1415 (1957).
J.I. Friedman and V.L. Telegdi,
  "Parity Nonconservation in $\pi^+ - \mu^+ - e^+$,"
    Phys.Rev. **105**,1681 (1957).
M. Goldhaber, L. Grodzins and A.W. Sunyar,
  "Helicity of Neutrinos,"
    Phys.Rev. **109**, 1015 (1958).
R.E. Marshak and E.C.G. Sudarshan,
  "Chirality and the Universal Fermi Interaction,"
    Phys.Rev. **109**, 1860 (1958).
R. Feynman and M. Gell-Mann,
  "Theory of the Fermi Interaction,"
    Phys.Rev. **109**, 193 (1958).
R. Christenson, J. Cronin, V.L. Fitch and R. Turlay,
  "2π Decay of the K_2° Meson,"
    Phys.Rev.Lett. **13**, 138 (1964).

---

6. SYMMETRY VIOLATIONS

Wu *et al.* investigated the decay of polarized cobalt nuclei, Co^{60}, into electrons. Because these nuclei are polarized they rotate parallel to each other. The experiment consisted of two setups, identical except that the directions of rotation of the initial nuclei were opposite; that is, each was a looking-glass image of the other. The experimenters found, however, that the patterns of the final electron distributions in these two setups are not mirror images of each other. In short, the initial states are mirror images, but the final configurations are not (see Figure 5). This established parity nonconservation, i.e., right-left asymmetry.

The β decay experiment by Wu *et al.* and the $\pi - \mu$ decay experiment by Garwin, Lederman and Weinrich and by Friedman and Telegdi proved not only parity P violation, but also asymmetry under particle-antiparticle conjugation C. This is illustrated in Figure 6. Both the neutrino and the antineutrino possess a spin (angular momentum). For a neutrino, if one aligns one's left thumb parallel to its momentum, then the curling of one's four fingers would always be in the direction of its spin, indicating P violation. Therefore, the spin and momentum direction of a neutrino defines a perfect left-hand screw, whereas the spin and momentum direction of an antineutrino defines a perfect right-hand screw, showing C violation in addition to P violation. This property holds for neutrinos and antineutrinos everywhere, independently of how they are produced.

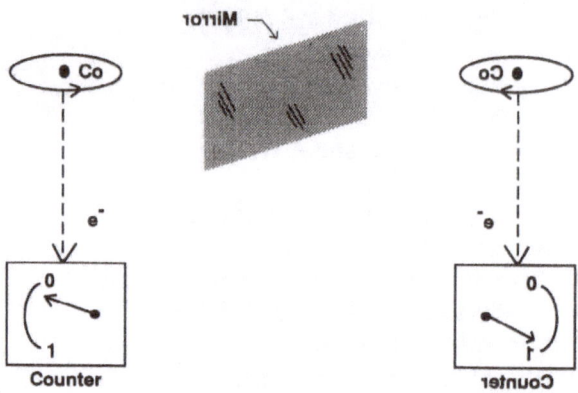

Figure 5. The initial setups of these two experiments on Co^{60} decay are exact mirror images but the final electron distributions are not, as indicated by the different readings on the counters.

In 1964, the experiment by Christenson *et al.*[37] established CP violation. An example of how such a violation can be established experimentally is to examine the decay[38] of K_L^0 into $e^+ \pi^- \nu$ and $e^- \pi^+ \overline{\nu}$. The long-lived neutral kaon K_L^0 is a spherically symmetric particle with zero spin; it carries no electric charge and no electromagnetic form factor of any kind. Yet these two decay modes have different rates

$$\frac{\text{rate}(K_L^0 \rightarrow e^+ + \pi^- + \nu)}{\text{rate}(K_L^0 \rightarrow e^- + \pi^+ + \overline{\nu})} \;=\; 1.00648 \pm 0.00035\,.$$

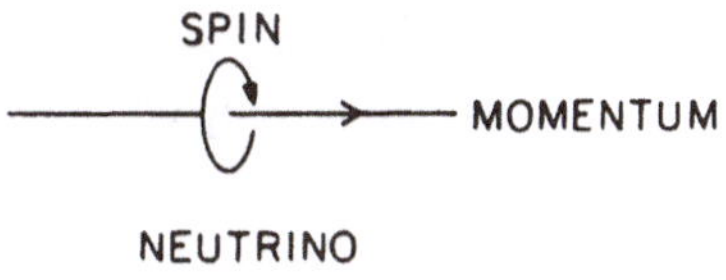

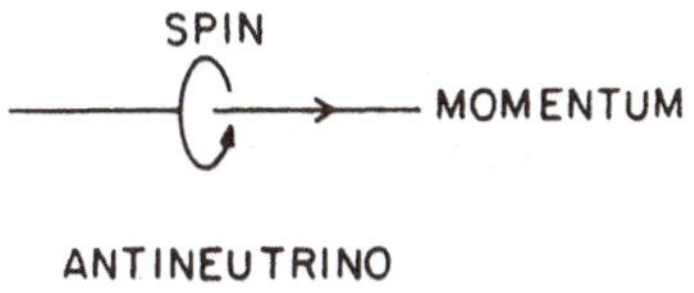

Figure 6. The spin momentum of a neutrino defines a left-hand screw; that of an antineutrino defines a right-hand screw.

Table 2

	Present Status
CPT	Good
P	X
C	X
CP	X
T	X ← without assuming CPT
CT	X
PT	X

Consequently, by using a time-clocking device it is possible to differentiate the positive sign of electricity vs. the negative sign of electricity, showing C violation. Because this rate difference remains true under mirror reflection, CP is also violated. The present status of these discrete symmetries and asymmetries is given in Table 2. Only the combined symmetries CPT remain valid. In other words, only when we interchange

$$\text{particle} \leftrightarrow \text{antiparticle},$$

$$\text{right} \leftrightarrow \text{left},$$

$$\text{past} \leftrightarrow \text{future},$$

do all physical laws appear to be invariant. From CPT symmetry, it follows that CP violation also implies time reversal T asymmetry.

7. TIME REVERSAL

Time reversal symmetry T means that the time-reversed sequence of any motion is also a possible motion. Some of you may think this absurd, since we are all getting older, never younger. So why should we even contemplate that the laws of nature should be time-reversal symmetrical?

In this sense we must distinguish between the evolution of a small system and a large system. Let me give an example. In Figure 7a each circle represents an airport, and a line indicates an air corridor. We assume that between any two of these airports the number of flights going both ways along any route is the same (this property will be referred to as microscopic reversability). Thus a person in Ajaccio can travel to Paris (for this discussion we assume the only air connection from Ajaccio is to Paris), then through Paris to New York or San Francisco. At any point in his travel, he can return to Ajaccio with the same ease.

But suppose that in every airport we were to remove all the signs and flight information, while maintaining exactly the same number of flights, as shown in Figure 7b. A person starting from Ajaccio would still arrive in Paris, since that is the only airport connected to Ajaccio. However, without the signs to guide him, it would be very difficult for him to pick out the return flight to Ajaccio from the many gates in the Paris airport. The plane he gets on may be headed for San Francisco. If, in San Francisco, he then tries another plane again without any guidance, he could perhaps arrive in New York. If he keeps on going this way, his chance of getting back to Ajaccio is very slim indeed. In this example, we see that microscopic reversibility is strictly maintained. When all the airport destination signs and other flight information are given clearly, then macroscopically we also have reversibility. On the other hand, if all such information is withheld, then the whole macroscopic process appears irreversible. Thus, macroscopic irreversibility is not in conflict with microscopic reversibility.

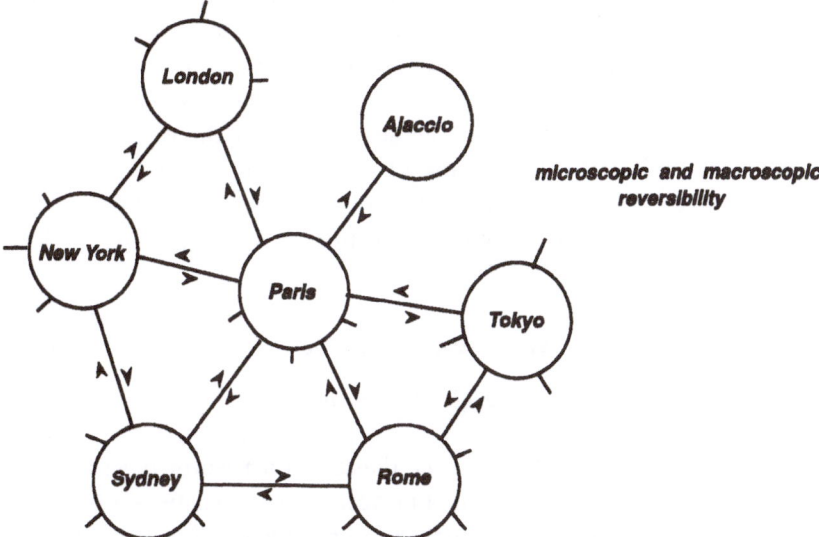

microscopic and macroscopic reversibility

Figure 7a. In this example, microscopic reversibility means an equal number of flights in either direction on every air route. When the names of the airports, the numbers of the gates, and all flight information are known, there is also macroscopic reversibility.

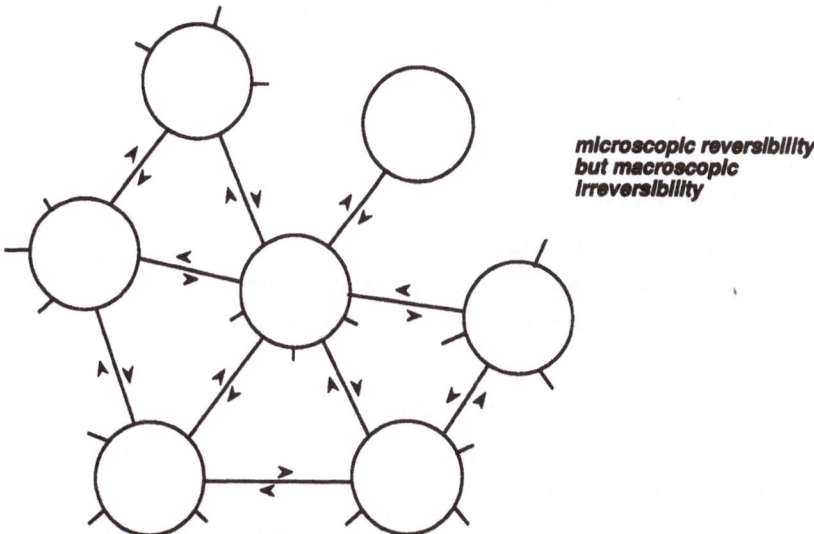

microscopic reversibility but macroscopic irreversibility

Figure 7b. If we maintain the same number of flights in each direction on any route (i.e., microscopic reversibility), but remove destination signs, gate numbers, and all other information, then it is nearly impossible to find our way back (i.e., macroscopic irreversibility).

Time reversal symmetry in physics refers to *microscopic reversiblity* between all molecular, atomic, nuclear and subnuclear reactions. Since none of these molecules, atoms, nuclei or subnuclear particles can be easily marked, any microscopic system in nature would exhibit irreversibility. This result is independent of microscopic reversibility. In any macroscopic process, we have to average over an immense number of unmarked microscopic units of atoms, molecules, and so forth (as in the example of unmarked airports and air routes), and that gives rise to *macroscopic irreversibility*. It is in this statistical sense of ever increasing disorder (entropy) that we define the direction of our macroscopic time flow. We may recall the words from *H.M.S. Pinafore*, by Gilbert and Sullivan,

What never? No, never!

What never? Well, hardly ever!

The existence of the macroscopic time direction then leaves open the important question whether time-reversal symmetry, or microscopic reversibility, is true or not. Since 1964, after the discovery of CP violation by Christenson *et al.*, through a series of remarkable experiments involving the kaons, it was found that the microscopic reversibility is indeed violated. Nature does not seem to respect time-reversal symmetry!

8. PRESENT STATUS

In 1967, the paper of S. Weinberg[39] unified the weak interaction with the electromagnetic force, called the electroweak interaction, which is mediated by the photon γ, the charged intermediate bosons $W^{\pm}$ and their neutral partner Z^0 (these particles were discovered at CERN in 1983 [40]).

At present, our theoretical structure of electroweak, strong and gravitational forces can be summarized as follows:

QCD (strong interaction)

$SU(2) \times U(1)$ **Theory** (electroweak)

General Relativity (gravitation).

However, in order to apply these theories to the real world, we need a set of about 17 parameters, all of unknown origins. Thus, this theoretical edifice cannot be considered complete.

The two outstanding puzzles that confront us today are:

i) **Missing symmetries** - All present theories are based on symmetry, but most symmetry quantum numbers are *not* conserved.

ii) **Unseen quarks** - All hadrons are made of quarks; yet, no individual quark can be seen.

As we shall see, the resolutions of these two puzzles probably are tied to the structure of our physical vacuum.

The puzzle of missing symmetries implies the existence of an entirely new class of fundamental forces, the one that is responsible for symmetry breaking. Of this new force, we know only of its existence, and very little else. Since the masses of all known particles break these symmetries, an understanding of the symmetry-breaking forces will lead to a comprehension of the origin of the masses of all known particles. One of the promising directions is the spontaneous symmetry-breaking mechanism in which one assumes that the physical laws remain symmetric, but the physical vacuum is not. If so, then the solution of this puzzle is closely connected to the structure of the physical vacuum; the excitations of the physical vacuum may lead to the discovery of Higgs-type mesons.

In some textbooks, the second puzzle is often "explained" by using the analogy of the magnet. A magnet has two poles, north and south. Yet, if one breaks a bar magnet in two, each half becomes a complete magnet with two poles. By splitting a magnet open one will never find a single pole (magnetic monopole). However, in our usual description, a magnetic monopole can be considered as either a fictitious object (and therefore unseeable) or a real object but with exceedingly heavy mass beyond our present energy range (and therefore not yet seen). In the case of quarks, we believe them to be real physical objects and of relatively low masses (except the top quark); furthermore, their interaction becomes extremely weak at high energy. If so, why don't we ever see free quarks? This is, then, the real puzzle.

The fact that quarks and gluons cannot be seen individually suggests an entirely new direction in our understanding of particles. Traditionally, any particle is either stable or unstable; the former is represented by a pole on the physical sheet and the latter by a pole on the "second" sheet. But now, we have a new third class: particles that cannot be seen individually. This is illustrated in Table 3, and its solution in Table 4. Following the KLN Theorem,[41] gluons and quarks are indeed discovered by observing high energy jets. The second method, of using entropy change, will be discussed shortly. Before doing that, we may recall what has happened in the past.

At the end of the last century, there were also two physics puzzles:

1. No absolute inertial frame (Michelson-Morley Experiment 1887),
2. Wave-particle duality (Planck's formula 1900).

These two seemingly esoteric problems struck classical physics at its very foundation. The first became the basis for Einstein's theory of relativity and the second laid the foundation for us to construct quantum mechanics. In this century, all the modern scientific and technological developments—*nuclear energy, atomic physics, molecular structure, lasers, x-ray technology, semiconductors, superconductors, supercomputers*—only exist because we have relativity and quantum mechanics. To humanity and to our understanding of nature, these are all-encompassing.

Now, near the end of the twentieth century, we must ask what will be the legacy we give to the next generation in the next century? At present, like the physicists at the end of the 1890's, we are also faced with two profound puzzles. It seems likely that the present two puzzles may bring us as important a change in the development of science and technology in the twenty-first century.

Table 3

Stable particles poles on the "physical" sheet	Unstable particles poles on the "second" sheet	Particles that cannot be seen individually
GRAVITON γ ν e p :	μ π K : n J / ψ Υ : W Z :	GLUONS QUARKS :

Table 4

HOW TO DETECT PARTICLES THAT
CANNOT BE SEEN INDIVIDUALLY?

---

1. JETS $\left(\text{KLN THEOREM ON MASS SINGULARITIES (1964)}\right)$

2. ENTROPY DENSITY CHANGE $\left(\text{VACUUM EXCITATIONS T.D.L. AND G.C.WICK (1974)}\right)$

9. PHYSICAL VACUUM

The current explanation of the quark confinement puzzle is again to invoke the vacuum. We assume the QCD vacuum to be a condensate of gluon pairs and quark-antiquark pairs so that it is a perfect color dia-electric[42] (i.e., color dielectric constant $\kappa = 0$). This is in analogy to the description of a superconductor as a condensate of electron pairs in BCS theory, which results in making the superconductor a perfect dia-magnet (with magnetic susceptibility $\mu = 0$). When we switch from QED to QCD we replace the magnetic field $\vec{H}$ by the color electric field $\vec{E}_{color}$, the superconductor by the QCD vacuum, and the QED vacuum by the interior of the hadron. Just as the magnetic field is expelled outward from the superconductor, the color electric field is pushed into the hadron by the QCD vacuum, and that leads to color confinement, or the formation of bags.[43] This situation is summarized below.

QED superconductivity as a perfect dia-magnet		QCD vacuum as a perfect color dia-electric
$\vec{H}$	$\longleftrightarrow$	$\vec{E}_{color}$
$\mu_{inside} = 0$	$\longleftrightarrow$	$\kappa_{vacuum} = 0$
$\mu_{vacuum} = 1$	$\longleftrightarrow$	$\kappa_{inside} = 1$
inside	$\longleftrightarrow$	outside
outside	$\longleftrightarrow$	inside

In the resolution of both puzzles, missing symmetry and quark confinement, the system of elementary particles no longer forms a self-contained unit. The microscopic particle physics depends on the coherent properties of the macroscopic world, represented by the appropriate operator averages in the physical vacuum state.

If we pause and think about it, this represents a rather startling conclusion, contrary to the traditional view of particle physics which holds that the microscopic world can be regarded as an isolated system. To a very good approximation it is separate and uninfluenced by the macroscopic world at large. Now, however, we need these vacuum averages; they are due to some long-range ordering in the state vector. At present our theoretical technique for handling such coherent effects is far from being developed. Each of these vacuum averages appears as an independent parameter, and that accounts for the large number of constants needed in the present theoretical formulation.

On the experimental side, there has hardly been any direct investigation of these coherent phenomena. This is because hitherto in most high-energy experiments, the higher the energy the smaller has been the spatial region we are able to examine. In order to explore physics in this fundamental area, relativistic heavy ion collisions offer an important new direction.[44] The basic idea is to collide heavy ions, say gold on gold, at an ultra-relativistic region. Before the collision, the vacuum between the ions is the usual physical vacuum; at a sufficiently high energy, after the collision almost all of the baryon numbers are in the forward and backward regions (in the center-of-mass system). The central region is essentially free of baryons and, for a short duration, it is of a much

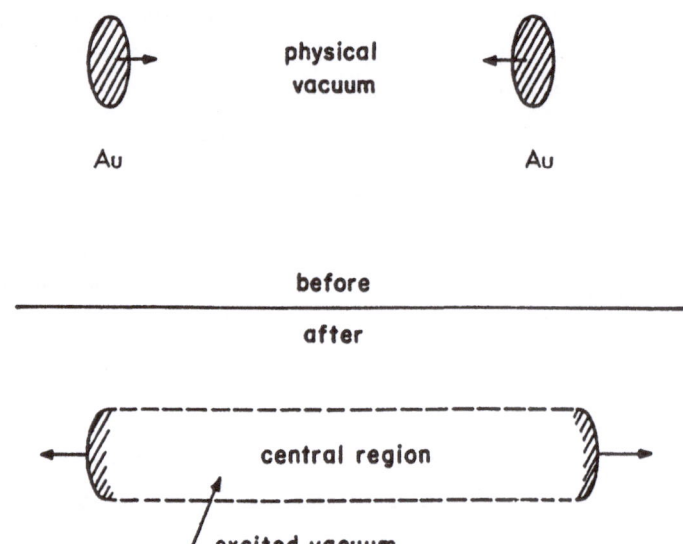

Figure 8. Vacuum excitation through relativistic heavy ion collisions.

higher energy density than the physical vacuum. Therefore, the central region represents the excited vacuum (Figure 8).

As we shall see, we need RHIC, the 100 GeV × 100 GeV (per nucleon) relativistic heavy ion collider at the Brookhaven National Laboratory, to explore the QCD vacuum.

10. PHASE DIAGRAM OF THE QCD VACUUM

A normal nucleus of baryon number A has an average radius $r_A \approx 1.2A^{\frac{1}{3}} fm$ and an average energy density

$$\mathcal{E}_A \approx \frac{m_A}{(4\pi/3)r_A^3} \approx 130 MeV/fm^3 .$$

Each of the A nucleons inside the nucleus can be viewed as a smaller bag which contains three relativistic quarks inside; the nucleon radius is $r_N \approx 0.8 fm$ and its average energy density is

$$\mathcal{E}_N \approx \frac{m_N}{(4\pi/3)r_N^3} \approx 440 MeV/fm^3.$$

Consequently, even without any sophisticated theoretical analysis we expect the QCD phase diagram to be of the form given by Figure 9.

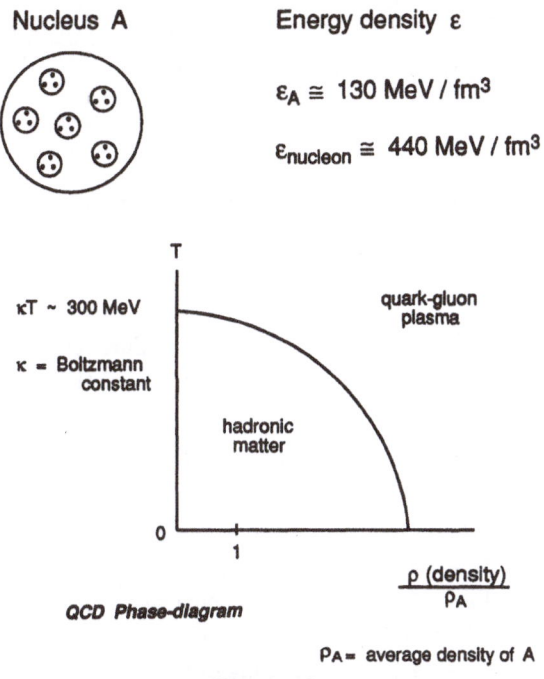

Figure 9

In Figure 9, the ordinate is κT (κ = Boltzmann constant, T = temperature), the abscissa is ρ/ρ_A (ρ = nucleon density, ρ_A = average nucleon density in a normal nucleus A). A typical nucleus A is when $\rho = \rho_A$. The scale can be estimated by noting that the critical $\kappa T \sim 300$ MeV is about the difference of $1 fm^3$ times $\mathcal{E}_N - \mathcal{E}_A$ and the critical $\rho/\rho_A \sim 4$ is just the nearest integer larger than $(1.2/0.8)^3$.

Accurate theoretical calculation exists only for pure lattice QCD (i.e., without dynamical quarks). The result is shown in Figure 10.

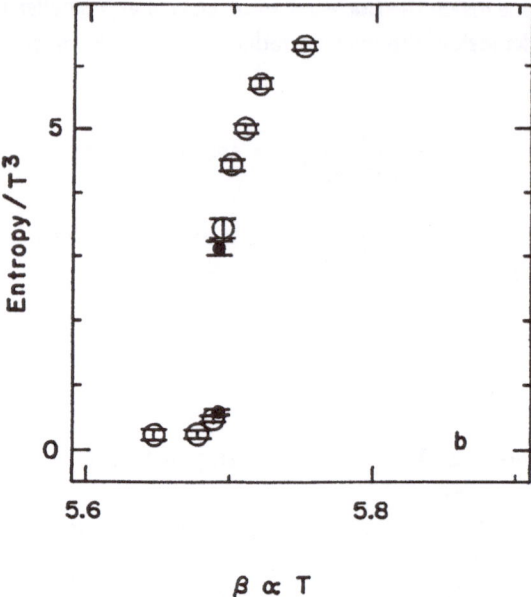

Figure 10. Phase transition[45] (pure QCD)

If one assumes scaling, then the phase transition in pure QCD (zero baryon number, $\rho = 0$) occurs at $\kappa T \sim 340$ MeV with the energy density of the gluon plasma

$$\mathcal{E}_P \sim 3 GeV/fm^3 .$$

To explore this phase transition in a relativistic heavy ion collision, we must examine the central region. Since only a small fraction of the total energy is retained in the central region, it is necessary to have a beam energy (per nucleon) at least an order of magnitude larger than $\mathcal{E}_N \times (1.2fm)^3 \sim 5$ GeV; this makes it necessary to have an ion collider of 100 GeV $\times$ 100 GeV (per nucleon) for the study of the QCD vacuum.

Another reason is that at 100 GeV $\times$ 100 GeV the heavy nuclei are almost transparent, leaving the central region (in Figure 10) to be one almost without any baryon number. As remarked before, this makes it an ideal situation for the study of the excited vacuum. Suppose that the central region does become a quark-gluon plasma. How can we detect it? This will be discussed in the following.

11. $\pi\pi$-INTERFEROMETRY

As shown in Figure 11, the emission amplitude of two pions of the same charge with momenta $\vec{k}_1$ and $\vec{k}_2$ from points $\vec{r}_1$ and $\vec{r}_2$ is proportional to

$$A \equiv e^{i\vec{k}_1 \cdot \vec{r}_1 + i\vec{k}_2 \cdot \vec{r}_2} + e^{i\vec{k}_2 \cdot \vec{r}_1 + i\vec{k}_1 \cdot \vec{r}_2},$$

because of Bose statistics. Let

$$\vec{q} \equiv \vec{k}_1 - \vec{k}_2 \qquad \text{and} \qquad \vec{r} \equiv \vec{r}_1 - \vec{r}_2 .$$

Since

$$|A|^2 \equiv 1 + \cos \vec{q} \cdot \vec{r}$$

changes from $|A|^2 = 2$ as $\vec{q} \to 0$, to $|A|^2 = 1$ as $\vec{q} = \infty$, a measurement of the $\pi\pi$ correlation gives a determination of the geometrical size R of the region that emits these pions, like the Hanbury-Brown/Twiss determination of the stellar radius.

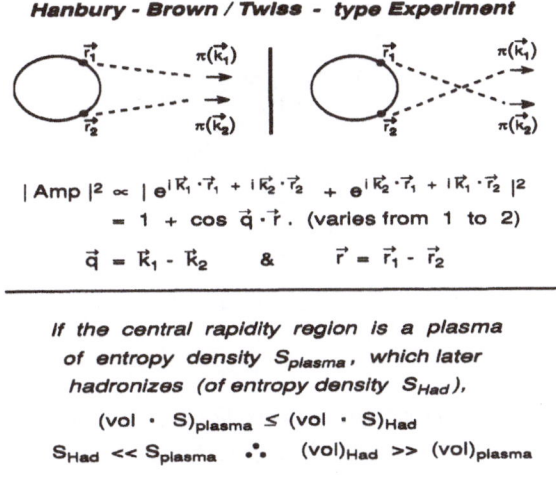

Figure 11

Now, if the central region is a plasma of entropy density S_P occupying a volume V_P, which later hadronizes to ordinary hadronic matter (of entropy density S_H and volume V_H), the total final entropy $S_H V_H$ must be larger than the total initial entropy $S_P V_P$. Since $S_P > S_H$, we have

$$V_H > V_P .$$

The experimental configurations and results[46] are given in Figure 12 and Table 5. One sees that the hadronization radius in the central region is indeed much larger than that in the fragmentation region. This is, at best, only indicative of the quark-gluon plasma. Much work and higher energy are needed for a more definitive proof. Nevertheless, it does show that relativistic heavy ions can be an effective means of exploring the structure of the vacuum.

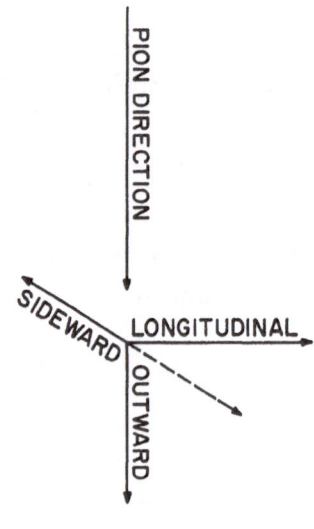

Figure 12

Table 5. $\pi\pi$-interference result[46] from the collision of an O beam (200 GeV/nucleon) on a stationary Au target

Rapidity Interval	Gaussian		
	R_T(fm)	R_L (fm)	Λ
$1 < y < 2$	4.3 ± 0.6	2.6 ± 0.6	$0.34^{+0.09}_{-0.06}$
	$R_T^{side} = 4.0 \pm 1.0$ fm	2.6 ± 0.6	$0.34^{+0.09}_{-0.06}$
	$R_T^{out} = 4.4 \pm 1.0$ fm	2.6 ± 0.6	$0.34^{+0.09}_{-0.06}$
$2 < Y < 3$	8.1 ± 1.6	$5.6^{+1.2}_{-0.8}$	0.77 ± 0.19
Central region (mid-rapidity)	$R_T^{side} = 6.6 \pm 1.8$ fm	$5.6^{+1.2}_{-0.8}$	0.77 ± 0.19
	$R_T^{out} = 11.2 \pm 2.3$ fm	$5.6^{+1.2}_{-0.8}$	0.77 ± 0.19

R (Oxygen) $\cong$ 3 fm

12. CONCLUDING REMARKS

To conclude we emphasize, once again, that the most challenging problems in physics are

(1) the symmetry-breaking force, and

(2) the structure of the vacuum.

It is quite likely that the answer to these two problems lies in the same direction. They can only be solved when we learn how to excite the vacuum. In the traditional way of thinking, our world is the world of particles. Larger units are made of small ones, which in turn are made of even smaller elements. The search for the smallest building block that everything is made of drives us to explore physical phenomena within smaller and smaller distances; that necessitates energies higher and higher in inverse proportion to the distance in question. On the other hand, the puzzles of missing symmetry and quark confinement have forced us to face the profound possibility that the vacuum could be a physical medium.

As we look into the future, the completion of RHIC in 1997 offers an unprecedented opportunity for physicists to explore the possibility of exciting the vacuum and to examine whether it is indeed a physical medium.

If the vacuum is the underlying cause for the strange phenomena in the microscopic world of particle physics, it must also have been actively responsive to the macroscopic distribution of matter and energy in the universe. Because the vacuum is everywhere and forever, these two, the micro- and the macro-, have to be linked together; neither can be considered a separate entity. Future history books will record that ours was a time when humankind was able to forge this bond on a scientific basis.

ACKNOWLEDGEMENTS

This research was supported in part by the U.S. Department of Energy

REFERENCES

[1] E. Rutherford, Philos.Mag. **42**, 392 (1898).

[2] M. and P. Curie, C.R. Acad.Sci. **130**, 647 (1900).

[3] A.S. Eve, *Rutherford* (Cambridge, Cambridge University Press, 1939), p. 80.

[4] A. Pais, *Inward Bound* (Oxford, The Clarendon Press, 1985).

[5] O. Hahn and L. Meitner, Phys.Z. **9**, 321, 697 (1908).

[6] W. Wilson, Proc.Roy.Soc. **A82**, 612 (1909).

[7] O. von Baeyer and O. Hahn, Phys.Z. **11**, 488 (1910).

[8] J. Chadwick, Verh.Dtsch.Phys.Ges. **16**, 383 (1914).

[9] L. Meitner, Z.Phys. **9**, 131, 145 (1922).

[10] E.D. Ellis, Proc.Cambridge Philos.Soc. **21**, 121 (1922).

[11] N.Bohr, J.Chem.Soc. **135**, 349 (1932).

[12] W. Pauli, American Physical Society Meeting in Pasadena, June 1931.

[13] E. Fermi, Ric.Scient. **4**, 491 (1934); Nuovo Cimento 11, 1 (1934); Z.Phys. **88**, 161 (1934).

[14] E.J. Konopinski and G.E. Uhlenbeck, Phys.Rev. **48**, 7 (1935).

[15] C.S. Wu and R.D. Albert, Phys.Rev. **75**, 315 (1949).

[16] E. Fermi and E. Teller, Phys.Rev. **72**, 399 (1947).

[17] M. Conversi, E. Pancini and O. Piccioni, Phys.Rev. **68**, 232 (1945).

[18] W. Heisenberg, Z.Phys. **101**, 533 (1936).

[19] C.M.G. Lattes, H. Muirhead, G.P.S. Occhialini and C.F. Powell, Nature **159**, 694 (1947).

[20] J. Steinberger, Phys.Rev. **75**, 1136 (1949).

[21] T.D. Lee, M. Rosenbluth and C.N. Yang, Phys.Rev. **75**, 905 (1949).

[22] O. Klein, Nature **161**, 897 (1948).

[23] G. Puppi, Nuovo Cimento **6**, 194 (1949).

[24] J. Tiomno and J.A. Wheeler, Rev.Mod.Phys. **21**, 153 (1949).

[25] L. Michel, Nuovo Cimento **10**, 319 (1953).

[26] M. Fierz, Z.Phys. **104**, 553 (1937).

[27] B.M. Rustad and S.L. Ruby, Phys.Rev. **89**, 880 (1953).

[28] R.H. Dalitz, Philos.Mag. **44**, 1068 (1953); Phys.Rev. **94**, 1046 (1954).

[29] T.D. Lee and J. Orear, Phys.Rev. **100**, 932 (1955).

[30] *High Energy Nuclear Physics*, Proc. 6th Annual Rochester Conference, April 3-7 1956 (New York, Interscience, 1956), p. VI-20.

[31] T.D. Lee and C.N. Yang, Phys.Rev. **102**, 290 (1956).

[32] R. Budde, M. Chrétien, J. Leitner, N.P. Samios, M. Schwartz and J. Steinberger, Phys.Rev. **103**, 1827 (1956).

[33] T.D. Lee and C.N. Yang, Phys.Rev. **104**, 254 (1956).

[34] C.S. Wu, E. Ambler, R.W. Hayward, D.D. Hoppes and R.P. Hudson, Phys.Rev. **105**, 1413 (1957).

[35] R.L. Garwin, L.M. Lederman and M. Weinrich, Phys.Rev. **105**, 1415 (1957).

[36] J.I. Friedman and V.L. Telegdi, Phys.Rev. **105**, 1681 (1957).

[37] J. Christenson, J. Cronin, V.L. Fitch and R. Turlay, Phys.Rev.Lett. **13**, 138 (1964).

[38] C. Alff-Steinberger *et al.*, Phys.Lett. **20**, 207 (1966). See also: *Particles and Detectors, Festschrift for Jack Steinberger*, eds. K. Kleinknecht and T.D. Lee (Berlin, Springer Verlag, 1986), p. 285ff for other references.

[39] S. Weinberg, Phys.Rev.Lett. **19**, 1264 (1967). See also S.L. Glashow, Nucl.Phys. **22**, 579 (1961); A. Salam and J.C. Ward, Nuovo Cimento 11, 568 (1959), Phys.Lett. **13**, 168 (1964); A. Salam, in *Proceedings of the 8th Nobel Symposium*, ed. N. Svartholm (Stockholm, Almqvist and Wiksell, 1968), p. 367.

[40] C. Rubbia, in *Proceedings of the International Conference on High Energy Physics*, eds. J. Guy and C. Costain (Didcot, Rutherford Appleton Laboratory, 1983), p. 880.

[41] T. Kinoshita, J.Math.Phys. **3**, 650 (1950). T.D. Lee and M. Nauenberg, Phys.Rev. **133**, B1549 (1964).

[42] R. Friedberg and T.D. Lee, Phys.Rev. **D18**, 2623 (1978).

[43] A. Chodos, R.J. Jaffe, K. Johnson, C.B. Thorn and V.F. Weisskopf, Phys.Rev. **D9**, 3471 (1974).

[44] T.D. Lee and G.C. Wick, Phys.Rev. **D9**, 2291 (1974); T.D. Lee, "Relativistic Heavy Ion Collisions and Future Physics," in *Symmetries in Particle Physics*, ed. I. Bars, A. Chodos and C.H. Tze (New York, Plenum Press, 1984), p. 93.

[45] F.R. Brown, N.H. Christ, Y.F. Deng, M.S. Gao and T.J. Woch, Phys.Rev.Lett. **61**, 2058 (1988).

[46] W. Willis, "Experimental Studies on States of the Vacuum," in *Relativistic Heavy-Ion Collisions*, ed. R.C. Hwa, C.S. Gao and M.H. Ye (New York, Gordon and Breach Science Publishers, 1990), p. 39.

PHYSICS AT LEP

Lorenzo Foà

Scuola Normale Superiore - Pisa
INFN Sezione di Pisa

1-INTRODUCTION

These three lectures are devoted to the description of some examples of quantitative physics, as experimentally studied at LEP. Emphasis will be put on the presentation of the experimental methods and to the discussion of their intrinsic precision, while no effort will be made to be complete and up to date with the results obtained by the four collaborations. The numerical values obtained with the 1991 data will be indeed presented at the Dallas Conference in few weeks and it will be one job of the rapporteurs to weigh them up and draw the final conclusions.

The three lectures are organized as follows:

i)   Short review of the parameters of LEP and of the measurement of its energy. Then some details are given on the measurement of the luminosity and the main characteristics of the experiments are recalled.
ii) Description of the measurements of the parameters of the Z resonance, of the axial and vector weak coupling constants and present status of the test of the Standard Model with limits on the top mass.
iii) Physics of the b-quark and of the τ-lepton as examples of the precision measurements now possible with the installation of the minivertex detectors with emphasis on the attempts to measure the lifetimes. In this case most of the results are taken from ALEPH and DELPHI.

2-LEP

For three years LEP has been running with 4 bunches of electrons and 4 bunches of positrons with initial currents of about 1 mA and a luminosity slightly below $10^{31}\text{cm}^{-2}\text{s}^{-1}$. In '90 and '91 the limit to the luminosity was due to the beam currents. In '92 the optics of the machine has been modified in order to be ready to use the Pretzel scheme with 8 bunches in both beams and to be consistent with the working mode of LEP 200. These new conditions have caused for the first part of the year a stronger beam-beam effect which has limited the initial luminosity to

.6 10^{31}cm^{-2}s^{-1}, a luminosity which then remained basically constant for many hours, the specific luminosity increase compensating the reduction of the currents. In the very last few days the performance of the machine has been improving and we can still hope to collect half a million Z per experiment before the end of the year.

3-LEP ENERGY MEASUREMENTS

The physics of LEP 100 consists basically in the determination of the parameters of the Z resonance, first of all of its mass. In the data of 1990 the uncertainty on M_z was largely dominated by the limited knowledge of the beam momentum which in turn depends on $\Delta B/B$, where B is the magnetic field of the ring and, to a much smaller extent, on $\Delta r/r$, where r is the machine radius.

In the past years several methods had been envisaged to measure B with high precision:

i) A flipping coil in a magnet in series with the dipoles of LEP. The intrinsic precision of the measurement is $\pm 2 \cdot 10^{-5}$, but the magnet is kept at constant temperature and humidity and has a different structure (no concrete) with respect to the ring magnets.

ii) Fixed coils is the LEP dipoles, cycled with a known frequency. The precision of the method is $\pm 10^{-4}$ and its weakness is the fact that it can be used only without beams.

iii) β of protons versus β of electrons injected into LEP at 20 GeV. This elegant method, with an intrinsic precision of $\pm 3 \cdot 10^{-5}$, requires an extrapolation at 45 GeV which reduces its accuracy.

As a result of measurements made with all these methods, the energies of LEP in 1990 have been determined with an error of $\pm 2.4 \cdot 10^{-4}$, resulting in ΔM_z=20 MeV.

During 1991 advantage has been taken of the natural transverse polarization of the electrons in LEP[1]. This polarization is determined by shooting a circularly polarized laser beam against the positrons and by measuring the angular distribution of the backscattered photons. The centre of the vertical distribution is displaced, by switching the photon helicity, by an amount

$$\Delta y = k \xi P$$

where k is the analysing power of the polarimeter, ξ is the photon helicity and P is the electron polarization.

With realistic figures this displacement amounts to 40 µm so that a clear asymmetry is visible in the difference between data taken with opposite helicities (see Fig. 1), corresponding to polarizations ranging between 10 and 20%.

This polarization can be used to measure the electron momentum with high precision since the precession frequency of the electrons in the magnetic field of LEP is given by

$$f_{prec} = \nu \, f_{Revol}$$

where

$$\nu = \frac{g_e - 2}{2} \frac{E_{beam}}{m_e c^2} = \frac{E_{beam}(GeV)}{0.4406486(1)} \; .$$

represents the number of spin precessions in one turn of the ring.

The precession frequency is found by sweeping the frequency of a kicker magnet acting on the electrons in the radial direction, because when this frequency and f_{prec} are equal the beam polarization is suddenly distroyed or reversed as shown in Fig. 2. This method has an intrinsic

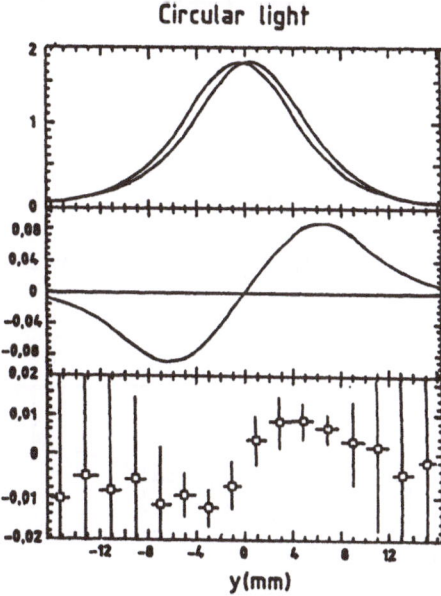

Circular light

Fig. 1- a) Angular distribution of backscattered photons, with opposite helicities, b) expected behaviour of the difference and c) experimental value.

precision of about 1 MeV (~10^{-5}) but a series of measurements[2] done throughout the year showed a dispersion of 6÷7 MeV.

Several attempts have been made to identify the origin of this dispersion and the most promising explanation seems to be the terrestrial tidal effect. According to current Earth models the joint moon and sun tides can modify the radius of LEP by about 1 mm, which can cause up to 6 or 7 MeV/c difference in beam momentum.

A compilation of all the depolarization results as a function of the tidal argument shows a satisfactory agreement with a dispersion of 1 MeV as expected. Further investigations on this effect are in progress and for the moment the values of E_{beam} are quoted with an error of 6 MeV, which still dominates over the statistical error on M_z, as shown later.

4-MEASUREMENTS OF THE MACHINE LUMINOSITY

The determination of the parameters of the Z resonance is based on the number of events collected in an identified channel and of the luminosity integrated by the experiment in the same period of time, with different settings of the beam energy, typically in steps of 1 GeV.

The luminosity measurement follows a common procedure in all experiments and its precision at present limits the determination of the peak cross-section.

The measurement is performed by counting the events of Bhabha scattering collected inside the acceptance (Acc) of small angle calorimeters positioned as in Fig. 3, and by evaluating

$$L = \frac{N_{events}}{\sigma_{Acc.}\left(e^+e^- \rightarrow e^+e^-(\gamma)\right)} \quad \text{where}$$

$$\sigma_{Acc.}(e^+e^- \rightarrow e^+e^-(\gamma)) = \int_{Acc.} \varepsilon \frac{d\sigma}{d\Omega} d\Omega$$

ε being the efficiency.

Since $\sigma_{Acc.}$ depends on $\dfrac{1}{\theta_{min}^2}$,

$$\frac{\Delta\sigma_{Acc.}}{\sigma_{Acc.}} \approx 2 \frac{\Delta\theta_{min}}{\theta_{min}} \approx 2 \left(\frac{\Delta r_{min}}{r_{min}} \otimes \frac{\Delta z}{z} \right)$$

The request $\Delta\sigma_{Acc.}/\sigma_{Acc.} \approx 0.5\%$ corresponds to an accuracy in Δr_{min} of 350 µm and in Δz of .5 cm.

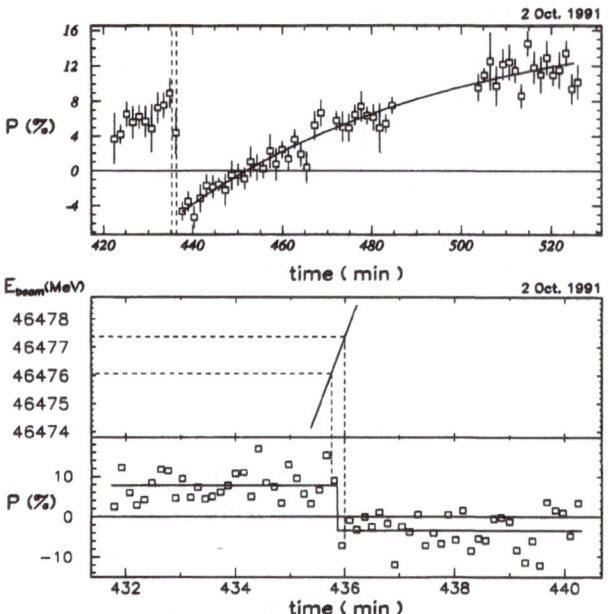

Fig. 2-Polarization as a function of time over 100 minutes and (enlarged scale below) around the moment when the polarization is reversed. The vertical scale shows the energy resolution achievable in the beam.

All small angle calorimeters are structured in cells, defined by the dimensions either of the counters (lead-glass or BGO) or of the towers in lead-gas detectors. With reference to Fig. 4a, the sums of the signals of the left and right cells allow the definition of an asymmetry

$$A = \frac{E_R - E_L}{E_R + E_L}$$

which depends on the x position as shown in Fig. 4b (Aleph data). By expressing the space resolution as

$$\delta x = \left(\frac{dA}{\frac{dE}{E}} \right) \frac{dx}{dA} \frac{dE}{E}$$

with dA/dE/E given by the MonteCarlo simulation, dx/dA taken from the test beam data, when the centre of gravity of the shower falls on the border between two cells, and dE/E from the standard performance of the calorimeters, it is easy to obtain $\delta x < 400\mu m$.

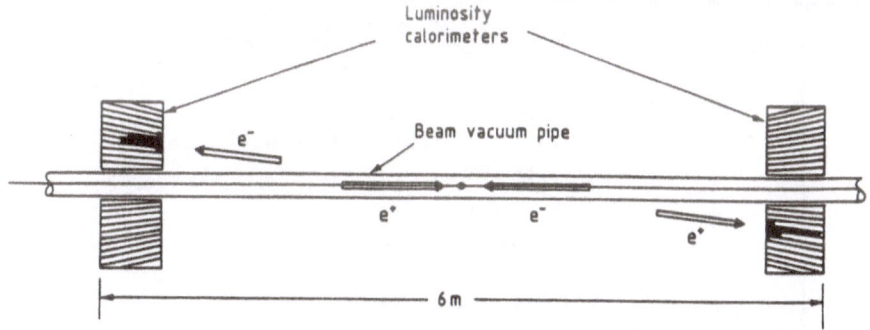

Fig. 3-Disposition of the luminosity calorimeters in the LEP experiments.

This shows that the requested precision on r_{min} can be reached if the inner edge of the fiducial region is defined by borders of cells which never belong to the innermost ring.

An important source of systematic error can be due to unexpected and unknown displacements of the interaction point, the position of which is used to define the e+ and e- scattering angles. The MonteCarlo simulation shows that displacements of the beam crossing point of the order of 1 mm with respect to the nominal position can produce losses up to 1% if symmetric θ_{min} cuts are applied to electrons and positrons. A procedure is then adopted which defines $\theta^{e^-}_{min}$ smaller then $\theta^{e^+}_{min}$ for odd-number events and viceversa for even-number events, as sketched in Fig. 5. In this way, unknown beam crossing point displacements induce the detection of an excess of events in one case and a loss of events in the other case with an average which is mostly insensitive to the beam position.

Once having taken this care in the definition of the acceptance, events are selected by requiring a minimum value[3] for E_{e^+}, E_{e^-}, $E_{e^+} + E_{e^-}$.

The final selection requires in addition that the electron and the positron be coplanar within some ±10°. Fig. 6 shows, in the case of ALEPH, the $\Delta\phi$ distribution which agrees on more than four orders of magnitude with the MonteCarlo predictions based on the Bhabha generator BHLUMI[4], which takes into account multiphoton-emission.

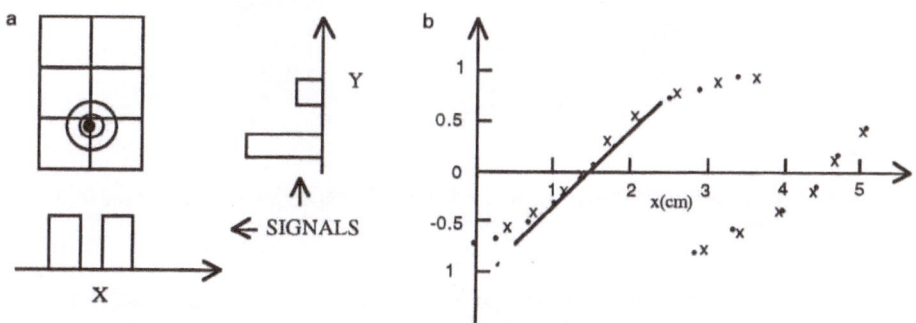

Fig. 4-a) Analog signals from towers organized in x, y coordinates
    b) asymmetry of the calorimeter response as a function of x. The maximum sensitivity (slope) corresponds to the counter edges, where the shower is split into two equal parts.

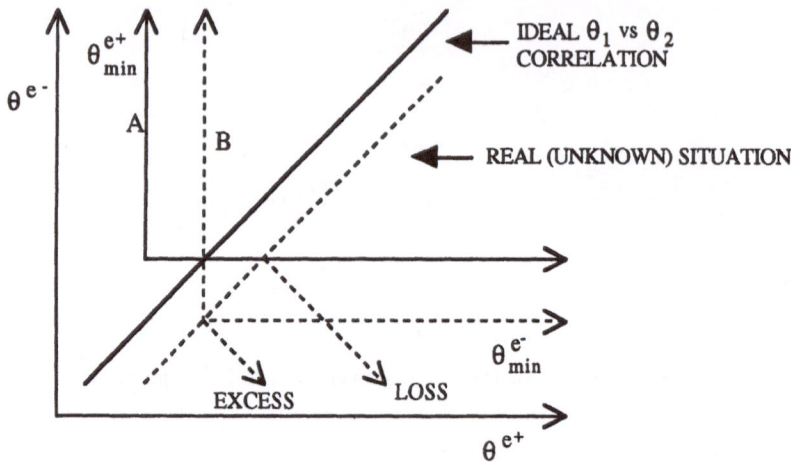

Fig. 5-θ^{e^-} versus θ^{e^+} for Bhabba events. The full line corresponds to the ideal situation, the dotted line to a realistic case with the interaction point displaced by an unknown amount.

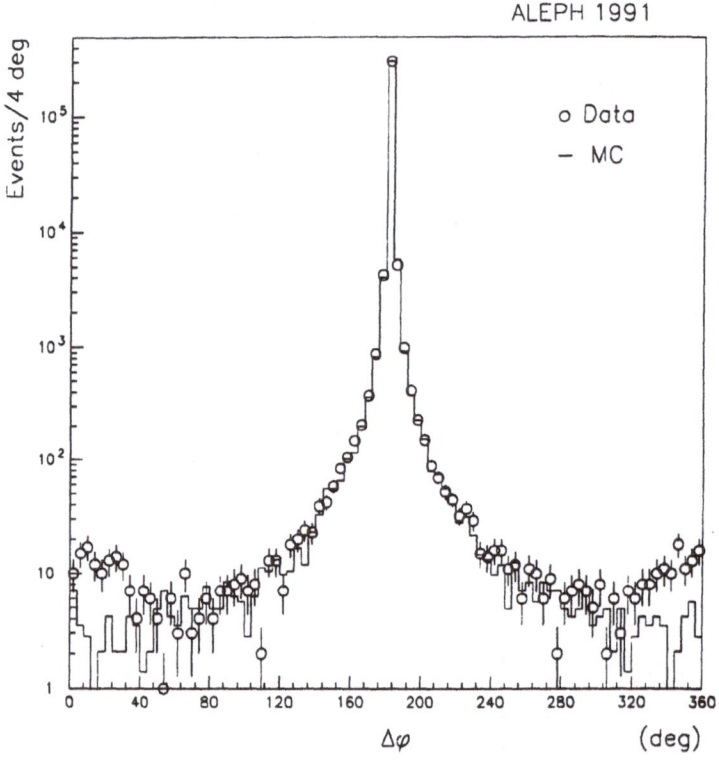

Fig. 6-Difference in azimuthal angle between electron and positron. Data are shown as points and the BHLUMI simulation as a histogram.

A worse agreement was obtained last year with the BABAMC generator, which takes into account radiative corrections only up to the first order.

A careful examination of the residual sources of systematic errors leads to the figures quoted in table 1.

Table 1

Source	Systematic error (%)
Background	0.03
Alignement	0.24
Fiducial side cut	0.38
Energy cut	0.23
Montecarlo statistics	0.18
others	<0.1
Total (experiment)	0.6
Theoretical error	0.3
Overall error	0.7

5-DETECTORS

No effort will be made here to describe in detail the four experiments ALEPH, DELPHI, L3 and OPAL since exaustive descriptions of them exist in the litterature[5-8].

The characteristics of the Z decay have guided the common design of all detectors, aiming to full coverage for detecting and identifying jets, electrons and muons, both isolated or imbedded in jets. Indeed the Z boson is expected to decay 6% of the times into each type of neutrino which escape detection, 3% of the times into each lepton pair $\left(e^+e^-, \mu^+\mu^-, \tau^+\tau^-\right)$ and the rest into two quarks, with a slight preference for down-type quarks (15% per family) with respect to up-type quarks (12% per family). These decay rates correspond to measurable quantities, Γ_{inv} for all neutrinos and unknown, undetectable new objects, $\Gamma_{e^+e^-}, \Gamma_{\mu^+\mu^-}, \Gamma_{\tau^+\tau^-}$ for the leptonic decays, $\Gamma_{hadrons}$ for the sum of all quark-antiquark final states, irrespective of the additional gluon jets radiated by the produced quarks.

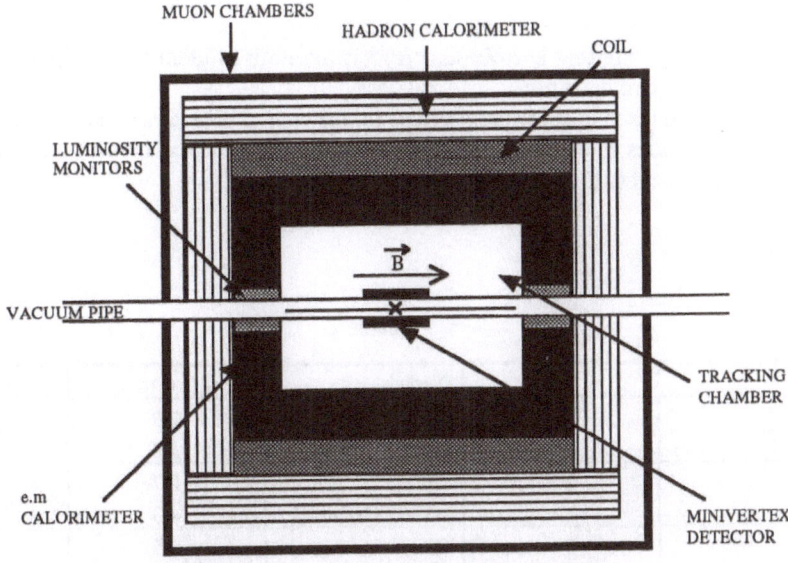

Fig. 7-General structure of LEP experiments with the sequence of detectors organized around the interaction point.

Each detector follows the basic design of Fig. 7, with small or large variations and distortions. A silicon minivertex detector surrounds the beam pipe at the interaction point, providing two or three points, for large angle tracks, in space or in the r-φ projection, with a resolution of about 10 μm. Around it a much longer inner drift chamber provides a trigger on charged prongs and several points to each track with resolutions ranging between 35 and 100 μm. Gaseous tracking detectors, with radii ranging between half a metre and two metres complete the tracking systems of all apparatus, with the exception of DELPHI which contains also a set of Ring Imaging Cherenkov Counters for π/K/p identification. Two of the central detectors are Time Projection chambers (TPC) (ALEPH and DELPHI), one is a cylindrical drift chamber with stereo view for the z coordinate (OPAL) and one is a chamber designed for high resolution (L3). Electromagnetic calorimeters precede the magnet coils in all experiments but OPAL for which the use of a warm coil has necessitated that the coil precede the calorimeter.

Two experiments have chosen sampling lead-gas calorimeters, ALEPH with small projective towers and DELPHI in the barrel with the new technique of the Heavy Projection Calorimeter (HPC), the other two have chosen crystal detectors (Lead-gas for OPAL and BGO for L3).

Hadron calorimeters absorb then all the residual energy of jets and track the path of muons. Three out of four experiments have imbedded Iarocci tubes in the return yoke of the magnet whereas L3 uses a compact hadron calorimeter working in proportional mode. Muon chambers complete the apparatus for all experiments, just outside the hadron calorimeters for ALEPH, DELPHI and OPAL whereas L3 fills the large volume inside the giantic aluminium coil with three layers of drift chambers, therefore measuring the momentum of the muons outside the hadron calorimeter.

Even if the designs of the experiments seem superficially similar, the detailed performance both in tracking and calorimetry are quite different.

The tracking capabilities are summarized in Table 2 which shows, for each experiment, the magnetic field B, the product BL^2 where L is the length of the measured tracks, proportional to the sagitta, and the coefficient k in the expression of the momentum resolution

$$\frac{\Delta p_\perp}{p_\perp^2} = k \ (Gev/c)^{-1} \ .$$

It is worth noticing that k=.6% $(GeV/c)^{-1}$ is the limit to identify the charge of a 50 GeV particle with 3 standard deviations.

Care has been taken to show these values for the central detectors alone and for the complex of all tracking devices since the use of several separate detectors allows a better resolution but requires the alignment of all detectors to be fully understood.

Table 2

Expt.	B(T)	Detector	Bl^2 (Tm^2)	k(%)(GeV/c)$^{-1}$
Aleph	1.5	TPC	2.9	0.11
		MV+TPC	4.8	0.08
Delphi	1.2	TPC	0.7	0.6
		MV+TPC+Out	4.3	0.11
L3	0.5	TEC	0.5	0.8
		TEC+μCH	4.3	0.04
OPAL	0.4	JET	1.0	0.18
		MV+JET	1.3	0.15

Table 3

Expt.	Detector	$\dfrac{\Delta E}{E} = \dfrac{K}{\sqrt{E}}$	$\Delta\theta$ (mr)	N° of towers
ALEPH	Lead-prop. tubes	18%	1	75000 (x3)
DELPHI	HPC	20%	1	3-D shower reconstruction
L3	BGO	~1%	40	12000
OPAL	Lead-glass	6%	50 2 with preshower	10600

Also in the case of calorimetry, the choice between gas detectors and crystals correponds to different priorities attributed to granularity, which is good in the gaseous detectors and very useful to identify electrons from random superposition of pions and photons, and to energy resolution, which is better in crystal detectors and helps in the identification of electrons from hadrons.

A comparison of the performance of the four electromagnetic calorimeters is given in Table 3.

A final comment on muon identification. The large volume spectrometer of L3 provides a very high degree of purity and a momentum resolution of ~2% at 50 GeV/c. ALEPH has implemented a bidimensional tracking system with 1 cm cells over 23 layers providing a detailed view of muon tracks inside the iron which allows an easy identification of muons also close to hadrons.

6-EVENT SELECTION

As described before, the main decay channels that each detector must identify are $q\bar{q}$, e^+e^-, $\mu^+\mu^-$, and $\tau^+\tau^-$. The main identification criteria are common to all experiments, even if they can differ in small details from one detector to another. In order to be quantitative, numerical examples will be taken from the ALEPH data.

a-Hadronic Events (≥2jets)

Two different analyses can be used to identify hadronic events, the first one based on the charged tracks seen by the tracking detectors, the second one based on calorimetry.

Since the two selections make use of different detectors, their results can be compared in order to evaluate possible systematic errors. The first selection makes use of the informations provided by the Inner Tracking Chamber (ITC) and by the TPC and requires the reconstruction of at least 5 charged tracks with a total energy of at least one tenth of E_{CM}. The efficiency, as evaluated by MonteCarlo simulation, amounts to (97.4±0.24)%, with a background, mainly $Z \to \tau^+\tau^-$ and $e^+e^- \to e^+e^-$ hadrons, smaller than 0.5%. A systematic error of 0.26% has been attribuited to the efficiency to take into account the uncertainty due to the absolute scale of the momenta measured by the TPC. A typical 2-jet event is shown in Fig. 8.

The second selection, based on the informations provided by the electromagnetic (ECAL) and hadronic (HCAL) calorimeters, requires

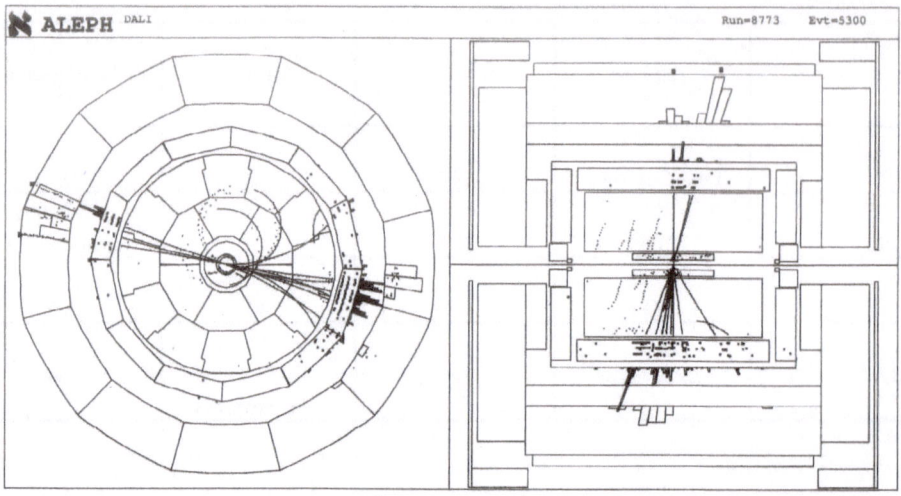

Fig. 8-Typical 2-jet event in ALEPH in the projection orthogonal to the beams (x, y) and along them (R, z). Tracks are visible in the ITC and TPC whereas energy deposits are shown in the calorimeters and muons are easy to recognize from their path in the hadron calorimeter and in the muon chambers.

$$E_{ECAL} + E_{HCAL} > 0.2 E_{CM}$$

with at least 7 GeV in the barrel of ECAL or 1.5 GeV in each end cap.

In addition a time window of 100 ns is selected around the crossing time to reject cosmic rays and leptons are excluded with the criteria outlined later. The efficiency is $(99\pm0.09)\%$ with a background of less than 0.7% and a systematic error of 0.2%.

As quoted before, a possible source of background is the two photon process $e^+e^- \to e^+e^-\gamma\gamma \to e^+e^-$ hadrons which is seen at low energy in Fig. 9, where the distribution of the sum of the energy of the charged tracks reconstructed by the detectors is shown. At low energy, the data show a fast rise not reproduced by the MonteCarlo simulation which contains only Z decay events. This background is mostly dangerous in the data taken when the energy of the machine is set on the tails of the Z resonance since its cross section does not depend appreciably on the energy between 88 and 93 GeV.

In order to evaluate the contribution of this process in the limit of zero resonance production, the graph in Fig. 10 shows the number of events in the window between $x = E_{ch}/E_{CM} = 0.10$ and $x=0.15$ as a function of the number of events detected above $x=0.3$. The latter variable is large when the machine energy is at the peak, being small on the tails. In this plot, a straight line passing through the origin means that no event remains in the small energy window when no Z event is detected, i.e. no background from two photon events is present in the data.

From a fit to the data the contamination results into the value $\sigma_{\gamma\gamma}^{(detected)} = (50\pm26)$pb. A similar result is obtained for the calorimetric selection. As a final check the ratio R of the cross-sections measured with the two selections is $R=1.0023\pm0.0004$ in good agreement with the quoted systematic errors of ~2‰, and shows no dependence on the energy of the machine.

b-Leptonic Events

As a first step, all leptonic events, irrespective of their flavour, are selected by requiring that only a few tracks are present. Defining as "good" the tracks which have at least 4 TPC points and which originate in a small cylinder around the interaction point, the requests $n_{charged} < 8$, with at least one track in each emisphere, no track at $\theta > 31.8°$ with respect to the thrust axis, and a acollinearity smaller than $20°$ between the total momenta in the two emispheres are sufficient to clean the leptonic event sample to better than a percent.

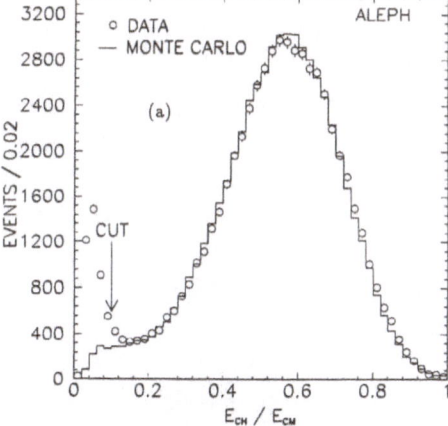

Fig. 9-Distribution of the energy of charged tracks. Points at low energy show the contribution of $\gamma\gamma$ events.

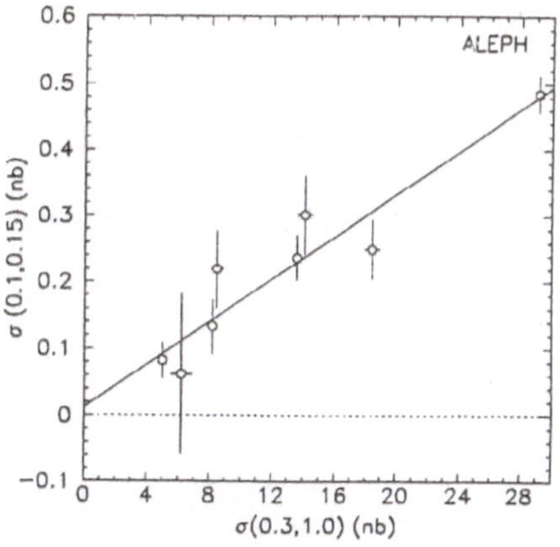

Fig. 10-Hadronic cross section in the charged energy interval (0.1 to 0.15) $\sqrt{s}$ as a function of the same quantity between (0.3 and 1)$\sqrt{s}$ at different center of mass energies.

A simple way to see the structure of the overall leptonic sample is shown in Fig. 11, where the event rate is plotted against the variable $x = \dfrac{\text{"E"}}{E_{CM}}$, with

$$\text{"E"} = \sum_i E_i + \sum_i |\vec{p_i}|.$$

In this expression E_i are the energies of the particles as measured by the electromagnetic calorimeter and $|\vec{p_i}|$ are the absolute values of their momenta, as measured by the TPC.

Muons show the full momenta in the TPC but deposit almost no energy in ECAL, and therefore cluster around $x=1$. Electrons deposit the full energy in ECAL and show part of their momenta in the TPC, often being accompanied by radiated photons, so that x peaks at 2 with a smooth tail towards $x=1$.

For τ's, only a fraction of the energy and of the momentum is visible because of the various neutrinos present in their decay. They are then mostly confined below x=1 in good agreement with the MonteCarlo simulation. Beyond this general selection for lepton channels more restrictive criteria are adopted to identify pure Z→e⁺e⁻, Z→μ⁺μ⁻ and Z→τ⁺τ⁻ samples.

c-e⁺e⁻ Events

The selection of this category of events is restricted to the angular region -0.9<cosθ<0.7 in order to avoid the forward cone where the non resonant e⁺e⁻ elastic scattering (Bhabha scattering) dominates. The residual Bhabha events at cosθ<0.7 are subtracted by means of a MonteCarlo simulation based on the ALIBABA[9] generator. The identification criteria for e⁺e⁻ events are:

$$\sum E_i \geq 0.2 \ E_{CM}$$
$$\sum |\vec{p_i}| \geq 0.05 \ E_{CM}$$

$$"E" > 1.2 \ E_{CM}.$$

An example of an e⁺e⁻ event is given in Fig. 12a.

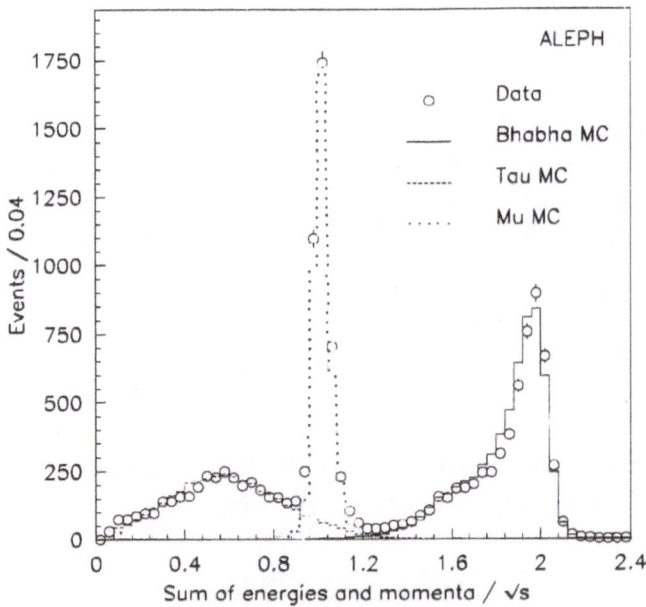

Fig. 11-"E" distribution for lepton pairs.

d-μ⁺μ⁻ Events

Events with only two muons in the final state are selected by requiring that the extrapolations of the two TPC tracks find a sufficient number of fired tubes in the hadron calorimeter, in particular in the last 10 planes, and some signal in the muon chambers compatible with the multiple scattering cone. The typical pattern of a μ⁺μ⁻ event is give in Fig. 12b.

e-τ⁺τ⁻ Events

The main characteristics of $\tau^+\tau^-$ events is that each τ has at least one neutrino in its decay. These neutrinos, propagating in almost opposite directions, build up a large missing mass as shown by data and MonteCarlo simulation in Fig. 13. A cut at missing mass larger than 20 GeV, requesting in addition $E_{measured}$<55 E_{CM}/M_Z GeV to reject the e⁺e⁻ events with two radiated photons lost in the pipe or in the cracks, is sufficient to select a pure $\tau^+\tau^-$ sample.

A typical example of $\tau^+\rightarrow\rho^+\bar{\nu}_\tau$, $\tau^-\rightarrow\rho^-\nu\nu_\tau$ is shown in Fig. 12c.

The $\tau^+\tau^-$ sample is the most critical since it contains the lepton events which are not identified as e⁺e⁻ or μ⁺μ⁻ and since, owing to its larger multiplicity, it can be contaminated by the low multiplicity tail of hadronic events.

A nice way to check its purity is the following: a sample of events in which one τ is identified with very stringent cuts is selected and the accompaning τ is taken. In this way we obtain a very pure and unbiased sample of single τ's which are then combined two by two by

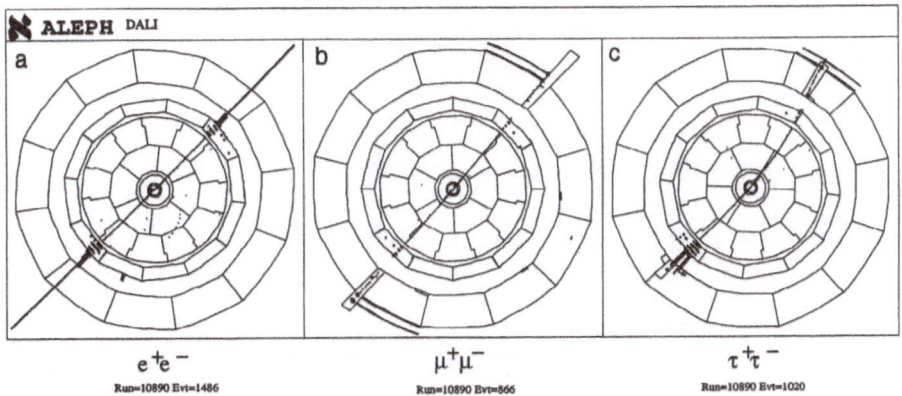

Fig. 12-Examples of a) e⁺e⁻, b) μ⁺μ⁻, c) τ⁺τ⁻ events in ALEPH.

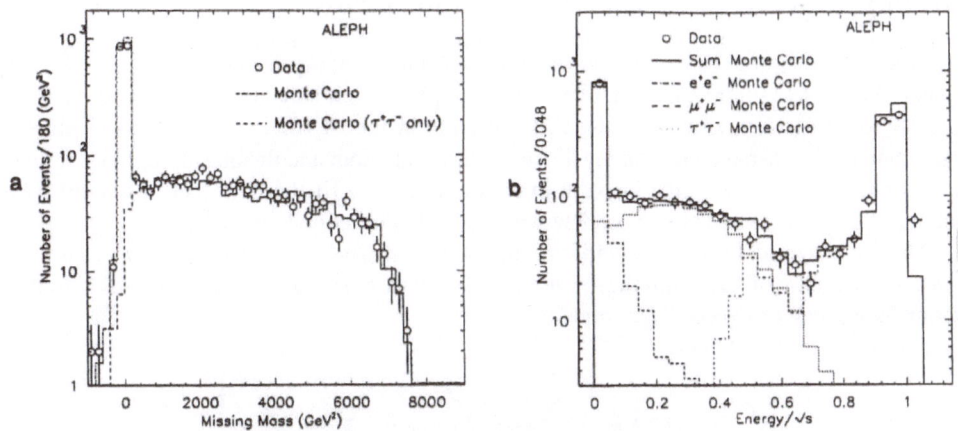

Fig. 13-a) Missing mass squared for the full lepton sample
b) total energy in the electromagnetic calorimeter for the lepton sample.

requesting that they have opposite charge and direction. The missing mass and visible energy distribution are then calculated and compared in Fig. 14 with the MonteCarlo predictions. The good agreement of the expected and seen distributions is a convincing check of the selection procedure.

Finally the inclusive "lepton" sample and the three separate channels can be compared as in Fig. 15, showing that the overlap between the different channels never reaches 1%.

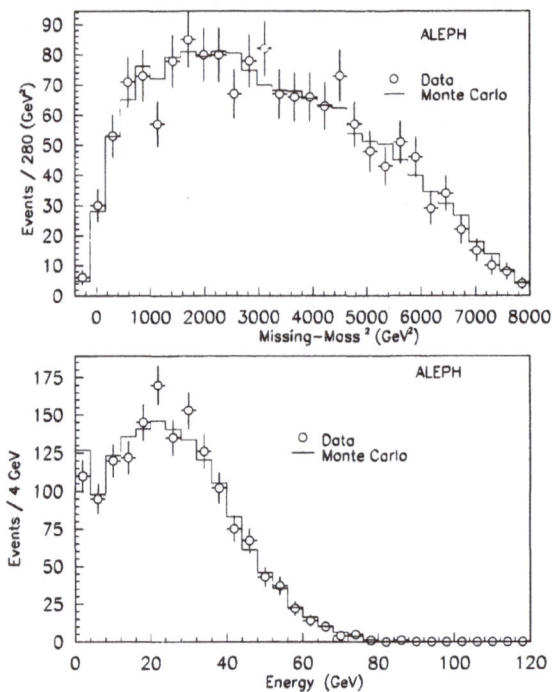

Fig. 14-Missing mass and visible energy in data and MonteCarlo predictions with the $\tau^+\tau^-$ sample, as described in the text.

7-MODEL INDEPENDENT FIT TO THE Z RESONANCE

We have seen how to isolate events coming from the decay of the Z boson and how to separate them into hadronic, e^+e^-, $\mu^+\mu^-$ and $\tau^+\tau^-$ samples. These data, taken at several energies, from 88 to 93 GeV in steps of 1 GeV, follow the shape of the Z resonance and allow a measurement of its parameters, namely the mass, the total width and the branching ratios into the various channels. In the following, the expression used to fit the data and to extract these parameters in the most model independent way is recalled for completeness[10].

The cross-section for fermion-antifermion production around '90 GeV is proportional to the square of the sum of the annihilation amplitudes with a virtual photon and a Z boson as intermediate states, as shown diagrammatically below

$$\sigma\left(e^+e^-\to f\bar{f}\right) \propto \left| \begin{matrix} e^+ & & f \\ & \text{">\!\!\!\sim\!\!\!<} & \\ e^- & "\gamma" & \bar{f} \end{matrix} + \begin{matrix} e^+ & & f \\ & \text{>\!\!\cdots\!\!<} & \\ e^- & Z & \bar{f} \end{matrix} \right|^2$$

At the tree level, using only standard quantum field theory, the cross-section can be written as

$$\sigma = \frac{4}{3}\pi\frac{\alpha^2}{s} + I \cdot \frac{s-M_z^2}{s} + \text{Resonance}.$$

The first term represents the electromagnetic annihilation process and α is the electromagnetic coupling constant, the second represents the interference term which vanishes at the central value of the resonance, and the third term gives the true contribution of the Z boson,

$$\text{Resonance} = \frac{12\pi}{M_z^2}\Gamma_{e^+e^-}\cdot\Gamma_{f\bar{f}}\frac{s}{\left(s-M_z^2\right)^2 + M_z^2\,\Gamma_z^2}$$

where the fraction corresponds to the propagator of a resonance with central mass value M_z and total width Γ_z. This expression, evaluated at the tree level, is not accurate enough to be used for fitting data which have an accurancy better than 1% and radiative corrections must be taken into account.

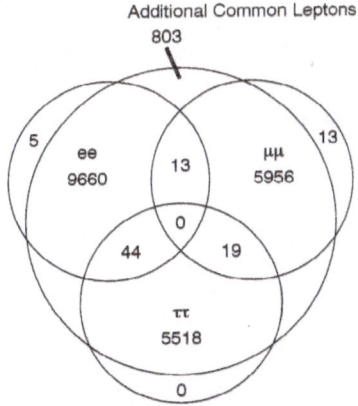

Fig. 15-Comparison of the separate flavour samples with the overall lepton selection.

A basic part of these corrections comes from the propagator self energy terms which correspond, in the case of the e.m. annihilation, to replacing the tree level diagram, $Q_e\,Q_f\frac{e^2}{s}$,

with the diagram

where

Calling $\Pi_\gamma(s)$ the amplitude for a single loop, the sum of all diagrams corresponds to a geometrical series and can be written as

$$Q_e\, Q_f \frac{e^2}{s}\frac{1}{1+\Pi_\gamma(s)}.$$

The s-dependence of the propagator corrections can be transferred to the electric charge with the definition

$$e^2(s)= \frac{e^2}{1+\Pi_\gamma(s)}$$

which introduces the concept of "running coupling constant". Indeed the coupling constant of electromagnetic interaction, α, is 1/137 at very small values of s and becomes 1/129 at the Z mass.

The advantage of this formalism lies in the fact that the expression of the amplitude is still $Q_e\, Q_f\, e^2(s)/s$ as in the tree level approximation.

The same approach is followed for the process with a Z intermediate state, by replacing the tree level diagram

where the propagator corresponds to $\frac{1}{s-M^2}$ and M is the "naked" Z mass, with the diagram

where

Calling $\Sigma(s)$ the single loop amplitude and summing the geometrical series we get for the propagator

$$\frac{1}{s-M^2+\text{Re } \Sigma(s)+i\text{Im}\Sigma(s)} .$$

We can now define the renormalized mass

$$M_z^2 = M^2 - \text{Re } \Sigma\left(M_z^2\right)$$

as the true, observable mass of the Z boson, expanding

$$\Sigma(s)=\Sigma\left(M_z^2\right)+\frac{\partial\Sigma}{\partial s}\Big|_{M_z^2}\left(s-M_z^2\right)+....$$

and

$$\text{Im}\Sigma(s)=M\Gamma(s)\left(1+\frac{\partial\Sigma}{\partial s}\Big|_{M_z^2}\right)$$

with $M\Gamma(s)=\Gamma_z\frac{s}{M_z} + ... $.

The propagator can now be written as

$$\frac{1}{1 + \frac{\partial\Sigma}{\partial s}\Big|_{M_z^2}} \; \frac{1}{s - M_z^2 + \frac{is}{M_z}\Gamma_z}$$

where the first factor can again be incorporated into the running coupling constant and the resonance width is also a running quantity.

Altogether, the cross-section is now expressed as

$$\sigma(s) = A\frac{4}{3}\pi\frac{\alpha^2(s)}{s} + I\frac{s - M_z^2}{s}\cdot B +$$

$$+ 12\pi\frac{\Gamma_{e^+e^-}\Gamma_{f\bar{f}}}{M_z^2}\frac{s}{\left(s-M_z^2\right)^2 + \frac{s^2}{M_z^2}\Gamma_z^2}\frac{1}{1+\delta_{QED}} .$$

In this expression all Γ quantities correspond to measurable observable, and contain vertex corrections to all orders, whilst the small corrections factors A and B can be found in table 4 according to the selected final state:

Table 4

channel	A	B
$e^+e^-\rightarrow\mu^+\mu^-$	$1+\delta_{QED}$	1
$e^+e^-\rightarrow q\bar{q}$	$\left(\frac{11}{3} + \frac{35}{27}\delta_{QCD}\right)\left(1+\delta_{QED}\right)$	$1+\delta_{QED}$

where $\delta_{QED} =\frac{3}{4}\frac{\alpha}{\pi}\simeq 0.17\%$ and $\delta_{QCD} =\frac{\alpha_s(M_z)}{\pi} + 1.4\left(\frac{\alpha_s}{\pi}\right)^2 = 4\%$

The factor I, describing the interference term, cannot be evaluated in a model independent way, but its value according to the Standard Model is extremely small so that this term is normally neglected. Also the non resonant box diagram contributions are very small, of the order 0.02%, and are neglected.

The only important correction which has not yet been applied is due to the initial state radiation. This is usually done by covoluting the production cross-section at all energies below the machine energy, $\sigma(zs)$, $z_0 < z < 1$, with a function $G(z)$ expressing the probability of radiating energy before the interaction so that the annihilation takes place at a real energy of $\sqrt{zs}$.

The value of z_0 is normally fixed by the lowest energy detectable by the apparatus:

$$\sigma(s) = \int_{z_0}^{1} dz\ \sigma(zs)\ G(z).$$

This correction is large and at the first order modifies the peak cross-section by 30%.

However the full treatment of $G(z)$ at the 2° order leaves only an uncertainty of less then 3.10^{-4}, quite adequate for the experimental data, as shown in table 5.

Table 5

	ΔM_z(Mev)	$\Delta\Gamma_z$(Mev)	$\Delta\sigma_0$(pb)
$e^+e^- \rightarrow \mu^+\mu^-$	1.6	0.7	0.2
$e^+e^- \rightarrow q^+q$	0.1	0.06	0.1

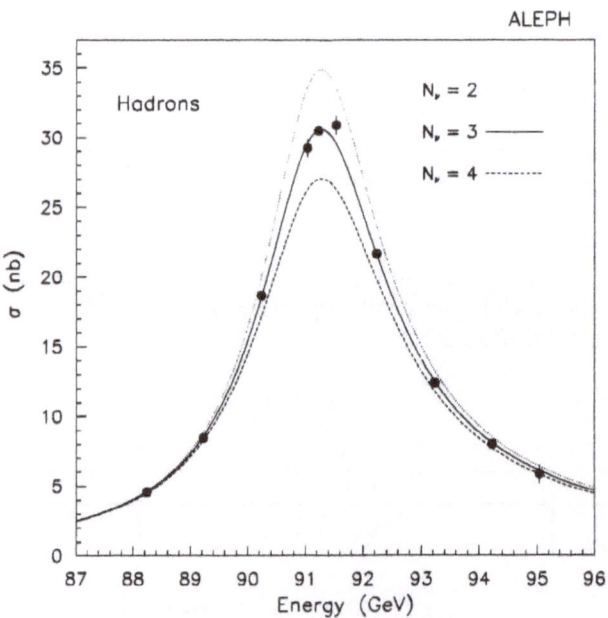

Fig. 16-Cross section for hadron events.

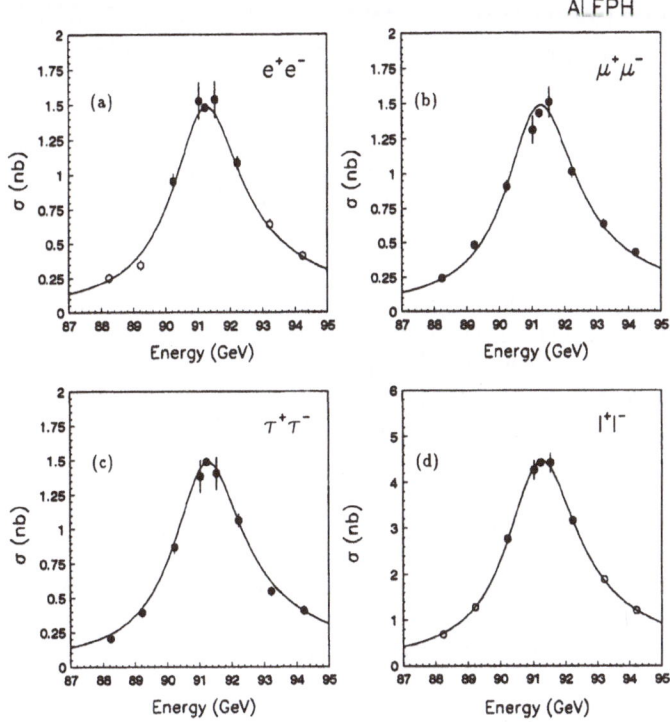

Fig. 17-Cross section for separate flavour events and for the lepton sample.

8-WEAK INTERACTION RESULTS

The experimental data on the shape of the Z resonance, as seen in the hadronic and in the leptonic channels, are shown, together with the fit based on the expression of paragraph 7, in figs. 16 and 17. The first conclusion that can be drawn from them is that the e^+e^-, $\mu^+\mu^-$ and $\tau^+\tau^-$ data agree among themselves to better than 1%, in full agreement with the hypothesis of lepton universality. In the following, very often the average of the three sets of data, called l^+l^- channel, will be used to increase the statistical accuracy. These data are extremely similar in all the four experiments and the fluctuations of individual measurements are normally contained inside one standard deviation. When the analysis of the data of 1990 has been finalized in all collaborations, an effort has been done to obtain the average values of the basic parameters of the Z resonance shown in the first column of table 6[11]. In the meanwhile the '91 data have been analysed separately in each collaboration, but they are not yet available and will be presented at the Dallas Conference in a few weeks time. In order to show the improvement that can be expected, the ALEPH data of '90 and '91 together are given in the second columns of table 6.

Table 6

	All expts (90)	ALEPH (90+91)
M_Z(MeV)	91˙175±21	91˙191±6±9
Γ_Z(MeV)	2˙487±10	2˙501±13
σ_{had}^0(nb)	41.33±0.23	41.53±0.33
Γ_{had}(MeV)	1740±12	1748±12
Γ_l(MeV)	83.24±0.42	84.16±0.54

Apart from the large improvement of the systematic error on M_Z, fully dominated by the measurement of the energy of LEP, the other parameters are determined, using the '90 and '91 data of a single experiment, with an error only 1.2 times larger than in the four experiments with the '90 data alone.

The global parameters of the Z resonance shown in table 6 are only a small fraction of the quantities that can be measured at LEP. The most interesting, together with the individual leptonic widths, are the forward-backward asymmetry in the angular distributions for the leptonic channels and, whenever possible, for the $q\bar{q}$ channels and the polarization of the τ leptons.

Using only the basic theory of weak interactions, all these quantities can be expressed in terms of the vector, g_v^f, and the axial, g_a^f, weak coupling constants.

Obviously these constants are now running quantities, in agreement with the procedure outlined in paragraph 7 and at LEP we can measure $g_v^f\left(M_Z^2\right)$ and $g_a^f\left(M_Z^2\right)$. Each of the measurements quoted above can be used to determine these quantities, as shown in the next paragraphs.

a)-Forward-backward asymmetry A_{FB}

Experimentally, the angular distribution for the leptonic channels is given by

$$\frac{d\sigma^l}{d\Omega} = A^l(s)\left(1+\cos^2\theta\right) + B^l(s)\cos\theta$$

where $B^l(s)=\frac{8}{3}A_{FB}^l$ and

$$A_{FB}^l = 3\left(1+\frac{1-\frac{M_Z^2}{s}}{2g_v^e\,g_v^l}\right)\frac{g_v^e\,g_a^e}{g_v^{e2}\,g_a^{e2}}\,\frac{g_v^l\,g_a^l}{g_v^{l2}\,g_a^{l2}}.$$

At the peak, neglecting g_v^2 with respect to g_a^2, this expression can be simplified to the form

$$A_{FB}^{l,peak} = 3\,\frac{g_v^e\,g_v^l}{g_a^e\,g_a^l}$$

which expresses the interference between the vector and axial parts of the weak current. Outside the peak, the second term dominates due to the smallness of $g_v^e\,g_v^l$ and the asymmetry becomes

$$A_{FB}^{l,out} \simeq 3\left(1-\frac{M_Z^2}{s}\right)\frac{1}{g_{ae}\,g_{al}},$$

an expression which is of no particular interest.

Examples of the asymmetry data obtained by ALEPH in '91 are shown in Fig. 18 and the results at the peak are summarized in table.7.

b)-τ polarization P_τ

The polarization of the τ leptons is given by

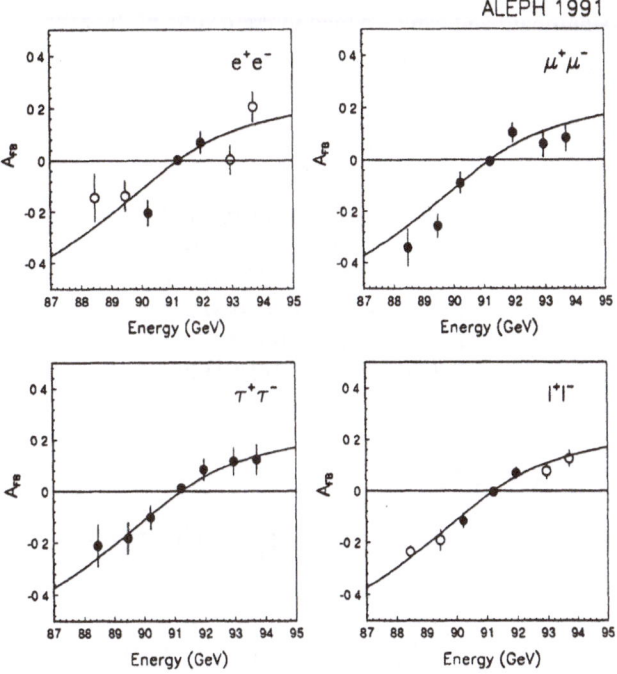

Fig. 18-Forward-backward asymmetry for e^-, μ^-, τ^- and l^- as a function of the centre of mass energy. Points with open circles are not used in the fit.

$$P_\tau = -2 \frac{g_v^\tau / g_a^\tau}{1 + \left(g_v^\tau / g_a^\tau \right)^2}$$

and can be measured in all relevant decay channels of τ's

i) $\tau^- \rightarrow e^- \bar{\nu}_e \nu_\tau$
ii) $\tau^- \rightarrow \mu^- \bar{\nu}_\mu \nu_\tau$
iii) $\tau^- \rightarrow \pi^- \nu_\tau$
iv) $\tau^- \rightarrow \rho^- \nu_\tau$ followed by $\rho^- \rightarrow \pi^- \pi^0$ and $\pi^0 \rightarrow \gamma\gamma$
v) $\tau^- \rightarrow a_1^- \nu_\tau$ followed by $a_1^- \rightarrow \pi^+ \pi^- \pi^-$

Channel iii) is the simplest, since it corresponds to a two body decay and the pion has spin zero.

Channels iv) and v) are again two body decays but the spin 1 of ρ^- and a^- dilutes the memory of the τ^- helicity. Channels i) and ii) have a smaller sensivity being three body decays.

In the first three cases the polarization is extracted from a fit to the energy distribution of the charged particle, e^-, μ^- and π^-. This is easily understood in the $\tau \rightarrow \pi \nu_\tau$ case since, according to its helicity, the τ^- emits the ν_τ preferentially in the forward or backward direction in its rest frame. This corresponds to a smaller or larger pion energy in the laboratory system due to the Lorentz boost.

In the case of channels iv) and v) the final state decay requires the use of additional quantities, such as the π^- angle in the ρ^- rest frame or the angle between the normal to the 3 pion plane in the a_1 rest frame with the a_1 direction.

Because of the limited statistics of each individual sample, the errors on the separate values of P_τ are very large as can be seen in table 8. However the average of all these results over the

'90 and '91 data is significant and can be added to the results of the asymmetry A_{FB}^l to extract the values of g_v^l and g_a^l.

Table 8

$$P_\tau(e) = -0.188 \pm 0.110 \pm 0.058$$
$$P_\tau(\mu) = -0.215 \pm 0.088 \pm 0.036$$
$$P_\tau(\pi) = -0.160 \pm 0.045 \pm 0.026$$
$$P_\tau(\rho) = -0.108 \pm 0.040 \pm 0.022$$
$$P_\tau(a_1) = -0.138 \pm 0.114$$
$$P_\tau(\text{average}) = -0.145 \pm 0.025$$

c)-Determination of g_v^l and g_a^l

Additional data for the determination of g_v^l, g_a^l are obtained from the leptonic Z decay widths which can be written as

$$\Gamma_{ll} = \frac{G_F M_Z^3}{6\sqrt{2}\pi} \left(g_v^{l2} + g_a^{l2} \right) \left(1 + \frac{3}{4}\frac{\alpha}{\pi} \right)$$

From the overall set of data, three values of A_{FB}, five values of P_τ and three values of Γ_{ll}, we get

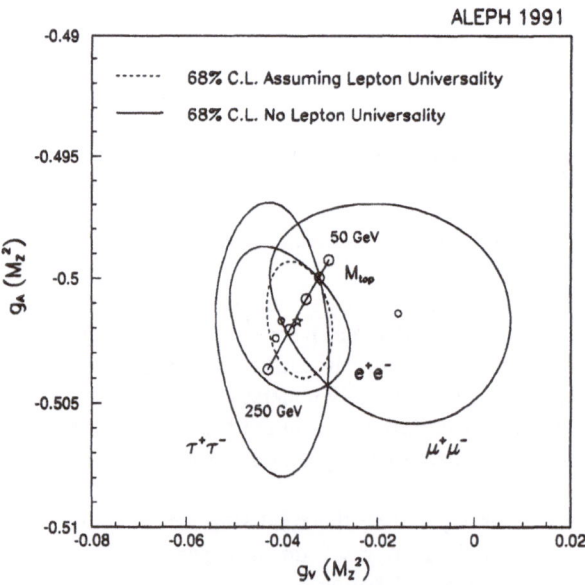

Fig. 19-g_a^l versus g_v^l for separate flavour samples (full lines) and assuming lepton universality (dotted line). The predictions of the Standard Model are shown as a function of M_{top}.

$$g_v^l \left(M_z^2 \right) = - 0.037 \, {}^{+0.006}_{-0.005}$$

$$g_a^l \left(M_z^2 \right) = - 0.501 \pm 0.002$$

averaged over all three leptonic channels, where the relativ sign of g_v^l with respect to g_a^l is provided by the measurement of P_τ and the value of g_v^l, even if small, is clearly different from zero.

This result is shown pictorially in Fig. 19, where the three full contours give the 1σ limits for g_v and g_a for individual leptons, whilst the dotted contour assumes lepton universality.

Without entering into any detail on the experimental procedure, it must be added that the asymmetry can be measured also in the $Z \to q\bar{q}$ events by using the fact that the charge of the leading particles of jets maintains a memory of the charge of the quark and later it will be shown that, in the particular case of $Z \to b\bar{b}$, the asymmetry can be measured by means of the semileptonic decays.

d)-Determination of N_v

We now have all the elements to evaluate the number of light neutrino species with a minimum of help from the predictions of the Standard Model.

The total Z width, Γ_z, can be written as

$$\Gamma_z = \Gamma_h + 3\,\Gamma_{ll} + \Gamma_{inv}$$

where Γ_{inv} corresponds to the decay of the Z boson into all known neutrino types and possibly into further neutrinos or unknown neutral weakly interacting objects with mass smaller than $M_Z/2$.

Dividing the previous expression by Γ_{ll} we get

$$\frac{\Gamma_{inv}}{\Gamma^{ll}} = N_v \frac{\Gamma_v}{\Gamma_{ll}} = \frac{\Gamma_z}{\Gamma_{ll}} - R - 3$$

where we have used $R = \Gamma_h/\Gamma_{ll}$ and we have introduced the number of neutrino families, N_v, which, in the absence of new generations of fermions and of unknown new objects, is expected to be equal to 3.

The value of the cross section into hadrons at the peak can be used to express Γ_z

$$\sigma_h^{peak} = \frac{12\pi}{M_z^2} \frac{\Gamma_{ll}\Gamma_h}{\Gamma_z^2}$$

obtaining

$$N_v \frac{\Gamma_v}{\Gamma_{ll}} = \sqrt{\frac{12\pi R}{M_z^2 \sigma_h^{peak}}} - R - 3$$

where all quantities are directly measured in the experiment, with the exception of the ratio $\frac{\Gamma_v}{\Gamma_{ll}}$.

This ratio must be taken from the Standard Model as

$$\frac{\Gamma_{ll}}{\Gamma_v} = \frac{1}{2}\left[1 + \left(\frac{g_v^l(M_z^2)}{g_a^l(M_z^2)}\right)^2\right]\left(1 + \frac{3}{4}\frac{\alpha}{\pi}\right)$$

Inserting the experimental values of R, σ_h^{peak}, g_v^l and g_a^l, the result of the four experiments with the data of 1990 is

$$N_v = 3.00 \pm 0.005$$

while the ALEPH result with the data of 1990 and 1991 is

$$N_v = 2.98 \pm 0.006$$

The striking agreement of N_v with the integer number 3 strongly supports the conclusion that no new generation of fermions with larger masses exist beyond the known ones.

9-TEST OF THE STANDARD MODEL

The numerical values of all the quantities that have been described in the previous sections are predicted by the Standard Model. If the calculations are extended to take into account one loop weak corrections, the agreement is quite satisfactory for all quantities, and an interesting upper limit can be derived for the unknown mass of the top quark.

At the one loop level, the Standard Model predictions are

$$g_a^l\left(M_z^2\right) = \frac{1}{2}\sqrt{1+\Delta\rho}$$

$$g_v^l\left(M_z^2\right) = g_a^l\left(M_z^2\right)\left[1-4\sin^2\theta_w\left(M_z^2\right)\right]$$

with $\Delta\rho = \frac{\alpha}{\pi}\frac{M_t^2}{M_z^2} - \frac{\alpha}{4\pi}\ln\frac{M_h^2}{M_z^2}$

and

$$\sin^2\theta_w = 1 - \frac{M_w^2}{(1+\Delta\rho)M_z^2}$$

The correction term due to the mass of the Higgs boson is very small and can be neglected at the present level of accuracy. On the contrary, the large value of the mass of the top quark, which is now known to be larger than '91 GeV, the lower limit being set by the CDF collaboration[12], makes the quadratic correction term $\alpha/\pi\, M_t^2/M_z^2$ appreciable, so that many of the values predicted by the Standard Model vary with M_t and the LEP data can be used to estimate the value of this unknown quantity.

We can therefore separate the quantities that are sensitive to the M_t dependence, as Γ_Z, Γ_h, Γ_l, M_Z, g_v^l, g_a^l and $\sin^2\theta_w$ from the quantities that, because of cancellation of the $(1+\Delta\rho)$ factor, are almost insensitive to M_t, as σ_h^{peak} and R.

The last quantities can be used to check the absolute value of the predictions of the Standard Model and the agreement is always better than 1%, as for instance in Fig. 20 where σ_h^{peak} is plotted against R.

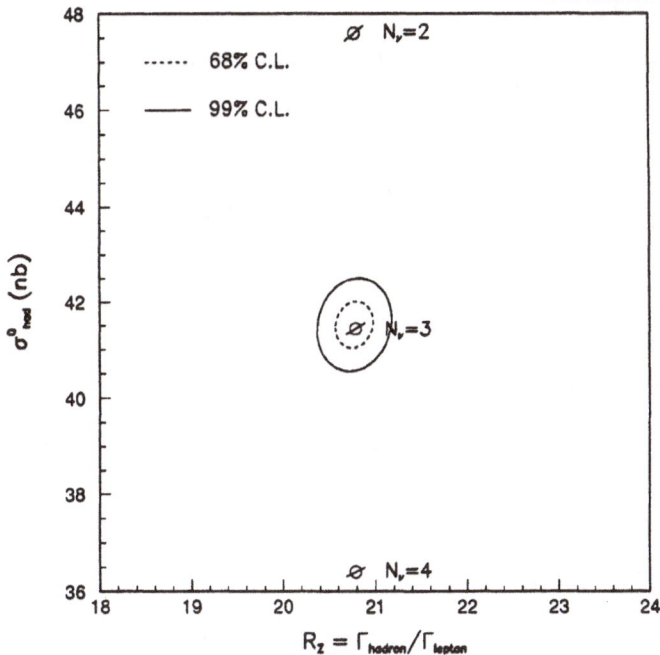

Fig. 20-Comparison of data and Standard Model predictions in the σ_{had}^{peak} versus R plot.

The first group of quantities can be used, on the contrary, to estimate the value of M_t, as for instance in the g_a^l versus g_v^l plot of Fig. 19, where the Standard Model prediction moves along a line passing through the experimental result and favours a M_t value between 150 and 200 GeV.

A very popular way of presenting the best estimate of M_t consists of expressing all measured quantities which are sensitive to M_t as functions of $\sin^2\theta_w$.

The most interesting are:

$$\Gamma_{ll} = \frac{\alpha\left(M_z^2\right)M_z}{48\sin^2\theta_w \cos^2\theta_w} \left[\left(1-4\sin^2\theta_w\right)^2+1\right] \left(1+\frac{3}{4}\frac{\alpha}{\pi}\right)$$

which is sensitive to $\Delta\rho$ through M_z, the values of Γ_z, Γ_h and

$$M_z^2 = \frac{\pi\alpha\left(M_z^2\right)}{\sqrt{2}\ G_F\ (1+\Delta\rho)\sin^2\theta_w\cos^2\theta_w}.$$

They can be compared with the constant value of $\sin^2\theta_w$ which is obtained from the P_τ and A_{FB}^l data.

In addition, $\sin^2\theta_w$ can be measured at other machines and in completely different experiments, as at

1) $p\bar{p}$ **Machines**, where the ratio M_w/M_z has been measured with high precision by CDF[13] and UA2[14]:

$$\sin^2\theta_w = 1 - \frac{M_w^2}{\left((1+\Delta\rho)M_z^2\right)}$$

and

2) **neutrino beams**, where the ratio of the neutrino neutral current to charged current cross sections can be expressed as a function of $\sin^2\theta_w$:

$$R = \frac{\sigma^\nu_{NC}}{\sigma^\nu_{CC}} = \left(\frac{M_w}{M_z}\right)^4 \frac{\frac{1}{2}\sin^2\theta_w + \frac{5}{4}(1\text{-}r)\sin^4\theta_w}{\cos^2\theta_w}$$

and the quantity $r = \sigma^{\bar\nu}_{CC}/\sigma^\nu_{CC}$ is measured at the same time[15].

Fig. 21 shows the results for $\sin^2\theta_w$, as obtained from all these measurements, as function of M_t, averaged over the 1990 data of the four collaborations.

The best fit value of M_t, with only LEP data, fixing $\alpha_s = 0.118 \pm 0.008$, is

$$M_t = 124\,{}^{+40+21}_{-56-21}$$

Using also $p\bar{p}$ and neutrino data the result is

$$M_t = 132\,{}^{+27+18}_{-31-19}$$

The ALEPH data alone, summed over 1990 and 1991, in conjunction with $p\bar{p}$ and neutrino data give

$$M_t = 162\,{}^{+25+16}_{-29-19}.$$

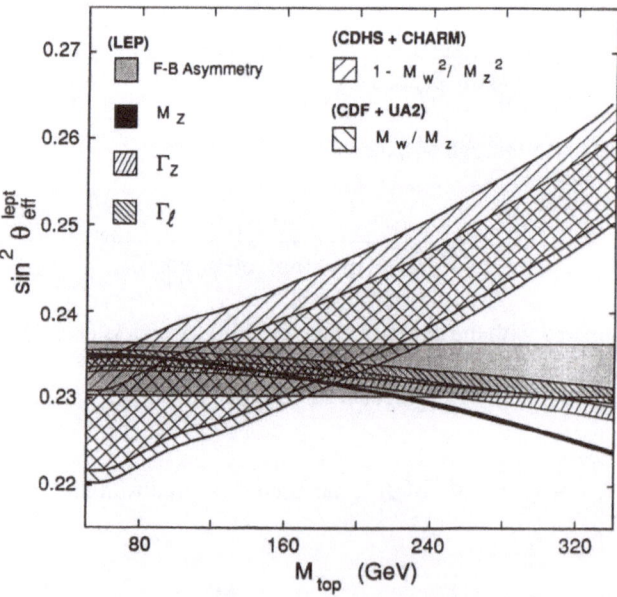

Fig. 21-$\sin^2\theta_w$ as a function of the mass of the top quark from various measurements as explained in the text.

10-B PHYSICS AT LEP

B physics is becoming, year after year, one of the main subjects of investigation at LEP. The possibility of identifying the Z→b$\bar{\text{b}}$ final state offered by the heavy mass of b-quarks allows a variety of measurements ranging from the Z decay properties to the characteristics of the b quark.

Up to now, Z→b$\bar{\text{b}}$ events have been selected by identifying the semileptonic decay of b quarks.

The main criteria are the following:

1) b-quarks, being heavy, hadronize into B hadrons which maintain a substantial fraction of the quark momentum (on average 0.7). Indeed the total energy transferred to hadronization particles is smaller than in the case of light quarks. This means that the lepton of the semileptonic decay, basically a 3 body process, is energetic.

2) The transverse momentum of the lepton, which is a Lorentz invariant, can reach the value of 2.5 GeV/c, because of the large mass of the b-quark, whilst the leptons from charm decay cannot even reach 1 GeV/c.

The standard procedure for selecting b$\bar{\text{b}}$ events proceds then through the following steps:

i) look for events with at least one identified lepton (e, μ);

ii) reconstruct charged particles and neutral clusters;

iii) construct jets according the rule

$$M^2_{12} = 2E_1E_2 \left(1-\cos\theta_{12}\right)$$

where 1,2 are the charged particles or neutral clusters providing the smallest invariant mass. If the jet contains the identified lepton, particle 1 is the lepton itself;

iv) if y= M^2_{12}/M^2_{vis} is smaller than a preselected cut, the two particles are merged into a single object and the procedure is repeated with a new particle or a new neutral object. As new particle 2 the one providing the smallest invariant mass is chosen again. The procedure continues until no new particle allows y to be smaller than the cut. For b-physics the normal choice of the cut is $y_{cut} = 0.0044$ corresponding to an invariant mass of 6 GeV, just above the b-quark mass;

v) define the jet axis as the direction of the total momentum of the particles associated in the jet, including (or excluding) the lepton. The question wether it is more efficient to include or exclude the identified lepton in the definition of the jet direction has been going on for three years and different methods of analysis prefer different choices. If we choose to include the lepton, the best cut on the transverse momentum of the lepton with respect to the jet is of the order of 1 GeV/c.

The main condition to select b$\bar{\text{b}}$ events will then be

$$P_{lepton} > 3 \text{ GeV/c}$$

$$P_{\perp \, lepton} > 1 \text{ GeV/c}$$

Fig. 22 shows the $p_\perp$ spectra of electrons and muons in the described selection, with the contributions, estimated with a MonteCarlo simulation, of the processes

b→l+X
b→c→l+X
c→l+X
q→hadron (π,κ)→l+X
q→hadron (misidentified as leptons)

among which only the first one, and sometimes the second is of real interest. It is clear that a cut at $p_\perp$=1 GeV/c provides a strong rejection against background processes, even if it is at the expenses of a small efficiency in the desired channel. Table 9 shows the contribution of the various channels to the selected sample and the final b tagging efficiency.

Table 9

channel	e(%)	μ(%)
b→l	83	75
b→c→l	7	7.5
c→l	4.6	5.3
hadrons (mis.)	4.3	11
Total efficiency for b	6	8

As an example of the measurements on Z decay properties allowed by this selection, Fig. 23 shows the angular distribution of leptons coming from b decay, providing the asymmetry A_{FB}^{b}.

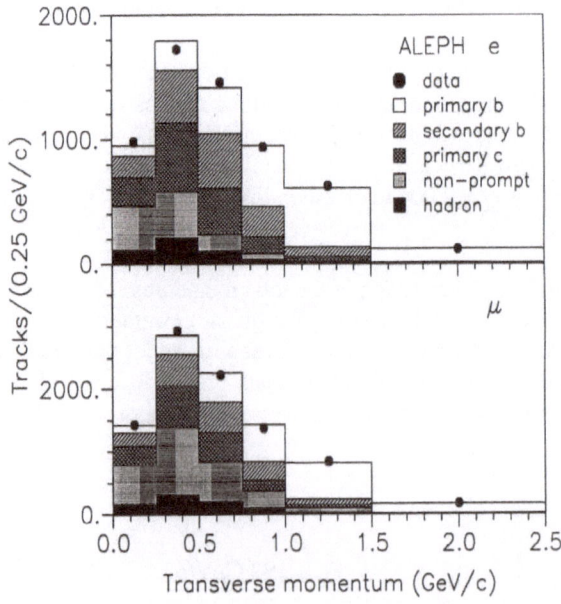

Fig. 22-$p_\perp$ spectra for electrons and muons, showing the contribution of the various processes which produce leptons.

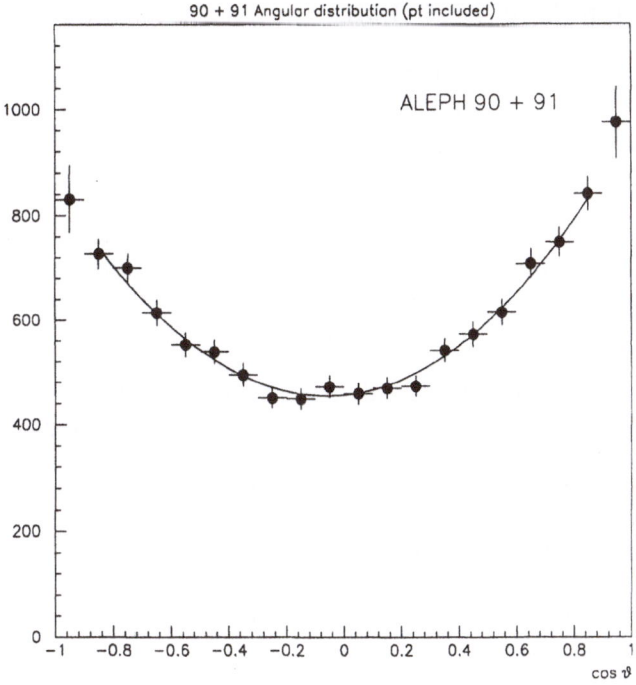

Fig. 23-Angular distribution for b⁻ quarks in the Z decay, showing the forward-backward asymmetry.

This asymmetry is expressed as

$$A_{FB}^b = 3 \, \frac{g_v^e \, g_a^e}{g_v^{e2} + g_a^{e2}} \, \frac{g_v^b \, g_a^b}{g_v^{b2} + g_a^{b2}}$$

and since $g_v^b/g_a^b = 1 - 4/3 \sin^2\theta_w \approx 0.69$ is about 10 times larger than g_v^l/g_a^l, owing to the fractionary charge of the quark, the b asymmetry is in principle more sensitive than the lepton asymmetry. Indeed the distortion in the angular distribution is perfectly visible in the data. However the smaller purity and the small efficiency of the sample compensate this larger sensitivity. In addition the true asymmetry must be corrected for the B-Bmixing which changes the charge of the decay lepton in about 15% of the cases. Even if direct measurements of the mixing are produced by all LEP experiments by studying di-lepton events, the error on this result is the main source of uncertainty in A_{FB}^b. Indeed, the row asymmetry is, in ALEPH '90+'91 data,

$$A_{FB}^b = (4.7 \pm 0.9)\%$$

which, after the correction for the mixing, becomes

$$A_{FB}^b = (6.7 \pm 2.1 \pm 0.54)\%.$$

The measurements of the mixing and of the asymmetry of B's is an interesting experimental challenge in itself. However they will not be described here in detail, since the next part of this paper will concentrate on the measurement of lifetimes, which is the best example of how the introduction of a new precision device such as the minivertex detector can change the quality of the results, reducing the resolution on the track points from ~150 μm to ~15 μm.

11-MEASUREMENTS OF LIFETIMES

At LEP the Lorentz boost is rather large so that the mean path of B hadrons is of the order of millimetres. The same is true also for τ's, which indeed have a shorter lifetime, compensated however by the larger momentum and the smaller mass.

Two main methods are used to measure the lifetime of unstable particles:

1) *Impact parameter method.* In the projection orthogonal to the beams the impact parameter is defined as the minimum distance between the trajectory of a charged particle and the production point of the event. A sign is attribuited to it by means of an ad hoc convention. It is positive if the crossing point of the charged particle and the flight direction of the parent is in the same hemisphere as the charged particle, negative otherwise. The definition is the same if the analysis is done in 3-dimensions or is restricted to the projection in the plane transverse to the beams. In the first case, however, "crossing point" must be replaced by "minimum approach distance".

2) *Decay length method.* The decay length is the distance between the production and the decay points of a short lived particle. Its measurement requires the identification of the primary and of the secondary vertices as the crossing points of several (at least two) charged tracks.

The main advantage of the impact parameter method lies in its independence, on an average, of the particle momentum, since

$$<\delta> \cong c\tau$$

This is particularly important for B's, because of the unknown momentum fraction subtracted by the hadronization particles. It loses however the advantage of the boost, which helps in measuring the decay length.

a-Inclusive measurement of τ_B

B-hadrons, irrespective of their charge or nature (bosons or barions), are selected as outlined before with a purity of 95% for the electron and 88% for the muon decay channel. In this case the cascade decay, $b \rightarrow c \rightarrow l$, is as good as the direct semileptonic decay. A probability function is then defined as

$$P = \text{Resolution function} \otimes \text{Physics function}$$

where the *physics function* is the true δ-distribution in the MonteCarlo simulation before detector smearing, already signed by means of the reconstructed jet direction, and the *resolution function* is the distribution taken from the data of δ/σ_δ for all hadrons. Only the negative part of the distribution is selected to avoid the memory of the decay time, is then corrected for the difference between leptons and hadrons calculated by the MonteCarlo program, and is fitted with two gaussians. This is repeated for all processes contributing to the data and a likelihood function is calculated

$$L = \Pi \left[f_b\, P_b(\tau_B) + f_{bc}\, P_{bc}(\tau_B,\tau_D) + f_c\, P_c(\tau_c) + f_m\, P_m + f_d\, P_d \right]$$

where f_b, f_{bc}, f_c are the weights given by the MonteCarlo programme for the decays $b \rightarrow l, b \rightarrow c \rightarrow l$, $c \rightarrow l$, f_m is the fraction of misidentified leptons and f_d is the fraction of π/K decays into muons.

Fig. 24a shows the final data and best fit curve, with the small contribution of the background terms. This must be compared with the equivalent data of 1990, without a minivertex detector, in Fig. 24b.

The ALEPH preliminary result from the 1991 data is

$$\tau_B = (1.49 \pm 0.03 \pm 0.06) \text{ ps.}$$

If confirmed, this figure is larger than the previous world average (1.24 ± 0.07 ps), and also of the results of DELPHI (1.30 ± 0.13 ps)[16] and ALEPH (1.29 ± 0.12) obtained in 1990.

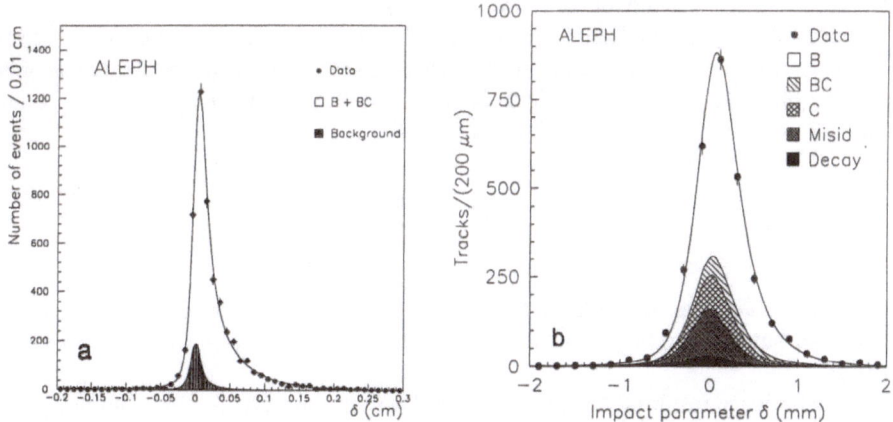

Fig. 24-Impact parameter distribution for a) 1991 data and b) 1990 data (without minivertex detector) with background contributions.

b-τ_B inclusive measurement in events with J/ψ

A different approach to the measurement of the lifetime of b-quarks, irrespective of the charge and of the nature of the produced hadrons (B^0, B^+, B_s, Λ_b, ...) consist in tagging the events with the presence of a J/ψ in the final state. This decay can be described as in the sketch below

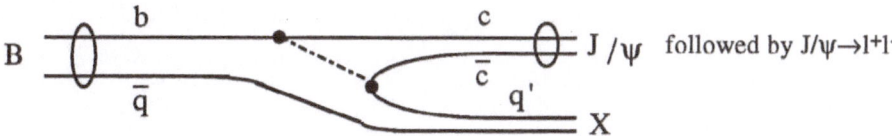

There are two advantages of this process:

1) The background, mainly due to b$\rightarrow$cl$^-\bar\nu$ followed by c$\rightarrow$l$^+\nu$x, is very small since the two leptons must by chance have an invariant mass in a narrow window around $m_{J/\psi}$. Also the hadronization of hard gluons into J/ψ's adds a negligible contribution to the background.
2) Since the J/ψ decays almost immediately into lepton pairs, the dilepton vertex coincides with the b-decay vertex and gives an ideal way to find and measure its position.

The price to pay for this favourable experimental situation is the small branching ratio BR(b→J/ψ+x)=(1.24±0.13)% which forbids the measurement of the lifetime for exclusive channels, even if a very clean peak can be reconstructed with few events in the $B^{\pm}\to J/\psi K^{\pm}$ channel. It should be noticed that this inclusive measurement of τ_B coincides with the previous one only if the ratio BR(b→lX)/BR(b→J/ψX) is equal for all B hadrons.

In this measurement the quantity that can be directly measured is the decay length, i.e. the distance between the b-decay point, defined as the point of minimum distance between the two Leptons of the J/ψ, and the b-jet production point. This is identified by the following procedure: jets are reconstructed with the method outlined at the beginning of the section and all tracks belonging to a jet are projected in a plane orthogonal to its axis. The vertex formed in this plane has no memory of the b lifetime and can move freely along the jet direction. This is repeated for all jets in the event. The jet production point is then defined as the point which is most consistent with these points and the centre of the average beam spot.

The decay length is not a Lorentz invariant as the impact parameter and the b momentum is needed, event by event, to evaluate the proper time. An estimate of the jet momentum is obtained by summing the momenta of the particles belonging to the jet which contains the J/ψ and by defining

$$(\gamma\beta)_{Jet} = \frac{\Sigma \vec{P_i}}{M_{Jet}}$$

A comparison of the quantity with the true value of $(\gamma\beta)_b$ at the MonteCarlo level shows that $(\gamma\beta)_{Jet}$ is distributed around $(\gamma\beta)_b$ with a very small overall correction (~1%) and a spread of about 10%.

The preliminary result of this analysis based on the time distribution shown in Fig. 25, which will improve with the increased statistics of 1992, is

$$\tau_B = \left(1.35 \; {}^{+0.19}_{-0.17} \pm 0.06\right) \text{ ps}$$

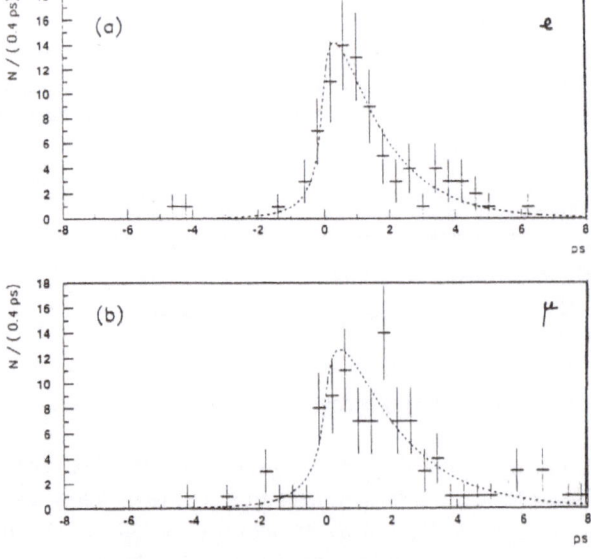

Fig. 25-Decay time distribution for J/ψ lX events.

c-Study of exclusive b channels

On the hypothesis that B-hadrons decay via the "spectator diagram"

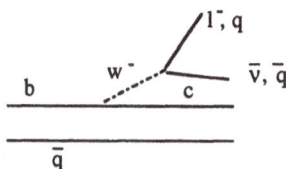

the lifetime of B^0, B^+, B_s, Λ_b and all other b states would be identical. However other diagrams as $b\bar{q}$ annihilation or W exchange contribute differently to different channels, as largely demostrated in the charmed particle family where the lifetimes can be different up to a factor of 4. The theoretical predictions suggest that these differences should be reduced to about 10% in the case of the B family. However these estimate are rather model dependent and direct measurements are needed. To move in this direction, several attempts are underway and look very promising for next year with more than doubled luminosity.

As a first example a systematic study of the b-hadron decays has been done in the channels

$$B_s \to D_s X l^+ \nu \qquad\qquad D_s^- \to \phi\pi^-$$
$$\Lambda_b \to \Lambda_c^+ X l^+ \nu \qquad\qquad \Lambda_c^+ \to pK^-\pi^+$$

The high momenta of the decay products and the large invariant mass of the Lepton-charmed hadron system are good ways to reject the background. However no mass peak can be reconstructed owing to the presence of the neutrino in the final state.

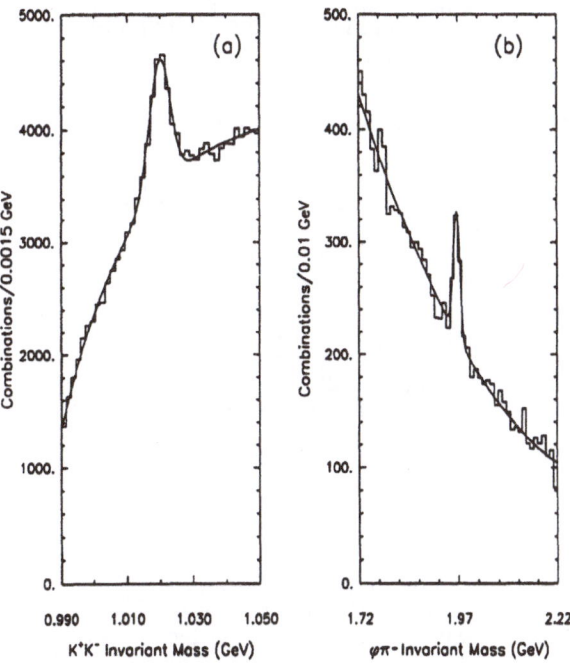

Fig. 26-a) invariant mass of the $K^+ K^-$ system with the ϕ peak
b) invariant mass of the $\phi\pi^-$ system with the D_s peak

The B_s search consists of the following steps:

a) reconstruct the φ as a peak in the K+K- invariant mass, using the dE/dx information provided by the TPC on each track (Fig. 26a)

b) combine the ϕ with every π^- with energy larger than 3 GeV to build the D_s candidate (Fig. 26b)

c) combine the D_s with the lepton requiring a lD_s invariant mass larger than 3 GeV

d) plot the D_s mass for right $\left(D_s^- l^+ \right)$ and wrong $\left(D_s^- l^- \right)$ sign combination, as in Fig. 27.

A clear peak is visible in the first case, nothing in the second, indicating that the source of the events in the peak is indeed the decay of a B_s boson.

A similar strategy allows the identification of Λ_b reconstructing at first the Λ_c^+, through the decay $\Lambda_c^+ \to pK^-\pi^+$ and combining it with the lepton again in the right and wrong sign combinations (Fig. 28).

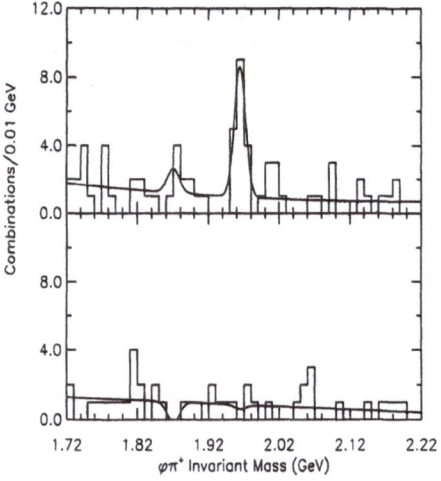

Fig. 27-Invariant mass spectrum of the $\varphi\pi$ system for the right sign $\left(D_s^- l^+ \right)$ and a wrong sign $\left(D_s^- l^- \right)$ combinations.

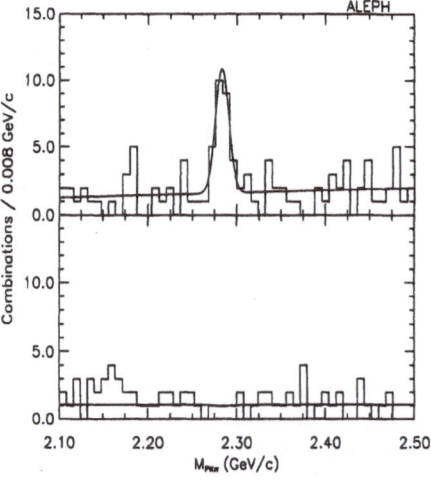

Fig. 28-Invariant mass spectrum of the $pK^-\pi^+$ system for the right sign $\left(\Lambda_c^+ l^- \right)$ and wrong sign $\left(\Lambda_c^+ l^+ \right)$ combinations.

The final samples of D_s (~15 events) and Λ_b (~20 events) obtained with these analyses are not yet sufficient for a measurement of the individual lifetimes. A richer sample can be selected by applying the same analysis to the process

$$\Lambda_b \rightarrow \Lambda_c^+ \, l^- \nu$$
$$\hookrightarrow \Lambda X$$
$$\hookrightarrow p\pi^-$$

and by studying the sign correlations of Λ's and Leptons as shown in Fig. 29. From these data a first value of the Λ_b lifetime

$$\tau_{\Lambda_b} = \left(1.12 \, {}^{+0.29}_{-0.26}\right) \text{ps}$$

has been derived.

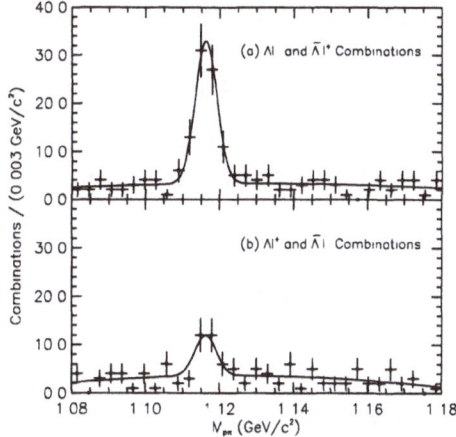

Fig. 29-Invariant mass spectrum of the $p\pi$ system for the right sign $\left(\Lambda l^-\right)$ and wrong $\left(\Lambda l^+\right)$ combinations.

A different example is given by the direct measurement of the lifetimes of $\overline{B}^0$ and B^+ through the reconstruction of the D boson in the final state:

$$\left\{ \begin{array}{l} \overline{B}^0 \rightarrow D^{*+} l^- \overline{\nu} \\ \overline{B}^0 \rightarrow D^+ l^- \overline{\nu} \\ \overline{B}^0 \rightarrow D^{(*)} \pi l^- \overline{\nu} \end{array} \right. \qquad \left\{ \begin{array}{l} B^- \rightarrow D^{*0} l^- \overline{\nu} \\ B^- \rightarrow D^0 l^- \overline{\nu} \\ B^- \rightarrow D^{(*)} \pi l^- \overline{\nu} \end{array} \right.$$

based on the identification of the decays

$$D^0 \rightarrow K^- \pi^+$$
$$D^0 \rightarrow K^- \pi^+ \pi^- \pi^-$$

and

$$D^{*+} \to \pi^+ D^\circ.$$

Once the D° in one of the two decay channels is reconstructed, the D^{*+} bosons are identified by means of the very soft π^+ by requiring

$$\left| m_{D^\circ \pi} - m_{D^\circ} - 0.145 \right| < 0.0015 \text{ GeV}.$$

The sample of events containing a D° which does not come from a D^* decay is selected by the condition that for every pion in the jet

$$\left| M_{D^\circ \pi} - m_{D^\circ} - 0.145 \right| > 0.0015 \text{ GeV}.$$

Again the selection of D states coming from a B decay is provided by the presence of the Lepton in the jet, with the requirement of an invariant Dl, D*l mass larger than 3 GeV.

We have then two samples, without overlap, one of $D^\circ l$, enriched in B^- parents, and one of $D^{*+}l$, enriched in $\bar{B}^\circ$ parents. For both of them the decay length distribution is fitted with the weighted sum of three functions, one containing τ_{B^-}, the second containing τ_{B° and the third the apparent lifetime of the background, weighted according to the branching ratios.

However, care must be taken of the fact that the weights depend also on the ratio $\tau_{B^-}/\tau_{B^\circ}$, since, for semileptonic events,

$$\Gamma(B^- \to l^- X) = \Gamma\left(\bar{B}^\circ \to l^- X\right)$$

and consequently

$$\frac{BR\left(B^- \to l^- x\right)}{\tau_{B^-}} = \frac{BR\left(\bar{B}^\circ \to l^- x\right)}{\tau_{B^\circ}}$$

For the evaluation of the B momentum the same procedure as in the previous section is used.

The proper time distribution of the two samples is shown in Fig. 30 and the preliminary result is

$$\frac{\tau_{B^-}}{\tau_{B^\circ}} = 0.83 \begin{smallmatrix} +0.18 & +0.13 \\ -0.15 & -0.15 \end{smallmatrix},$$

which is consistent with unity and of no interest at the moment, but which will become significant for the data collected in 1992.

d-τ lifetimes

The measurement of the lifetime of τ's, thanks to the very clean topology of their decays, is a good example of the precision studies which are possible at LEP. Both methods described before, namely the impact parameter and the decay length, have been extensively used, the first one on one prong decays (1-1 events) the second on three body final states (1-3, 3-3 events).

An important difference between τ and b lifetime measurement lies in the fact that the b production vertex can be identified by means of the particles produced in the quark hadronization, whilst the production vertex of the τ⁺τ⁻ pairs is invisible and the measurements (impact parameter or decay length) must be made with respect to the centre of the interaction volume.

A second difference is given by the uncertainty in the τ direction, which for one prong decays can only be identified by the single charged particle. In order to cope with one or the other of these difficulties two new methods have been developed at ALEPH which make use of the full event, assuming that the τ⁺ and τ⁻ are produced back to back. This approximation is quite acceptable if the measurement is done in the projection orthogonal to the beam line, which does not feel the effect of initial state radiation. They consist in measuring the distribution of the sum and of the difference of the impact parameters.

The first method requires the determination of the distance between the tracks at their point of closest approach to the beam axis, as shown in Fig. 31.

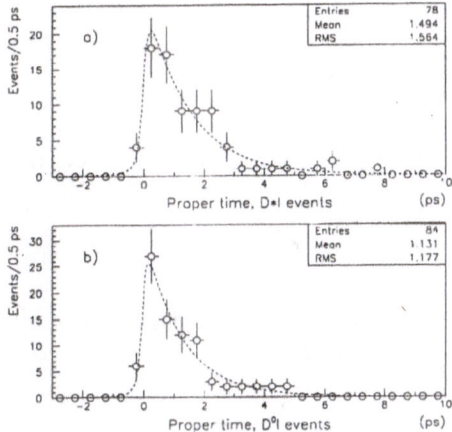

Fig. 30-Proper time distributions for a) the D*l and b) the D⁰l samples fitted with the expected mixture of B̄° and B⁻ decays.

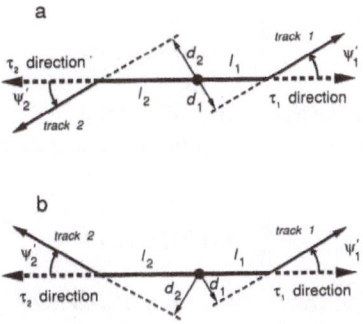

Fig. 31-τ⁺ and τ⁻ production and decay with the definition of the kinematical variables.

Each individual impact parameter with respect to the beam axis is signed according to the convention that the sign is positive if the track moves counterclockwise with respect to the beam, negative in the opposite case. For collinear tracks the distribution of the sum is completely independent of the position of the production vertex, if the angles ψ_i are not equal a small contribution depending on the acollinearity and on the beam size appears. This method suffers however from the uncertainty on the $\tau^+\tau^-$ axis defined as the event sphericity. The distribution of the sum of impact parameters for single prong τ decays is shown in Fig. 32, and is compared with the same quantity for 45 GeV $\mu^+\mu^-$ from Z decays.

The second method makes use of the fact that the average impact parameter difference $<d_+ - d_->$ is proportional to the average decay length and therefore to τ_τ. This method is independent of the determination of the τ direction, while the beam size contributes doubly to the smearing on the impact parameter difference with a corresponding reduction of statistical precision.

The three methods give the following results

impact parameter sum:	$\tau_\tau=288.3\pm7.5\pm5.1$ fs
impact parameter difference:	$\tau_\tau=310.5\pm13.0\pm3.5$ fs
decay length:	$\tau_\tau=298.0\pm10.6\pm4.5$ fs.

The average of these results is

$$\tau_\tau=295.5\pm5.9\pm3.1 \text{ fs}$$

which, combined with the previous ALEPH results, adjusted to the same, new, value of the τ mass, gives

$$\tau_\tau=294\pm5.4\pm3.0 \text{ fs}.$$

This number can be used to test the μ-τ universality, together with the measurement of the $BR\left(\tau^-\to e^-\bar{\nu}_e\nu_\tau\right)$ through the relationship

$$\left(\frac{g_\tau}{g_\mu}\right)^2 = \left(\frac{m_\mu}{m_\tau}\right)^5 \frac{\tau_\mu}{\tau_\tau} BR\left(\tau^-\to e^-\bar{\nu}_e\nu_\tau\right).$$

The new revised value of m_τ recently available[17,18] and the new result of ALEPH on the τ branching ratios into electrons and muons (averaged together on the basis of the e-μ universality) give as a result

$$\frac{g_\tau}{g_\mu} = 0.997\pm0.016$$

in good agrement with the hypothesis of e-μ-τ universality.

Just before these new results and before the recent new measurements of the mass of the τ, the value was

$$\frac{g_\tau}{g_\mu} = 0.974\pm0.013$$

suggesting a possible deviation from universality at the level of 2σ, which seems now very unlikely.

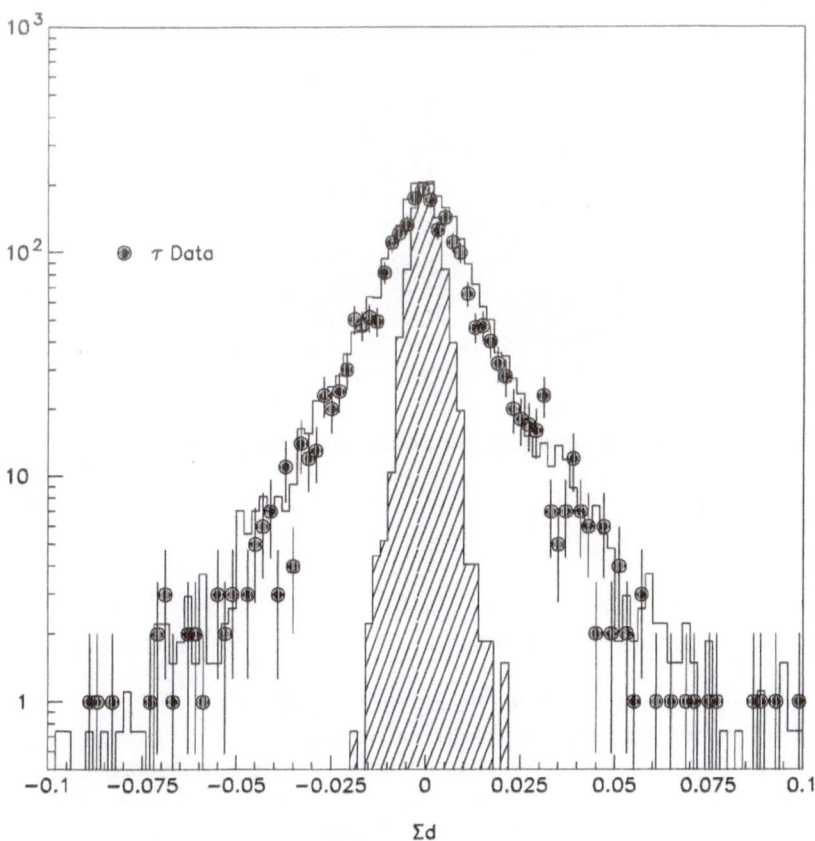

Fig. 32-Distribution of the sum of the impact parameters for the single prongs in the τ decay, showing the broadening due to the τ-lifetime, compared with the same quantity for $Z \rightarrow \mu^+\mu^-$ events.

12-BRANCHING RATIO FOR THE DECAY B→τ⁻ν̄$_\tau$ X

To complete this section which was devoted to the physics of B's and τ's, I am pleased to present a new result of ALEPH on the determination of the decay B→τν̄$_\tau$ X branching ratio, the largest of the decay modes not yet seen, which is predicted to be

$$BR = (2.83 \pm 0.31)\%.$$

This channel is difficult to identify since the presence of the ν_τ in the final state forbids the identification of individual events and its study must be performed by using the large missing mass characteristic of these events. The analysis therefore uses the "Energy flow" algorithm which reconstructs the visible energy as the sum of the charged track momenta, of the energy of the identified photons and of the energy of neutral hadrons as seen by the calorimeters, correcting for the double counting due to electrons or muons. The missing energy of events is shown in Fig. 33 for light quarks with e, μ rejection together with the MonteCarlo predictions.

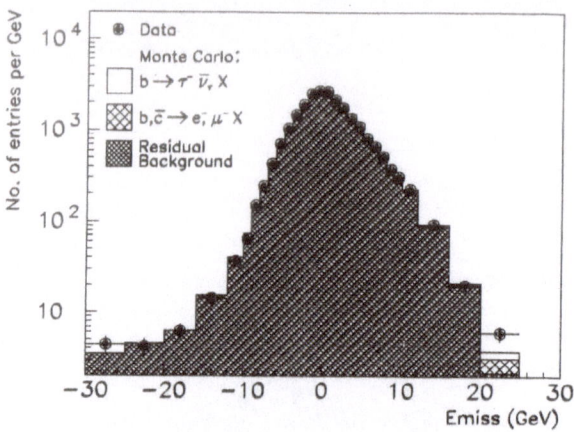

Fig. 33-Missing energy reconstructed for light quark events.

The analysis proceds in the following way:

i) measure the missing energy with the "Energy flow" algorithm
ii) use only events well contained in the detector ($\cos\theta<.7$)
iii) divide the events in two halves, orthogonally to the thrust axis, and define

$$E_{miss}^{ev.} = E_{miss}^{1half} + E_{miss}^{2half}$$

which allows a gain of a factor $\sqrt{2}$ since

$$\sigma^2\left(E_{miss}^{ev.}\right) = 2\sigma^2\left(E_{miss}^{half}\right)$$

iv) throw away the half events having electrons or muons, since they have associate neutrinos
v) select $B\bar{B}$ events searching for large impact parameter tracks with the minivertex detector.

This selection provides a good efficiency (77%) with an acceptable purity (81%).

Fig. 34 shows the missing energy for the selected events and the contribution of the events due to $B\rightarrow\tau X$ decay visible in the tail on the right hand side. The last value of BR is

$$BR\left(B^-\rightarrow\bar{\tau}\,\bar{\nu}_\tau X\right) = \left(4.20 \,^{+0.72}_{-0.68}\pm0.52\right)\%$$

It has been evaluated that the contribution of events with $D_s\rightarrow\tau X$, both coming from B decay or direct charm production is negligible compared to the present accuracy.

This is the first example of an analysis in which the b identification relies fully on the minivertex informations, a first hint of what will be the basic strategy at LEP 200 when searching for Higgs bosons.

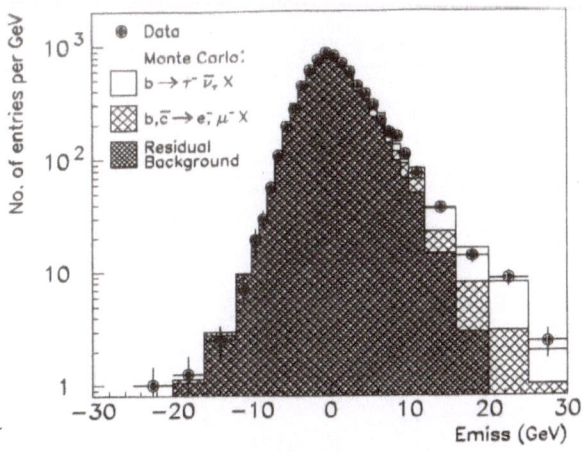

Fig. 34-Missing energy for selected events compared with the MonteCarlo predictions including the B→τX decay.

References

1. L. Knudsen et al., Phys. Lett. **B270** (1991), 97.
2. L. Arnaudon et al., Phys. Lett. **284B** (1992), 431.
3. D. Decamp et al., CERN -PPE-91-129.
4. S. Jadach et al., Phys. Lett. **B268** (1991), 253; S. Jadach and B.F.L. Ward, Phys. Rev. **D40** (1989), 3582.
5. D. Decamp et al., Nucl. Instrum. Methods **A294** (1990), 121.
6. P. Aarnio et al., Nucl. Instrum. Methods **A303** (1991), 233.
7. B. Adeva et al., Nucl. Instrum. Methods **A289** (1990), 35.
8. K. Ahmet et al., Nucl. Instrum. Methods **A305** (1991), 275.
9. W. Beenakker et al., Nucl. Phys. **B355** (1991), 281.
10. A. Borrelli et al., Nucl. Phys. **B333** (1990), 357; M. Martinez et al., Zeit. Phys. **C49** (1991), 645; W. Hollik, 1989 CERN-JINR School of Physics, Egmond-aan-Zee (CERN 91-07, 1991) p. 50.
11. The LEP Collaborations Phys. Lett. **276B** (1992), 247.
12. F. Abe et al., (CDF Collaboration) Phys. Rev. Lett. **68** (1992), 447 and Phys. Rev. **D45** (1992), 3921.
13. F. Abe et al. (CDF Collaboration) Phys. Rev. **D43** (1991), 2070.
14. J. Alitti et al. (UA2 Collaboration) Phys. Lett. **B241** (1990), 150.
15. A. Blondel et al. (CDHS Collaboration) Z. Phys. **C45** (1990), 361; J.V. Allaby et al (CHARM Collaboration) Z. Phys. **C36** (1987), 611.
16. P. Abreu et al., Z. Phys. **C53** (1992), 567.
17. J.Z. Bai et al. (BES Collaboration) SLAC-PUB-5870 and BEPC-EP-92-01 (1992).
18. H. Albrecht at al. (ARGUS Collaboration) DESY-92-086 (1992).

ELECTROWEAK PRECISION TESTS: WHAT DO WE LEARN?

R. Barbieri

CERN, Geneva, Switzerland
Dept. of Physics, Univ. Pisa, and
INFN, Sezione di Pisa, Italy

1. INTRODUCTION

The experimentation at the Z resonance at LEP makes it possible, for the first time, to study with great precision some properties of the electroweak interactions at distances of the order of the inverse Fermi scale, $G_F^{1/2}$. An immediate question arises: What is it that we learn from these studies? The answer is not an easy one. It may in fact require some time and some complementary studies (both experimental and theoretical) to be able to fully appreciate the significance of these precision tests of the electroweak interactions. What is sure is that a judgment of the impact of these tests requires an appreciation of several different aspects of the theory of the electroweak interactions themselves. In particular, this appreciation is not achieved, I believe, by only making a comparison of the Standard Model [1] (SM) predictions with the experimental results.

This is the physical background that motivates these lectures. It is actually my opinion that the experimental program being carried out at LEP 1, especially in association with the direct search for the top quark at Fermilab, although indirectly, may shed light on the symmetry breaking issue in electroweak physics. My purpose, rather than trying to convince the reader of this statement, is to make him able to judge by himself.

To discuss this subject, one has to master the problem of calculating the radiative corrections in the SM, as well as in alternative more or less defined theories, a technically heavy and often boring task. For this reason, when discussing these issues, one frequently ends up by making reference to other people's work for detailed calculations and similar technical things. This is an attitude that I would like to avoid here, aiming rather at being as self-contained as possible. The reader is however expected to be familiar with the basic properties and techniques of gauge field theories as well as with the Lagrangian of the SM. During the course, I have actually devoted one lecture to this last issue, but its content is not described here. In this respect, it may be useful to consult the lectures of Luciano Maiani [2] in Cargèse 90 or those of Guido Altarelli [3] in Les Houches 91.

Quantitative Particle Physics, Edited by
M. Lévy *et al.*, Plenum Press, New York, 1993

2. DEFINITION OF THE THEORY - BASIC OBSERVABLES

As I have made it clear in the Introduction, we do not want to restrict our attention to the SM only. Rather I consider a generic theory that fulfils the following requirements:

i) The gauge group is SU(2)×U(1);

ii) The spectrum includes the standard three generations of fermions with the usual SU(2)×U(1) assignments;

iii) At the tree level, the vector boson masses are related to the gauge couplings and to a scale parameter v by the relations

$$M_W^2 = \frac{g^2 v^2}{2} \ , \quad M_Z^2 = \frac{(g^2 + g'^2) v^2}{2} \ . \tag{2.1}$$

The first two requirements do not need any comment, except perhaps for saying that I will not consider extensions of the gauge group as, e.g., extra U(1) factors. The third condition is dictated by the experimental observation that the ρ-parameter, to be precisely defined later on, equals one within few per mille. In this way, among other things, I certainly include the SM itself, a generic multi-Higgs doublet model, the Minimal Supersymmetric Standard Model [4] (MSSM) and QCD-like TechniColour (TC) models [5]. In an SU(2)×U(1) gauge theory with any number of Higgs doublets H_i, like the MSSM itself where $i = 1, 2$, it is always possible to define the normalized linear combination

$$H = \frac{\sum_i v_i H_i}{(\sum_i v_i^2)^{1/2}} \ , \quad \langle H_i^0 \rangle = v_i \ , \tag{2.2}$$

which gets a vacuum expectation value

$$v = \left(\sum_i v_i^2 \right)^{1/2} \tag{2.3}$$

and plays the same role as the SM Higgs boson in giving rise to the W and Z masses (2.1). In the technicolour case, it is well known that the relations (2.1) hold, in fact to all orders in the technicolour interactions, as a consequence of an extra "custodial" SU(2)$_C$ symmetry [6], which is only broken by the gauging of the U(1) factor or by the interactions responsible for the top-bottom mass splitting (see Section 13).

In this way I am led to consider a model described by a Lagrangian $\mathcal{L}(g, \ g', \ v; \ ...)$, which fixes, at the tree level, in terms of g, g' and v only, a series of "precision observables". The dots in $\mathcal{L}(g, \ g', \ v; \ ...)$ stand for the many other possible parameters, e.g. the top or the Higgs masses. The "precision observables", by definition, can be influenced by these extra parameters only via radiative corrections.

Given this framework, even sticking only to the tree level predictions, three measurements are needed to determine the basic parameters, whereas any extra measurement can be used to check the theory (or determine, via radiative corrections, the remaining parameters). The basic observables that are used to determine the theory (rather than checking it) must be precisely measured on one side, and theoretically calculable in a clean way on the other side. At present, three quantities neatly emerge:

i) The electromagnetic fine structure constant, α, as measured by the Josephson effect or the electron $(g - 2)$

$$\alpha^{-1} = 137.035 \ 9895 \ (61) \ . \tag{2.4}$$

ii) The Fermi constant G as determined from the muon lifetime τ_μ and the theoretical formula [7]

$$\frac{1}{\tau_\mu} = \frac{G^2 m_\mu^5}{192\pi^3}\left(1 - 8\frac{m_e^2}{m_\mu^2}\right)\left[1 + \left(\frac{\alpha}{\pi}\right)\left(\frac{25}{4} - \pi^2\right)\left(1 + \frac{2\alpha}{3\pi}\lg\frac{m_\mu}{m_e}\right)\right] , \tag{2.5}$$

which includes the QED corrections to the 4-fermion interaction:

$$G = 1.16637(2) \times 10^{-5} \text{ GeV}^{-2} \tag{2.6}$$

iii) The Z-mass [8]

$$M_Z = 91.187 \pm 0.007 \text{ GeV} . \tag{2.7}$$

The uncertainty on M_Z, which is the less precisely known among the basic observables, is far smaller (by more than one order of magnitude at least) than the relative uncertainty in any of the other observables that are currently measured and that we shall consider.

At the tree level the basic observables are expressed in terms of the parameters g, g', v as

$$\alpha_0 = \frac{g^2 s^2}{4\pi}, \quad G_0 = \frac{\sqrt{2}}{4v^2}, \quad M_{Z0}^2 = \frac{g^2 v^2}{2c^2} , \tag{2.8}$$

where $s^2 \equiv (g'/g)^2/(1 + (g'/g)^2)$ and $c^2 \equiv 1 - s^2$. [1)]

3. DERIVED OBSERVABLES. TREE LEVEL VALUES

Equations (2.8) can be solved in favour of g, g' and v as functions of α_0, G_0, M_{Z0}. In this way, at the tree level, the precision observables that we shall use to check the theory can all be expressed in terms of α_0, G_0, M_{Z0}. To this end, one needs:

i) the W-mass

$$M_{W0}^2 = M_{Z0}^2 c_0^2 , \quad c_0^2 \equiv 1 - s_0^2 \equiv \frac{1}{2}\left(1 + \left(1 - \frac{4\pi\alpha_0}{\sqrt{2}G_0 M_{Z0}^2}\right)^{1/2}\right) ; \tag{3.1}$$

ii) the photon coupling to the fermion f of charge Q_f

$$V_\mu(\gamma - f\bar{f}) = -\sqrt{4\pi\alpha_0}Q_f\gamma_\mu ; \tag{3.2}$$

iii) the Z coupling to the fermion f of weak isospin T_{3f}

$$V_\mu(Z - f\bar{f}) = (\sqrt{2}G_0 M_{Z0}^2)^{1/2}\gamma_\mu\left(g_{Vf}^0 - g_{Af}^0\gamma_5\right) \tag{3.3}$$

$$g_{Vf}^0 = T_{3f} - 2Q_f s_0^2 , \quad g_{Af}^0 = T_{3f} ;$$

iv) the W-coupling to the fermion doublet (f, f')

$$V_\mu(W - f\bar{f}') = \frac{\sqrt{4\pi\alpha_0}}{2\sqrt{2}s_0}\gamma_\mu(1 - \gamma_5) . \tag{3.4}$$

Of special interest to us are the Z widths into a pair of fermions $\Gamma(Z \to f\bar{f})$, the forward-backward asymmetries at the Z pole

$$A_{FB}^f \equiv \frac{\sigma_F^f - \sigma_B^f}{\sigma_F^f + \sigma_B^f} \tag{3.5}$$

---

1) The reader should be aware of the fact that, although keeping the same symbol, the precise meaning of s^2 (or c^2) will change as we progress.

$$\sigma^f_{F(B)} = \int_0^1 \left(\int_{-1}^0 \right) d\cos\theta \frac{d\sigma}{d\cos\theta} (e^+e^- \to f\bar{f}) \, ,$$

and the τ-polarization asymmetry

$$A^\tau_{\text{Pol}} \equiv \frac{\sigma_L - \sigma_R}{\sigma_L + \sigma_R}$$
$$\sigma_{L(R)} = \sigma(e^+e^- \to \tau^+_{R(L)}\tau^-_{L(R)}) \, . \tag{3.6}$$

It is a simple matter to obtain, from these definitions, the following well-known tree-level expressions[2]

$$\Gamma^{(0)}\left(Z \to f\bar{f}\right) = N_c \frac{G_0 M^3_{Z0}}{6\pi\sqrt{2}} \left(g^{0}_{Vf}{}^2 + g^{0}_{Af}{}^2\right) \, , \tag{3.7}$$

where $N_c = 1, 3$ for leptons and quarks respectively, and

$$A^{f(0)}_{FB} = 3\eta_e\eta_f \tag{3.8}$$
$$A^{\tau(0)}_{\text{Pol}} = 2\eta_e \tag{3.9}$$

where

$$\eta_f = \frac{g^0_{Vf}g^0_{Af}}{g^{0}_{Vf}{}^2 + g^{0}_{Af}{}^2} \, . \tag{3.10}$$

If we do not care at all about radiative corrections, it is already possible to obtain the predictions for some of the "derived observables", starting from the input values of the basic observables given in (2.4-7). This is shown in Table 1, column A.

Table 1. Comparison of theory (in different approximations) with experiment

Observable	A "Bare" relations	B e.m. corrections	C e.m. + large m_t	Experiment	
$\Gamma(Z \to \ell\bar{\ell})$/MeV	84.99	83.56	86.36	83.33 ± 0.30	[8]
A^ℓ_{FB}	0.0657	0.0188	0.0371	0.0174 $\pm$ 0.0027	[8]
A^τ_{Pol}	0.296	0.150	0.216	0.140 ± 0.018	[8]
A^b_{FB}	0.210	0.105	0.152	0.098 ± 0.012	[8]
M_W/M_Z	0.8876	0.8768	0.8935	0.8807 ± 0.0030	[9]

2) For simplicity I do not write the trivial fermion mass corrections, which are however not completely negligible for the τ and the b quark.

4. ELECTROMAGNETIC CORRECTIONS

At the level of precision that is of interest here, it is not possible to doubt the effectiveness of QED in describing the radiative corrections from photon exchanges. Always having in mind the required precision, it is less obvious but nevertheless true, with some provisos, that one can speak of pure QED corrections in isolation from the full electroweak corrections. For these reasons, I only describe the two main QED corrections that affect the LEP observables at the Z pole. In particular I refer to the works [10] that deal with the initial-state radiation and with the QED corrections to Bhabha scattering.

For observables at the Z-pole, the largest effect, by far, from pure QED radiative corrections is the change in the electric charge when going from $q^2 = 0$, where the fine structure constant defined in (2.4) is measured, to $q^2 = M_Z^2$. This change from α to $\alpha(M_Z^2)$ is related to the photon vacuum polarization function

$$\Pi_{\mu\nu}^{\gamma\gamma}(q) = -ig_{\mu\nu}\left[A^{\gamma\gamma}(0) + q^2 F^{\gamma\gamma}(q^2)\right] + q_\mu q_\nu \text{ terms} \tag{4.1}$$

via

$$\alpha(M_Z^2) = \frac{\alpha}{1 + F^{\gamma\gamma}(M_Z^2) - F^{\gamma\gamma}(0)} \equiv \frac{\alpha}{1 - \Delta\alpha} . \tag{4.2}$$

In $F^{\gamma\gamma}$ we only include the contributions from the lepton and the light quark loops:

$$\Delta\alpha = \Delta\alpha_\ell + \Delta\alpha_q , \tag{4.3}$$

that is to say we will conventionally include in the remainder of the corrections both the top quark and the W-boson loops. The lepton loops give $\Delta\alpha_\ell = -0.0314$, whereas, for the quark contribution, a perturbative calculation is not possible, because of strong interaction effects at low q^2. The way out consists [11] in relating $\Delta\alpha_q$ to the measured, and properly normalized, hadronic cross-section

$$R(s) = \frac{3s}{4\pi\alpha^2}\sigma(e^+e^- \to \gamma \to \text{hadrons}; s) \tag{4.4}$$

via a dispersive representation

$$\Delta\alpha_q = \frac{\alpha}{3\pi}M_Z^2 \int_{4m_{\pi^2}}^{\infty} \frac{ds\, R(s)}{s(s - M_Z^2)} . \tag{4.5}$$

This gives [12] $\Delta\alpha_q = -0.0282 \pm 0.0009$, with the error dominated by the uncertainties in the measured cross-section around the $b\bar{b}$ threshold. Putting together Eqs. (4.2-5), one then obtains [12]

$$\alpha(M_Z)^{-1} = 128.87 \pm 0.12 , \tag{4.6}$$

which replaces Eq.(2.4) as the reference value of the fine structure constant. The uncertainty of this new value is not such that we can obviously neglect it in the following.

The importance of the charge renormalization effect is manifest from column B of Table 1, which is obtained in the same way as the first was, except for the use of (4.6) instead of (2.4) and the introduction in A_{FB}^ℓ of the other electromagnetic correction, that we now describe.

Such correction, which is practically only relevant to A_{FB}^ℓ, arises from the interference in the $e^+e^- \to \ell^+\ell^-$ amplitude between the Z and the photon exchange contributions. More precisely, since at the pole the Z-exchange amplitude is purely imaginary, the most significant interference effect arises from the imaginary part of the photon propagator [13]. The calculation goes as follows. The four helicity amplitudes for the process $e^+e^- \to \ell\bar{\ell}$ are given by

$$G_{AB}(s) = \sqrt{2}G g_A g_B \frac{M_Z^2}{s - M_Z^2 + iM_Z\Gamma_Z} + \frac{4\pi\alpha}{s}(1 + F^{\gamma\gamma}(s)) , \qquad (4.7)$$

where A, $B = L$, R are the helicities of the electrons (A) and of the final lepton (B),

$$\begin{aligned} g_L &= g_{V\ell} + g_{A\ell} \\ g_R &= g_{V\ell} - g_{A\ell} , \end{aligned} \qquad (4.8)$$

and the one-loop correction term, $F^{\gamma\gamma}(s)$, to the photon exchange amplitude has been introduced to make it complex. At the pole one has in fact

$$Im G_{AB}(M_Z^2) = \frac{\sqrt{2}G M_Z}{\Gamma_Z}\left(g_A g_B + 4\pi\alpha \frac{Im F^{\gamma\gamma}(M_Z^2)}{\sqrt{2}G M_Z^3}\Gamma_Z \right) \qquad (4.9)$$

or, since

$$Im F^{\gamma\gamma}(M_Z^2) = \frac{\alpha}{3}\sum Q^2 = \frac{20}{9}\alpha \qquad (4.10)$$

with the sum extended to all leptons and light quarks,

$$Im G_{AB}(M_Z^2) = \frac{\sqrt{2}G M_z}{\Gamma_Z}(g_A g_B + \delta) \qquad (4.11)$$

$$\delta = 4\pi\alpha \frac{20\alpha}{9}\frac{\Gamma_Z}{\sqrt{2}G M_Z^2} = 3.2 \times 10^{-4} . \qquad (4.12)$$

In spite of being so small, δ contributes significantly to A_{FB}^ℓ. Since every helicity amplitude contributes to the cross-section with a definite angular distribution, from the very definition (3.5) one gets

$$A_{FB}^\ell = \frac{3}{4}\frac{|G_{LL}|^2 + |G_{RR}|^2 - |G_{LR}|^2 - |G_{RL}|^2}{|G_{LL}|^2 + |G_{RR}|^2 + |G_{LR}|^2 + |G_{RL}|^2}. \qquad (4.13)$$

Therefore, expanding in δ, one obtains

$$A_{FB}^\ell \simeq A_{FB}^\ell(\text{uncorrected}) + 6\delta , \qquad (4.14)$$

which amounts to a more than 10% relative shift ($6\delta = 2 \times 10^{-3}$) of the uncorrected asymmetry. Such correction is included in column B of Table 1. An analogous calculation would show that the same effect is totally negligible in the case of the τ-polarization asymmetry and very small ($\delta A_{FB}^b/A_{FB}^b \simeq 2\%$) for the forward-backward b asymmetry. In Table 1, a factor $(1+(3/4)\alpha/\pi)$ is also included in the leptonic width, which accounts for the final state radiation, again a pure electromagnetic effect.

5.   RENORMALIZATION

If we now want to deal with the full electroweak corrections, we need to go through the renormalization procedure. In principle, this opens the way to many possible different schemes. In a weak coupling theory, however, as the one that we are dealing with, one procedure emerges above all, which avoids useless intermediate steps and speeds up the necessary calculations in a significant way. Technically, it is itself a renormalization scheme.

First, we focus on the basic observables

$$(\alpha_0, \ G_0, \ M_{Z0}) \equiv a_0^{\iota} \tag{5.1}$$

and we compute the radiative corrections to them

$$a_0^{\iota} \overset{RC}{\to} a_0^{\iota} + \delta a^{\iota}(a_0^{\iota}) \equiv a^{\iota}(a_0^{\iota}) \ . \tag{5.2}$$

In this way we obtain the renormalized basic observables $a^{\iota}(a_0^{\iota})$ as (divergent) power series in $\hbar$, which can be formally inverted to $a_0^{\iota} \equiv a_0^{\iota}(a^{\iota})$.

In the same way we have to consider the loop corrections to the derived observables, or rather to the S-matrix elements that allow to compute the derived observables. As an effect of these corrections, any tree level S-matrix element $S_0(a_0^{\iota})$ goes into the renormalized one

$$S_0(a_0^{\iota}) \overset{RC}{\to} S_0(a_0^{\iota}) + \delta S(a_0^{\iota}) \equiv S(a_0^{\iota}) \ , \tag{5.3}$$

again as a formal (and divergent) power series in $\hbar$.

Finally we get the desired (finite) connection between the renormalized S-matrix elements, needed to compute the derived observables, and the renormalized basic observables a^{ι}, by considering the expansion in $\hbar$ of $S(a_0^{\iota}(a^{\iota}))$ both in $S(a_0^{\iota})$ and in the $a_0^{\iota}(a^{\iota})$ themselves. For example, at one loop, one will have

$$
\begin{aligned}
S(a_0^{\iota}(a^{\iota})) &\simeq S_0(a_0^{\iota}) + \delta^{(1)}S(a_0^{\iota}) \\
&\simeq S_0(a^{\iota}) + \delta^{(1)}S(a^{\iota}) - \sum_{\iota} \frac{\partial S_0}{\partial a^{\iota}} \delta^{(1)} a^{\iota} \\
&\equiv S_0(a^{\iota}) + \Delta S^{(1)}(a^{\iota})
\end{aligned}
\tag{5.4}
$$

which shows explicitly that the finite one-loop shift $\Delta S^{(1)}(a^{\iota})$ gets both a direct contribution $\delta^{(1)}S(a^{\iota})$ and an indirect one, $-\sum_{\iota}(\partial S_0/\partial a^{\iota})\delta^{(1)}a^{\iota}$, from the shifts of the basic parameters. I shall follow this procedure consistently throughout these lectures. The fact that, in a two-loop calculation, other parameters come in which need to be renormalized, simply requires an enlargement of the basic observables.

6. RADIATIVE CORRECTIONS IN THE "GAUGE-LESS" LIMIT

We have already seen how significant the electromagnetic corrections are. With the purpose of dealing with the other corrections in order of importance, we now consider the radiative corrections to the various precision observables that grow like powers of the top-quark mass. From the early work of Veltman [14], it is well known that such effects do arise from top/bottom loop corrections to the W and Z vacuum polarization amplitudes, making the ρ-parameter deviate from one. Relatively more recently, an analogous effect has been pointed out [15], which leads to the presence of a GIM-violating $Z \to b\bar{b}$ vertex, also violently growing with m_t. Following Ref. 16, I propose to deal with these effects by means of a Lagrangian that knows nothing about the gauge couplings. The point is that, at least for a very heavy top, the leading corrections both to the ρ-parameter and to the GIM violation $Z \to b\bar{b}$ vertex can be viewed as power series in the "top fine structure constant" $\alpha_t = g_t^2/4\pi$, where g_t is the top Yukawa coupling. In this sense, these corrections have nothing to do with the gauge couplings.

With this in mind, let us consider the SM Lagrangian for the third generation of quarks, with the vector bosons treated as external classical currents and the bottom Yukawa coupling neglected:

$$\mathcal{L} = i\bar{Q}_L \not{D} Q_L + i\bar{t}_R \not{D}\, t_R + i\bar{b}_R \not{D}\, b_R + |D_\mu \varphi|^2 + V(\varphi) + (g_t \bar{Q}_L \varphi t_R + \text{h.c.})\,, \qquad (6.1)$$

where $Q_L = (t, b)_L$ and φ is the Higgs doublet.

As usual, to minimize $V(\varphi)$, we set

$$\varphi = \begin{pmatrix} v + \frac{1}{\sqrt{2}}(H + i\chi) \\ i\varphi^- \end{pmatrix}\,, \qquad (6.2)$$

so that we obtain

$$\mathcal{L} = \left| \partial_\mu \varphi^- - \frac{gv}{\sqrt{2}} W_\mu^- \right|^2 + \frac{1}{2}\left| \partial_\mu \chi - \frac{gv}{\sqrt{2}c} Z_\mu \right|^2 + \frac{1}{2}(\partial_\mu H)^2$$
$$i\bar{t}\not{\partial}\, t + (g_t v \bar{t}_L t_R + \text{h.c.}) + i\bar{b}\not{\partial}\, b + \dots \qquad (6.3)$$

The dots in (6.3) stand for all the interactions among the Goldstone bosons φ^-, χ, the physical Higgs fields H, and the t, b quarks, dependent upon the top Yukawa coupling g_t and the quartic Higgs coupling λ. Suppose now that we perform loop corrections with this Lagrangian. We can compactly describe their results in terms of the following effective Lagrangian

$$\mathcal{L}^{eff} = Z_2^\varphi \left| \partial_\mu \varphi^- - \frac{gv}{\sqrt{2}} W_\mu^- \right|^2 + \frac{Z_2^\lambda}{2}\left| \partial_\mu \chi - \frac{gv}{\sqrt{2}c} Z_\mu \right|^2$$
$$+ i Z_{2L}^t \bar{t}_L \not{\partial}\, t_L + i Z_{2R}^t \bar{t}_R \not{\partial}\, t_R + i Z_{2L}^b \bar{b}_L \not{\partial}\, b_L + i \bar{b}_R \not{\partial}\, b_R \qquad (6.4)$$
$$+ g_t v Z_m^t \bar{t}_L t_R + \text{h.c.}$$
$$+ i\frac{Z_1}{v}\left(\partial_\mu \chi - \frac{gv}{\sqrt{2}c} Z_\mu \right) \bar{b}_L \gamma_\mu b_L + \dots$$

which contains several (divergent) constants $Z_i = Z_i(g_t, \lambda)$. The important points about (6.4) are:

i)   the presence of the $\chi b\bar{b}$ coupling, which must be of the derivative type because the b-quark is massless (the b_R does not interact);

ii) the fact that the derivative terms in the Goldstone boson fields keep the covariant form in terms of the classical fields W_μ and Z_μ.

In terms of the constants Z_i's appearing in (6.4) we can now express the renormalized observables. The gauge couplings do not get any corrections, so that, in particular

$$\alpha = \frac{g^2 s^2}{4\pi}\,. \qquad (6.5)$$

The Z and the W masses, from (6.4), become

$$M_Z^2 = Z_2^\chi \frac{g^2 v^2}{2c^2}\,, \quad M_W^2 = Z_2^\varphi \frac{g^2 v^2}{2}\,. \qquad (6.6)$$

The μ-decay, or the Fermi constant, is only affected through the corrections to the W mass, namely

$$G = \frac{\sqrt{2} g^2}{8 M_W^2} = \frac{\sqrt{2}}{4v^2 Z_2^\varphi}\,. \qquad (6.7)$$

One may also be interested in the neutral-current Fermi constant G^{NC} defined in an analogous way to G, for example, from the elastic electron-neutrino amplitude at $q^2 = 0$

$$G^{NC} = \frac{\sqrt{2} g^2}{8 M_Z^2 c^2} = \frac{\sqrt{2}}{4v^2 Z_2^\lambda} \qquad (6.8)$$

Finally, for the GIM-violating $Z \to b\bar{b}$ vertex, always from (6.4) and taking into account the renormalization of the b_L field, one has

$$V_\mu^{GIM}(Z \to b\bar{b}) = \frac{g}{2c} \frac{Z_1}{Z_2^b} \gamma_\mu \frac{1-\gamma_5}{2} \equiv \frac{g}{2c} \tau \gamma_\mu \frac{1-\gamma_5}{2} \; . \tag{6.9}$$

According to the procedure outlined in Section 5, we have to express all the derived observables in terms of α, G and M_Z. We have therefore, from Eqs. (6.5-8) [17]:

$$G^{NC} = G \frac{Z_2^\varphi}{Z_2^\chi} \equiv G\rho \tag{6.10}$$

$$s^2 c^2 = \frac{\pi \alpha}{\sqrt{2} G M_Z^2} \frac{Z_2^\chi}{Z_2^\varphi} = \frac{\pi \alpha}{\sqrt{2} G M_Z^2 \rho} \tag{6.11}$$

$$M_W^2 = M_Z^2 \rho c^2 \tag{6.12}$$

Knowing $\rho \equiv Z_2^\varphi / Z_2^\chi$ and $\tau \equiv Z_2 / Z_2^b$ only, we can in fact express all the radiative effects that we are dealing with in this section. Since, from (6.6-7),

$$\frac{g^2}{2c^2} = \sqrt{2} G M_Z^2 \rho \; , \tag{6.13}$$

for the Z widths into any pair of fermions other than b one has

$$\Gamma(Z \to f\bar{f} \neq b\bar{b}) = N_c \rho \frac{G M_Z^3}{6\pi\sqrt{2}} (g_{Vf}^2 + g_{Af}^2) \; , \tag{6.14}$$

where g_{Vf}, g_{Af} have their tree level value with s^2 given by (6.11). On the other hand, for the $Z \to b\bar{b}$ width, taking into account of the anomalous contribution (6.9), one has

$$\Gamma(Z \to b\bar{b}) = \rho \frac{3 G M_Z^3}{6\pi\sqrt{2}} \left[g_{Vb}^2 + g_{Ab}^2 \right] \tag{6.15}$$

$$g_{Vb} = \frac{1}{2} \left(1 - \frac{4}{3} s^2 + \tau \right) \; , \quad g_{Ab} = \frac{1}{2} (1 + \tau) \; . \tag{6.16}$$

Finally, for the asymmetries at the Z-pole again the tree level expressions hold with s^2 given by (6.11). (The inclusion of τ in A_{FB}^b according to (6.16) gives a numerically irrelevant correction due to the smallness of g_{Ve}).

7. HOW HEAVY CAN THE TOP BE?

Now that we have seen how the large m_t effects spread out, via ρ and τ, in all the precision observables, we can compute ρ and τ themselves in the SM using the Lagrangian (6.1). At one-loop level, the only diagrams that contribute to ρ and τ are shown in Figs. 1a and b respectively. From them it is immediate to get the well-known leading-order contributions [14,15]

$$\rho - 1 \simeq 3x \; , \quad \tau = -2x \; , \tag{7.1}$$

where we have set

$$x = \frac{G m_t^2}{8\pi^2 \sqrt{2}} \; , \tag{7.2}$$

which, in leading order, coincides with $\alpha_t / 8\pi$.

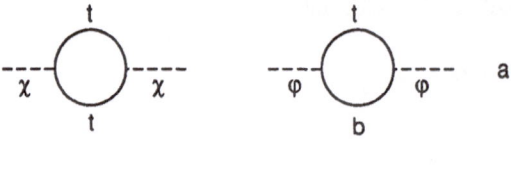

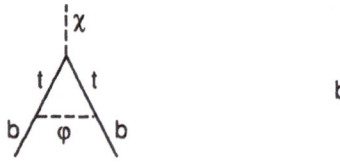

Fig. 1: Relevant one-loop diagrams contributing to ρ (1a) and τ (1b).

Quite a few more diagrams contribute at the two-loop level. Nevertheless, the simplification achieved by working with the Goldstone Lagrangian rather than the gauge Lagrangian itself makes even the two-loop calculation quite manageable [18]. At this order, also the Higgs quartic coupling λ comes in, which can be traded for the Higgs mass m_H by means of the lowest-order relation $m_H^2 = 4\lambda v^2$. In other words, expanding in x, the second-order coefficients of both ρ and τ are functions of m_H^2/m_t^2. For $m_H^2/m_t^2 \ll 1$, they are[3]

$$\rho - 1 \simeq 3x(1 + x(22 - 2\pi^2)) \tag{7.3}$$

$$\tau \simeq -2x(1 + x(9 - \pi^2/3)) \tag{7.4}$$

whereas, for $r \equiv (m_t^2/m_H^2) \ll 1$,

$$\rho - 1 \simeq 3x\left(1 + \frac{x}{4}(61 + 4\pi^2 + 54\lg r + 6\lg^2 r)\right) \tag{7.5}$$

$$\tau \simeq -2x\left(1 + \frac{x}{144}(311 + 24\pi^2 + 282\lg r + 90\lg^2 r)\right) \tag{7.6}$$

Needless to say, we have obtained finite second-order coefficients because we have expressed α_t in favour of the basic observables G and m_t. It is

$$\alpha_t = \frac{g_t^2}{4\pi} = \frac{Gm_t^2}{\pi\sqrt{2}}\frac{Z_2^\varphi Z_{2L}^t Z_{2R}^t}{(Z_m^t)^2} , \tag{7.7}$$

where the top renormalization constants have the usual meaning, as in (6.4), except that they are computed at the pole of the top propagator, since m_t is defined as the position of the pole itself.

Figures 2 show ρ and τ as functions of m_t for different values of m_H [18]. The expansion up to second order is well convergent for all $m_t \leq 300$ GeV, $m_H \leq 2$ TeV. This gives us confidence in the use of the perturbative calculation to get an upper bound on m_t from the comparison with the experimental data, since the bound that one obtains is well inside this region. If we use, say, $m_t = 300$ GeV (and any $m_H \leq 2$ TeV), where ρ and τ give the largely dominating corrections, and plug their values in the expressions for any at the observables given in the previous section, we obtain in all cases striking deviations from the data (see Table 1, column C). To get an accurate

---

3) Eq. (7.3) confirms a previous result obtained in Ref.19.

determination of the bound itself, one actually needs to take into account all the other electroweak corrections that do not grow like powers of m_t (see Section 10).

If view of the importance of the bound on m_t, let me insist a little more on it and play the devil's advocate. Since τ only affects the b-quark, it is ρ that plays the main role here. What then if, for some reason, the perturbative calculation of ρ goes wrong or simply if ρ does not deviate from 1 even for some large values of m_t, where perturbation theory cannot be trusted anyhow? Unlikely as it may be, this possibility cannot be excluded. It becomes therefore important to study the role of τ independently, which brings the focus on $\Gamma(Z \to b\bar{b})$.

Apart from the direct determination of this width, which has at present a 2.5% relative uncertainty [8], an indirect information on $\Gamma(Z \to b\bar{b})$ comes from the total Z width and from the ratio $R_h \equiv \Gamma(Z \to h)/\Gamma(Z \to \ell\bar{\ell})$, which are both very precisely measured [8]. A two-parameter fit, in terms of ρ and τ, can be made of the various observables, as described in the previous section. This is essentially what has been recently done in Ref. 20, with the result

$$\tau = (0.4 \pm 0.9) \; 10^{-2}$$

which can be directly turned into an upper bound on m_t, using Fig. 2b. More important for us is the conclusion that, to get around the bound on m_t, the perturbative analysis, done here at two-loop level, should have to go wrong both in the case of ρ and in the case of τ, which does not seem likely.

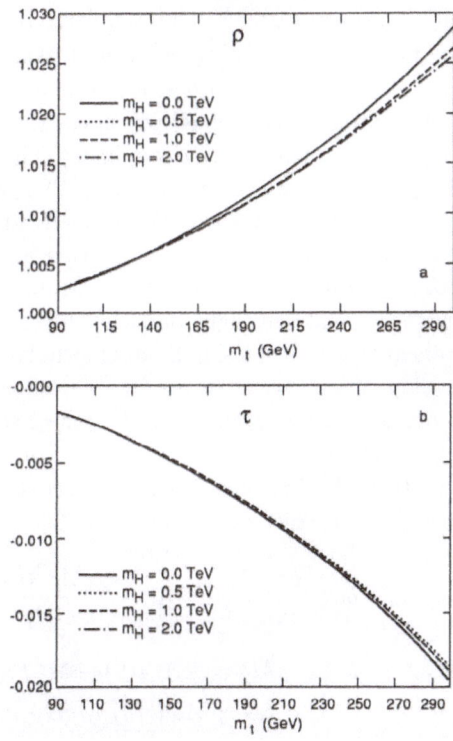

Fig 2: The ρ parameter (Fig 2a) and the GIM-violating $Z \to b\bar{b}$ coupling τ (Fig 2b) as functions of the top mass, for different values of the Higgs mass m_H

8. GENERAL ONE-LOOP EXPRESSIONS OF THE ELECTROWEAK RADIATIVE CORRECTIONS

It is time that we discuss the full electroweak radiative correction effects. To this purpose, I give in the following the general expressions of the derived observables in terms of the one-loop corrected Green functions, in such a way that one will no longer have to worry about the renormalization procedure. In doing this, I closely follow the scheme outlined in Section 5.

The relevant quantities, or the needed ingredients, are

i) the vacuum polarization amplitudes for the W, Z and γ

$$\Pi_{\mu\nu}^{ij}(q) = -ig_{\mu\nu}\left[A^{ij}(0) + q^2 F^{ij}(q^2)\right] + q_\mu q_\nu \text{ terms } ; \tag{8.1}$$

where $i, j = W, \gamma, Z$ or possibly $i, j = 0, 3$ for the W_3 or the B-boson respectively;

ii) the contributions to the vector and the axial form factors at $q^2 = M_Z^2$ in the $Z \to \ell^+\ell^-$ vertex $(\ell = e, \mu, \tau)$ from proper vertex diagrams and fermion self energies only

$$- i\frac{e}{2sc}\bar{v}\gamma_\mu(\delta g_V - \gamma_5\delta g_A)u \; ; \tag{8.2}$$

iii) all the one-loop corrections except the vacuum polarization (boxes, vertices and fermion self-energies) to the μ-decay amplitude at zero external momenta

$$- i\delta G_{V,B}(\bar{e}\gamma_\mu(1 - \gamma_5)\nu_e)(\bar{\nu}_\mu\gamma_\mu(1 - \gamma_5)\mu) \; . \tag{8.3}$$

Before proceeding further, several comments are in order. There is, in principle, no special reason to isolate the vector boson vacuum polarization amplitudes (8.1) from the remaining terms: the motivation for doing this will be clear in the following. In the expression (8.2) for the $Z \to \ell^+\ell^-$ vertex, also a magnetic form factor might be introduced. We assume that the lepton chiral symmetries, in the theory under consideration, are controlled by their masses (or their Yukawa couplings). This assumption makes the magnetic form factors negligible to the present purposes. The various functions defined in (8.1-3) have in general some imaginary parts. We have already discussed in Section 4 their effect in the case of the photon vacuum polarization amplitude. In view of the present phenomenological constraints, we assume that possible further contributions to the imaginary parts from new relatively light particles give negligible effects. Finally, there is the obvious comment that any of the individual form factors defined in (8.1-3) is in general neither finite nor gauge-invariant. It is only when their different contributions are grouped together in the derived observables that finiteness and gauge-invariance will be restored.

In terms of the quantities defined in (8.1-3), we can first express the shifts of the basic observables (or the input parameters) as defined in Section 5. They are:

$$\alpha = \alpha_0 + \delta\alpha; \quad \frac{\delta\alpha}{\alpha} = -F^{\gamma\gamma}(0) - 2\frac{s}{c}\frac{A^{\gamma Z}(0)}{M_Z^2} \; ; \tag{8.4}$$

$$M_Z^2 = M_{Z0}^2 + \delta M_Z^2 \; ; \quad \delta M_Z^2 = -A_{ZZ}(0) - M_Z^2 F_{ZZ}(M_Z^2) \; ; \tag{8.5}$$

$$G = G_0 + \delta G \; ; \quad \frac{\delta G}{G} = \frac{A_{WW}(0)}{M_W^2} + \frac{\delta G_{V,B}}{G} \; . \tag{8.6}$$

The derived observables of interest to us are: the Z width into a pair of charge leptons $\Gamma(Z \to \ell^+\ell^-)$, the asymmetries at the Z pole and the W mass. At one loop, these observables receive direct contributions from the amplitudes (8.1-3) as well as corrections due to the shifts (8.4-6) of the input parameters. For the W mass one immediately obtains

$$M_W^2 = M_Z^2 c^2 + \delta M_W^2 - \frac{M_Z^2 c^2}{s^2 - c^2}\left(s^2\frac{\delta\alpha}{\alpha} - s^2\frac{\delta G}{G} - c^2\frac{\delta M_Z^2}{M_Z^2}\right) , \qquad (8.7)$$

where

$$s^2 = 1 - c^2 = \frac{1}{2}\left[1 - \left(1 - \frac{4\pi\alpha}{\sqrt{2}GM_Z^2}\right)^{\frac{1}{2}}\right] \qquad (8.8)$$

and

$$\delta M_W^2 = -A_{WW}(0) - M_W^2 F_{WW}(M_W^2) . \qquad (8.9)$$

To obtain the Z leptonic widths and the asymmetries, one first has to write down the $e^+e^- \to \ell^+\ell^-$ amplitude close to the Z pole

$$\begin{aligned}
A(e^+e^- \to \ell^+\ell^-) &= \sqrt{2}G_0 M_Z^2(1 - F_{ZZ}(M_Z^2) - M_Z^2 F'_{ZZ}(M_Z^2))\frac{1}{q^2 - M_Z^2}\\
&\times \bar{v}_e\gamma_\mu\left(\left(-\frac{1}{2} + 2s_0^2 + 2scF_{Z\gamma}(M_Z^2) + \delta g_V\right) - \gamma_5\left(-\frac{1}{2} + \delta g_A\right)\right)u_e\\
&\times \bar{v}_\ell\gamma_\mu\left(\left(-\frac{1}{2} + 2s_0^2 + 2scF_{Z\gamma}(M_Z^2) + \delta g_V\right) - \gamma_5\left(-\frac{1}{2} + \delta g_A\right)\right)u_\ell
\end{aligned}$$
$$(8.10)$$

from which, using the shifts (8.1-3), one gets

$$\begin{aligned}
\Gamma(Z \to \ell^+\ell^-) &= \\
&= \frac{GM_Z^3}{6\pi\sqrt{2}}\left(1 - \frac{\delta G}{G} - \frac{\delta M_Z^2}{M_Z^2} - F_{ZZ}(M_Z^2) - M_Z^2 F'_{ZZ}(M_Z^2)\right)\\
&\quad \times \left((g_V + \Delta g_V)^2 + (g_A + \Delta g_A)^2\right)
\end{aligned} \qquad (8.11)$$

$$A_{FB}^\ell = 3\frac{(g_V + \Delta g_V)^2(g_A + \Delta g_A)^2}{[(g_V + \Delta g_V)^2 + (g_A + \Delta g_A)^2]^2} \qquad (8.12)$$

where

$$\begin{aligned}
g_V &= -\frac{1}{2} + 2s^2, \quad g_A = -\frac{1}{2}\\
\Delta g_V &= -2\delta s^2 + \delta g_V + 2scF_{Z\gamma}(M_Z^2), \quad \Delta g_A = \delta g_A\\
\delta s^2 &= -\frac{s^2 c^2}{c^2 - s^2}\left(\frac{\delta G}{G} + \frac{\delta M_Z^2}{M_Z^2} - \frac{\delta\alpha}{\alpha}\right) .
\end{aligned} \qquad (8.13)$$

Both in the amplitude (8.10) and in the forward-backward asymmetry (8.12) we have neglected possible box diagram contributions, which is legitimate for observables at the Z pole. Analogous expressions hold for A_{Pol}^τ and A_{FB}^b, with the lepton couplings g_V, g_A replaced by $g_V + \Delta g_V$ and $g_A + \Delta g_A$ as in (8.12). We have already mentioned that the forward-backward b-asymmetry is practically insensitive to (reasonable) corrections to the $Z - b\bar{b}$ couplings. For them one can then simply take the tree level expressions with s^2 given in (8.8).

It is customary, and useful for later purposes, to define three auxiliary dimensionless parameters $\Delta r_W, \Delta\rho, \Delta k'$ which are in direct correspondence with three derived observables $M_W, \Gamma(Z \to \ell^+\ell^-)$ and A_{FB}^ℓ via [21,22]

$$\left(1 - \frac{M_W^2}{M_Z^2}\right)\frac{M_W^2}{M_Z^2} \equiv \frac{\pi\alpha}{\sqrt{2}GM_Z^2(1 - \Delta r_W)} \qquad (8.14)$$

$$\Gamma(Z \to \ell^+ \ell^-) \equiv \frac{G_F M_Z^3}{6\pi\sqrt{2}} (g_V^2 + g_A^2) \qquad (8.15)$$

$$A_{FB}^\ell \equiv \frac{3g_V^2 g_A^2}{(g_A^2 + g_A^2)^2} \qquad (8.16)$$

$$g_A \equiv -\frac{1}{2}\left(1 + \frac{\Delta\rho}{2}\right) \qquad \frac{g_V}{g_A} \equiv 1 - 4s^2(1 + \Delta k') \qquad (8.17)$$

and s^2 given in (8.8). Notice that g_V and g_A defined in (8.15-16) have only an auxiliary role, being related to $\Delta\rho, \Delta k'$ through (8.17). By looking at Eqs. (6.14) and (8.15-17), notice also the relation between $\Delta\rho$ and the ρ parameter defined in Section 6, $\rho \simeq 1+\Delta\rho$. From these definitions one obtains [23]

$$\Delta r_W = \frac{\delta G}{G} + \frac{c^2}{s^2}\frac{\delta M_Z^2}{M_Z^2} - \frac{\delta\alpha}{\alpha} + \frac{s^2 - c^2}{c^2}\frac{\delta M_W^2}{M_W^2}$$

$$\Delta\rho = -\frac{\delta G}{G} - \frac{\delta M_Z^2}{M_Z^2} - F_{ZZ}(M_Z^2) - M_Z^2 F'_{ZZ}(M_Z^2) - 4\Delta g_A$$

$$\Delta k' = \frac{1}{2s^2}(\Delta g_V - (1 - 4s^2)\Delta g_A) \qquad (8.18)$$

This is as much as one can do in general, without specifying the actual theory that one wants to deal with: the SM or something else.

9.   THE ϵ-PARAMETERS: "OBLIQUE" OR NON "OBLIQUE"

In any given theory of the type defined in Section 2, it becomes now a rather straightforward matter to compute explicitly from the various equations in Section 8 the one-loop electroweak corrections, without having to worry about the different subtleties of the renormalization procedure, since everything has already been taken care of. Let us take however a further general step, still with the purpose of being able to deal with different models in an efficient, unified way.

There are two independent reasons that suggest the introduction of three basic phenomenological parameters (observables) to study the electroweak radiative corrections. The first one [22,24] is of a purely phenomenological nature: if one concentrates on the LEP observables at the Z pole and, among them, on those not affected by the strong interaction uncertainties, one remains with two physical observable quantities, which are the effective g_V and g_A couplings defined in Eqs. (8.15-16). In fact, as will be made more precise later on, not only $\Gamma(Z \to \ell^+\ell^-)$ and A_{FB}^ℓ, but also A_{pol}^τ and A_{FB}^b, are determined by g_V and g_A. One then ends up with three effective observables if one adds M_W, as motivated by the foreseen precision on its experimental determination.

The usefulness of introducing three effective variables has also some more theoretical motivations based on the assumption that new physics effects might especially occur in the electroweak precision test through vector boson vacuum polarizations (the so-called "oblique" corrections) [25,26].

This can be readily seen as follows. In a gauge SU(2)×U(1) theory the kinetic vector boson Lagrangian has the form

$$L_{VB}^{kin} = -\frac{1}{4}\vec{W}_{\mu\nu}\vec{W}_{\mu\nu} - \frac{1}{4}B_{\mu\nu}B_{\mu\nu} . \qquad (9.1)$$

After spontaneous symmetry breaking down to U(1)$_{em}$, the most general effective kinetic Lagrangian, allowing for a rescaling of the fields that can be re-absorbed in a redefinition of the couplings g, g', has the form

$$
\begin{aligned}
L_{VB}^{kin;eff} = & -\frac{1}{2}W_{\mu\nu}^+ W_{\mu\nu}^- - \frac{1}{4}(1 - e_2)W_{\mu\nu}^3 W_{\mu\nu}^3 \\
& -\frac{1}{4}B_{\mu\nu}B_{\mu\nu} + \frac{1}{2}\frac{s}{c}e_3 B_{\mu\nu}W_{\mu\nu}^3 \\
& +\frac{1}{2}M_Z^2(cW_3^\mu - sB^\mu)^2 + M_Z^2 c^2(1 + e_1)W_\mu^+ W_\mu^-
\end{aligned}
\tag{9.2}
$$

where s it the sine of the same angle as the one that appears in the neutral current. It is in the constants e_i, $i = 1, 2, 3$, that all possible "oblique" corrections enter, needless to say apart from the coefficients in front of higher- dimensional operators, related to higher derivatives of the vacuum polarizations, where non-standard effects are assumed to be negligible.

With all this in mind, we define [22] our phenomenological parameters ϵ_i, $i = 1, 2, 3$ by relating them directly to Δr_W, $\Delta\rho$, $\Delta k'$ via a triangular linear system as follows

$$
\begin{aligned}
\epsilon_1 &= \Delta\rho , \\
\epsilon_2 &= c^2\Delta\rho + \frac{s^2\Delta r_W}{c^2 - s^2} - 2s^2\Delta k' , \\
\epsilon_3 &= c^2\Delta\rho + (c^2 - s^2)\Delta k' .
\end{aligned}
\tag{9.3}
$$

Notice that $\Delta\rho$ and $\Delta k'$, and therefore also ϵ_1 and ϵ_3, only depend on the LEP observables, whereas ϵ_2 is affected by Δr_W, (or M_W) as well.

Although this is only a matter of convention, these definitions are actually complete only after it is made clear which of the purely electromagnetic effects discussed in Section 4 we want to introduce in the ϵ_i's. First of all we avoid including in the ϵ_i's the very important, but also rather trivial, electric charge renormalization effect due to light quarks and leptons. This is simply achieved by taking, in the defining Eqs. (8.14-17) for $\Delta\rho, \Delta r_W$ and $\Delta k'$, $\alpha(M_Z)$ in place of α. In particular, for s^2 in (8.17), from Eqs. (8.8), (4.6) and (2.7), one must take

$$
s^2 = 0.2312(3) ,
\tag{9.4}
$$

this error being determined almost exclusively by the uncertainty on $\alpha(M_Z)$. For practical (and historical) reasons we shall on the contrary include in the definition of the ϵ_i's both the final-state radiation factor $(1 + 3\alpha/4\pi)$ in the leptonic widths and the $\gamma - Z$ interference correction to the forward-backward μ asymmetry that was described in Section 5. This last point requires in fact a brief discussion. Were it not for this effect, to determine $\Delta k'$ (or g_V/g_A from (8.17) we could have equally well used the τ-polarization and the b-asymmetries together with

$$
A_{Pol}^\tau = \frac{2g_V/g_A}{1 + (g_V/g_A)^2}
\tag{9.5}
$$

and

$$
A_{FB}^b = 3\frac{g_V/g_A}{1 + (g_V/g_A)^2}\frac{1 - \frac{4}{3}s^2}{1 + (1 - \frac{4}{3}s^2)^2} .
\tag{9.6}
$$

The presence of the $\gamma - Z$ interference effect in A_{FB}^μ makes it necessary to correct for it before averaging over the three different determinations of g_V/g_A or $\Delta k'$. The fact that we have used A_{FB}^ℓ to define, through (8.16), g_V/g_A requires that $\Delta k'$, obtained from both A_{Pol}^τ and A_{FB}^b through Eqs. (9.5-6) and (8.17) be shifted by a fixed amount $\delta(\Delta k') \cong -5 \times 10^{-3}$. Alternatively one can take the average value of g_V/g_A from experimental papers that use (9.5) as its very definition and shift this value to obtain our g_V/g_A, Eq. (8.16), by $\delta(g_V/g_A) = 4.7 \times 10^{-3}$.

All this procedure to determine the experimental values of the ϵ_i's can be numerically summarized by the following equations:

$$\Gamma = 83.41 \ (1 - 0.26\epsilon_3 + 1.20\epsilon_1) \ \text{MeV} \tag{9.7}$$

$$\frac{g_V}{g_A}\bigg|_{av} = 0.0752 \pm 0.0012 - 1.72\epsilon_3 + 1.32\epsilon_1 \tag{9.8}$$

$$\frac{M_W}{M_Z} = 0.8768 + 0.63\epsilon_1 - 0.44\epsilon_2 - 0.38\epsilon_3 \ . \tag{9.9}$$

where the error in the right-hand side of (9.8) reflects the uncertainty on $\alpha(M_Z)$.

Once the ϵ_i's are experimentally determined, they can be compared with the predictions of any given theory. From the definitions (9.3) and the expressions obtained for Δr_W, $\Delta\rho$ and $\Delta k'$ in the previous section, we get [23], in terms of the amplitudes defined in (8.1-3),

$$\epsilon_1 = e_1 - e_5 - \frac{\delta G_{V,B}}{G} - 4\delta g_A \tag{9.10}$$

$$\epsilon_2 = e_2 - s^2 e_4 - c^2 e_5 - \frac{\delta G_{A,B}}{G} - \delta g_V - 3\delta g_A \tag{9.11}$$

$$\epsilon_3 = e_3 + c^2 e_4 - c^2 e_5 + \frac{c^2 - s^2}{2s^2}\delta g_V - \frac{1 + 2s^2}{2s^2}\delta g_A \ , \tag{9.12}$$

where, for the contributions of the vacuum polarization amplitudes, we introduce the combinations [23]

$$e_1 = \frac{A_{33}(0) - A_{WW}(0)}{M_W^2} \tag{9.13}$$

$$e_2 = F_{WW}(M_W^2) - F_{33}(M_Z^2) \tag{9.14}$$

$$e_3 = \frac{c}{s}F_{30}(M_Z^2) \tag{9.15}$$

$$e_4 = F_{\gamma\gamma}(0) - F_{\gamma\gamma}(M_Z^2) \tag{9.16}$$

$$e_5 = M_Z^2 F_{ZZ}'(M_Z^2) \ . \tag{9.17}$$

Notice that for the first three of these coefficients we have used the same notation as for the parameters defined in the effective Lagrangian (9.2). Indeed they coincide, which clearly illustrates, via (9.10-12), the connection between the ϵ_i's and the e_i's. The ϵ_i's are directly measurable quantities, which receive all sorts of contributions from different kinds of corrections: first and second (e_1, e_2, e_3) and higher derivatives (e_4, e_5) of the vacuum polarization amplitudes, as well as vertex and box corrections $(\delta g_{V,A}, \ \delta G_{V,B})$.

What has been done so far is general and exact, in terms of the leptonic observables at the Z pole and of the W mass. Still in a model-independent way, other observables could be introduced, following the previous discussion, if we knew that the only deviations from the SM were contained in e_1, e_2, e_3. In this case, for any given observable, we could explicitly compute the remainder from the SM and parametrize the possible deviations in terms of e_1, e_2, e_3 themselves. This is the attitude taken by Peskin and Takeuchi [26] and others [27,28]. Their S, T, U variables are proportional to the deviations Δe_1, Δe_2, Δe_3 from the SM values at a given reference point, with fixed m_t and m_H, according to the definitions

$$S = \frac{4s^2}{\alpha}\Delta e_3 \ , \quad T = \frac{1}{\alpha}\Delta e_1 \ , \quad U = -\frac{4s^2}{\alpha}\Delta e_2 \tag{9.18}$$

Since the m_t dependence of the radiative corrections is a major effect, at least until the top mass will not be independently measured, I prefer to analyse the data in terms of variables that can be defined in an m_t-independent way [22]. In turn, this is perfectly possible for all the observables that do not involve the $Z \to b\bar{b}$ vertex, since they are only affected by the top-quark mass via e_1, e_2, e_3 (e_4 and e_5 have a negligible top-mass dependence for $m_t \geq 90$ GeV). This allows us to include in the analysis the information coming from low-energy experiments as well. If all the possible deviations from the SM are contained in the vacuum polarization functions (more precisely in e_1, e_2, e_3), together with Eqs.(9.7-9), one has for the ratio R_ν of neutral- to charged-current processes in deep inelastic neutrino scattering on nuclei [27,29]:

$$R_\nu = 0.3025 + 8.27\epsilon_1 - 2.74\epsilon_3 \qquad (9.19)$$

or, for the weak charge Q_W measured in atomic parity violation in caesium:

$$Q_W = -72.84 \pm 0.13 - 102\epsilon_3 , \qquad (9.20)$$

where the error is determined by the uncertainties in the atomic wave functions.

A final comment concerns the term e_4, Eq. (9.16), which enters the expressions for ϵ_2 and ϵ_3, Eqs. (9.11 and 12), to account for the running of the electromagnetic coupling constant from $q^2 = 0$ to $q^2 = M_Z^2$. Since the contribution to this running from light quarks and leptons has already been included in the reference value of s^2, Eq. (9.4), e_4 should only contain the remaining contributions, such as the one from W-exchange and t-exchange or from any new charged particle.

10. COMPARISON WITH THE STANDARD MODEL

Equations (9.7-9) allow a direct determination of the ϵ parameters from the leptonic observables at LEP and the W-mass measurements at hadron colliders. The relevant experimental information is given in Table 1. As we have discussed in the previous section, the LEP observables include some electromagnetic corrections (final-state photon radiation and $\gamma - Z$ interference effects). The corresponding results are

$$\epsilon_1 = (-0.9 \pm 3.7)10^{-3} , \quad \epsilon_2 = (-9.9 \pm 8.0)10^{-3} , \quad \epsilon_3 = (-0.9 \pm 4.1)10^{-3} . \quad (10.1)$$

One comment is worth making: the consistency with zero of all the ϵ's is a reflection of the fact, illustrated in Table 1, that the theoretical predictions, after inclusion of the electric charge renormalization, are in agreement with the observations. Of course this implies that the remaining corrections have to be small enough.

Since the SM fulfils all the requirements discussed in Section 2, its predictions can be compared with experiment for any given value of the top and Higgs mass. [4] Such a comparison is shown in Figs. 3 for representative values of m_t and m_H. In Fig. 3a, we show the $\epsilon_1 - \epsilon_3$ correlation from LEP data alone, so that ϵ_2 is not entering. The value of ϵ_2 obtained by including also the W-mass determination is consistent with the SM prediction, with a still large error (Fig. 3b).

---

4) Equations (9.10-12) give the basis for the explicit calculations. Ref. 30 contains an explicit analytic description of the one-loop calculation in the SM, with full reference to the original works.

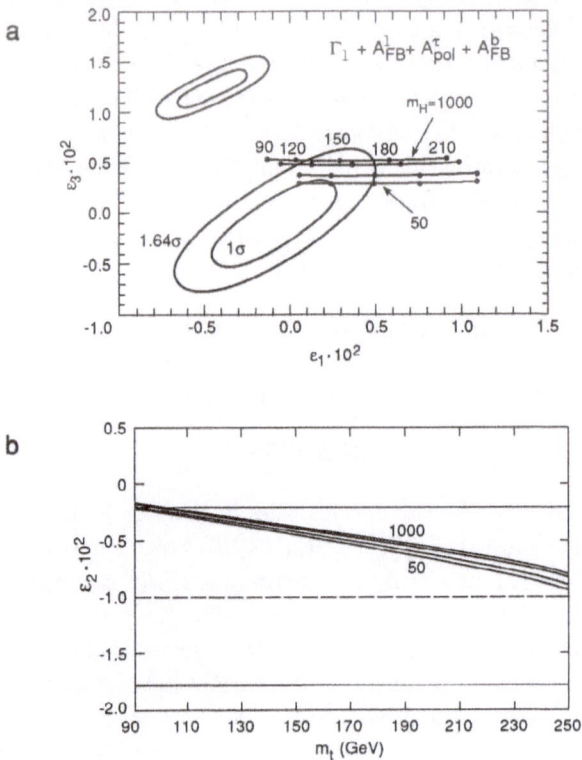

Fig. 3: Data on ϵ_1, ϵ_2 (a) and ϵ_3 (b) from LEP and M_W. In Fig. 3a the 1σ and 1.64σ ellipses are shown. In Fig. 3b the 1σ error is indicated. The Standard Model predictions are also given for reference purposes (the four solid lines are for different values of the Higgs mass, $m_H = 50, 100, 500, 1000$ GeV, and the dots mark values of m_t in the range 90–210 GeV). The ellipses in the right upper corner of Fig. 3a indicate the foreseen precision as described in the text.

As a way to discriminate among theories that deviate from the SM only in vacuum polarizations, one can include also the low-energy precision tests according to Eqs. (9.19-20). Using, together with Table 1, the present average value from νN scattering [31], [32]

$$\langle R_\nu \rangle = 0.312 \pm 0.030 \tag{10.2}$$

(which includes the recent CCFR results [32]) and, from Atomic Parity Violation (APV) in caesium [33],

$$Q_W = -71.04 \pm 1.81 , \tag{10.3}$$

one obtains the overall result shown in Fig. 4 with the corresponding values of the ϵ-parameters

$$\epsilon_1 = (0.6 \pm 2.8) \times 10^{-3} , \quad \epsilon_2 = (-8.6 \pm 7.8) \times 10^{-3} , \quad \epsilon_3 = (0.0 \pm 3.6) \times 10^{-3} . \tag{10.4}$$

Of considerable importance is to know which sensitivity should be reachable in the determination of the ϵ-parameters in a not too distant future. A clear improvement is expected at the end of LEP 1 phase. With about 2M Z events per experiment, it should

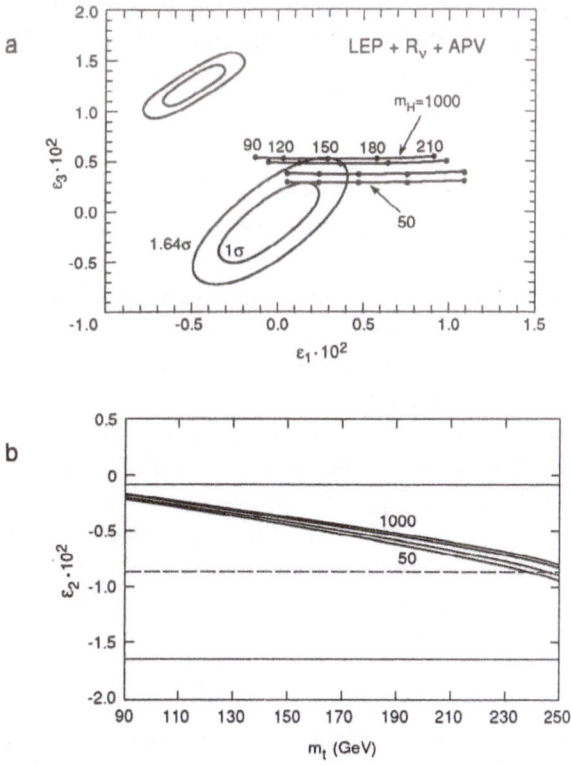

Fig. 4: Data on ϵ_1, ϵ_2 (a) and ϵ_3 (b) from LEP, M_W, R_ν and APV under the assumption specified in the text.

be possible to determine the leptonic widths and the ratio of the lepton Z couplings with an error

$$\Delta\Gamma_\ell/\Gamma_\ell = \pm 2^\circ/_{\circ\circ}, \quad \Delta(g_V/g_A) = \pm 0.0015 . \tag{10.5}$$

A significant improvement is also expected on the W-mass measurement, first at the Tevatron and ultimately at LEP 2, with an estimated error

$$\Delta M_W = \pm 50 \text{ MeV} . \tag{10.6}$$

The errors in Eq. (10.5) give rise to the "future" 1σ and 1.64σ ellipses also shown in Figs. 3a and 4a.

11. COMPARISON WITH THE MINIMAL SUPERSYMMETRIC STANDARD MODEL

The Minimal Supersymmetric Standard Model [4] is a well-motivated alternative to the Standard Model. From the point of view of the precision electroweak tests, however, the significant features of the MSSM make it clearly resemble the SM itself: i) in both cases one has a fundamental Higgs; ii) both the MSSM and the SM (with a not too heavy Higgs) are perturbative theories.

The MSSM is really defined by two assumptions: i) it has minimal supersymmetric particle content; ii) the supersymmetry breaking occurs in a hidden sector and is transmitted to standard matter by universal supergravity couplings [34]. (R-parity conservation plays no role in the following considerations.) Two basic properties are in turn implied by the previous assumptions:

i)   The MSSM has the same but one-dimensionless parameters as the SM: the quartic Higgs self-coupling is not a free parameter but it is actually a gauge coupling. On the contrary, the MSSM has five parameters with dimension of mass, m_i, in place of the only mass parameter appearing in the quadratic term of the SM scalar potential. In a standard notation [4]

$$m_i \equiv m, M, \mu, m_A, m_B , \qquad (11.1)$$

all of them being related to the scale of effective supersymmetry breaking[5];

ii)   By taking $m_i \gg M_Z$, the light spectrum of the MSSM is identical to the SM one with a light Higgs. As is by now well known, the exact upper bound on the light Higgs depends critically on the top-quark mass [35]. All the remaining particles predicted by the MSSM can be made arbitrarily heavy, although at the price of an increasingly unnatural fine tuning.

These two properties imply that, in the $m_i \gg M_Z$ limit, all the superparticles "decouple" from radiative-correction effects and the MSSM becomes undistinguishable from the SM itself with a light Higgs [36]. Notice that to reach this conclusion, both properties i) and ii) are essential. Differently stated, it is essential that the mass terms of the particles that decouple be compatible with unbroken SU(2)×U(1). Needless to say, the top quark is the best example of a particle that does not "decouple" from radiative corrections: also the top-quark mass can be made large, but, since the mass term is not consistent with unbroken SU(2)×U(1), a heavy top requires a correspondingly large Yukawa coupling.

How heavy need the superparticles be in order to make the MSSM undistinguishable from the SM with a light Higgs, as far as radiative correction effects are concerned? Not much, it turns out. Figure 5 shows the range of values that ϵ_1 and ϵ_3 can attain in the parameter space of the MSSM, with the dotted area being by far the most probable one [23]. Notice that the SM with a Higgs of about 100 GeV sets the upper value of ϵ_3 in the MSSM. On the contrary, ϵ_3 might attain a lower value by having charginos and neutralinos, the fermionic partners of the vector and Higgs bosons, just around the corner of LEP 1 (dashed area). This is a threshold effect in the Z wave-function renormalization, contributing to a positive e_5-term, Eq. (9.17), mostly due to the vector coupling of charginos and (off-diagonal) neutralinos to the Z itself. Notice that this effect on e_5 propagates in all the ϵ_i's according to Eqs. (9.10-12) and reduces all of them by a similar amount. Physically, all the Z widths would be reduced without affecting the asymmetries [23].

In the MSSM, one further possible effect has to be mentioned, which is not included in Fig. 5. The scalar partners of quarks and leptons in a given generation are expected, from the universality of the supergravity couplings, to have splittings proportional to the splittings of the corresponding fermions. In turn, if all the squarks are not too heavy, the scalar top ($\tilde{t}$) and bottom ($\tilde{b}$) partners could have a mass difference comparable to their mean mass. In such a case, their virtual exchange could contribute to e_1, and therefore to ϵ_1, by a positive amount at most equal to the corresponding $t-b$ contribution discussed in Section 7 [36], [37]. This contribution, which requires one light stop, would not affect ϵ_3 in a significant way.

---

5) Even the so-called μ parameter, which, in the low-energy theory, appears in a supersymmetric term, may very well originate in a more complete theory as a result of supersymmetry breaking, as many examples show.

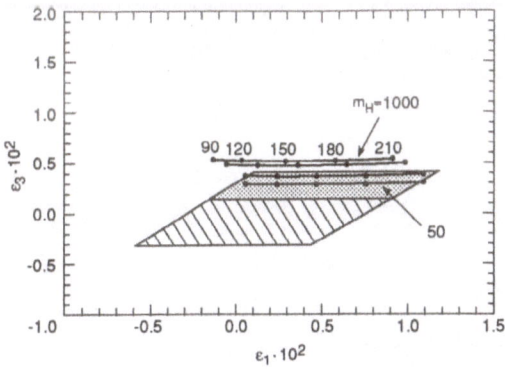

Fig. 5: Ranges of values attainable by ϵ_1 and ϵ_3 in the parameter space of the MSSM, compared with the SM predictions. The dotted area is the most probable one. The dashed area can be attained with electroweak gaugino-higgsinos close to the Z threshold. The top mass ranges from 90 to 210 MeV. Not shown is the possible effect of a light stop, as discussed in the text.

12. "HIGGS-LESS" THEORIES: a) THE SM FOR INFINITE HIGGS MASS

In this section we start trying to tackle the hard problem of estimating the size of the radiative corrections in theories without a Higgs particle, that is to say in theories which deviate from the SM in the sector that is supposed to break both the local and the global symmetries of the pure $SU(2) \times U(1)$ gauge Lagrangian. Since the local symmetry has to be broken spontaneously, this sector must contain in general, among other things, the charged and neutral Goldstone bosons that provide the longitudinal modes of the W and the Z, respectively called φ and χ in Section 6. These Goldstone bosons could be composite states arising from some new strongly interacting dynamics.

In this situation, it makes sense to assume that all possible deviations from the SM radiative correction effects be confined to the vector boson vacuum polarization amplitudes, in particular to their low derivatives only. We then concentrate on e_1, e_2 and e_3 defined in Eqs. (9.13-15).

To deviate from zero, both e_1 and e_2 require a breaking of the (custodial) symmetry that is responsible, at the tree level, for $\rho = 1$, or the mass relations (2.1). This can be seen, for example, from the global $SU(2)$ symmetry that is recovered in the effective Lagrangian (9.2) by taking $e_1 = e_2 = 0$ and neglecting mixing effects ($s = 0$ or $g' = 0$). With reference to Fig. 6 we then set

$$e_1 = e_1^\Delta + e_1^B \; ; \quad e_2 = e_2^\Delta + e_2^B \; , \tag{12.1}$$

where $e_{1,2}^B$ are due to the B-boson exchange and $e_{1,2}^\Delta$ to the remaining sources of isospin breaking, like the top-bottom mass difference. The blobs in this figure may have to be computed non-perturbatively in the symmetry breaking sector of the theory. In the case of e_1, or $\Delta\rho$, we have actually seen in Section 6 how it can be viewed as the deviation from 1 of the ratio of the wave function renormalization constants of the charged and neutral Goldstone bosons, as explicitly indicated in Fig. 6. In other words, e_1^Δ will be independent from the gauge couplings (in leading order), whereas e_1^B will be proportional to g'^2. This is unlike the case for e_2, which requires the couplings to the external gauge bosons: e_2^B in particular is proportional to the fourth power of the

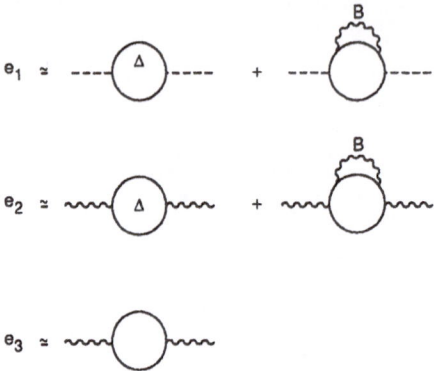

Fig. 6: Contributions to e_1, e_2, e_3 in an expansion in custodial isospin-breaking (Δ) and in the gauge couplings. The dotted lines in e_1 stand for the Goldstone bosons φ, χ. The blobs are to all orders in the non-perturbative couplings of the symmetry-breaking sector

gauge couplings and is generally small. Unlike e_1 and e_2, e_3 does not need any custodial isospin breaking, which therefore can hopefully be neglected.

A first step towards a "Higgs-less" theory can be made by considering the various e_i's in the SM for an infinitely heavy Higgs [38], namely when the Higgs is never excited. The relevant diagrams at one loop are shown in Fig. 7. In this approximation, there is in fact no diagram at all with a Higgs boson internal line that contributes to e_1^Δ and e_2^Δ; these are therefore independent from the Higgs mass. There are, on the contrary, diagrams with internal Higgs boson lines that contribute to e_1^B and e_3, which would in fact be needed to make the pure Goldstone diagrams of Fig. 7 ultraviolet-convergent. In terms of an ultraviolet and infrared cut-off, Λ and μ respectively, the diagrams of Figs. 7b and 7e, both computed at $q^2 = 0$, immediately give [39]

$$e_1^B \simeq -\frac{3\alpha}{8\pi c^2} \lg \frac{\Lambda}{\mu} \simeq -1.2 \times 10^{-3} \lg \frac{\Lambda}{\mu} \tag{12.2}$$

$$e_3 \simeq \frac{\alpha}{24\pi s^2} \lg \frac{\Lambda}{\mu} \simeq 0.5 \times 10^{-3} \lg \frac{\Lambda}{\mu} \tag{12.3}$$

The result for e_1^B is actually obtained by taking the B-boson propagator in the Landau gauge. This is the gauge that has to be used if one wants to compute e_1 in terms of Goldstone boson properties only; the corresponding gauge-fixing term, although breaking the local symmetry, respects in fact the global symmetry that is responsible for the existence of the massless Goldstone bosons in the first place.

This is at least a quick way to obtain, in the SM, the leading logarithmic dependence [40] of e_1 and e_3 as the Higgs mass becomes infinitely large, since for that it is enough to replace Λ in Eqs. (12.2) and (12.3) with the Higgs mass. (The infrared cut-off in a complete calculation is replaced by the gauge-boson masses.) What the calculation actually shows is that, to compute the ϵ parameters (or the ϵ's) at the per mille level, which is the foreseen precision in their determination (see Section 10), it is necessary to specify how the divergences in Eqs. (12.2) and (12.3) are regularized in a theory without the Higgs as an explicit degree of freedom. Needless to say, it is quite clear what the counterpart of the divergences (12.2) and (12.3) is in an effective Lagrangian description of the Goldstone boson interactions. Unlike the case of the SM itself, where both the local SU(2)$\times$U(1) and the approximate global custodial SU(2) symmetries are linearly realized in terms of the full Higgs doublet, the same symmetries can be non-linearly

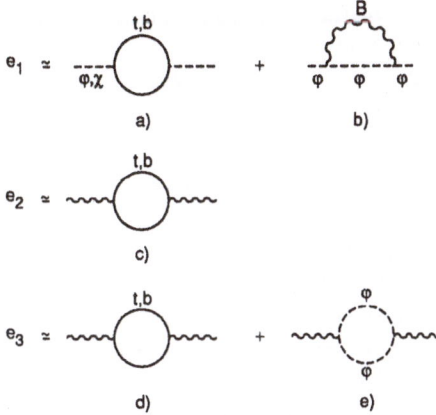

Fig. 7: One-loop contributions to e_1, e_2, e_3 in the SM without Higgs boson internal lines in the approximation discussed in the text. The B-boson propagator in Fig. 7b is in the Landau gauge.

realized using the Goldstone bosons only [39]. The point now is that the corresponding (non-renormalizable) effective Lagrangian realizing these symmetries will have terms that contribute directly to e_1 and e_3 with unknown coefficients. Without independent theoretical or experimental inputs, this makes the ϵ coefficients unpredictable.

13. "HIGGS-LESS" THEORIES: b) TECHNICOLOUR MODELS

The divergences due to the Golstone boson exchanges that we have computed in the previous section are certainly cut off in models where the breaking of the gauge symmetry is produced by the condensation of a fermion bilinear, as in technicolour theories [5]. In such theories, it is therefore a challenge to compute the finite residual contributions to the ϵ parameters.

Simply for orientation, let us compute the contributions to the e's from the virtual exchange of one "technicolour" SU(2) doublet with mean mass $m_T \gg M_Z$ and with mass difference, inside the doublet, $\Delta m_T \ll m_T$. From simple one-loop diagrams, one readily obtains

$$e_1^\Delta \simeq N_{TC} \frac{G(\Delta m_T)^2}{6\pi^2 \sqrt{2}} \simeq 10^{-3} N_{TC} \left(\frac{\Delta m_T}{100 \text{ GeV}} \right)^2 \tag{13.1}$$

$$e_1^B = 0 \tag{13.2}$$

$$e_2^\Delta \simeq -N_{TC} \frac{\alpha}{30\pi s^2} \left(\frac{\Delta m_T}{m_T} \right)^2 = -N_{TC} \, 0.3 \times 10^{-3} \left(\frac{\Delta m_T}{m_T} \right)^2 \tag{13.3}$$

$$e_3 \simeq \frac{\alpha}{24\pi s^2} N_{TC} = 0.5 \times 10^{-3} \, N_{TC} \tag{13.4}$$

for N_{TC} technicolours. What should be done with these equations is unclear, to say the least. Nevertheless we learn something from them. The heavy fermions do not "decouple" in the large mass limit. Furthermore, to make a numerical estimate of some sort, it is crucial that we know how many heavy fermions there are.

The non-decoupling property of these effects is of course due to the fact that the mass of the technifermions (left-handed SU(2) doublets and right-handed singlets) is not compatible with unbroken SU(2)×U(1). (The discussion of the top-quark-mass effects given in Section 6 is an illustrative example in this sense.) As to the needed number of technifermions, one should not forget the gigantic global symmetry of the SM Lagrangian in the absence of Higgs interactions, which is supposedly broken by extra gauge interactions, shared between the standard fermions and the technifermions. In actual model building, this is what generally requires a proliferation of technifermions. Finally, the parameter Δm_T hides the problem of generating, in technicolour models, the top-bottom mass difference.

At least this problem of the custodial isospin breaking can be avoided if we stick to the parameters e_3 (or to e_1^B, although this contribution will always appear together with the uncertain e_1^A contribution). It is therefore interesting to see how an estimate of e_3 can be attempted if one makes the extra assumption that the technicolour dynamics mimics QCD as much as it can [26,41,426)].

From the definition in (9.15), one sees that e_3 is related to the Fourier transform of the correlation function (Lorentz and space indices omitted)

$$\langle 0|J_{3L}J_0|0\rangle \tag{13.5}$$

between the current coupled to the neutral W boson

$$J_{3L} = \frac{1}{2}(J_{3V} - J_{3A}) \tag{13.6}$$

and the current coupled to the B boson

$$J_0 = \frac{1}{2}(J_{3V} + J_{3A}) + \frac{1}{2}J_Y , \tag{13.7}$$

where J_{3V}, J_{3A} and J_Y are the third components of the vector and axial isospin currents and of the (conventional) hypercharge current respectively. Since, as in QCD, technicolour is assumed to conserve parity and isospin, one has therefore

$$\langle 0|J_{3L}J_0|0\rangle = \frac{1}{4}\left(\langle |J_{3V}J_{3V}|0\rangle - \langle 0|J_{3A}J_{3A}|0\rangle\right) \tag{13.8}$$

Precisely, taking the proportionality factors into account one has

$$e_3 = \frac{\pi\alpha}{s^2}\left(F_{VV}(0) - F_{AA}(0)\right) , \tag{13.9}$$

where, in analogy with (8.1),

$$\Pi^{\mu\nu}_{VV,AA} = \int d^4x\, e^{-iqx}\langle 0|J^\mu_{V,A}(x)J^\nu_{V,A}(0)|0\rangle \tag{13.10}$$

$$= -ig^{\mu\nu}\left[A_{VV,AA}(0) + q^2 F_{VV,AA}(q^2)\right] + q_\mu q_\nu \text{ terms}$$

In the strict QCD case, the right-hand side of (13.9) can be estimated reliably in several different ways. The quickest one [26,41], consists in using Vector Meson Dominance, together with the pion pole and current conservation, for the vector and axial vacuum polarization functions

$$-i\Pi^{\mu\nu}_{VV} = \frac{M_V^2}{f_V^2}\frac{1}{q^2 - M_V^2}(q^2 g_{\mu\nu} - q_\mu q_\nu) \tag{13.11}$$

---

6) For an alternative view, see [43]

$$- i\Pi_{AA}^{\mu\nu} = \frac{M_A^2}{f_A^2} \frac{1}{q^2 - M_A^2} (q^2 g_{\mu\nu} - q_\mu q_\nu) + F_\pi^2 \left(g_{\mu\nu} - \frac{q_\mu q_\nu}{q^2} \right) \qquad (13.12)$$

and the convergence requirement [44]

$$\Pi_{\mu\nu}^V - \Pi_{\mu\nu}^A \overset{q^2 \to \infty}{\simeq} O \left(\frac{1}{(q^2)^2} \right) . \qquad (13.13)$$

which give

$$\frac{M_V^2}{f_V^2} = \frac{M_A^2}{f_A^2} + F_\pi^2 ; \qquad \frac{M_V^4}{f_V^2} = \frac{M_A^4}{f_A^2} . \qquad (13.14)$$

From (13.9-14), one then gets

$$e_3 = \frac{\pi \alpha}{s^2} \frac{F_\pi^2}{M_V^2} \left(1 + \frac{M_V^2}{M_A^2} \right) . \qquad (13.15)$$

where F_π, M_V and M_A are the analogues, in the QCD case, of the π decay constant, the ρ mass and the A_1 mass respectively.

How can this QCD result be extrapolated to the technicolour case? Suppose that we continue to make the VMD assumption. The problem is not in the difference between the QCD and the TC scales, since Eq. (13.15) is scale-independent. Rather it comes from the generally different numbers of colours and flavours. For N_D doublets and N_{TC} technicolours, planar QCD (or TCD) [45] gives

$$\frac{F_\pi^2}{M_V^2} = N_D \frac{N_{TC}}{3} \frac{f_\pi^2}{m_\rho^2} \qquad (13.16)$$

where now we use the lower case letters for the real π decay constant and ρ mass. Other considerations, that give more weight to the threshold behaviour of the vacuum polarization amplitudes, or the Goldstone boson contributions [41,42], suggest different estimates, such as [46]

$$\frac{F_\pi^2}{M_V^2} \simeq \frac{N_D N}{16\pi^2} , \qquad (13.17)$$

where N is the total number of technicolour flavours, so $N \geq 2N_D$. Together with

$$\frac{M_V^2}{M_A^2} \simeq \frac{m_\rho^2}{m_{A_1}^2} \qquad (13.18)$$

and (13.15), the two estimates (13.16) and (13.17) lead respectively to

$$e_3 = \frac{\pi\alpha}{3s^2} \frac{f_\pi^2}{m_\rho^2} \left(1 + \frac{m_\rho^2}{m_{A_1}^2} \right) N_D N_{TC} = 0.7 \times 10^{-3} \, N_D N_{TC} \qquad (13.19)$$

$$e_3 = \frac{\alpha}{16\pi s^2} \left(1 + \frac{m_\rho^2}{m_{A_1}^2} \right) N_D N = 0.9 \times 10^{-3} \, N_D N \qquad (13.20)$$

The comparison of (13.19) and (13.20) is a bit embarrassing. For a "realistic" technicolour model with $N_{TC} = 4$ and 1 technifamily ($N_D = 4$, $N = 8$), the two results would differ by about a factor of 2, which is perhaps not an untolerable discrepancy, but is also a rather fortuitous "agreement". On the other hand, only the estimate (13.19) is in accord with the free techniquark estimate, (13.4), and this for quite obvious reasons. The fact that none of these estimates shows the infrared behaviour required from the

Goldstone boson exchanges, [Eqs. (12.2) and (12.3)] is at this point only a relatively minor drawback.

I personally draw the following conclusions from these considerations. It is obviously difficult to make a sensible estimate of the radiative correction effects in a technicolour model. The difficulty is partly technical (the necessity of dealing with a strongly interacting field theory) and partly fundamental (the undefiniteness of technicolour models in solving the flavour problem). Nevertheless, I believe that the estimates (13.19) and (13.20) make it hard to maintain that, in view of the relatively weak dependence on the ultra-violet cut-off of the radiative correction effects in a "Higgs-less" model [Eqs. (12.2) and (12.3)], the electroweak precision tests are insensitive to the symmetry- breaking sector of the theory. At least in technicolour models, for "realistic" choices of N_D, N and N_{TC}, one finds corrections close to or bigger than the per cent level, to be compared with the per mille level attainable by the experimental determination of the relevant quantities. Furthermore, one can rather clearly identify the source of these potentially large corrections in the way models without the Higgs-Yukawa couplings try to solve the flavour problem, namely by a proliferation of extra degrees of freedom that do not decouple in their radiative correction effects. Finally, it is also significant to notice that the sign of the estimates (13.19) and (13.20) indicate a pattern of the corrections in technicolour not favoured by present data. For a "realistic" technicolour model ($N_{TC} = 4, N_D = 4, N = 8$), if one adds (13.19) or (13.20) to the value of ϵ_3 in the SM with a Higgs mass in the range of the typical technicolour mass scales (500 GeV to 1 TeV), one finds positive values of about $(1 \div 2\%)$, to be compared with the results shown in Figs. 3 and 4.

14. SUMMARY AND CONCLUSIONS

The following is a summary of the main physics points that I have discussed in the course of the lectures, more or less in order of presentation.

i) The bulk of the radiative correction effects that are clearly seen at LEP is still of pure electromagnetic origin (Section 4).

ii) The dominant top quark radiative correction effects, in an approximation which gets better as the top gets heavier, can be discussed without ever making reference to the gauge theory (Section 6). The two quantities most sensitive to the top mass, the ρ parameter and the $Z \to b\bar{b}$ vertex, have been studied independently; both are shown not to be compatible with too heavy a top. To circumvent this conclusion, the perturbative analysis, done at the two-loop level, would have to go wrong in two distinct cases (Sections 6 and 7).

iii) By introducing suitable parameters, it is possible to make a rather model-independent analysis of the "genuine" electroweak radiative correction effects, not of pure electromagnetic origin (Section 9). The LEP 1 programme may show clear evidence for these effects, which at the moment, are constrained to be small enough. The SM is at present compatible with these constraints (Section 10).

iv) All the extra particles introduced in the MSSM fulfil a decoupling property in their virtual contributions to the electroweak precision tests. The SM with a light Higgs is mimicked in most of the MSSM parameter space. Two possibly significant effects have been identified, requiring either a light stop or light gaugino-higgsinos (Section 11).

v) For intrinsic and technical reasons, it is difficult to make a reliable estimate of the radiative correction effects in theories without fundamental scalars, such as technicolour models. In the considered examples, however, these effects attain values that are about one order of magnitude larger than the sensitivity expected at the end of the LEP 1 programme (Section 13). To the extent that technicolour models are representative cases, these semi-quantitative considerations indicate no reason why theories with or without a Higgs should show the same pattern of radiative correction effects at the level of precision attainable in the experiments. In view of all this, I believe that the experimental programme that is being carried out at LEP 1 may shed light on the symmetry-breaking issue in electroweak physics. To this end, it would be very important to have also a direct determination, with sufficient precision, of the top-quark mass, a goal within reach at Fermilab.

ACKNOWLEDGEMENTS

I have benefited from many useful conversations and discussions on these matters with G. Altarelli, D. Bardin, F. Caravaglios, M. Frigeni, G. Curci, M. Beccaria, A. Viceré, M. Giuliani, H. Haber, L. Okun, L. Maiani, P. Ciafaloni, S. Jadach and many others. It is a pleasure to thank Raymond Gastmans and Jean-Marc Gérard for the effective and pleasant organization of the School.

REFERENCES

[1] S.L. Glashow, *Nucl. Phys.* 22:579 (1961);
S. Weinberg, *Phys. Rev. Lett.* 19:1264 (1967);
A. Salam, Proc. 8th Nobel Symposium, Aspenäsgården, 1968, ed. N. Svartholm (Almqvist and Wiksell, Stockholm, 1968), p. 367.

[2] L. Maiani, in Z^0 Physics, Cargèse 1990, eds. M. Levy et al. (Plenum Press, New York, 1991), p. 237.

[3] G. Altarelli, Lectures given at the LVth Les Houches Summer School, France, July 1991.

[4] For reviews, see for example:
H.P. Nilles, *Phys. Rep.* 110:1 (1984);
H. Haber and G. Kane, *Phys. Rep.* 117:75 (1985);
R. Barbieri, *Riv. N. Cimento* 11:1 (1988).

[5] For a review, see for example:
E. Farhi and L. Susskind, *Phys. Rep.* 74:277 (1981).

[6] P. Sikivie, L. Susskind, M. Voloshin and V. Zakharov, *Nucl. Phys.* B173:189 (1980).

[7] R. Behrends, R. Finkelstein and A. Sirlin, *Phys. Rev.* 101:866 (1956);
T. Kinoshita and A. Sirlin, *Phys. Rev.* 113:1652 (1959).

[8] LEP Collaborations, as summarized by G. Rolandi, at the 1992 Int. Conf. on High Energy Physics, Dallas, August 1992.

[9] CDF Collaboration, F. Abe et al., *Phys. Rev. Lett.* 65:2243 (1990);
UA2 Collaboration, G. Unal, Proc. XXVI Rencontre de Moriond, Les Arcs, France (1991).

[10] See F. Berends in Z Physics at LEP1, eds. G. Altarelli et al. (CERN 89-08, Geneva, 1989), Vol.1, p. 89;
M. Caffo and E. Remiddi, ibidem, Vol. 1, p. 171.

[11] H. Burkhardt, F. Jegerlehner, G. Penso and C. Verzegnassi, *Z. Phys.* C43:497 (1989).

[12] See F. Jegerlehner, Villigen preprint PSI-PR-91-08 (1991).

[13] M. Böhm and W. Hollik, in Z Physics at LEP1, eds. G. Altarelli et al. (CERN 89-08, Geneva, 1989), Vol. 1, p. 203.

[14] M. Veltman, *Nucl. Phys.* B123:89 (1977).

[15] A.A. Akhundov, D.Yu. Bardin and T. Riemann, *Nucl. Phys.* B276:1 (1988); F. Diakonos and W. Wetzel, preprint HD-THEP-88-21 (1988); W. Beenakker and W. Hollik, *Z. Phys.* C40:141 (1988); J. Bernabéu, A. Pich and A. Santamaria, *Phys. Lett.* B200:569 (1988); B. Lynn and R. Stuart, *Phys. Lett.* B252:676 (1990).

[16] R. Barbieri, M. Beccaria, P. Ciafaloni, G. Curci and A. Viceré, *Phys. Lett.* B288:95 (1992).

[17] A. Cohen, H. Georgi and B. Grinstein, *Nucl. Phys.* B232:61 (1984); M. Consoli, W. Hollik and F. Jegerlehner, *Phys. Lett.* B227:167 (1989); H. Georgi, *Nucl. Phys.* B363:301 (1991).

[18] R. Barbieri, M. Beccaria, P. Ciafaloni, G. Curci and A. Viceré, CERN preprint TH. 6713/92 (1992).

[19] J. van der Bij and F. Hoogeveen, *Nucl. Phys.* B283:477 (1987).

[20] A. Blondel, A. Djouadi and C. Verzegnassi, DESY preprint DESY 92-112 (August 1992).

[21] G. Burgers, in Polarization at LEP, eds. G. Alexander et al. (CERN 80-06, Geneva 1988), p. 121.

[22] G. Altarelli and R. Barbieri, *Phys. Lett.* B253:161 (1990).

[23] R. Barbieri, F. Caravaglios and M. Frigeni, *Phys. Lett.* B279:169 (1992).

[24] G. Altarelli, R. Barbieri and S. Jadach, *Nucl. Phys.* B369:3 (1992).

[25] D. Kennedy and B. Lynn, *Nucl. Phys.* B322:1 (1989); B. Lynn, SLAC-PUB-5077 (1989); D. Kennedy, *Nucl. Phys.* B351:81 (1991).

[26] M. Peskin and T. Takeuchi, *Phys. Rev. Lett.* 65:964 (1990).

[27] W. Marciano and J. Rosner, *Phys. Rev. Lett.* 65:2963 (1990), E:66:395 (1991); *Phys. Rev.* D44:1591 (1991).

[28] D. Kennedy and P. Langacker, *Phys. Rev. Lett.* 65:2967 (1990).

[29] W. Marciano, Brookhaven preprint BNL-45999 (1991).

[30] V.A. Novikov, L.B. Okun and M.I. Vytsosky, CERN preprint TH.6538/92 (1992).

[31] U. Amaldi et al., *Phys. Rev.* D26:1385 (1987); A. Blondel et al., CDHSW Collaboration, *Z. Phys.* C45:361 (1990); J.V. Allaby et al., CHARM Collaboration, *Z. Phys.* C36: 611 (1987); P.G. Reutens et al., CCFRR Collaboration, *Z. Phys.* C45:539 (1990); T.S. Mattison et al., FMM Collaboration, *Phys. Rev.* D42:1311 (1990).

[32] CCFR Collaboration, as presented by T. Bolton, 1992 International Conference on High Energy Physics, Dallas, August 1992.

[33] M.C. Noecker et al., *Phys. Rev. Lett.* 61:310 (1988); M.A. Bouchiat, Proc. 12th Int. Atomic Physics Conference (1990); S.A. Blundell et al., *Phys. Rev. Lett.* 65:1411 (1990); V. Dzuba et al., *Phys. Lett.* A141:147 (1989).

[34] R. Barbieri, S. Ferrara and C. Savoy, *Phys. Lett.* 119B:343 (1982); P. Nath, R. Arnowitt and A. Chamseddine, *Phys. Rev. Lett.* 49:970 (1982).

[35] H. Haber and R. Hempfling, *Phys. Rev. Lett.* 66:1815 (1991);
J. Ellis, G. Ridolfi and F. Zwirner, *Phys. Lett.* B257:83 (1991);
Y. Okado, M. Yamaguchi and T. Yanagida, *Progr. Theor. Phys. Lett.* 85:1 (1991);
R. Barbieri, F. Caravaglios and M. Frigeni, *Phys. Lett.* B258:167 (1991).

[36] R. Barbieri and L. Maiani, *Nucl. Phys.* B224:32 (1983).

[37] L. Alvarez-Gaumé, J. Polchinski and M. Wise, *Nucl. Phys.* B221:495 (1983).

[38] M. Veltman, *Acta Phys. Polon.* B8 (1977) 475.

[39] A. Longhitano, *Phys. Rev.* D22:1166 (1980); *Nucl. Phys.* B188:118 (1981).

[40] G. Passarino and M. Veltman, *Nucl. Phys.* B160:151 (1979).

[41] R. Cahn and M. Suzuki, *Phys. Rev.* D44:3641 (1991);
M. Peskin and T. Takeuchi, *Phys. Rev.* D46:381 (1991).

[42] R. Renken and M. Peskin, *Nucl. Phys.* B211:93 (1983);
M. Golden and L. Randall, *Nucl. Phys.* B361:3 (1990);
B. Holdom and J. Terning, *Phys. Lett.* B247:88 (1990);
A. Dobado, D. Espriu and M.J. Herrero, *Phys. Lett.* B255 (1990) 405.

[43] R. Casalbuoni et al., *Phys. Lett.* B258 (1991) 161.

[44] S. Weinberg, *Phys. Rev. Lett.* 18:507 (1967).

[45] G. 't Hooft, *Nucl. Phys.* B72:461 (1974).

[46] R.S. Chivukula, M.J. Dugan and M. Golden, Harvard preprint HUTP-92/A033 (July 1992).

CHIRAL PERTURBATION THEORY

Gerhard Ecker

Inst. Theor. Physik
Univ. Wien
Boltzmanng. 5
A-1090 Wien

1   INTRODUCTION

Advances in particle physics are achieved on two experimental frontiers:

- Experiments using the highest available energies search for new degrees of freedom which are either expected in the standard theory (top, Higgs) or would point to new physics.
- At the low–energy frontier, the main goal is to perform precision tests of the standard theory. To make up for the smaller energies, high sensitivity is an indispensable requirement (factories).

Effective Field Theories (EFT) are the appropriate theoretical tool for 'low–energy' physics. There are two different categories:

- An EFT can be used to parametrize new physics in cases where the underlying fundamental theory is unknown. The Higgs sector of the standard model (SM) may be of this type.
- Even if the fundamental theory is known, it may not be directly applicable in the region of interest. This is the case of the SM in the confinement regime. The corresponding EFT of the SM at low energies is called Chiral Perturbation Theory (CHPT) [1, 2, 3] and it is the subject of these lectures.

It is useful to distinguish three energy domains:

1. **Short distances** $(E \gtrsim 1 GeV)$
   This is the domain of perturbative QCD.
2. **Intermediate distances** $(M_K \lesssim E \lesssim 1 GeV)$
   In this region, theoretical improvements are needed most, such as the non-perturbative implementation of unitarity, incorporation of resonances beyond the low–energy expansion, etc. The general theme is to match the short–distance and

the long–distance descriptions of the SM. That such a matching is possible is by no means obvious a priori, but previous successes of duality give support to such attempts.

3. **Long distances** ($E \overset{<}{\sim} M_K$)

The confinement regime is the domain of CHPT.

The theoretical basis of CHPT was formulated by Weinberg [1] as a 'theorem': for a given set of asymptotic states (hadrons rather than quarks and gluons in our case), perturbation theory with the most general Lagrangian containing all terms allowed by the assumed symmetries will yield the most general S-matrix elements consistent with analyticity, perturbative unitarity and the assumed symmetries. The 'theorem' carries quotation marks because it is at the same time almost self–evident and extremely difficult to prove. In fact, the theorem was in 1979, and still is today, only 'proved' to the extent that no counterexamples are known.

Anticipating the contents of these lectures, let me list the main features of CHPT right at the beginning:

- CHPT is a direct consequence of the SM and its assumed ground state.

- Even though it is non–renormalizable, it is a perfectly well–defined QFT with all its general properties (analyticity, unitarity).

- CHPT is a low–energy expansion in terms of momenta and meson masses. It incorporates the crucial property of Goldstone bosons that their interactions become weak at small momenta.

- There is no problem of double–counting (hadronic vs. quark and gluon degrees of freedom).

- The softly broken chiral symmetry of the SM is manifest at each stage. There is no need to check chiral Ward identities, PCAC and the like: they are automatically fulfilled.

- It is ideally suited for the physics of the light pseudoscalar bosons which are the pseudo–Goldstone bosons of spontaneous chiral symmetry breaking (SCSB).

- The chiral anomaly is unambiguously incorporated at the hadronic level.

- The price to pay for all these assets is substantial: at each order of the chiral expansion, certain low–energy constants appear which are not restricted by the symmetries of the SM. They can be interpreted as being due to the loss of information in going from the fundamental to the effective theory. The SM is singled out by specific values of these constants which must be determined either phenomenologically or with additional theoretical input.

- However, the structure of amplitudes is fixed once and for all. Further theoretical progress in this field will predict some of the low–energy constants, but it cannot modify the structure of the SM at low energies as given by CHPT.

- Finally, CHPT affords a fascinating theoretical challenge: it is the first time in particle physics that we derive experimentally measurable predictions from a non–renormalizable QFT in a systematic way.

CHPT as discussed in these lectures is based on the most general effective Lagrangian (à la Weinberg) with all the symmetries of the SM, but without additional dynamical assumptions. The advantage of such an approach is that the SM is certainly included. The drawback is that this applies equally well to many possible alternative underlying theories with the same symmetries as the SM. For this reason, the trademark CHPT is frequently (mis)used for what should more properly be called chirally–inspired models. A few examples in decreasing order of theoretical reliability are:

1. Models for the low–energy constants
   Additional dynamical assumptions are used to determine those constants: resonance contributions, $1/N_C$ expansion, models of SCSB, sum rules, lattice gauge theory, etc. Some of these approaches will be discussed in Sect.2.

2. Linear σ model [4]
   The model has the correct chiral structure, but the wrong phenomenology at the next–to–leading order in the chiral expansion for any value of M_σ [2].

3. Skyrme model [5]
   The baryons are viewed as solitons of a truncated effective Lagrangian. Interesting as it may be from a theoretical point of view, one must not forget the inherent uncertainties of the model. Since the $O(p^4)$ terms are needed to stabilize the solitons, the soliton energy in particular receives comparable contributions from the terms of $O(p^2)$ (non–linear σ model) and $O(p^4)$ (Skyrme term). Thus, the Skyrme model lies outside the domain where a perturbative expansion in the momenta can be expected to give reliable results. One should not expect predictions from the Skyrme model (although some people do) to better than 50 %.

4. Chiral handwaving
   Attractive especially for linguistic reasons, it often allows for illustrative order–of–magnitude estimates. However, it must not be confused with genuine CHPT predictions. There is a rich literature based on such a misunderstanding. A useful attitude towards the chiral literature is not to believe predictions from a non–renormalizable QFT without a systematic analysis.

The purpose of these lectures is to give a rather self–contained introduction to CHPT on the one hand and to review some of the recent applications, successes and problems of the method. Since CHPT has not yet acquired the status of textbook physics (see however Refs. [6, 7]), I have decided to go into quite some details in describing the structure of CHPT in Sect. 2. In the applications to the strong, electromagnetic and semileptonic weak interactions in Sect. 3 and to the non–leptonic weak interactions in Sect. 4, on the other hand, only results will be presented in most cases, referring for details to the original literature.

Among the many topics not treated in these lectures, mostly for reasons of space but sometimes also for lack of competence, are CHPT at finite temperature and volume, the odd–intrinsic parity sector at $O(p^6)$, CHPT and CP violation (see the lectures of J.-M. Gérard in these Proc.) and the applications of CHPT to baryons. For those topics and also for the ones included here, I refer to the already mentioned books [6, 7] and to several reviews and Proceedings of specialized workshops [8,..,16].

2 THE STRUCTURE OF CHPT

2.1 Chiral Symmetry

Our starting point is the QCD Lagrangian

$$
\begin{aligned}
\mathcal{L}^0_{QCD} &= \bar{q}i\gamma^\mu\left(\partial_\mu + ig_s\frac{\lambda_\alpha}{2}G^\alpha_\mu\right)q - \frac{1}{4}G^\alpha_{\mu\nu}G^{\alpha\mu\nu} + \mathcal{L}_{\text{heavy quarks}} \qquad (2.1) \\
&= \bar{q}_L i\, \slashed{D} q_L + \bar{q}_R i\, \slashed{D} q_R + \ldots \\
q &= \text{column}(u,d,s) \quad , \quad q_{R,L} = \frac{1}{2}(1\pm\gamma_5)q
\end{aligned}
$$

with massless light quarks u, d, s. The global symmetry of this Lagrangian is

$$
\underbrace{SU(3)_L \times SU(3)_R}_{\text{chiral group G}} \times U(1)_V \times U(1)_A ,
$$

but only the chiral group G will be relevant in the following. The quark number symmetry $U(1)_V$ is trivially realized in the meson sector and $U(1)_A$ is not a symmetry at the quantum level. The corresponding Abelian anomaly [17] gives in particular a mass to the η' meson even in the chiral limit. The Noether currents associated with the chiral group G are

$$
J_X^{a\mu} = \bar{q}_X\gamma^\mu\frac{\lambda_a}{2}q_X \qquad X = L, R \qquad a = 1,\ldots,8 . \qquad (2.2)
$$

The local form of the Lie algebra of the Noether charges

$$
\begin{aligned}
[Q_X^a, Q_X^b] &= if_{abc}Q_X^c \qquad (X = L, R) \\
[Q_L^a, Q_R^b] &= 0 \\
Q_X^a &= \int d^3x\, J_X^{a0}
\end{aligned} \qquad (2.3)
$$

was the starting point of the current algebra methods of the Sixties [18].

How can we realize a symmetry like G in QFT ? Let us consider an arbitrary global internal symmetry with an associated charge

$$
Q = \int d^3x\, J^0(x) . \qquad (2.4)
$$

As an internal symmetry it commutes with space–time translations,

$$
\begin{aligned}
[Q, P_\mu] &= 0 \qquad (2.5) \\
J^0(x) &= e^{iPx}J^0(0)e^{-iPx} . \qquad (2.6)
\end{aligned}
$$

Operating with Q on the vacuum state $(P_\mu|0\rangle = 0)$, we obtain

$$
\begin{aligned}
\||Q|0\rangle\|^2 &= \langle 0|QQ|0\rangle = \int d^3x \langle 0|QJ^0(x)|0\rangle \\
&= \int d^3x \langle 0|Qe^{iPx}J^0(0)e^{-iPx}|0\rangle = \int d^3x \langle 0|QJ^0(0)|0\rangle . \qquad (2.7)
\end{aligned}
$$

Observing that the integrand in the last integral is independent of x, we arrive at the

$Q\|0\rangle = 0$	$\|Q\|0\rangle\| = \infty$
Wigner–Weyl	Nambu–Goldstone
linear representation	non–linear realization
degenerate multiplets	massless Goldstone bosons
exact symmetry	spontaneously broken symmetry

There is ample phenomenological and theoretical evidence that the chiral group G is actually spontaneously broken :

1. Absence of parity doublets in the hadron spectrum.

2. The pseudoscalar mesons are by far the lightest hadrons.

3. The consistency conditions of 't Hooft [19] require the spontaneous breaking of G for $N = 3$ light quarks.

4. In the large–N_C limit, where N_C is the number of colours, but for any number N_{fl} of flavours, the global symmetry group $U(N_{fl})_L \times U(N_{fl})_R$ is either unbroken or spontaneously broken to the vectorial subgroup $SU(N_{fl})_V \times U(1)_V$ [20].

5. In vector–like gauge theories like QCD, $SU(N_{fl})_V$ cannot be spontaneously broken (for the vacuum angle $\theta_{QCD} = 0$) [21].

6. There is now rather solid evidence from lattice gauge theories (see below).

All this evidence puts the basic assumption of CHPT on rather firm grounds :

$$G = SU(3)_L \times SU(3)_R \xrightarrow{SCSB} SU(3)_V . \tag{2.8}$$

The simplest explicit example for SCSB is provided by the linear σ model [4] with the Lagrangian

$$\mathcal{L}_\sigma = \frac{1}{2} \left[(\partial_\mu \sigma)^2 + (\partial_\mu \vec{\pi})^2 \right] - \frac{\lambda}{4} \left(\sigma^2 + \vec{\pi}^2 - v^2 \right)^2 , \qquad \lambda > 0 . \tag{2.9}$$

For $v^2 < 0$, the global symmetry $O(4) \sim SU(2) \times SU(2)$ is realized à la Wigner–Weyl with degenerate states $\sigma, \vec{\pi}$ with mass $m^2 = -\lambda v^2$. For the more interesting case $v^2 > 0$, the potential

$$V(\sigma, \vec{\pi}) = \frac{\lambda}{4} \left(\sigma^2 + \vec{\pi}^2 - v^2 \right)^2 \tag{2.10}$$

has degenerate minima for all $\sigma, \vec{\pi}$ with $\sigma^2 + \vec{\pi}^2 = v^2$. In QFT, these minima correspond to degenerate ground states (vacua), each of which is not invariant under G; these ground states are rather transformed into each other under a chiral rotation. The symmetry is realized à la Nambu–Goldstone in this case. With the choice

$$\langle 0|\sigma(x)|0\rangle = v , \quad \langle 0|\vec{\pi}(x)|0\rangle = 0 \qquad (v^* = v) \tag{2.11}$$

and rewriting $\mathcal{L}_\sigma$ in terms of the shifted scalar field $\hat{\sigma} = \sigma - v$,

$$\mathcal{L}_\sigma = \frac{1}{2} \left((\partial_\mu \hat{\sigma})^2 - 2\lambda v^2 \hat{\sigma}^2 \right) + \frac{1}{2} (\partial_\mu \vec{\pi})^2 \tag{2.12}$$

$$- \lambda v \hat{\sigma} (\hat{\sigma}^2 + \vec{\pi}^2) - \frac{\lambda}{4} \left(\hat{\sigma}^2 + \vec{\pi}^2 \right)^2 + \text{const.} ,$$

one finds immediately the following states (particles) in the classical approximation:

$$\hat{\sigma} = \sigma - v \qquad\qquad M_{\hat{\sigma}} > 0 \qquad\qquad \text{positive curvature of } V$$
$$\vec{\pi} \qquad\qquad\qquad M_{\pi} = 0 \qquad\qquad \text{flat directions in } V$$

The Lagrangian (2.12) suggests the following two questions :

- Does the mechanism of SCSB depend on the existence of a massive scalar particle ?

- How can the non–derivative couplings of the pions in Eq. (2.12) be reconciled with the statement that the interactions of Goldstone bosons become arbitrarily weak at low momenta ?

Following a procedure reminiscent of the unitary gauge in the SM, we will now perform field transformations to bring (2.12) into the standard form of the lowest–order chiral Lagrangian. The additional terms appearing in this transition are of higher order in the chiral expansion and they reduce the linear σ model to a toy model: except at lowest order p^2, where all chiral Lagrangians are equivalent once the pion decay constant and the quark condensate are fixed, the linear σ model is not a phenomenologically viable EFT of QCD [2].

Introducing the matrix notation

$$\Sigma(x) = \sigma(x)\mathbf{1} + i\vec{\tau} \cdot \vec{\pi}(x) \ , \qquad\qquad (2.13)$$

we rewrite $\mathcal{L}_\sigma$ as

$$\mathcal{L}_\sigma = \frac{1}{4}\langle \partial_\mu \Sigma \partial^\mu \Sigma^\dagger \rangle - \frac{\lambda}{16}\left(\langle \Sigma^\dagger \Sigma \rangle - 2v^2\right)^2 \ , \qquad\qquad (2.14)$$

where $\langle A \rangle$ stands for the trace of the matrix A. In this formulation, a chiral transformation ($G = SU(2)_L \times SU(2)_R$ in this case) is simply [1]

$$\Sigma \xrightarrow{\ G\ } g_R \Sigma g_L^\dagger \qquad , \qquad g_{L,R} \in SU(2)_{L,R} \ . \qquad\qquad (2.15)$$

We now perform a polar decomposition of the matrix field $\Sigma(x)$

$$\Sigma(x) = (v + S(x))\, U(\vec{\phi}(x)) \qquad\qquad (2.16)$$

in terms of a hermitian scalar field S and pseudoscalar [2] fields $\vec{\phi}$. By construction, U is unitary, but a calculation of the determinant of Σ shows that U is also unimodular ($\det U = 1$), so that we can parametrize it in the familiar form

$$U(\phi) = e^{i\vec{\tau} \cdot \vec{\phi}/v} \ . \qquad\qquad (2.17)$$

The non–linear field transformations

$$\hat{\sigma}, \vec{\pi} \longrightarrow S, \vec{\phi}$$

imply the chiral transformation properties

$$S \ \rightarrow \ S$$
$$U(\phi) \ \rightarrow \ g_R U(\phi) g_L^\dagger$$

---

[1] Adopting the so–called inverse Bernese notation in the ordering of left and right.
[2] The parity transformation $\Sigma \rightarrow \Sigma^\dagger$ is a symmetry of the Lagrangian (2.14) .

and lead to the following form of the linear σ model Lagrangian

$$\mathcal{L}_\sigma = \frac{v^2}{4}\left(1 + \frac{S}{v}\right)^2 \langle \partial_\mu U^\dagger \partial^\mu U \rangle \tag{2.18}$$

$$+ \frac{1}{2}\left(\partial_\mu S \partial^\mu S - 2\lambda v^2 S^2\right) - \lambda v S^3 - \frac{\lambda}{4}S^4 + \text{const} .$$

Comments:

1. The interaction of the Goldstone bosons takes the form of the non–linear σ model, the unique chiral Lagrangian of lowest order in the derivative expansion. The Lagrangian (2.18) allows a clear separation between the model–independent Goldstone boson interaction induced by SCSB and the model–dependent part involving the scalar field S. To lowest order in CHPT, we may as well forget about S altogether.

2. Expanding $U(\phi)$ in a power series in ϕ, one obtains a properly normalized kinetic term for the fields $\vec{\phi}$. This ensures that the fields $S, \vec{\phi}$ produce the same S–matrix elements as the original $\hat{\sigma}, \vec{\pi}$.

3. As announced, but remarkable nevertheless, the Goldstone fields have purely derivative couplings. It is worthwhile to discuss the origin of this fundamental property of Goldstone bosons. Let us assume for a moment that we have constant fields

$$\vec{\phi}(x) = \vec{\phi}(x_0) = \text{const} .$$

In that case, we can perform a global axial transformation of the type

$$g_L = g_R^\dagger = e^{i\vec{\tau} \cdot \vec{\phi}(x_0)/2v}$$

to eliminate the fields $\vec{\phi}$ altogether in the matrix field Σ [cf. Eq.(2.15)].

Therefore, constant Goldstone fields cannot influence the dynamics which implies derivative couplings among the Goldstone bosons, at least in a properly chosen basis of fields. This basic feature of Goldstone bosons, which is at the heart of CHPT, is not at all manifest in the original form (2.12) of the Lagrangian. Nevertheless, all S–matrix elements are the same whether one uses $\hat{\sigma}, \vec{\pi}$ or $S, \vec{\phi}$, as a fundamental theorem [22] of QFT guarantees. Thus, although the couplings in Eq. (2.12) seem to be at variance with the general theorem, all measurable amplitudes with external Goldstone bosons are of course perfectly in accordance with it.

Since the rigorous quantum field theorists insist that a spontaneously broken generator does not 'exist' [cf. Eq. (2.7)], we are advised to exercise some care in formulating the Goldstone theorem [23] in a model–independent way. Let us therefore define the Noether charge (2.4) in a finite volume V as

$$Q^V(x^0) = \int_V d^3x J^0(x)$$

and assume the existence of an operator A such that

$$\lim_{V \to \infty} \langle 0|[Q^V(x^0), A]|0\rangle \neq 0 , \tag{2.19}$$

which is clearly only possible if

$$Q|0\rangle \neq 0 .$$

Then the Goldstone theorem tells us that there exists a state $|G\rangle$ with

$$\langle 0|J^0(0)|G\rangle\langle G|A|0\rangle \neq 0 \qquad \text{and} \qquad P^2|G\rangle = 0 . \tag{2.20}$$

Each spontaneously broken (continuous) symmetry implies the existence of a corresponding massless Goldstone boson, with its other quantum numbers dictated by those of J^0 or A.

The quantity on the left–hand side of Eq. (2.19) is called the order parameter of the spontaneous symmetry breakdown. What is the order parameter of QCD ? From the previously given arguments, the broken generators of the chiral group G are axial. Thus, $|G^a>$ must be pseudoscalar states and, consequently, the A^a must be pseudoscalar operators. The simplest possibility in QCD are the quark bilinear operators

$$A^a = \bar{q}\gamma_5\lambda_a q \qquad (a = 1,\ldots,8) \tag{2.21}$$

with the relevant commutators

$$[Q_A^a, \bar{q}\gamma_5\lambda_b q] = -\frac{1}{2}\bar{q}\{\lambda_a, \lambda_b\}q , \tag{2.22}$$

where $Q_A^a = Q_R^a - Q_L^a$ are the axial generators (even though they are not allowed to exist !). Taking vacuum expectation values and using the conservation of the diagonal subgroup $SU(3)_V$, one finds as a candidate for the order parameter of SCSB the quark condensate

$$< 0|\bar{u}u|0 >=< 0|\bar{d}d|0 >=< 0|\bar{s}s|0 > . \tag{2.23}$$

Until recently, it was necessary to concede the hypothetical nature of this simplest possibility for the order parameter of SCSB. Recent work with dynamical staggered fermions [24] has now produced very convincing evidence 'directly' from (lattice) QCD that the quark condensate is indeed non–vanishing. The authors of Ref. [24] have extracted the (renormalization group invariant) quark condensate from their lattice data to be

$$< 0|\bar{q}q|0 >_{RGI}= (215 \pm 25 MeV)^3 \tag{2.24}$$

in the chiral limit. In fact, they even verify the Gell–Mann, Oakes, Renner relation [25] for non–vanishing quark masses [see Eq. (2.70)]. The impressive results of Altmeyer et al. [24] establish the standard scenario of CHPT beyond reasonable doubt.

2.2 Non–Linear Realizations

How can a symmetry be realized on quantum fields ? As already mentioned in connection with the Goldstone alternative, there are two possibilities : linear representation (Wigner–Weyl) and non–linear realization (Nambu– Goldstone).

Let us concentrate first on the transformation of the Goldstone fields under a given symmetry G :

$$\vec{\phi} \xrightarrow{g \in G} \vec{\phi}' = \vec{f}(g, \vec{\phi}) . \tag{2.25}$$

It is a remarkable fact [26] that for a given group G the (non–linear) transformation functions $\vec{f}$ are unique up to field transformations of the $\vec{\phi}$. Consider first elements $h \in G$ such that

$$\vec{f}(h, \vec{0}) = \vec{0} . \tag{2.26}$$

All elements h that leave the origin in $\vec{\phi}$ space invariant obviously form a group, the conserved subgroup $H \subset G$. For all other elements $g \in G$, but $g \notin H$, the group property

$$\vec{f}(gh, \vec{0}) = \vec{f}(g, \vec{f}(h, \vec{0})) = \vec{f}(g, \vec{0}) \qquad h \in H \tag{2.27}$$

tells us that $\vec{f}(g, \vec{0})$ is an element of the coset space G/H. Since this mapping between the Goldstone fields and coset space is invertible, the Goldstone fields parametrize G/H. A field transformation of the $\vec{\phi}$ is just a coordinate transformation in G/H. In particular, we have derived the well–known result that the number of Goldstone fields is equal to the dimension of the coset space. Note also that a non–linear transformation of the type (2.27) can only be a symmetry if the fields $\vec{\phi}$ are massless; a mass term in the ϕ's could not possibly be invariant under (2.27) which creates something out of nothing (inhomogeneous transformation).

The non–linear realization of a symmetry is related to the geometry of G/H. We decompose the Lie algebra of the symmetry group G into elements H_i corresponding to the conserved subgroup H (the generators of $SU(3)_V$ in the chiral case) and the remaining spontaneously broken generators X_a (the axial generators in our case). The Lie algebra has the form [3]

$$[H_i, H_j] = ic_{ijk}H_k \qquad \text{subgroup } H \qquad (2.28)$$
$$[H_i, X_a] = ic_{iab}X_b \qquad X_a \text{ representation of } H \qquad (2.29)$$
$$[X_a, X_b] = ic_{abi}H_i \qquad (2.30)$$

with structure constants c. Denoting an element of coset space by $L(\phi)$, which in the standard parametrization [26] would have the form

$$L(\phi) = e^{i\phi^a X_a} , \qquad (2.31)$$

the (left) action of a symmetry transformation $g \in G$ on G/H is given by

$$L(\phi) \rightarrow gL(\phi) = L(\phi')h(\phi, g) \qquad (2.32)$$

where the so–called compensator $h(\phi, g) \in H$ accounts for the fact that the transformed coset space element will have the chosen parametrization only up to an H transformation. Specializing first to $g \in H$, one finds in accordance with the previous discussion

$$gL(\phi) = e^{i\alpha^i H_i}e^{i\phi^a X_a} = \underbrace{e^{i\alpha^i H_i}e^{i\phi^a X_a}e^{-i\alpha^i H_i}}_{L(\phi')=e^{i\phi'^a X_a}}\underbrace{e^{i\alpha^i H_i}}_{h(g)=g} . \qquad (2.33)$$

The compensator h is trivial in this case, i.e. independent of ϕ, and the ϕ themselves transform linearly because of (2.29) :

$$e^{i\alpha^i H_i}X_a e^{-i\alpha^i H_i} = X_b R_X(\alpha)_{ba} \qquad (2.34)$$
$$\phi'^a = R_X(\alpha)_{ab}\phi^b .$$

On the other hand, if $g \notin H$, the compensator field $h(\phi, g)$ depends non–trivially on ϕ and the ϕ transform non–linearly.

The geometry of coset space is characterized by the Cartan–Maurer one–form

$$L(\phi)^{-1}dL(\phi) = \omega + e = (\omega_a + e_a) d\phi^a = \left(\omega_a^i H_i + e_a^b X_b\right) d\phi^a . \qquad (2.35)$$

From the transformation property of $L(\phi)$ one derives

$$e \xrightarrow{g} heh^{-1} \qquad (2.36)$$
$$\omega \xrightarrow{g} h\omega h^{-1} + hdh^{-1} . \qquad (2.37)$$

---

[3]The Lie bracket (2.30) is not the most general one, but the structure constants $c_{abc} = 0$ for the chiral group.

The one–form $e(\phi)$ is called the vielbein and $\omega(\phi)$ is the natural connection on coset space. The connection has a very important practical application for constructing G–invariant couplings of the Goldstone fields to other matter fields. A generic multi-component matter field ψ will transform in the usual linear way under the conserved subgroup H:

$$\psi \xrightarrow{h} \psi' = h_\psi(h)\psi \qquad (2.38)$$

where h denotes both the group element and its linear representation on the ψ. For a general transformation $g \in G$ one defines

$$\psi \xrightarrow{g} \psi' = h_\psi(\phi, g)\psi \qquad (2.39)$$

in terms of the compensator field $h_\psi(\phi, g)$ in the ψ representation. Although the group property is readily checked, there is a problem in constructing G–invariant Lagrangians: since $h_\psi(\phi(x), g)$ is a field depending on space–time, the normal derivative $\partial_\mu \psi$ does not transform covariantly. The remedy is known from gauge theories and the place of the gauge potential is now taken by the connection ω. The covariant derivative

$$\nabla\psi = (d + \omega)\psi = \left(\partial_\mu + \omega_a \frac{\partial \phi^a}{\partial x_\mu}\right)\psi dx^\mu = \nabla_\mu \psi dx^\mu \qquad (2.40)$$

transforms as it should and supplies the missing ingredient for constructing G–invariant Lagrangians in a completely general way.

Specializing now to the chiral symmetry

$$G = SU(3)_L \times SU(3)_R \xrightarrow{SCSB} H = SU(3)_V , \qquad (2.41)$$

we employ the notation

$$
\begin{aligned}
g &= (g_L, g_R) & g \in H \leftrightarrow g_L = g_R \\
L(\phi) &= (l_L(\phi), l_R(\phi))
\end{aligned}
$$

to write the chiral transformation as

$$L(\phi) \to L(\phi') = gL(\phi)h(\phi, g)^{-1} = (g_L, g_R)(l_L(\phi), l_R(\phi))(h(\phi, g)^{-1}, h(\phi, g)^{-1}) . \quad (2.42)$$

The chiral group has the special property of an additional symmetry, (the outer automorphism) parity. This allows the construction of a field

$$U(\phi) = l_R(\phi)l_L(\phi)^{-1} \qquad (2.43)$$

transforming linearly under chiral transformations in view of (2.42) :

$$U(\phi) \xrightarrow{g} g_R l_R h^{-1}(g_L l_L h^{-1})^{-1} = g_R l_R l_L^{-1} g_L^{-1} = g_R U(\phi) g_L^{-1} . \qquad (2.44)$$

We have gone to quite some efforts to convince ourselves that the matrix field $U(\phi)$, which appeared first in the polar decomposition (2.16) in the discussion of the linear σ model, is in fact an intrinsic object of SCSB. One may still wonder if we are not omitting some information by considering only the special product (2.43) of the coset space elements l_L, l_R. However, we can always use a parametrization of the chiral coset space (sometimes called a choice of gauge [26]) where

$$
\begin{aligned}
l_R(\phi) &= l_L(\phi)^{-1} = e^{i\phi^a Q_A^a / F} \doteq u(\phi) \\
U(\phi) &= u(\phi)^2
\end{aligned}
\qquad (2.45)
$$

where I have inserted a dimensionful quantity F to give the fields ϕ^a the proper mass dimension of scalar fields. If we choose the fundamental representation with $Q_A^a = \lambda_a/2$, we recover the formula (2.17) except that we have $SU(3)$ now instead of $SU(2)$ and that v has been replaced by F.

For the connection and the vielbein we will use the notation

$$\omega = \tfrac{1}{2}\left(udu^\dagger + u^\dagger du\right) = \Gamma_\mu(\phi)dx^\mu \tag{2.46}$$

$$e = \tfrac{1}{2}\left(udu^\dagger - u^\dagger du\right) = \frac{i}{2}u_\mu(\phi)dx^\mu \ .$$

The hermitian vielbein fields $u_\mu(\phi)$ transform as [Eq. (2.36)]

$$u_\mu \xrightarrow{G} hu_\mu h^{-1} \tag{2.47}$$

and they can also be expressed as

$$u_\mu = iu^\dagger \partial_\mu U u^\dagger = -iu\partial_\mu U^\dagger u \ . \tag{2.48}$$

There is clearly only one chiral invariant of lowest order p^2, i.e. with two derivatives,

$$\langle u_\mu u^\mu \rangle = \langle \partial_\mu U^\dagger \partial^\mu U \rangle \tag{2.49}$$

and we recover the non-linear σ model, first encountered in (2.18), in a completely model-independent way. It is trivial to check that the only chiral invariant of $O(p^0)$ is an irrelevant constant.

2.3   Effective Chiral Lagrangian of Lowest Order

In the approach of Gasser and Leutwyler [2, 3] one couples the quarks to external hermitian matrix-valued fields v_μ, a_μ, s, p via the Lagrangian

$$\mathcal{L} = \mathcal{L}_{QCD}^0 + \bar{q}\gamma^\mu(v_\mu + \gamma_5 a_\mu)q - \bar{q}(s - i\gamma_5 p)q. \tag{2.50}$$

The external field method offers several advantages:

- The electromagnetic and semileptonic weak interactions are automatically included with

$$r_\mu = v_\mu + a_\mu = -eQA_\mu \tag{2.51}$$

$$l_\mu = v_\mu - a_\mu = -eQA_\mu - \frac{e}{\sqrt{2}\sin\theta_W}(W_\mu^+ T_+ + h.c.)$$

$$Q = \frac{1}{3}\mathrm{diag}(2, -1, -1), \qquad T_+ = \begin{pmatrix} 0 & V_{ud} & V_{us} \\ 0 & 0 & 0 \\ 0 & 0 & 0 \end{pmatrix}$$

where the V_{ij} are Kobayashi–Maskawa matrix elements.

- Not only S–matrix elements, but general Green functions of quark currents can be obtained from the generating functional $Z[v, a, s, p]$ to be defined below.

- The approach provides an extremely useful way to incorporate explicit chiral symmetry breaking through the quark masses. The physical Green functions (and S–matrix elements) are obtained as functional derivatives of $Z[v, a, s, p]$ at

$$v_\mu = a_\mu = p = 0 \ ,$$

but at

$$s = \mathcal{M} = \text{diag}(m_u, m_d, m_s) \,. \tag{2.52}$$

The tremendous advantage is that we can calculate $Z[v, a, s, p]$ in a manifestly chiral invariant way till the very end before taking the appropriate functional derivatives for the actual Green functions.

- The method avoids the problem of non–multiplicative renormalization of ϕ^a or $U(\phi)$. In the calculation of $Z[v, a, s, p]$, the Goldstone fields are just coordinates in the functional integral corresponding to their role as coordinates in coset space in the classical case. Nevertheless, for most practical purposes there is no harm in calculating amplitudes directly with the ϕ^a corresponding to physical states.

The Lagrangian (2.50) exhibits a local $SU(3)_L \times SU(3)_R$ symmetry

$$
\begin{aligned}
q &\rightarrow g_R \frac{1}{2}(1 + \gamma_5)q + g_L \frac{1}{2}(1 - \gamma_5)q \\
r_\mu &\rightarrow g_R r_\mu g_R^\dagger + i g_R \partial_\mu g_R^\dagger \\
l_\mu &\rightarrow g_L l_\mu g_L^\dagger + i g_L \partial_\mu g_L^\dagger \\
s + ip &\rightarrow g_R(s + ip)g_L^\dagger \\
g_{R,L} &\in SU(3)_{R,L}.
\end{aligned}
\tag{2.53}
$$

The local nature of the chiral symmetry requires the introduction of a covariant (now with respect to the external gauge fields v, a) derivative

$$D_\mu U = \partial_\mu U - i r_\mu U + i U l_\mu \tag{2.54}$$

and of non–Abelian field strength tensors

$$
\begin{aligned}
F_R^{\mu\nu} &= \partial^\mu r^\nu - \partial^\nu r^\mu - i[r^\mu, r^\nu] \\
F_L^{\mu\nu} &= \partial^\mu l^\nu - \partial^\nu l^\mu - i[l^\mu, l^\nu].
\end{aligned}
\tag{2.55}
$$

The definition of the generating functional $Z[v, a, s, p]$ and Weinberg's 'theorem' [1] can be assembled in the master formula

$$e^{iZ[v, a, s, p]} = < 0|Te^{i \int d^4 x \mathcal{L}}|0> = \int [dU(\phi)] e^{i \int d^4 x \mathcal{L}_{eff}} \tag{2.56}$$

where $\mathcal{L}$ is defined in Eq. (2.50). The effective chiral Lagrangian

$$\mathcal{L}_{eff} = \mathcal{L}_2 + \mathcal{L}_4 + \ldots \tag{2.57}$$

is written as an expansion in derivatives and external fields. The chiral counting rules are :

$$
\begin{array}{ll}
U & O(p^0) \\
D_\mu U, v_\mu, a_\mu & O(p) \\
s, p, F_{L,R}^{\mu\nu} & O(p^2)
\end{array}
\tag{2.58}
$$

which are obvious except for s, p, to be explained shortly.

CHPT in the formulation of Gasser and Leutwyler [2, 3] employs the following two methods:

1. The quantum field theory underlying CHPT is defined via the path integral (2.56). This approach avoids the pitfalls of canonical quantization for a non–polynomial non–renormalizable QFT. All the usual axioms of QFT are guaranteed in this way.

2. The second ingredient to define the theory is dimensional regularization. It respects all the symmetries it should and it does not respect those it shouldn't (anomalies). It also avoids spurious (quadratic) divergences.

The lowest order (locally) chiral invariant Lagrangian

$$\mathcal{L}_2 = \frac{F^2}{4} \langle D_\mu U D^\mu U^\dagger + \chi U^\dagger + \chi^\dagger U \rangle \tag{2.59}$$

is the generalization of the non–linear σ model in the presence of external fields with

$$\chi = 2B_0(s + ip) . \tag{2.60}$$

The corresponding generating functional $Z_2[v, a, s, p]$ is just the classical action

$$Z_2[v, a, s, p] = \int d^4x \mathcal{L}_2(U, v, a, s, p) \tag{2.61}$$

where

$$U = U[v, a, s, p]$$

is to be understood as a functional of the external fields via the equations of motion for $\mathcal{L}_2$:

$$\Box U U^\dagger - U \Box U^\dagger = \chi U^\dagger - U \chi^\dagger - \frac{1}{3} \langle \chi U^\dagger - U \chi^\dagger \rangle \mathbf{1} . \tag{2.62}$$

For most practical purposes, like calculating pseudoscalar meson amplitudes, one can directly insert the expansion of $U(\phi) = \exp\left(i\lambda_a \phi^a / F\right)$ in the chiral Lagrangian (2.59) with

$$\Phi = \frac{1}{\sqrt{2}} \lambda_a \phi^a = \begin{pmatrix} \frac{\pi^0}{\sqrt{2}} + \frac{\eta_8}{\sqrt{6}} & \pi^+ & K^+ \\ \pi^- & -\frac{\pi^0}{\sqrt{2}} + \frac{\eta_8}{\sqrt{6}} & K^0 \\ K^- & \overline{K^0} & -\frac{2\eta_8}{\sqrt{6}} \end{pmatrix} . \tag{2.63}$$

The parameters F and B_0 are the only free constants at $O(p^2)$. To understand their meaning, let us first calculate the chiral Noether currents from the lowest–order Lagrangian $\mathcal{L}_2$ in (2.59):

$$V^\mu \doteq \tfrac{1}{\sqrt{2}} \lambda_a J_V^{a\mu} = -\frac{iF^2}{2\sqrt{2}} \left(U^\dagger D^\mu U + U D^\mu U^\dagger \right)$$

$$A^\mu \doteq \tfrac{1}{\sqrt{2}} \lambda_a J_A^{a\mu} = \frac{iF^2}{2\sqrt{2}} \left(U^\dagger D^\mu U - U D^\mu U^\dagger \right) = -F D^\mu \Phi + O(\Phi^3) . \tag{2.64}$$

The Goldstone matrix element for the axial current

$$\langle 0 | J_A^{a\mu}(x) | \phi^b(p) \rangle = i\delta_{ab} e^{-ipx} F p^\mu \tag{2.65}$$

shows that F equals the pion decay constant to lowest order in CHPT :

$$F = F_\pi = 93.2 \text{ MeV} . \tag{2.66}$$

The second constant B_0 is related to the quark condensate :

$$\frac{\delta Z}{\delta s_{ij}}\Big|_{v=a=s=p=0} = -\langle 0|\overline{q}_i q_j|0\rangle = F^2 B_0 \delta_{ij} \ . \tag{2.67}$$

We can now extract the pseudoscalar meson masses to lowest order in CHPT by expanding $\mathcal{L}_2$ in (2.59) to second order in Φ :

$$\mathcal{L}_2^{kin} = \frac{1}{2}\langle\partial_\mu\Phi\partial^\mu\Phi\rangle - B_0\langle\mathcal{M}\Phi^2\rangle \ . \tag{2.68}$$

Performing the trace, we obtain for the masses

$$
\begin{aligned}
M_{\pi^+}^2 &= 2\hat{m}B_0 \\
M_{\pi^0}^2 &= 2\hat{m}B_0 + O\left[\frac{(m_u - m_d)^2}{m_s - \hat{m}}\right] \\
M_{K^+}^2 &= (m_u + m_s)B_0 \\
M_{K^0}^2 &= (m_d + m_s)B_0 \\
M_{\eta_8}^2 &= \frac{2}{3}(\hat{m} + 2m_s)B_0 + O\left[\frac{(m_u - m_d)^2}{m_s - \hat{m}}\right] \\
\hat{m} &\doteq \frac{1}{2}(m_u + m_d) \ .
\end{aligned}
\tag{2.69}
$$

From these relations we can immediately infer several well–known equations, which hold to lowest order in the chiral expansion :

- Gell-Mann–Oakes–Renner relation [25]

$$F_\pi^2 M_\pi^2 = -2\hat{m} < 0|\overline{q}q|0 > \ ; \tag{2.70}$$

- Current algebra mass ratios [25, 27]

$$B_0 = \frac{M_\pi^2}{2\hat{m}} = \frac{M_{K^+}^2}{m_u + m_s} = \frac{M_{K^0}^2}{m_d + m_s} \ ; \tag{2.71}$$

- Gell-Mann–Okubo mass relation [28]

$$3M_{\eta_8}^2 = 4M_K^2 - M_\pi^2 \ . \tag{2.72}$$

Let me emphasize here that the success of the Gell-Mann–Okubo mass formula is another strong argument for a non–vanishing quark condensate ($B_0 \neq 0$). If B_0 were zero, there would be no natural way to understand the quadratic mass relation (2.72). The mass formulas (2.69) also show why the scalar fields s and χ are counted as $O(p^2)$ in the chiral expansion. Since $B_0 \neq 0$ in the chiral limit, the squares of the meson masses are proportional to the quark masses contained in s or χ.

Having determined the two free constants F, B_0, we can now use $Z_2[v, a, s, p]$ to derive all the current algebra results for mesonic Green functions and amplitudes in terms of F_π and of the meson masses M_a^2. The constant B_0 is hidden in the meson masses and will never appear explicitly.

As an example, we can easily obtain the Weinberg amplitude for $\pi\pi$ scattering [29] [cf. Sect. (3)] :

$$A(s,t) = \frac{s - M_\pi^2}{F_\pi^2} \ . \tag{2.73}$$

The remarkable thing about this amplitude is that we can make an absolute prediction. How is that possible from symmetry arguments only? In fact, we could never get such an absolute prediction from a linearly realized symmetry (Wigner–Weyl). However, the non–linearly realized chiral symmetry relates amplitudes with different numbers of (pseudo–) Goldstone bosons. Therefore, the $\pi\pi$ scattering amplitude can be related to the Goldstone matrix element (2.65) allowing for an 'absolute' prediction in terms of F_π.

2.4 Generating Functional to $O(p^4)$

We now compute $Z[v,a,s,p]$ to next–to–leading order p^4. There are three parts entering $Z_4[v,a,s,p]$ [2, 3]:

- The general effective chiral Lagrangian $\mathcal{L}_4$ of $O(p^4)$ to be considered at tree level.

- The one–loop functional for the lowest–order Lagrangian $\mathcal{L}_2$ in (2.59).

- The Wess–Zumino–Witten functional [30] to account for the chiral anomaly [31].

2.4.1 Chiral Lagrangian of $O(p^4)$

The general chiral Lagrangian $\mathcal{L}_4(U,v,a,s,p)$, invariant with respect to $G = SU(3)_L \times SU(3)_R$, parity and charge conjugation,

$$
\begin{aligned}
\mathcal{L}_4 &= \sum_i L_i P_i \qquad\qquad\qquad\qquad\qquad\qquad (2.74)\\
&= L_1\langle D_\mu U^\dagger D^\mu U\rangle^2 + L_2\langle D_\mu U^\dagger D_\nu U\rangle\langle D^\mu U^\dagger D^\nu U\rangle\\
&\quad + L_3\langle D_\mu U^\dagger D^\mu U D_\nu U^\dagger D^\nu U\rangle + L_4\langle D_\mu U^\dagger D^\mu U\rangle\langle \chi^\dagger U + \chi U^\dagger\rangle\\
&\quad + L_5\langle D_\mu U^\dagger D^\mu U(\chi^\dagger U + U^\dagger\chi)\rangle + L_6\langle \chi^\dagger U + \chi U^\dagger\rangle^2 + L_7\langle \chi^\dagger U - \chi U^\dagger\rangle^2\\
&\quad + L_8\langle \chi^\dagger U\chi^\dagger U + \chi U^\dagger\chi U^\dagger\rangle - iL_9\langle F_R^{\mu\nu} D_\mu U D_\nu U^\dagger + F_L^{\mu\nu} D_\mu U^\dagger D_\nu U\rangle\\
&\quad + L_{10}\langle U^\dagger F_R^{\mu\nu} U F_{L\mu\nu}\rangle + L_{11}\langle F_{R\mu\nu} F_R^{\mu\nu} + F_{L\mu\nu} F_L^{\mu\nu}\rangle + L_{12}\langle \chi^\dagger\chi\rangle,
\end{aligned}
$$

depends on 10 a priori arbitrary low-energy constants $L_1,\ldots,L_{10}$ (L_{11} and L_{12} are not directly accessible experimentally).

The L_i parametrize the loss of information in passing from the fundamental theory QCD to the effective CHPT level. Expressed more drastically, the SM is a point in the 10-dimensional space of the L_i containing all possible underlying theories with the same symmetries G, P and C. The theoretical challenge consists of finding this needle in a 10-dimensional stack of hay!

2.4.2 One–loop functional

The master formula (2.56) shows that all internal lines of Feynman diagrams to arbitrary loop order correspond to (pseudo-) Goldstone bosons with propagators

$$
\frac{i}{k^2 - M_a^2} \qquad a = 1,\ldots,8. \qquad\qquad (2.75)
$$

A general diagram contains N_d vertices of $O(p^d)$ ($d = 2, 4, \ldots$). After the loop integrations, the amplitude is a homogeneous function in the external momenta and the pseudoscalar masses M_a. The chiral dimension of an L–loop amplitude is essentially the primitive degree of divergence:

$$D_L = 4L - 2I + \sum_d dN_d, \tag{2.76}$$

where L is the number of loops and I the number of internal lines. Using the topological relation

$$L = I - V + 1 \tag{2.77}$$
$$V = \sum_d N_d \tag{2.78}$$

valid for any simply connected diagram, we obtain [1]

$$D_L = 2L + 2 + \sum_d (d-2)N_d, \qquad d = 2, 4, \ldots \tag{2.79}$$

and therefore, up to $O(p^4)$:

$$D_L = 2: \quad L = 0, \ d = 2 \qquad Z_2 = \int d^4x \mathcal{L}_2 \tag{2.80}$$

$$D_L = 4: \quad L = 0, \ d = 4 \qquad Z_4^{\text{tree}} = \int d^4x \mathcal{L}_4 \tag{2.81}$$

$$L = 1, \ d = 2 \qquad Z_4^{L=1} \text{ for } \mathcal{L}_2. \tag{2.82}$$

One must not be misled by the chiral counting: the amplitudes of CHPT are, of course, not just polynomials in momenta and masses. In fact, in order to guarantee analyticity and (perturbative) unitarity, threshold factors and logarithms are never expanded in CHPT. It may be helpful to think of CHPT amplitudes in terms of dispersion relations: CHPT produces the correct dispersion integrals (perturbatively) as required by the general principles of QFT and it organizes the subtraction polynomials in a derivative expansion.

For a given S-matrix element, the chiral dimension D_L increases with L according to Eq. (2.79). In order to reproduce the (fixed) physical dimension of the amplitude, each loop produces a factor $1/F^2$. Together with the geometric loop factor $(4\pi)^{-2}$, the loop expansion suggests

$$4\pi F_\pi = 1.2 \text{ GeV} \tag{2.83}$$

as the natural scale for the chiral expansion [32]. We will have to check whether the coefficients of the local Lagrangian $\mathcal{L}_4$ in (2.74) respect this scale. Anticipating the positive answer, we expect that in the domain of applicability of CHPT (cf. Sect. 1) with $|p| \lesssim O(M_K)$ the expansion parameter of chiral amplitudes should not exceed

$$\frac{M_K^2}{16\pi^2 F_\pi^2} = 0.18 . \tag{2.84}$$

This is the basis for the statement that higher–order chiral corrections can be expected to be around 20% in amplitude.

For the actual calculation of the one–loop functional $Z_4^{L=1}[v, a, s, p]$, Gasser and Leutwyler employ [2, 3] the method of functional integration around the classical solution $U(\phi_{\text{cl}}[v, a, s, p])$, which is a solution of the equation of motion (2.62) with the boundary condition $U(0) = 1$. The procedure consists of expanding $\mathcal{L}_{\text{eff}}(U, v, a, s, p)$ around $\phi = \phi_{\text{cl}}$ and functionally integrating over the fluctuations $\phi - \phi_{\text{cl}}$.

Table 1. Phenomenological values and source for the renormalized coupling constants $L_i^r(M_\rho)$ taken from Ref. [36].

i	$L_i^r(M_\rho) \times 10^3$	source	Γ_i
1	0.7 ± 0.5	$K_{e4}, \pi\pi \to \pi\pi$	$3/32$
2	1.2 ± 0.4	$K_{e4}, \pi\pi \to \pi\pi$	$3/16$
3	-3.6 ± 1.3	$K_{e4}, \pi\pi \to \pi\pi$	0
4	-0.3 ± 0.5	Zweig rule	$1/8$
5	1.4 ± 0.5	$F_K : F_\pi$	$3/8$
6	-0.2 ± 0.3	Zweig rule	$11/144$
7	-0.4 ± 0.2	Gell-Mann-Okubo,L_5, L_8	0
8	0.9 ± 0.3	$M_{K^0} - M_{K^+}, L_5$, $(2m_s - m_u - m_d) : (m_d - m_u)$	$5/48$
9	6.9 ± 0.7	$<r^2>_{em}^\pi$	$1/4$
10	-5.5 ± 0.7	$\pi \to e\nu\gamma$	$-1/4$
11			$-1/8$
12			$5/24$

For $L = 1$, it is sufficient to expand $\mathcal{L}_2$ to $O[(\phi - \phi_{cl})^2]$:

$$S_2 = \int d^4x \mathcal{L}_2 = S_2[\phi_{cl}] - \frac{F^2}{2}(\xi, D\xi) + O(\xi^3) \tag{2.85}$$

$$\xi^a = \phi^a - \phi_{cl}^a, \tag{2.86}$$

where D is an explicitly known differential operator acting on the fluctuations ξ. The one–loop functional $Z_4^{L=1}[v, a, s, p]$ is then given by [33]

$$Z_4^{L=1} = \frac{i}{2} \log \det D = \frac{i}{2} \operatorname{tr} \log D. \tag{2.87}$$

In accordance with another theorem of Weinberg [34], the divergent part $Z_{4,\mathrm{div}}^{L=1}$ has the form of a local action with all the symmetries of $\mathcal{L}_2$. It was first calculated in Ref. [35] for the $SU(2)$ non–linear σ model without external fields. The complete expression for chiral $SU(3)$ in the presence of external fields and with dimensional regularization [3] is of the form (2.74),

$$\mathcal{L}_{4,\mathrm{div}}^{L=1} = -\Lambda \sum_{i=1}^{12} \Gamma_i P_i \tag{2.88}$$

$$\Lambda = \frac{\mu^{d-4}}{(4\pi)^2}\left\{\frac{1}{d-4} - \frac{1}{2}[\log 4\pi + 1 + \Gamma'(1)]\right\}$$

with divergent coefficients. The relative coefficients Γ_i are listed in Table 1.

Although it may sound paradoxical, we can now renormalize a non–renormalizable QFT by defining renormalized, measurable coupling constants $L_i^r(\mu)$ as

$$L_i = L_i^r(\mu) + \Gamma_i \Lambda. \tag{2.89}$$

The sum

$$Z_{4,\mathrm{div}}^{L=1} + \int d^4x \mathcal{L}_4 = \int d^4x \mathcal{L}_4(L_i^r(\mu)) \tag{2.90}$$

is then finite, but depends on the arbitrary scale μ via the $L_i^r(\mu)$. However, because of the definition of $\mathcal{L}_{4,\text{div}}^{L=1}$ in (2.88) the finite part $Z_{4,\text{fin}}^{L=1}$ is, of course, also μ-dependent since the total one–loop functional $Z_4^{L=1} = Z_{4,\text{div}}^{L=1} + Z_{4,\text{fin}}^{L=1}$ knows nothing about the scale μ. Therefore, the complete functional (except for the anomalous part to be discussed soon) of $O(p^4)$

$$Z_4 - Z_4^{\text{anom}} = Z_4^{L=1} + Z_4^{\text{tree}} = Z_{4,\text{fin}}^{L=1}(\mu) + \int d^4x \mathcal{L}_4(L_i^r(\mu)) \tag{2.91}$$

is both finite and μ-independent. The scale dependence of the loop amplitude and the local counterterm amplitude due to $\mathcal{L}_4$ cancel in any physical, measurable amplitude. The phenomenological values of the renormalized coupling constants $L_i^r(M_\rho)$ are listed in Table 1.

Remarks

1. Unlike in a generic regularization scheme producing spurious quadratic divergences, the lowest–order coupling constants F, B_0 are not renormalized in dimensional regularization.

2. The scale dependence of the $L_i^r(\mu)$ can be deduced from Eq. (2.89):

$$L_i^r(\mu_2) = L_i^r(\mu_1) + \frac{\Gamma_i}{(4\pi)^2} \log \frac{\mu_1}{\mu_2}. \tag{2.92}$$

Let me emphasize once more that this scale dependence is always cancelled by the finite part of the one–loop functional in a physical amplitude.

3. Mathematics vs. physics

   - Following the suggestive proposition that a non–renormalizable QFT should not be renormalized, one might be tempted to ignore the constants L_i altogether and use a 'physical' momentum cutoff $p_{\max}$ instead to define the amplitudes. A favourite example of Leutwyler [8] shows that this proposal is nonsense: in order to reproduce the experimental value of the pion charge radius, one would need a ridiculously high cutoff $p_{\max} \simeq 60$ GeV. Instead, the pion charge radius is predominantly determined by $L_9^r(\mu)$ for any reasonable value of μ like $\mu = M_\rho$.

   - The linear σ model as a renormalizable QFT allows the explicit calculation of the L_i (or rather their counterparts for chiral $SU(2)$) in terms of the parameters λ, v^2 in the Lagrangians (2.9) or (2.18). The conclusion [2] is that the linear σ model is phenomenologically ruled out as an EFT of QCD.

   - To renormalize the non–linear σ model Lagrangian (2.59) in a consistent way, it would be sufficient to include only terms in (2.74) with $\Gamma_i \neq 0$. However, a glance at Table 1 shows that this would be another "interesting way not to understand this world" [37]: the low–energy constants L_3 and L_7 are definitely non–zero although $\Gamma_3 = \Gamma_7 = 0$.

2.4.3 Chiral anomaly

The chiral anomaly is a fundamental property of the SM as a chiral QFT [31]. Although the anomaly has many facets, its origin can probably be best understood as a clash between a classical symmetry and the impossibility of defining the QFT (via a regularization procedure) in a way respecting this symmetry.

The problem for CHPT arises because we have so far made use of manifest chiral symmetry to construct the effective chiral Lagrangian $\mathcal{L}_2 + \mathcal{L}_4$. On the other hand, the chiral symmetry is violated by the anomaly. Writing an infinitesimal chiral transformation in the form

$$g_{L,R} = \mathbf{1} + i\alpha \mp i\beta + \dots \tag{2.93}$$

and requiring that the vector currents are conserved to guarantee electromagnetic gauge invariance, the anomaly is expressed by Bardeen's equations [31]

$$\delta_\alpha Z[v, a, s, p] = 0 \tag{2.94}$$

$$\delta_\beta Z[v, a, s, p] = -\frac{N_C}{4\pi^2} \int \langle \beta\{(Dv)^2 + \frac{1}{3}(Da)^2$$
$$+ \frac{4i}{3}(Dva^2 + aDva + a^2Dv) - \frac{8}{3}a^4\}\rangle. \tag{2.95}$$

I have used again the compact notation of differential forms with

$$\begin{array}{ll} v = v_\mu dx^\mu & a = a_\mu dx^\mu \\ Dv = dv - i(v^2 + a^2) & Da = da - i(va + av). \end{array} \tag{2.96}$$

Recalling the chiral counting rules (2.58), we recognize that the anomalous variation $\delta_\beta Z$ in (2.95) is of $O(p^4)$. Thus, the anomaly appears first at $O(p^4)$ requiring a contribution Z_4^{anom} to the generating functional Z_4.

The remarkable finding of Wess and Zumino [30] is that Z_4^{anom} can be given in explicit form, of course always up to a chiral invariant piece already contained in Z_4^{tree}. In the explicit formulation of Witten [30] and others [38], the anomaly functional Z_4^{anom} takes the form of the anomalous action

$$S[U, l, r]_{WZW} = -\frac{N_C}{240\pi^2} \int_{M_5} \langle L^5 \rangle - \frac{iN_C}{48\pi^2} \int_{M_4} [W(U, l, r) - W(\mathbf{1}, l, r)]$$

$$W(U, l, r) = \langle l^3 U^\dagger r U + \frac{1}{4} l U^\dagger r U l U^\dagger r U + i d l l U^\dagger r U + i U^\dagger d r U l U^\dagger r U$$
$$- L l U^\dagger r U l - L l^3 - i L U^\dagger d r U l - i L l d l - i L d l l$$
$$+ L^2 U^\dagger r U l - \frac{1}{2} L l L l + L^3 l \rangle - (l \leftrightarrow r, U \leftrightarrow U^\dagger) \tag{2.97}$$

$$L = iU^\dagger dU = L_\mu dx^\mu, \qquad L_\mu = iU^\dagger \partial_\mu U, \qquad N_C = 3.$$

Since the first integrand in Eq. (2.97) is a total divergence, the five-dimensional manifold M_5 is arbitrary except that its boundary ∂M_5 must be Minkowski space M_4. Although most phenomenological applications of the anomaly involve the local term $W(U, l, r)$, the five-dimensional surface term $\langle L^5 \rangle$ bears the distinctive mark of an anomaly. If $S[U, l, r]_{WZW}$ were a local action in 4-dimensional Minkowski space, we would be free to define a regularization procedure by subtracting the corresponding Lagrangian of $O(p^4)$ from $\mathcal{L}_4$. The resulting functional Z_4 would be completely chiral invariant and there would be no anomaly at all.

In spite of its considerable complexity, the anomalous action (2.97) has no free parameters at all ! Compare this remarkable property with the 10 arbitrary parameters L_i in the Lagrangian $\mathcal{L}_4$. The special status of the chiral anomaly is reflected in the unambiguous transition from the fundamental to the effective level.

2.5 Low–Energy Constants

The present state of the phenomenological determination [3, 39, 40, 41] of the constants L_i is summarized in Table 1 [36]. Comparing the Lagrangians $\mathcal{L}_2$ and $\mathcal{L}_4$ and recalling the chiral expansion parameter

$$\frac{p^2}{(4\pi F_\pi)^2} \tag{2.98}$$

suggested by the loop expansion, we expect

$$L_i \lesssim \frac{1}{(4\pi)^2} = 6.3 \cdot 10^{-3} \tag{2.99}$$

in agreement with the phenomenological values $L_i^r(M_\rho)$ in Table 1.

In accordance with the three energy domains of Sect. 1, there could in principle be contributions to the L_i both from short and intermediate distances. In particular, we expect all states which can couple to the pseudoscalar mesons to contribute. It is then a rather well–founded conjecture that meson resonances should play an important rôle for understanding the L_i.

A systematic analysis of the couplings between meson resonances of the type V, A, S, P and the pseudoscalar mesons was performed[4] in Ref. [42]. We use the general procedure of coupling matter fields to the Goldstone modes discussed in connection with the non–linear realization of G. For instance, an octet

$$R = \frac{1}{\sqrt{2}} \lambda_a R^a \tag{2.100}$$

of resonance fields transforms as

$$R \xrightarrow{G} h(\phi, g) R h(\phi, g)^{-1} \tag{2.101}$$

where $h(\phi, g)$ is the compensator field in the fundamental triplet representation. It turns out [42] that for V and A resonances only the octets can contribute to the L_i with the relevant couplings to the pseudoscalar mesons given by

$$
\begin{aligned}
\mathcal{L}_2[V(1^{--})] &= \frac{F_V}{2\sqrt{2}} \langle V_{\mu\nu} f_+^{\mu\nu} \rangle + \frac{iG_V}{\sqrt{2}} \langle V_{\mu\nu} u^\mu u^\nu \rangle \\
\mathcal{L}_2[A(1^{++})] &= \frac{F_A}{2\sqrt{2}} \langle A_{\mu\nu} f_-^{\mu\nu} \rangle \\
f_\pm^{\mu\nu} &= u F_L^{\mu\nu} u^\dagger \pm u^\dagger F_R^{\mu\nu} u \ .
\end{aligned}
\tag{2.102}
$$

For S and P, both octets and singlets can contribute.

The coupling constants F_V, G_V and F_A (and the corresponding ones for S, P resonances) can be estimated from resonance decays, e.g.

$$
\begin{aligned}
F_V &: \ \rho \to e^+ e^- \\
G_V &: \ \rho \to \pi\pi \\
F_A &: \ a_1 \to \pi\gamma.
\end{aligned}
\tag{2.103}
$$

Resonance exchange then immediately gives the corresponding contributions to the L_i, e.g.

$$L_9^V = \frac{F_V G_V}{2M_V^2}, \qquad L_{10}^A = \frac{F_A^2}{4M_A^2} \tag{2.104}$$

$$M_V \simeq M_\rho, \qquad M_A \simeq M_{a_1}.$$

The results of Ref. [42] can be summarized by two catchwords:

---

[4]See also Ref. [43] for related work.

Chiral duality:

The $L_i^r(M_\rho)$ are practically saturated by resonance exchange. There is very little room left for direct short–distance or additional intermediate–distance contributions.

Chiral VMD:

Whenever spin-1 resonances can contribute at all ($i = 1, 2, 3, 9, 10$), the $L_i^r(M_\rho)$ are almost completely dominated by V- (and for L_{10} only, also A-) exchange.

Subsequently, it was shown [44] that the high–energy structure of QCD gives additional information:

- Imposing the QCD constraints at high energies via dispersion relations, all phenomenologically successful models for V,A resonances were shown to be equivalent to $O(p^4)$: tensor field description used in Eq. (2.102), massive Yang-Mills [45], hidden gauge formulations [46], etc.

- With additional QCD–inspired assumptions of high–energy behaviour, like an unsubtracted dispersion relation for the pion form factor, all V and A couplings could be expressed in terms of F_π and M_ρ only. The results are in good agreement with the phenomenological values of the $L_i^r(M_\rho)$ for $i = 1, 2, 3, 9, 10$ and can be summarized in the relations

$$
\begin{aligned}
L_2 &= 2L_1 \quad \text{(large } N_C) \\
L_3^V &= -6L_1 = L_{10} \\
L_9 &= 4L_2 = \frac{F_\pi^2}{2M_V^2},
\end{aligned}
\tag{2.105}
$$

all to be taken at $\mu \simeq M_\rho$. Although L_3 is dominated by V exchange, there is an additional contribution from scalar octet exchange [42] in this case.

The theoretical prediction $G_V = F_\pi/\sqrt{2}$ implies

$$
\Gamma(\rho \to 2\pi) = \frac{M_\rho^3 (1 - 4M_\pi^2/M_\rho^2)^{3/2}}{96\pi F_\pi^2} = 141 \text{ MeV}
\tag{2.106}
$$

which is one version of the so–called KSFR relation [47]. This is a non–trivial result: higher–order terms in the chiral expansion could a priori influence the decay $\rho \to 2\pi$, but would be negligible at small momenta $|p| \overset{<}{\sim} M_K$. The underlying reason for the success of the KSFR relation is once again the intimate connection between the high–energy structure predicted by QCD and the low–energy structure of CHPT via dispersion relations [48].

There have been several more ambitious attempts to calculate the L_i 'directly' from QCD. The relevant literature is collected and discussed in a recent analysis of the extended Nambu–Jona-Lasinio model by Bijnens, Bruno and de Rafael [49]. In the remainder of this section, I will sketch some of the salient features of Ref. [49] referring to that paper for all details.

Since we don't know how to integrate out the gluons in a functional integral representation of the generating functional $Z[v, a, s, p]$ defined in Eq. (2.56), one needs an explicit model to implement SCSB. The authors of Ref. [49] conjecture that integrating out quarks and gluons down to some intermediate energy scale Λ_χ produces an effective interaction of the form

$$
\mathcal{L}_{QCD}^{E<\Lambda_\chi} + \mathcal{L}_{ENJL} + O(1/\Lambda_\chi^4)
\tag{2.107}
$$

with the extended Nambu–Jona-Lasinio Lagrangian

$$\mathcal{L}_{ENJL} = \frac{8\pi^2 G_S(\Lambda_\chi)}{N_C \Lambda_\chi^2} \sum_{i,j} (\overline{q_R^i q_L^j})(\overline{q_L^j q_R^i})$$

$$- \frac{8\pi^2 G_V(\Lambda_\chi)}{N_C \Lambda_\chi^2} \sum_{i,j} [(\overline{q_L^i \gamma^\mu q_L^j})(\overline{q_L^j \gamma_\mu q_L^i}) + (L \to R)] \qquad (2.108)$$

$$G_S, G_V = O(1) \qquad \text{for } N_C \to \infty.$$

By introducing so–called collective fields H, L_μ, R_μ, one can transform $\mathcal{L}_{ENJL}$ into a Lagrangian bilinear in the quark fields $q, \bar{q}$ which can therefore be integrated out. Minimizing the resulting effective action with respect to the vacuum expectation value $\langle H \rangle_0$ of the scalar collective field $H(x)$, one arrives at the well–known gap equation of the Nambu–Jona-Lasinio model [50]. For sufficiently big scalar coupling G_S, the ground state minimizing the effective action corresponds to

$$\langle H(x) \rangle_0 = M_Q \mathbf{1}$$
$$M_Q \neq 0 \quad \leftrightarrow \quad \langle \bar{q}q \rangle_0 \neq 0 \qquad (2.109)$$

implying SCSB, with $M_Q \simeq 300$ MeV playing the rôle of a constituent quark mass. In the mean–field approximation characterized by

$$H(x) = \langle H(x) \rangle_0 = M_Q \mathbf{1}, \qquad L_\mu = R_\mu = 0 \qquad (2.110)$$

one recovers essentially (for $M_Q/\Lambda_\chi \to 0$) the QCD effective action approach of Espriu, de Rafael and Taron [51]. In that limit, the resulting L_i obey almost all the relations (2.105):

$$8L_1 = 4L_2 = L_9 = \frac{N_C}{48\pi^2}[1 + O(1/M_Q^6)] = 6.3 \cdot 10^{-3} \qquad (2.111)$$

$$L_3 = L_{10} = -\frac{N_C}{96\pi^2}[1 + \frac{\pi^2}{5N_C} \frac{\langle \frac{\alpha}{\pi} GG \rangle}{M_Q^4} + O(1/M_Q^6)]$$

$$= -3.2 \cdot 10^{-3}[1 + c], \qquad c > 0.$$

It is quite remarkable that the model seems to 'know' enough about the high–energy structure of QCD responsible for Eqs. (2.105). Comparing, e.g., the predictions for L_9 in the chiral VMD approach of Eq. (2.105) and in the QCD effective action framework of Eq. (2.111), one arrives at the 'prediction' (see also Ref. [52])

$$M_V \simeq M_\rho = 2\sqrt{2}\,\pi F_\pi = 830 \text{ MeV}. \qquad (2.112)$$

Bijnens, Bruno and de Rafael have extended their analysis beyond the mean–field approximation by generating kinetic terms for the collective fields H, L, R (to be identified with S, V, A resonance fields) and their couplings to the pseudoscalar mesons. In the end, they obtain predictions for some 20 measurable quantities including the L_i in terms of the three basic parameters of the model G_S, G_V and Λ_χ and a possible fourth parameter measuring the leading $O(\alpha_s N_C)$ corrections. The quality of their fits [49] is quite impressive. As far as the L_i are concerned, this is maybe not too surprising since the model contains all the resonances which are known to saturate the L_i. However, the analysis of Ref. [49] clarifies in a transparent way an obvious problem of double–counting: in certain limits, the model approaches either the pure quark loop results of Eqs. (2.111) or the chiral VMD predictions (2.105), but in general it interpolates between these two cases. In addition, the model represents a definite improvement for the constants L_5, L_8 sensitive only to scalar exchange compared to the mean-field results of Ref. [51].

3 STRONG, ELECTROMAGNETIC AND SEMILEPTONIC WEAK INTERACTIONS

Starting with the classical papers of Gasser and Leutwyler [2, 3, 53, 54], there has been a wealth of applications of CHPT to next–to–leading order p^4 in recent years. I will concentrate here on a few recent applications and discuss some open problems.

3.1 $\pi\pi$ Scattering

The scattering amplitude for

$$\pi^a(p_a) + \pi^b(p_b) \rightarrow \pi^c(p_c) + \pi^d(p_d) \tag{3.1}$$

is determined by a single scalar function $A(s,t,u)$ defined by the isospin decomposition

$$
\begin{aligned}
T_{ab,cd} &= \delta_{ab}\delta_{cd}A(s,t,u) + \delta_{ac}\delta_{bd}A(t,s,u) + \delta_{ad}\delta_{bc}A(u,t,s) \\
A(s,t,u) &= A(s,u,t) \quad \text{[crossing]}
\end{aligned}
\tag{3.2}
$$

$$s = (p_a + p_b)^2, \qquad t = (p_a - p_c)^2, \qquad u = (p_a - p_d)^2.$$

The amplitudes of definite isospin ($I = 0, 1, 2$) are decomposed into partial wave amplitudes $T_l^I(s)$:

$$
\begin{aligned}
A^I(s, \cos\Theta) &= \frac{16\pi\sqrt{s}}{k} \sum_{l=0}^{\infty}(2l+1)P_l(\cos\Theta)T_l^I(s) \\
t &= -2k^2(1 - \cos\Theta) \\
k^2 &= \frac{s}{4} - M_\pi^2.
\end{aligned}
\tag{3.3}
$$

In the region of interest for CHPT, the amplitudes $T_l^I(s)$ can be expressed in terms of the phase shifts $\delta_l^I(s)$ because of elastic unitarity:

$$T_l^I(s) = \frac{1}{2i}(e^{2i\delta_l^I(s)} - 1) \tag{3.4}$$

$$\text{Im}\, T_l^I = |T_l^I|^2. \tag{3.5}$$

Bose symmetry requires that even (odd) l correspond to even (odd) I.

To lowest order p^2, the Weinberg amplitude (2.73) [29] produces only partial waves with $l \le 1$. At $O(p^4)$, the $\pi\pi$ scattering amplitude was first calculated in the early 70's [55, 35, 56] and in the modern framework in Ref. [2]. More recent analyses have been performed in Refs. [57, 58]. The scattering amplitude of $O(p^4)$

$$A_{(4)}(s,t,u) = B(s,t,u) + C(s,t,u) \tag{3.6}$$

is as usual the sum of a loop amplitude B and of a counterterm amplitude C depending on the coupling constants L_i in (2.74). The chiral expansion for the phase shifts is [58]

$$\delta_l^I = \arctan(\text{Re}\, T_l^I) + O(p^6) = \underbrace{\text{Re}\, T_{l(2)}^I}_{\delta_{l(2)}^I} + \underbrace{\text{Re}\, T_{l(4)}^I}_{\delta_{l(4)}^I} + O(p^6), \qquad s \ge 4M_\pi^4. \tag{3.7}$$

Comparison with the available data shows the definite improvement of CHPT to $O(p^4)$ [58] compared to the current algebra phase shifts due to the Weinberg amplitude

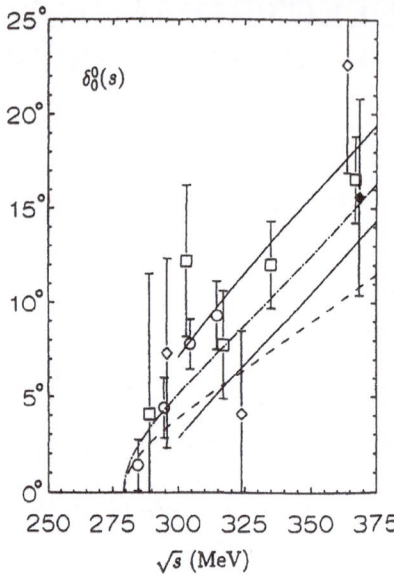

Figure 1. The $I = 0$ s-wave phase shift $\delta_0^0(s)$ in the threshold region taken from Ref. [58]. The dashed line is the tree–level result and the dash–dotted curve corresponds to the amplitude of $O(p^4)$.

(2.73). As an example, the $l = I = 0$ phase shift is shown in Fig. 1 [58]. Beyond $\sqrt{s} \simeq 450$ MeV, higher-order contributions make the $O(p^4)$ predictions unreliable. However, Gasser and Meißner have emphasized that those uncertainties cancel to a large extent in the difference $\delta_0^0 - \delta_0^2$ which governs the phase of the CP violation parameter ε' [59] for $s = M_K^2$. They find [58]

$$\delta_0^0(M_K^2) - \delta_0^2(M_K^2) = \begin{cases} 37^0 & O(p^2) \\ 45^0 \pm 6^0 & O(p^4) \end{cases} \tag{3.8}$$

which is much more trustworthy than the result from $K \to 2\pi$ decays subject to big isospin violating corrections which remain to be calculated [60].

Although statistically not very significant, there is a tendency in the experimental phase shift δ_0^0 to lie above the CHPT prediction near threshold (see Fig. 1). This is also borne out by the scattering length a_0^0 measuring the slope of δ_0^0 at threshold:

$$a_0^0 = \begin{cases} 0.16 & O(p^2) \ [29] \\ 0.20 & O(p^4) \ [2, 58] \\ 0.26 \pm 0.05 & \text{expt.} \ [61]. \end{cases} \tag{3.9}$$

How should one proceed at higher energies? Truong has suggested [52] applying the simplest Padé approximation to the CHPT amplitude to guarantee elastic unitarity non–perturbatively. A more modest, but less ambiguous approach was recently adopted by Bernard, Kaiser and Meißner [62]. They propose to take chiral duality [42, 43] seriously not only to $O(p^4)$, but in a certain sense to all orders by including the resonance poles. In the spirit of Ref. [42], they take the loop amplitude at $\mu = M_\rho$, include the resonance poles in the real part of the counterterm amplitude, e.g. with

$$L_1 = \frac{G_V^2}{8M_\rho^2} \to \frac{G_V^2}{8(M_\rho^2 - s)} \tag{3.10}$$

in the s-channel, and unitarize finally via the relation

$$\delta_l^I(s) = \arctan(\operatorname{Re} T_l^I(s)). \tag{3.11}$$

In this manner, good agreement with (most of) the data is obtained at least up to $\sqrt{s} = 700$ MeV where the ρ and a broad enhancement in the $l = I = 0$ channel dominate the phase shifts. It is hardly a surprise that the possible threshold enhancement of δ_0^0 is not reproduced in the treatment of Ref. [62]. To investigate if there is indeed a discrepancy between CHPT and experiment, we have started the ultimate analysis of $\pi\pi$ scattering at low to intermediate energies [63] combining all the axiomatic information (unitarity, crossing) encoded in the Roy equations [64] with the chiral structure of QCD.

3.2 Semileptonic K Decays

We have recently completed a comprehensive review of semileptonic kaon decays [36] motivated by the physics program of the Φ factory DAFNE under construction in Frascati [65]. Starting in 1995, DAFNE is expected to produce very high kaon fluxes of the order of $9 \cdot 10^9$ $(1 \cdot 10^9)$ tagged $K^\pm$ (K_L) per year [66].

From the theoretical point of view, semileptonic K decays offer a wide range of interesting tests of the SM at $O(p^4)$ in the chiral expansion. Since all the low–energy constants L_i have already been determined from other processes, these tests are essentially parameter free. Most of the decays, in particular all radiative semileptonic K decays, are sensitive to the chiral anomaly. All the anomalous amplitudes can be obtained from the WZW functional (2.97) by projecting out the relevant vertices. In addition to some non–local meson pole contributions [67], those amplitudes can be taken directly from the anomalous Lagrangian

$$\mathcal{L}_{\mathrm{anom}}(W) = \frac{e}{16\sqrt{2}\,\pi^2 \sin\Theta_W} \varepsilon^{\mu\nu\rho\sigma} W_\sigma^+ \tag{3.12}$$

$$\langle T_+(D_\mu U^\dagger D_\nu U D_\rho U^\dagger U + ieF_{\mu\nu}\{U^\dagger D_\rho U, Q + \tfrac{1}{2}U^\dagger QU\}\rangle + h.c.$$

The big experimental challenge will be to isolate the amplitudes (form factors) sensitive to specific aspects of CHPT to $O(p^4)$ such as the chiral anomaly or the loop contributions. In the following, I will sketch a few of the predictions and refer for all details and a comprehensive list of references to Ref. [36].

3.2.1 $K_{l2\gamma}$

In addition to inner bremsstrahlung, the decay

$$K^+(p) \to l^+(p_l)\nu_l(p_\nu)\gamma(q) \tag{3.13}$$

is characterized by the tensor amplitude

$$
\begin{aligned}
H^{\mu\nu} &= iV(W^2)\varepsilon^{\mu\nu\rho\sigma} q_\rho p_\sigma - A(W^2)(qW g^{\mu\nu} - W^\mu q^\nu) \\
W &= p - q = p_l + p_\nu.
\end{aligned}
\tag{3.14}
$$

To lowest $O(p^2)$, the form factors $V(W^2)$ and $A(W^2)$ vanish: the amplitude is pure bremsstrahlung. At next–to–leading order, the loops only renormalize F to F_K and the form factors are given by

$$
\begin{aligned}
A &= -\frac{4}{F}(L_9^r + L_{10}^r) \\
V &= -\frac{1}{8\pi^2 F}.
\end{aligned}
\tag{3.15}
$$

Table 2. Predictions for the decays $K^+ \rightarrow e^+ \nu_e l^+ l^-$ and comparison with experiment [69]

decay mode	$O(p^2)$	$O(p^4)$	expt.
$K^+ \rightarrow e^+ \nu_e \mu^+ \mu^-$	$3 \cdot 10^{-12}$	$1.1 \cdot 10^{-8}$	$-$
$K^+ \rightarrow e^+ \nu_e e^+ e^-$ $[m_{e^+e^-} > 140 \text{ MeV}]$	$2 \cdot 10^{-12}$	$3.4 \cdot 10^{-8}$	$\left(2.8 \begin{smallmatrix} +2.8 \\ -1.4 \end{smallmatrix} \right) \cdot 10^{-8}$

Both form factors are constant to $O(p^4)$. The prediction for the vector form factor can be immediately obtained from the anomalous Lagrangian (3.12) and it agrees within very large errors with the scarce experimental information. A real test of the anomaly in this decay should be possible at DAFNE. The anomaly is definitely seen in the corresponding pionic decay mode $\pi \rightarrow e\nu_e \gamma$ [68]:

$$\frac{V_{\text{exp}}}{V_{CHPT}} = 0.86 \begin{smallmatrix} +0.56 \\ -0.48 \end{smallmatrix} .$$

(3.16)

3.2.2 $K_{l2\gamma^*}$

The decay modes with a virtual photon,

$$K^+ \rightarrow l^+ \nu_l l'^+ l'^- \qquad l, l' = e, \mu,$$

(3.17)

can provide additional information. As in the decays with a real photon, the leading–order bremsstrahlung amplitudes are suppressed for $l = e$. These channels are therefore especially sensitive to the non–leading effects of $O(p^4)$. At this order, the non–trivial (in addition to the renormalization $F \rightarrow F_K$) loop effects only appear via the kaon electromagnetic form factor in one of the axial amplitudes. The vector form factor is again given by the anomaly only. The branching ratios [67, 36] shown in Table 2 give an indication of the experimental challenge.

3.2.3 K_{l3}

The K_{l3} decays $K^+ \rightarrow \pi^0 l^+ \nu_l$, $K^0 \rightarrow \pi^- l^+ \nu_l$ are described by two form factors, the vector form factor $f_+(t)$ and the scalar form factor $f_0(t)$. While there is perfect agreement between theory [53] and experiment for $f_+(t)$, there are still conflicting experimental results for the slope parameter λ_0 defined by the linear approximation

$$f_0(t) = f_+(0)[1 + \lambda_0 \, t/M_\pi^2]$$

(3.18)

to the scalar form factor. DAFNE, with an expected $3 \cdot 10^8$ events/yr, will certainly be able to settle the issue and to allow for a precise comparison with CHPT.

Once isospin violation is taken into account, one has to distinguish between the two charge modes. It was emphasized by Leutwyler and Roos [70] that the isospin violating corrections of $O(m_u - m_d)$ to $f_+(0)$ are enhanced by small denominators and can therefore compete with the suppressed $SU(3)$ breaking corrections [71]. The result [70]

$$f_+^{K^0 \pi^-}(0) = 0.961 \pm 0.008$$

(3.19)

Table 3. Branching ratios and expected number of events at DAFNE for $K_{l3\gamma}^0$.

$K_{e3\gamma}^0$	BR	#events/yr
full $O(p^4)$ amplitude	3.8×10^{-3}	4.2×10^6
tree level	3.6×10^{-3}	4.0×10^6
$O(p^4)$ without loops	4.0×10^{-3}	4.4×10^6

$K_{\mu3\gamma}^0$	BR	#events/yr
full $O(p^4)$ amplitude	5.6×10^{-4}	6.1×10^5
tree level	5.2×10^{-4}	5.7×10^5
$O(p^4)$ without loops	5.9×10^{-4}	6.5×10^5

allows for the most precise determination of the Kobayashi-Maskawa matrix element V_{us} and of the ratio F_K/F_π:

$$|V_{us}| = 0.2196 \pm 0.0023$$
$$\frac{F_K}{F_\pi} = 1.22 \pm 0.01. \tag{3.20}$$

3.2.4 $K_{l3\gamma}$

The full calculation of radiative K_{l3} decays to $O(p^4)$ has been completed only recently [67]. As always for radiative decays, there is a less interesting bremsstrahlung amplitude and a structure–dependent part which contains new information. The loop corrections are sizeable in this case and they interfere destructively with the counterterm amplitude with coupling constants L_9, L_{10}. A detailed study of various distributions will be necessary to extract the contribution of the chiral anomaly to these decay modes. To illustrate the expected rates, the CHPT predictions for $K_{l3\gamma}^0$ ($K_L \to \pi^\pm l^\mp \nu \gamma$) [67] are reproduced in Table 3.

3.2.5 K_{l4}

The K_{l4} decays $K \to \pi\pi l\nu_l$ (3 channels) are the cleanest source for $\pi\pi$ s-wave scattering near threshold. The recent CHPT analysis to $O(p^4)$ [40, 41] has led to an improved determination of the low–energy constants L_1, L_2, L_3. In particular, it allows for a direct test of the Zweig rule $L_2 = 2L_1$:

$$(L_2^r - 2L_1^r)/L_3 = -0.19 \, {}^{+0.16}_{-0.27} \, . \tag{3.21}$$

Rather surprisingly, the best evidence for the chiral anomaly in K decays stems from a high–statistics experiment on $K^+ \to \pi^+\pi^- e^+ \nu_e$ [72]. The vector form factor H of the K_{l4} decay amplitude [40, 41, 36] can be extracted from Eq. (3.12) as

$$H = -\frac{\sqrt{2}\,M_K^3}{8\pi^2 F^3} = -2.7, \tag{3.22}$$

in impressive agreement with the experimentally determined value at threshold [72]

$$H_{\text{expt}} = -2.68 \pm 0.68. \tag{3.23}$$

Note that the sign of H is a measurable quantity due to V,A interference.

3.3 Successes and Problems

3.3.1 Meson masses to $O(p^4)$

The current algebra relations (2.69) for the pseudoscalar meson masses are modified at $O(p^4)$ [3]. The additional terms depend on the low–energy constants L_i. There is however a relation between the quark and meson masses where the dependence on the L_i drops out [3]:

$$\frac{m_d^2 - m_u^2}{m_s^2 - \hat{m}^2} = \underbrace{\frac{M_\pi^2}{M_K^2} \frac{M_{K^0}^2 - M_{K^+}^2 - M_{\pi^0}^2 + M_{\pi^+}^2}{M_K^2 - M_\pi^2}}_{1/Q^2}[1 + O(m_q^2)]. \qquad (3.24)$$

To a good approximation, Eq. (3.24) can also be written as an ellipse

$$\left(\frac{m_u}{m_d}\right)^2 + \frac{1}{Q^2}\left(\frac{m_s}{m_d}\right)^2 = 1 \qquad (3.25)$$

in the m_u/m_d vs. m_s/m_d plane.

A few years ago, Kaplan and Manohar made the interesting observation [73] that the effective chiral Lagrangian $\mathcal{L}_2 + \mathcal{L}_4$ exhibits an accidental symmetry which does not allow a separate determination of m_u/m_d and m_s/m_d, using only information from mesonic S-matrix elements. This raises the question, relevant for the solution of the strong CP problem [59], whether

$$m_u = 0 \qquad (3.26)$$

is a possible value for the lightest quark mass. Of course, this would require a huge correction of $O(p^4)$ to compensate the non–vanishing m_u at $O(p^2)$ and would cast serious doubts on the whole idea of the chiral expansion.

The situation was recently clarified by Leutwyler [37]. The Kaplan–Manohar symmetry is most certainly not a symmetry of QCD, but an accidental symmetry of $\mathcal{L}_2 + \mathcal{L}_4$. Nevertheless, since it is a symmetry of the chiral Lagrangian, one has to turn to Green functions rather than S-matrix elements for additional information. The clue of Leutwyler's resolution of the ambiguity is the assumption of resonance dominance for the 2-point function of pseudoscalar currents, an assumption that has proved to be very successful in predicting all the low–energy constants L_i [2, 42].

The Kaplan–Manohar symmetry is a non–linear transformation of the quark masses accompanied by corresponding transformations of the L_i, in particular L_7. Fixing L_7 via η' exchange [37] thus 'freezes' the Kaplan–Manohar symmetry and the standard current algebra mass ratios are rescued. In Leutwyler's formulation, "$m_u = 0$ is an interesting way not to understand this world." Using different arguments involving matrix elements of the winding number density $G\widetilde{G}$, Donoghue and Wyler [74] arrive at essentially the same conclusion.

3.3.2 Radiative pion transitions

There is in general good agreement between theory and experiment in this sector [2, 3, 75], with the possible exceptions of the pion polarizabilities [75] and the process $\gamma\gamma \to \pi^0\pi^0$. Of course, the latter does not occur to lowest order (neutral mesons). At $O(p^4)$, none of the low–energy constants L_i can contribute so that the loop amplitude must be finite. Thus, one obtains a parameter–free absolute prediction to $O(p^4)$ [39, 76].

In Fig. 2, the data of the Crystal Ball Collaboration [77, 78] for the integrated cross-section for $\gamma\gamma \to \pi^0\pi^0$ as a function of the center-of-mass energy are compared with the

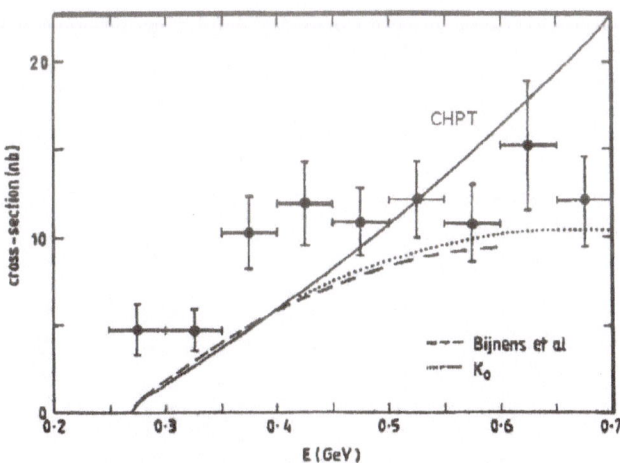

Figure 2. The integrated cross–section for $\gamma\gamma \to \pi^0\pi^0$ as a function of the center-of-mass energy taken from Ref. [78]. The data are from the Crystal Ball Collaboration [77]. The solid line is the prediction of CHPT to $O(p^4)$ [39, 76]. The dashed and dotted curves include quark loop and vector meson contributions as calculated in Refs. [79, 80].

CHPT prediction suggesting an excess near threshold. In the theoretical amplitude, the pion loop contribution dominates by far. We can write the invariant amplitude [there is only one at $O(p^4)$] in the form [76]

$$A(\gamma\gamma \to \pi^0\pi^0) = \frac{\alpha}{2\pi}\frac{s - M_\pi^2}{F^2} \cdot F(s/M_\pi^2) + K \text{ loop}, \qquad (3.27)$$

where

$$A(\pi^+\pi^- \to \pi^0\pi^0) = \frac{s - M_\pi^2}{F^2} \qquad (3.28)$$

is the lowest–order scattering amplitude [compare Eq. (2.73)] and $F(s/M_\pi^2)$ is a steeply rising loop function shown in Fig. 3.

How can we understand the possible discrepancy between CHPT and experiment ? After all, CHPT as a low–energy expansion should be expected to be especially reliable near threshold. Neither the inclusion of vector mesons nor the chiral quark model can modify the cross–section substantially near threshold [79, 80], as is also shown in Fig. 2. The most straightforward approach [81] is to extend the calculation to $O(p^6)$, in particular to the two–loop amplitude.

There is also a more pragmatic, albeit less rigorous way to proceed[5]. CHPT yields all contributions that the general principles of QFT demand and it parametrizes in a systematic manner the remainder in the form of local amplitudes which are polynomials in momenta and meson masses. The pragmatic approach to go beyond $O(p^4)$ consists of calculating those amplitudes which unitarity and analyticity allow us to calculate in the form of dispersion relations and to estimate the rest by chiral dimensional analysis. Applying this procedure to $\gamma\gamma \to \pi^0\pi^0$, we rewrite the amplitude (3.27) as [76]

$$A(\gamma\gamma \to \pi^0\pi^0) = \frac{\alpha}{2\pi} A(\pi^+\pi^- \to \pi^0\pi^0)F(s/M_\pi^2) + K \text{ loop} \qquad (3.29)$$

and use instead of the lowest–order amplitude (3.28) either the amplitude to $O(p^4)$ or the 'exact' amplitude in terms of the experimental phase shifts. The strategy of

---

[5]The following arguments have emerged in several discussions with Andy Cohen.

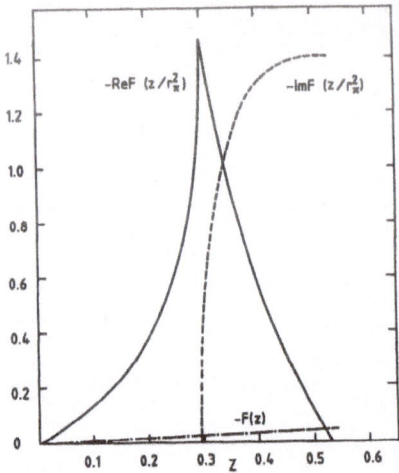

Figure 3. Real (solid line) and imaginary (dashed) parts of the pion loop function $F(s/M_\pi^2)$ with $z/r_\pi^2 = s/M_\pi^2$. The much smaller kaon loop contribution is represented by the dash–dotted curve.

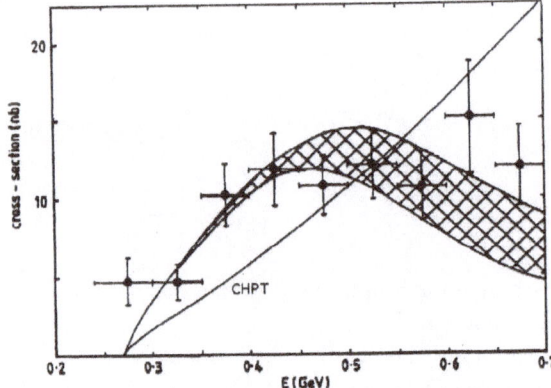

Figure 4. The results of the dispersive analysis of Morgan and Pennington [78] for $\gamma\gamma \to \pi^0\pi^0$ are shown as a band depending on various assumptions.

combining dispersion relations with the experimental $\pi\pi$ phase shifts has in fact been pursued in recent years by Morgan and Pennington [78] and the results of a recent analysis are shown in Fig. 4. The main source of the significant enhancement of the cross–section near threshold is the corresponding excess of the $\pi\pi$ s-wave amplitude over the chiral prediction of $O(p^4)$ discussed previously. Although the experimental evidence is far from compelling, it is intriguing that the effect seems to point in the same direction in both cases. In any case, the key to this problem lies in the s-wave of $\pi\pi$ scattering.

Let me emphasize that there are of course contributions to the amplitude already at the two–loop level which cannot be written in the form (3.29). The reason for the seeming success of the pragmatic approach to the $\gamma\gamma \to \pi^0\pi^0$ scattering amplitude is probably related to the exceptionally 'big' function $F(s/M_\pi^2)$. We shall encounter a similar situation in non–leptonic K decays in Sect. 4.

3.3.3 η decays

There are many open questions and unsolved problems in η decays (see the second review of Ref. [13]), both for the dominant decay modes $\eta \to 3\pi$ [54] and for rare decays. There are two special features characterizing η decays:

- Due to isospin violation, all decay amplitudes involve a factor $m_u - m_d$ or e or both.

- The η-η' mixing angle $\Theta \simeq -20^0$ is bigger [3] than previously thought. To $O(p^4)$, the effect is contained in the low–energy constant L_7. Because of the rather large mixing, one usually gets substantially bigger amplitudes by taking the η' into account explicitly. After all, in the chiral limit the η' becomes a Goldstone boson for $N_C \to \infty$.

Let me finish this section with an amusing example of a dramatic 'failure' of next–to–leading order CHPT. The decay $\eta \to \pi^0\gamma\gamma$ has obviously many things in common with $\gamma\gamma \to \pi^0\pi^0$: there are no contributions from either $\mathcal{L}_2$ or $\mathcal{L}_4$ and the amplitude is given exclusively by the finite one–loop contribution. However, the corresponding rate [82]

$$\Gamma(\eta \to \pi^0\gamma\gamma)_{loop} = 0.35 \cdot 10^{-2} \text{ eV} \tag{3.30}$$

is about a factor 230 lower than the (single) experimental measurement [83]

$$\Gamma(\eta \to \pi^0\gamma\gamma)_{exp} = (0.82 \pm 0.17) \text{ eV}. \tag{3.31}$$

The reason for this dramatic discrepancy is well understood [82]: the usually dominating pion loop contribution is suppressed by the notorious factor $m_u - m_d$, while the kaon loop is small anyway. Not even the two–loop amplitude is of much help [82]. The dominant contribution arises at $O(p^6)$ due to vector meson exchange and gives at least the right order of magnitude [84]. In a recent thorough analysis, Ametller et al. [85] have collected all relevant contributions including effects of $O(p^8)$ and higher. Assuming positive interference between the significant contributions, they conclude that the theoretical answer for the rate is probably still lower than the experimental result (3.31), but the discrepancy is no more than two standard deviations. A new measurement of $\eta \to \pi^0\gamma\gamma$ would be highly welcome in any case.

4 NON–LEPTONIC WEAK INTERACTIONS

For momenta much smaller than M_W, the $\Delta S = 1$ non–leptonic weak interactions are described by an effective Hamiltonian [86]

$$\mathcal{H}_{\text{eff}}^{\Delta S=1} = \frac{G_F}{\sqrt{2}} V_{ud} V_{us}^* \sum_i C_i(\mu)Q_i + h.c. \tag{4.1}$$

in terms of Wilson coefficients $C_i(\mu)$ depending on the QCD renormalization scale μ and of local operators

$$Q_1 = 4\overline{s_L}\gamma^\mu d_L \overline{u_L}\gamma_\mu u_L \qquad\qquad Q_2 = 4\overline{s_L}\gamma^\mu u_L \overline{u_L}\gamma_\mu d_L$$

$$Q_3 = 4\overline{s_L}\gamma^\mu d_L \sum_{q=u,d,s} \overline{q_L}\gamma_\mu q_L \qquad Q_4 = 4\sum_q \overline{s_L}\gamma^\mu q_L \overline{q_L}\gamma_\mu d_L \tag{4.2}$$

$$Q_5 = 4\overline{s_L}\gamma^\mu d_L \sum_q \overline{q_R}\gamma_\mu q_R \qquad Q_6 = -8\sum_q \overline{s_L}q_R\overline{q_R}d_L.$$

Only five of the operators Q_i are linearly independent. The operator product expansion in (4.1) is limited to the leading dim $= 6$ operators and to those terms that survive in the limit $\alpha_{em} = 0$.

4.1 Chiral Lagrangian of $O(p^2)$

We are looking for an effective chiral Lagrangian for the effective Hamiltonian (4.1). Of course, the weak interactions break the chiral symmetry. Instead of a chiral invariant Lagrangian as in Sect. 3, we now have to construct an effective Lagrangian that has the same transformation properties

$$(8_L, 1_R) + (27_L, 1_R) \tag{4.3}$$

under chiral rotations as the Hamiltonian (4.1).

To lowest order p^2, there are exactly two such terms[6], one octet and the other one a 27-plet with respect to $SU(3)_L$:

$$\mathcal{L}_2^{\Delta S=1} = G_8 F_\pi^4 (L_\mu L^\mu)_{23} + G_{27} F_\pi^4 (L_{\mu 23} L_{11}^\mu + \frac{2}{3} L_{\mu 21} L_{13}^\mu) + h.c. \tag{4.4}$$

$$L_\mu = iU^\dagger D_\mu U \sim \text{left-chiral current of } O(p).$$

The octet part of (4.4) was first written down by Cronin [87]. In the limit $N_C \to \infty$ ($\alpha_s \sim 1/N_C \to 0$), the Lagrangian (4.4) must assume the factorized form (see Ref. [88] and earlier work quoted therein)

$$\mathcal{L}_2^{\Delta S=1} \overset{N_C \to \infty}{\longrightarrow} -\frac{G_F}{\sqrt{2}} V_{ud} V_{us}^* F_\pi^4 L_{\mu 21} L_{13}^\mu + h.c. \tag{4.5}$$

In terms of dimensionless coupling constants g_A ($A = 8, 27$) defined by

$$G_A = -\frac{G_F}{\sqrt{2}} V_{ud} V_{us}^* g_A, \qquad (A = 8, 27) \tag{4.6}$$

the factorizable Lagrangian (4.5) corresponds to

$$g_8^{N_C \to \infty} = g_{27}^{N_C \to \infty} = \frac{3}{5}. \tag{4.7}$$

Determining g_8 and g_{27} from $K \to 2\pi$ decays, one finds instead

$$|g_8| = 5.1$$
$$|g_{27}| = 0.29. \tag{4.8}$$

The octet enhancement together with the 27-plet suppression are usually referred to as the $\Delta I = 1/2$ rule.

Let me try to summarize briefly the present theoretical status of the $\Delta I = 1/2$ rule:

1. Short-distance corrections [89, 90, 91, 92] raise $|g_8|$ and decrease $|g_{27}|$ from the values (4.7), but they are not big enough to explain the experimental values (4.8).

2. Long-distance effects ($\pi\pi$ rescattering) also point in the right direction. The analysis of Kambor, Missimer and Wyler [93] of $K \to 2\pi, 3\pi$ to $O(p^4)$ suggests an experimental value for $|g_8|$ that is about 30% smaller than the lowest-order value in (4.8). This result confirms earlier estimates by Bel'kov et al. [94].

3. The remaining missing factor of about 2 to 3 in the ratio g_8/g_{27} is by far the toughest part. Although there is little reason to doubt that QCD can account for the $\Delta I = 1/2$ rule, one must invoke additional model assumptions for the remaining corrections occurring at intermediate distances.

---

[6]Weak mass terms are omitted because they don't contribute to on-shell amplitudes.

In a series of papers [95], Bardeen, Buras and Gérard have estimated the intermediate–distance effects by matching QCD to a truncated chiral Lagrangian for large N_C. Identifying the QCD scale μ with the momentum cutoff for the chiral Lagrangian, they achieve the desired matching at $\mu \simeq 600$ MeV. Although the cutoff procedure on the low–energy side with its quadratic dependence on the cutoff is probably too crude, they obtain about 75% of the observed octet enhancement and practically all of the 27-plet suppression.

Another very interesting approach to understand the $\Delta I = 1/2$ rule is the diquark model of Stech and collaborators [96]. The phenomenological success of the model, not only for non–leptonic K decays, is quite intriguing, but the connections to QCD on the one hand and to the effective chiral Lagrangian on the other hand are not straightforward.

Motivated by the successful calculation of the strong low–energy constants L_i [51], Pich and de Rafael have adapted the method of the QCD effective action to the non-leptonic weak sector. The final results for g_8, g_{27} and the scale–invariant parameter $\widehat{B}$ relevant for K^0-$\bar{K}^0$ mixing are [88]:

$$
\begin{aligned}
g_8 &= \frac{1}{2}C_-(M_Q)(1-\Delta) + \frac{1}{10}C_+(M_Q)(1+\Delta) \\
&\quad + C_4(M_Q) - 16C_6(M_Q)L_5\left(\frac{\langle\bar{q}q\rangle_0}{F_\pi^3}\right)^2 + O(\alpha_s N_C^{-1}) \\
g_{27} &= \frac{3}{5}C_+(M_Q)(1+\Delta) \\
\widehat{B} &= \alpha_s(M_Q)^{-2/9}\frac{3}{4}(1+\Delta) \\
\Delta &\doteq \frac{1}{N_C}\left(1 - \frac{N_C}{2}\frac{\langle\frac{\alpha_s}{\pi}GG\rangle_0}{16\pi^2 F_\pi^4}\right) < 0.
\end{aligned}
\tag{4.9}
$$

$C_\pm$ are the Wilson coefficients for the operators $Q_\pm = Q_2 \pm Q_1$ and $M_Q \simeq 300$ MeV is to be identified with a constituent quark mass (cf. Sect. 2). Since $\Delta < 0$ with the standard values for the gluon condensate, the corrections of $O(\alpha_s N_C)$ exhibit the right tendency of increasing g_8 and lowering both g_{27} and $\widehat{B}$ compared to the large–N_C predictions. Since the $O(\alpha_s N_C)$ correction in Δ is about twice as big as the first term one may of course wonder about terms of $O(\alpha_s^2 N_C^2)$. Moreover, $M_Q \simeq 300$ MeV is a rather low scale for a perturbative calculation of the Wilson coefficients $C_i(\mu)$.

It is probably a fair conclusion that there is still room for further theoretical progress on the $\Delta I = 1/2$ rule. For CHPT, the coupling constants g_8, g_{27} are free parameters to be determined phenomenologically[7].

Due to the Goldstone nature of the pseudoscalar mesons, the Lagrangian $\mathcal{L}_2^{\Delta S=1}$ in Eq. (4.4) relates the processes $K \to 3\pi, 2\pi, \pi$. The bilinear vertices give rise to the so–called pole contributions and they are often a pain in the neck for loop calculations. Thus, it is of some practical advantage to rediagonalize the bilinear terms in $\mathcal{L}_2 + \mathcal{L}_2^{\Delta S=1}$ even though this weak diagonalization [97] breaks manifest chiral invariance. The corresponding transformations [97]

$$
\pi^+ \to \pi^+ - \frac{2M_K^2 F_\pi^2 G_8}{M_K^2 - M_\pi^2}K^+ + O(G_8^2) \qquad [G_{27} = 0] \qquad \text{etc.} \tag{4.10}
$$

induce weak terms in the strong chiral Lagrangian $\mathcal{L}_2 + \mathcal{L}_4 + \mathcal{L}_{WZW}$. Those terms sum up all the pole contributions of the standard procedure.

---

[7]Because of the large disparity in size, it is very often a good approximation to set $g_{27} = 0$.

4.2 $\Delta S = 1$ Non–Leptonics at $O(p^4)$

Except for the chiral anomaly, the procedure for constructing weak amplitudes of $O(p^4)$ exactly parallels the discussion in Sect. 3 for the strong interactions.

4.2.1   Effective Lagrangian of $O(p^4)$

There is a discouragingly large number of local terms of $O(p^4)$ with the proper transformation properties of $(8_L, 1_R)$ and $(27_L, 1_R)$, respectively [98]. By use of the Cayley-Hamilton theorem and the equations of motion (2.62), 37 independent terms remain in the octet sector alone [98, 99, 100]. A further significant reduction can be achieved by restricting attention to those terms which contribute to non–leptonic amplitudes where the only external gauge fields are photons (and Z bosons, but no W bosons). This leaves 22 linearly independent octet operators W_i of $O(p^4)$, four of which contain an ε tensor [101]. The effective Lagrangian has the general form

$$\mathcal{L}_4^{\Delta S=1} = \sum_i N_i W_i, \qquad (4.11)$$

where the dimensionless coupling constants N_i are the analogues of the L_i in the strong sector. For a complete list of the operators W_i I refer to Ref. [101].

The interpretation of the low–energy constants N_i is not as straightforward as in the strong case (chiral duality). In contrast to the L_i, there are genuine short–distance contributions such as the electroweak penguin operator contributing to N_{14}, N_{16} and N_{18} [97, 102]. In addition, the chiral anomaly contributes to and probably dominates the constants $N_{28}, \ldots, N_{31}$ [103]. Nevertheless, the observed resonance saturation of the L_i suggests a corresponding analysis of resonance exchange in the weak sector [101].

In the most general case where only symmetry constraints are used there are too many independent couplings between the meson resonances of spin 0, 1 and the pseudoscalar mesons. Therefore, we have considered two model approaches which both rest on the assumption that the strong chiral Lagrangian or the associated left–chiral current already determine the dominant features of the $\Delta S = 1$ effective Lagrangian: the factorization model [104, 105, 88] and the weak deformation model [106]. As we have shown [101], the two models are actually equivalent in predicting the same values for the N_i except for an overall scale factor.

Although these models are quite constrained, the results of a comparison with experimental data are not completely conclusive at this time. In the case of $K \to 2\pi, 3\pi$ decays where the data are rather precise, the normally dominant V and A exchange does not contribute as observed previously by Isidori and Pugliese [107]. Although there is evidence in the data for direct weak counterterm contributions ($N_i \neq 0$), the remaining S and P contributions predicted by the factorization or weak deformation model do not seem to be sufficient [101].

The situation is more promising for radiative K decays where only relatively few of the coupling constants N_i contribute, which are moreover sensitive only to V and A exchange. Here, the lack of data prevents a comprehensive comparison between theory and experiment. However, the destructive interference between the strong and weak couplings predicted by the two models in many cases is definitely seen in $K^+ \to \pi^+ e^+ e^-$ [108]. The N_i relevant for radiative K decays have also been calculated recently by Bruno and Prades [102] within the effective action approach [88].

4.2.2 One–loop functional

The divergences of the one–loop functional for the total Lagrangian $\mathcal{L}_2 + \mathcal{L}_2^{\Delta S=1}$ of lowest order are again of the form of a local Lagrangian [98, 99, 100]. The coefficients of this Lagrangian govern the scale dependence of both the $L_i^r(\mu)$ and the $N_i^r(\mu)$, where the renormalization procedure is exactly the same as in Sect. 2. The final result is the generalization of Eq. (2.91): the sum

$$Z_4^{L=1} + \int d^4x [\mathcal{L}_4(L_i) + \mathcal{L}_4^{\Delta S=1}(N_i)] =$$
$$= Z_{4,\text{fin}}^{L=1}(\mu) + \int d^4x [\mathcal{L}_4(L_i^r(\mu)) + \mathcal{L}_4^{\Delta S=1}(N_i^r(\mu))] \qquad (4.12)$$

gives rise to finite and scale–independent amplitudes in terms of the low–energy constants L_i^r, N_i^r.

The formalism was applied to $K = 2\pi, 3\pi$ decays by Kambor, Missimer and Wyler [93]. In addition to the two isospin amplitudes $A_0(\Delta I = 1/2)$ and $A_2(\Delta I = 3/2)$ for $K \to 2\pi$, there are 10 measurable quantities in the standard expansion of the $K \to 3\pi$ amplitudes to fourth order in the momenta [109]:

$$\Delta I = \frac{1}{2} \quad : \qquad \alpha_1, \beta_1, \zeta_1, \xi_1$$
$$\Delta I = \frac{3}{2} \quad : \qquad \alpha_3, \beta_3, \gamma_3, \zeta_3, \xi_3, \xi_3'. \qquad (4.13)$$

The current algebra analysis of $O(p^2)$ predicts 7 observable quantities in terms of the two parameters G_8, G_{27} (the quadratic slope parameters ζ_i, ξ_i, ξ_3' vanish to $O(p^2)$). It has been known for a long time that these predictions show the right qualitative trend only. There are sizeable discrepancies between theory and experiment at the current algebra level.

At $O(p^4)$, the agreement between CHPT and experiment is improved substantially [93]. One may worry that this improvement is mainly due to the many free coupling constants N_i entering the amplitudes. A nice way to illustrate the predictive power of CHPT to $O(p^4)$ has recently been presented in Ref. [110]. Treating the strong couplings L_i as known input, there are 7 unknown combinations of the weak constants N_i and the corresponding 27-plet couplings in addition to G_8, G_{27}. Since two of those combinations can be absorbed in G_8, G_{27}, there are altogether 7 parameters for 12 observables. The resulting 5 relations can be formulated as predictions for the quadratic slope parameters [110]. The predictions are compared to the experimental values in Table 4. Except for ζ_3, which is expected to be rather sensitive to isospin violating radiative corrections, the agreement is very satisfactory. CHPT has passed the test successfully.

4.3 Rare K Decays

Most rare K decays are radiative decays with real or virtual photons. In contrast to $K \to 2\pi, 3\pi$, relatively few of the weak couplings N_i contribute. With few exceptions ($K^+ \to \pi^+\nu\bar{\nu}$, $K_L \to \pi^0\nu\bar{\nu}$), the rare K decays are dominated by long–distance physics.

Many transitions do not occur at lowest $O(p^2)$. Thus, they offer non–trivial tests of the genuine QFT aspects of CHPT. The absence of $O(p^2)$ amplitudes is due in many cases to a mismatch between the available momenta at $O(p^2)$ and the constraints of electromagnetic gauge invariance [97]. In particular, any K decays into at most one pion, but any number of photons and leptons ($K^+ \to \pi^+ l^+ l^-$, $K^+ \to \pi^+\gamma\gamma$), vanish at $O(p^2)$ [97].

Table 4. Predicted and measured values of the quadratic slope parameters in the $K \to 3\pi$ amplitudes, all given in units of 10^{-8}. The table is taken from Ref. [110] and is based on the $O(p^4)$ CHPT calculation of Ref. [93].

parameter	prediction	expt. value
ζ_1	-0.47 ± 0.18	-0.47 ± 0.15
ξ_1	-1.58 ± 0.19	-1.51 ± 0.30
ζ_3	-0.011 ± 0.006	-0.21 ± 0.08
ξ_3	0.092 ± 0.030	-0.12 ± 0.17
ξ'_3	-0.033 ± 0.077	-0.21 ± 0.51

Non–leptonic K decays can be classified in three categories on the basis of the amplitudes of $O(p^4)$:

1. Chiral symmetry forbids counterterms for the transition. The amplitude of $O(p^4)$ is given entirely by the one–loop amplitude which is necessarily finite ($K_S \to \gamma\gamma$, $K_L \to \pi^0\gamma\gamma$, $K_S \to \pi^0\pi^0\gamma\gamma$).

2. The loop amplitude is finite but there is in addition a non–vanishing scale–independent counterterm amplitude ($K^+ \to \pi^+\gamma\gamma$, $K^+ \to \pi^+\pi^0\gamma$).

3. The loop amplitude diverges, requiring a scale–dependent counterterm amplitude ($K \to 2\pi, 3\pi$, $K \to \pi l^+ l^-$, $K_L \to \pi^0\pi^0 l^+ l^-$).

Here, I will only discuss two examples of the first category ($K_S \to 2\gamma$, $K_L \to \pi^0\gamma\gamma$) characterized by one–loop amplitudes depending only on the lowest–order coupling G_8 (for $G_{27} = 0$).

4.3.1 $K_S \to \gamma\gamma$

Although each of the four diagrams in Fig. 5 is quadratically divergent, chiral symmetry and electromagnetic gauge invariance enforce finiteness of the total amplitude. The chiral prediction compares well with experiment,

$$BR(K_S \to \gamma\gamma) = \begin{cases} 2.0 \cdot 10^{-6} & [111, 112] \\ (2.4 \pm 1.2) \cdot 10^{-6} & [113], \end{cases} \tag{4.14}$$

but the experimental error certainly leaves room for higher–order corrections.

A typical local coupling from $\mathcal{L}_6^{\Delta S=1}$ contributing to $K_S \to 2\gamma$ at $O(p^6)$ is [114]

$$\frac{c_6 e^2 G_8 F_\pi^4}{(4\pi F_\pi)^4} F_{\mu\nu} F^{\mu\nu} \frac{\Box K^0}{2F_\pi} \tag{4.15}$$

where the chiral counting rules would suggest $c_6 = O(1)$. On the other hand, in order to reproduce the experimental rate with the coupling (4.15) one would need

$$|c_6| \simeq 20. \tag{4.16}$$

How does the one–loop amplitude of CHPT [111, 112] account for this factor 20 enhancement? One factor is simply due to chiral power counting: the natural ratio between an $O(p^4)$ and an $O(p^6)$ K decay amplitude is

$$\frac{(4\pi F_\pi)^2}{M_K^2} \simeq 5. \tag{4.17}$$

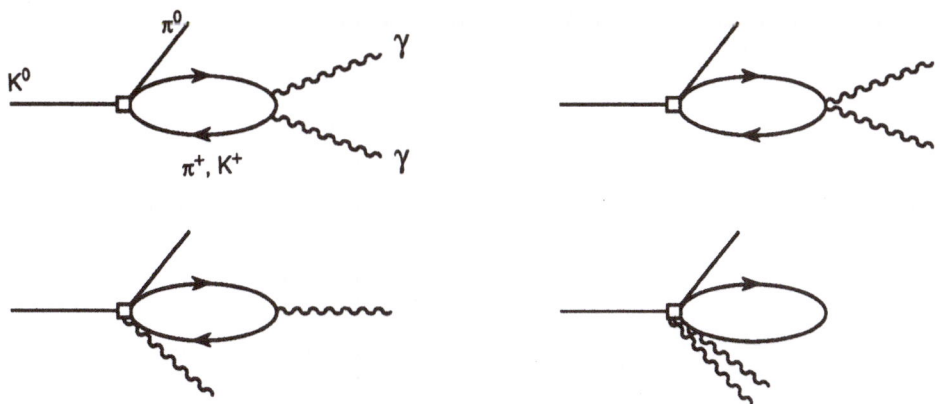

Figure 5. One–loop diagrams for $K_S \to \gamma\gamma$ (without the pion line in the final state) and for $K_L \to \pi^0\gamma\gamma$.

The remaining factor 4 is due to the special properties of the pion loop producing the 'big' loop function $F(s/M_\pi^2)$ ($s = M_K^2$ in this case) already encountered in the discussion of $\gamma\gamma \to \pi^0\pi^0$. The final conclusion is that the $O(p^6)$ corrections are expected to be small.

4.3.2 $K_L \to \pi^0\gamma\gamma$

To $O(p^4)$, the amplitude is again determined only by the one–loop diagrams of Fig. 5. The CHPT prediction [115, 116] for the normalized spectrum in the 2γ–invariant mass is in excellent agreement with the most recent results of the NA31-Collaboration shown in Fig. 6. On the other hand, with the canonical value of g_8 in Eq. (4.8) the CHPT prediction for the branching ratio seems to lie definitely below the experimental value:

$$BR(K_L \to \pi^0\gamma\gamma) = \begin{cases} 0.67 \cdot 10^{-6} & [115, 116] \\ (1.7 \pm 0.3) \cdot 10^{-6} & [117] \\ (1.7 \pm 0.7) \cdot 10^{-6} & [118]. \end{cases} \qquad (4.18)$$

The new experimental branching ratio of NA31 is precise enough to warrant a careful investigation of possible higher–order effects. Several authors [119, 80] have suggested that in analogy to $\eta \to \pi^0\gamma\gamma$ discussed in Sect. 3 vector meson exchange produces an important, if not dominant contribution to $K_L \to \pi^0\gamma\gamma$ (and to $K_L \to \pi^0 e^+ e^-$, for that matter). In the chiral language, V exchange appears first at $O(p^6)$. The relevant strong Lagrangian is [106]

$$\mathcal{L}(VP\gamma) = eh_V \tilde{F}_{\mu\nu} \langle V^\mu \{u^\nu, uQu^\dagger + u^\dagger Qu\} \rangle \qquad (4.19)$$

with $|h_V| = (3.7 \pm 0.3) \cdot 10^{-2}$ taken from $\rho, \omega \to \pi\gamma$ decays. V exchange and weak diagonalization (the pole diagrams in the standard parlance) give rise to a weak amplitude proportional to

$$a_V = \frac{512\pi^2 h_V^2 M_K^2}{9M_V^2} = 0.32. \qquad (4.20)$$

However, there is absolutely no reason not to expect direct weak contributions from $\mathcal{L}_6^{\Delta S=1}$ in analogy to the coupling constants N_i of $O(p^4)$. Based on the weak deformation

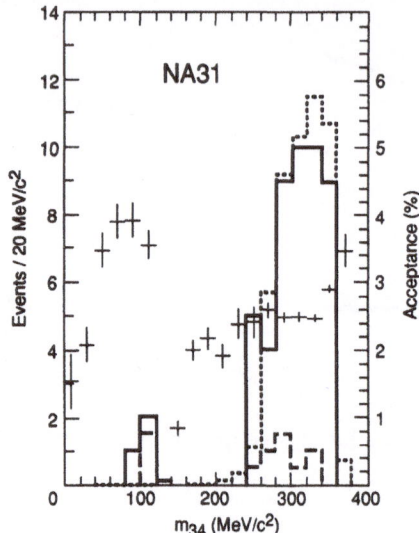

Figure 6. Distribution of unambiguous $K_L \rightarrow \pi^0\gamma\gamma$ candidates in the two–photon invariant mass from Ref. [117]. The dashed histogram corresponds to the normalized distribution for the one–loop amplitude of CHPT. The experimental acceptance is indicated by the crosses.

model, we estimated [106] $|a_V| \stackrel{<}{\sim} 0.3$ for the overall effective coupling in agreement with the recent bounds of NA31 [117]:

$$-0.32 < a_V < 0.19 \qquad (90\% \text{ c.l.}). \qquad (4.21)$$

For such an allowed range of the effective coupling a_V, V exchange in addition to the $O(p^4)$ amplitude cannot by itself explain the missing factor 2 in rate [106].

Let us entertain some chiral handwaving for a change. After all, a 40% correction in amplitude is not at all unusual in CHPT. Donoghue has offered a trial estimate of higher–order effects with the formula [9]

$$\Gamma(K_L \rightarrow \pi^0\gamma\gamma)_{\text{trial}} = \Gamma_{O(p^4)} \cdot \frac{\Gamma(K_L \rightarrow \pi^+\pi^-\pi^0)}{\Gamma(K_L \rightarrow \pi^+\pi^-\pi^0)_{O(p^2)}} \cdot \frac{\sigma(\gamma\gamma \rightarrow \pi^0\pi^0)}{\sigma(\gamma\gamma \rightarrow \pi^0\pi^0)_{O(p^4)}}. \qquad (4.22)$$

However, both enhancement factors on the right–hand side of (4.22) are slightly misleading. The first one suggests an enhancement of about 1.7 in rate due to the 30% change in G_8 found in Ref. [93]. However, the theoretical prediction in Eq. (4.18) is based on G_8 extracted from $K \rightarrow 2\pi$ decays (see below) at tree level, being larger instead of smaller than the value at $O(p^4)$. Secondly, the dispersive analysis implicit in Donoghue's argument involves $\gamma\gamma \rightarrow \pi^+\pi^-$ rather than $\gamma\gamma \rightarrow \pi^0\pi^0$.

Nevertheless, a dispersive analysis of the higher–order corrections is certainly called for. Both chiral dimensional analysis (recall the discussion for $K_S \rightarrow 2\gamma$) and the natural size of V exchange suggest that the local $O(p^6)$ corrections are not unusually large. In accordance with the discussion of $\gamma\gamma \rightarrow \pi^0\pi^0$ in Sect. 3, we write the invariant decay amplitude in the form

$$A(K_L \rightarrow \pi^0\gamma\gamma) = \frac{\alpha}{2\pi} A(K_L \rightarrow \pi^+\pi^-\pi^0) F(q^2/M_\pi^2) + K \text{ loop}$$
$$A(K_L \rightarrow \pi^+\pi^-\pi^0) = G_8(q^2 - M_\pi^2) \qquad (4.23)$$

where q^2 is the invariant mass^2 of the two photons. The simplest correction in accordance with Donoghue's proposal [9] amounts to extracting G_8 directly from $\Gamma(K_L \to \pi^+\pi^-\pi^0)$ instead of $\Gamma(K \to 2\pi)$, always keeping the $O(p^2)$ amplitude for $K_L \to \pi^+\pi^-\pi^0$. One finds

$$\left| \frac{G_8(K_L \to \pi^+\pi^-\pi^0)}{G_8(K \to 2\pi)} \right|^2 = 1.32 \tag{4.24}$$

implying a 30% increase in rate (not in amplitude !) compared to the CHPT prediction in (4.18).

A more refined analysis uses the $O(p^4)$ amplitude [93] or directly the experimentally determined amplitude for $K_L \to \pi^+\pi^-\pi^0$ in the dispersion relation. Due to the quadratic slope terms, the amplitude for $K_L \to \pi^0\gamma\gamma$ is then no longer of the factorized form as in Eq. (4.23). Moreover, as recently shown by Cappiello, D'Ambrosio and Miragliuolo [120], the second invariant amplitude of $K_L \to \pi^0\gamma\gamma$, which appears first at $O(p^6)$, develops a non–zero absorptive part. The complete dispersive analysis will soon be available [121].

4.4   The Chiral Anomaly in Non–Leptonic K Decays

As a fundamental property of the SM, the chiral anomaly [31] should be tested experimentally wherever possible. In non–leptonic K decay amplitudes it appears again at $O(p^4)$, but in two different manifestations.

The first type of anomalous amplitudes is characterized by tree–level Feynman diagrams where one vertex is due to the WZW action (2.97) and a second one to the weak Lagrangian (4.4) of $O(p^2)$. Examples are $K_S \to \pi^0\pi^0(\eta) \to \pi^0\gamma\gamma$ [115] and $K_L \to \pi^0\pi^0\pi^0 \to \pi^0\pi^0\gamma\gamma$ [123, 124]. The pole diagram contributions due to $\mathcal{L}_2^{\Delta S=1}$ can be summed up as usual by a weak diagonalization [97]. The resulting anomalous Lagrangian [122] contributes only to $K^+ \to \pi^+\pi^0\gamma(\gamma)$ and $K_L \to \pi^+\pi^-\gamma\gamma$.

There is however an additional direct weak anomaly functional [105, 103] due to the effective chiral realization of $\mathcal{H}_{\text{eff}}^{\Delta S=1}$ in (4.1) in the odd–intrinsic parity sector at $O(p^4)$. Let us first recall that the left–chiral current $\overline{q_L}\gamma^\mu q_L$ at the fundamental level corresponds to a functional derivative of the chiral action at the effective level:

$$\overline{q_L}\gamma^\mu q_L \leftrightarrow \frac{\delta S}{\delta l_\mu}. \tag{4.25}$$

The chiral realization of the 4-quark operators in (4.2) in $\mathcal{H}_{\text{eff}}^{\Delta S=1}$ is not as straightforward. The formal limit

$$\lim_{x \to y} \overline{q_L}(x)\gamma^\mu q_L(x)\overline{q_L}(y)\gamma_\mu q_L(y) \tag{4.26}$$

of a product of two currents is meaningless without introducing a regularization procedure as indicated in particular by the anomalous dimensions of four-quark operators [88]. To implement the limit (4.26), a model for SCSB must be chosen. The QCD effective action [51] adapted for the non–leptonic weak interactions [88] provides such a model.

It was recently shown [103] that the so–called factorizable contributions to the effective realization of 4-quark operators are actually model independent. Moreover, in contrast to the even–intrinsic parity sector [88, 102], the result of $O(p^4)$

$$\overline{q_L}\gamma^\mu q_L \overline{q_L}\gamma_\mu q_L \leftrightarrow \left\{ \frac{\delta S_{WZW}}{\delta l^\mu}, \frac{\delta S_2}{\delta l_\mu} \right\} \tag{4.27}$$

is exact to all orders in α_s [103]. This important property is a consequence of the non–renormalization of the chiral anomaly [125]. The anomalous current can be derived from the WZW action (2.97):

$$\frac{\delta S_{WZW}}{\delta l_\mu} = \frac{1}{16\pi^2}\,\varepsilon^{\mu\nu\rho\sigma}J_{\nu\rho\sigma}$$

$$J_{\nu\rho\sigma} = iL_\nu L_\rho L_\sigma + \{F^L_{\nu\rho} + \frac{1}{2}U^\dagger F^R_{\nu\rho}U, L_\sigma\}$$

$$+ \text{ a non–covariant polynomial in } l,r$$

$$L_\mu = iU^\dagger D_\mu U. \tag{4.28}$$

The non–covariant polynomial in the gauge fields l,r is physically irrelevant [103].

Summing up all contributions yields an effective Lagrangian of $O(p^4)$:

$$\begin{aligned}
\mathcal{L}^{\Delta S=1}_{eff} = & \frac{G_8 F^2}{16\pi^2}\Big(2a_1 i\varepsilon^{\mu\nu\alpha\beta}\langle\lambda L_\mu\rangle\langle L_\nu L_\alpha L_\beta\rangle \\
& + a_2\langle\lambda[U^\dagger\tilde{F}^{\mu\nu}_R U, L_\mu L_\nu]\rangle \\
& + 3a_3\langle\lambda L_\mu\rangle\langle(\tilde{F}^{\mu\nu}_L + U^\dagger\tilde{F}^{\mu\nu}_R U)L_\nu\rangle \\
& + a_4\langle\lambda L_\mu\rangle\langle(\tilde{F}^{\mu\nu}_L - U^\dagger\tilde{F}^{\mu\nu}_R U)L_\nu\rangle\Big) + h.c. \tag{4.29} \\
& \lambda = \frac{1}{2}(\lambda_6 - i\lambda_7)
\end{aligned}$$

The first observation is that this effective Lagrangian is in fact contained in the general expression (4.11) for $\mathcal{L}^{\Delta S=1}_4$. Although due to the chiral anomaly, the Lagrangian (4.29) has the normal chiral transformation property of a left–chiral octet (having set $G_{27} = 0$). The factorizable contributions from the dominant octet operator $Q_- = Q_2 - Q_1$ correspond to $a_i = 1$ ($i = 1,\ldots,4$). Both the non–factorizable contributions from Q_i, which cannot be calculated in a model–independent manner, and contributions from the other octet operators in $\mathcal{H}^{\Delta S=1}_{eff}$ are expected to be smaller [103] so that $a_i \simeq 1$ should be a reasonable first approximation.

The Lagrangian (4.29) contributes only to radiative K decays. An interesting example where both manifestations of the chiral anomaly appear is provided by the decay

$$K^+(P) \to \pi^+(p_1)\pi^0(p_2)\gamma(q). \tag{4.30}$$

The decay amplitude

$$A = \frac{\varepsilon^*_\mu(q)}{M^3_K}[E(x_i)(p_1 q p^\mu_2 - p_2 q p^\mu_1) + M(x_i)\varepsilon^{\mu\nu\rho\sigma}p_{1\nu}p_{2\rho}q_\sigma], \qquad x_i = \frac{P p_i}{M^2_K} \tag{4.31}$$

is characterized by an electric amplitude E and a magnetic amplitude M. In the realistic case where the photon helicity is not measured there is no interference between E and M.

At leading $O(p^2)$, the bremsstrahlung amplitude

$$E_B = -\frac{2eA(K^+ \to \pi^+\pi^0)}{M_K(1 - x_1 - x_2)(1 - 2x_2)} \sim G_{27} \tag{4.32}$$

is suppressed by the $\Delta I = 1/2$ rule. At $O(p^4)$, there are two contributions to the electric amplitude E_4 [122]:

- A scale–independent counterterm amplitude

$$E_4^{\text{tree}} = \frac{2ie}{F_\pi} G_8 M_K^3 (-N_{14} + N_{15} + N_{16} + N_{17}) \qquad (4.33)$$

in terms of the weak low–energy constants N_i.

- A finite, but small loop amplitude E_4^{loop} which can be neglected in first approximation.

The existing measurements [126] are not accurate enough to determine a non-vanishing electric direct emission amplitude, i.e. the combination $-N_{14} + N_{15} + N_{16} + N_{17}$ of weak coupling constants. However, they clearly establish that the direct emission amplitude is dominantly magnetic. Assuming $E = E_B$ (no electric direct emission) and applying certain cuts, the experimental branching ratio is found [126] to be

$$BR(K^+ \to \pi^+ \pi^0 \gamma)_{\text{exp}} = (1.8 \pm 0.4) \cdot 10^{-5}. \qquad (4.34)$$

At $O(p^4)$, the magnetic amplitude is exclusively due to the chiral anomaly [122, 103] in both its manifestations discussed above:

$$
\begin{aligned}
M_4 &= M_4^{(\text{i})} + M_4^{(\text{ii})} \\
M_4^{(\text{i})} &= \frac{eG_8 M_K^3}{2\pi^2 F_\pi} \\
M_4^{(\text{ii})} &= \frac{3eG_8 M_K^3}{2\pi^2 F_\pi} \left(a_3 - \frac{a_2}{2} \right).
\end{aligned} \qquad (4.35)
$$

With the same assumptions as in the experimental analysis, the CHPT prediction is

$$BR(K^+ \to \pi^+ \pi^0 \gamma)_{CHPT} = 2.2 \cdot 10^{-5} \left(\frac{2 + 6a_3 - 3a_2}{5} \right)^2 \left(\frac{G_8}{9 \cdot 10^{-6} \text{ GeV}^{-2}} \right)^2. \qquad (4.36)$$

The positive interference between $M_4^{(\text{i})}$ and $M_4^{(\text{ii})}$ for $2a_3 - a_2 > 0$ is supported by the data which are in fact fully consistent with the theoretical expectation

$$2a_3 - a_2 \simeq 1. \qquad (4.37)$$

A more precise measurement of the photon spectrum will allow for a separate determination of both $E - E_B$ and M.

The chiral anomaly can also be tested in other non–leptonic radiative K decays such as in $K_L \to \pi^+ \pi^- \gamma$ [122, 127].

5 Conclusions

1. CHPT is the effective field theory of the standard model at low energies.

2. Although the effective chiral Lagrangian is non–renormalizable, all axiomatic properties of a quantum field theory are incorporated in CHPT.

3. All symmetries of the SM are manifest by construction.

4. CHPT is a systematic expansion in momenta and meson masses. The relevant expansion scale is $4\pi F_\pi = 1.2$ GeV.

5. There is no–double counting, since only hadronic degrees of freedom appear in the chiral Lagrangian.

6. The transition from the SM to the effective chiral Lagrangian generates a number of low–energy constants. In the strong effective Lagrangian of $O(p^4)$, those constants are dominated by resonance exchange (chiral duality). More work is needed, both experimentally and theoretically, to determine and to understand the corresponding weak coupling constants.

7. The chiral anomaly has a special status in CHPT. It is unambiguously given by the WZW functional without any free parameters.

8. A wide range of applications has been reviewed in these lectures, but much remains to be done. In view of the forthcoming high–intensity kaon beams, K decays in particular will remain a promising field for CHPT.

Acknowledgements

I am indebted to J. Bijnens, A. Cohen, J. Gasser, J. Kambor, H. Leutwyler, H. Neufeld, A. Pich, E. de Rafael and D. Wyler for their collaboration and for their advice on many aspects of CHPT. I want to thank A. Irbäck for pointing out Ref. [24] to me. The questions and comments of the participants of the Cargèse and Corfu Summer Schools have been very helpful in improving the final version of these lecture notes. Last but not least, my thanks are due to R. Gastmans, J.-M. Gérard and G. Zoupanos for having invited me to lecture in such beautiful surroundings.

References

[1] S. Weinberg, Physica 96A (1979) 327.

[2] J. Gasser and H. Leutwyler, Ann. Phys. 158 (1984) 142.

[3] J. Gasser and H. Leutwyler, Nucl. Phys. B250 (1985) 465.

[4] J. Schwinger, Ann. Phys. 2 (1957) 407 ;
M. Gell–Mann and M. Lévy, Nuovo Cim. 16 (1960) 705.

[5] T. Skyrme, Proc. Royal Soc. A260 (1961) 127 ;
A.P. Balachandran, V.P. Nair, S.G. Rajeev and A. Stern, Phys. Rev. Lett. 49 (1982) 1124; Phys. Rev. D27 (1983) 1153;
E. Witten, Nucl. Phys. B223 (1983) 433.

[6] H. Georgi, "Weak Interactions and Modern Particle Theory",
Benjamin/Cummings (Menlo Park, 1984).

[7] J.F. Donoghue, E. Golowich and B.R. Holstein, "Dynamics of the Standard Model", Cambridge Univ. Press (Cambridge, 1992).

[8] Proc. of the Ringberg Workshop on Hadronic Matrix Elements and Weak Decays, Ringberg Castle, Germany, April 1988, Eds. A.J. Buras, J.-M. Gérard and W. Huber, Nucl. Phys. B(Proc. Suppl.)7A (1989).

[9] Proc. of the Workshop on Effective Field Theories of the Standard Model, Dobogókö, Hungary, Aug. 1991, Ed. U.-G. Meißner, World Scient. Publ. Co. (Singapore, 1992).

[10] H. Leutwyler, Chiral effective Lagrangians, Schladming and Boulder Lectures, Bern preprint BUTP–91/26 (1991).

[11] J. Gasser, The QCD vacuum and chiral symmetry, in: "Hadrons and Hadronic Matter", Eds. D. Vautherin et al. (Plenum Press, 1990);
Chiral dynamics, Proc. Workshop on Physics and Detectors for DAFNE, Ed. G. Pancheri (Frascati,1991).

[12] E. de Rafael, CHPT approach to hadronic weak amplitudes, Nucl. Phys. B(Proc. Suppl.)10A (1989) 37.

[13] A. Pich, Rare K decays and CP violation, Proc. DPF Meeting, Storrs 1988, DPF Conf. 1988, 306;
η decays and chiral Lagrangians, Proc. Workshop on Rare Decays of Light Mesons, Gif–sur–Yvette, 1990;
Rare decays in CHPT, Nucl. Phys. B(Proc. Suppl.)23A (1991) 399.

[14] J.F. Donoghue, Chiral symmetry as an experimental science, Int. School of Low–Energy Antiprotons, Erice, 1990, preprint CERN-TH.5667/90.

[15] J. Bijnens, Chiral perturbation theory and anomalous processes, Habilitations-schrift Univ. München, 1992.

[16] G. Ecker, CP violation in rare K decays, Proc. of Topical Conference on CP Violation, Calcutta, 1990, Ed. P. Ghose (in print);
Chiral realization of the non-leptonic weak interactions, Proc. 24th Int. Symposium on the Theory of Elementary Particles, Gosen (Berlin), Ed. G. Weigt (Zeuthen, 1991).

[17] G. 't Hooft, Phys. Rev. Lett. 37 (1976) 8;
C.G. Callan, R.F. Dashen and D.J. Gross, Phys. Lett. 36B (1976) 334;
R.J. Crewther, Phys. Lett. 70B (1977) 349.

[18] S. Adler and R.F. Dashen, "Current Algebras", Benjamin (New York, 1968);
V. de Alfaro, S. Fubini, G. Furlan and C. Rossetti, "Currents in Hadron Physics", North–Holland (Amsterdam, 1973).

[19] G. 't Hooft, in "Recent Developments in Gauge Theories", Eds. G. 't Hooft et al., Plenum Press (New York, 1980).

[20] S. Coleman and E. Witten, Phys. Rev. Lett. 45 (1980) 100.

[21] C. Vafa and E. Witten, Nucl. Phys. B234 (1984) 173.

[22] H.J. Borchers, Nuovo Cim. 15 (1960) 784.

[23] J. Goldstone, Nuovo Cim. 19 (1961) 154.

[24] R. Altmeyer et al. (MT_C Collaboration), The hadron spectrum in QCD with dynamical staggered fermions, Jülich preprint HLRZ 92-17, April 1992.

[25] M. Gell–Mann, R.J. Oakes and B. Renner, Phys. Rev. 175 (1968) 2195.

[26] S. Coleman, J. Wess and B. Zumino, Phys. Rev. 177 (1969) 2239;
C. Callan, S. Coleman, J. Wess and B. Zumino, Phys. Rev. 177 (1969) 2247.

[27] S. Weinberg, in "A Festschrift for I.I. Rabi", New York Acad. of Sciences, 1977, p.185.

[28] M. Gell–Mann, Phys. Rev. 106 (1957) 1296;
S. Okubo, Prog. Theor. Phys. 27 (1962) 949.

[29] S. Weinberg, Phys. Rev. Lett. 17 (1966) 616.

[30] J. Wess and B. Zumino, Phys. Lett. 37B (1971) 95;
E. Witten, Nucl. Phys. B223 (1983) 422.

[31] S.L. Adler, Phys. Rev. 177 (1969) 2426;
J.S. Bell and R. Jackiw, Nuovo Cim. 60A (1969) 47;
W.A. Bardeen, Phys. Rev. 184 (1969) 1848.

[32] A. Manohar and H. Georgi, Nucl. Phys. B234 (1984) 189.

[33] For a comprehensive review, see R.D. Ball, Phys. Reports 182 (1989) 1.

[34] S. Weinberg, Phys. Rev. 118 (1960) 838.

[35] G. Ecker and J. Honerkamp, Nucl. Phys. B35 (1971) 481.

[36] J. Bijnens, G. Ecker and J. Gasser, Semileptonic kaon decays, in Ref. [65].

[37] H. Leutwyler, Nucl. Phys. B337 (1990) 108.

[38] K. Chou et al., Phys. Lett. B134 (1984) 67;
H. Kawai and S. Tye, Phys. Lett. B140 (1984) 403;
Ö. Kaymakcalan, S. Rajeev and J. Schechter, Phys. Rev. D30 (1984) 594;
N.K. Pak and P. Rossi, Nucl. Phys. B250 (1985) 279.

[39] J. Bijnens and F. Cornet, Nucl. Phys. B296 (1988) 557.

[40] J. Bijnens, Nucl. Phys. B337 (1990) 635.

[41] C. Riggenbach, J. Gasser, J.F. Donoghue and B.R. Holstein, Phys. Rev. D43 (1991) 127.

[42] G. Ecker, J. Gasser, A. Pich and E. de Rafael, Nucl. Phys. B321 (1989) 311.

[43] J.F. Donoghue, C. Ramirez and G. Valencia, Phys. Rev. D39 (1989) 1947;
M. Praszalowicz and G. Valencia, Nucl. Phys. B341 (1990) 27.

[44] G. Ecker, J. Gasser, H. Leutwyler, A. Pich and E. de Rafael, Phys. Lett. B223 (1989) 425.

[45] U.-G. Meißner, Phys. Reports 161 (1988) 213.

[46] M. Bando, T. Kugo and K. Yamawaki, Phys. Reports 164 (1988) 115.

[47] K. Kawarabayashi and M. Suzuki, Phys. Rev. Lett. 16 (1966) 255;
Riazuddin and Fayyazuddin, Phys. Rev. 147 (1966) 1071.

[48] G. Ecker, Vector mesons and chiral symmetry, in Proc. of the 12th Warsaw Symposium on Elementary particle Physics, Kazimierz, Poland, June 1988, Eds. Z. Ajduk et al., World Scient. Publ. Co. (Singapore, 1990).

[49] J. Bijnens, C. Bruno and E. de Rafael, Nambu–Jona-Lasinio like models and the low–energy effective action of QCD, preprint CERN-TH. 6521/92, May 1992, to appear in Nucl. Phys. B.

[50] Y. Nambu and G. Jona-Lasinio, Phys. Rev. 122 (1961) 345; ibid. 124 (1961) 246.

[51] D. Espriu, E. de Rafael and J. Taron, Nucl. Phys. B345 (1990) 22; Err. ibid. B355 (1991) 278.

[52] T.N. Truong, Phys. Rev. Lett. 67 (1991) 2260; Phys. Lett. B273 (1991) 292.

[53] J. Gasser and H. Leutwyler, Nucl. Phys. B250 (1985) 517.

[54] J. Gasser and H. Leutwyler, Nucl. Phys. B250 (1985) 539.

[55] H. Lehmann, Phys. Lett. 41B (1972) 529; Acta Phys. Austriaca Suppl. XI (1973) 139.

[56] G. Ecker and J. Honerkamp, Nucl. Phys. B52 (1973) 211.

[57] J.F. Donoghue, C. Ramirez and G. Valencia, Phys. Rev. D38 (1988) 2195.

[58] J. Gasser and U.-G. Meißner, Phys. Lett. B258 (1991) 219.

[59] J.-M. Gérard, these Proceedings.

[60] E. de Rafael, in Ref. [8].

[61] J.-L. Basdevant, C.D. Froggatt and J.L. Petersen, Nucl. Phys. B72 (1974) 413.

[62] V. Bernard, N. Kaiser and U.-G. Meißner, Nucl. Phys. B364 (1991) 283.

[63] J. Bijnens et al., in preparation.

[64] S.M. Roy, Phys. Lett. 36B (1971) 353.

[65] The DAFNE Physics Handbook, Eds. L. Maiani, G. Pancheri and N. Paver, INFN-Frascati (to appear).

[66] P. Franzini, in Ref. [65].

[67] J. Bijnens, G. Ecker and J. Gasser, Radiative semileptonic kaon decays, preprint CERN-TH. 6625/92, Aug. 1992.

[68] S. Egli et al., Phys. Lett. B175 (1986) 97; ibid. B222 (1989) 533.

[69] A.M. Diamant-Berger et al., Phys. Lett. 62B (1976) 485.

[70] H. Leutwyler and M. Roos, Z. Phys. C25 (1984) 91.

[71] M. Ademollo and R. Gatto, Phys. Rev. Lett. 13 (1964) 264.

[72] L. Rosselet et al., Phys. Rev. D15 (1977) 574.

[73] D. Kaplan and A. Manohar, Phys. Rev. Lett. 56 (1986) 1994.

[74] J.F. Donoghue and D. Wyler, Phys. Rev. D45 (1992) 892.

[75] J.F. Donoghue and B.R. Holstein, Phys. Rev. D40 (1989) 2378.

[76] J.F. Donoghue, B.R. Holstein and Y.C. Lin, Phys. Rev. D37 (1988) 2423.

[77] H. Marsiske et al. (Crystal Ball Coll.), Phys. Rev. D41 (1990) 3324.

[78] D. Morgan and M.R. Pennington, Phys. Lett. B272 (1991) 134;
M.R. Pennington, Predictions for $\gamma\gamma \to \pi\pi$: what photons at DAFNE will see, in Ref. [65].

[79] J. Bijnens, S. Dawson and G. Valencia, Phys. Rev. D44 (1991) 3555.

[80] P. Ko, Phys. Rev. D41 (1990) 1531.

[81] S. Bellucci, J. Gasser and M.E. Sainio, in preparation.

[82] G. Ecker, J. Gasser, A. Pich and E. de Rafael, unpublished work reported by G. Ecker in Ref. [8].

[83] F. Binon et al., Nuovo Cim. 71A (1982) 497; Sov. J. Nucl. Phys. 36 (1982) 391.

[84] T.P. Cheng, Phys. Rev. 162 (1967) 1734.

[85] L. Ametller, J. Bijnens, A. Bramon and F. Cornet, Phys. Lett. B276 (1992) 185.

[86] F.J. Gilman and M.B. Wise, Phys. Rev. D20 (1979) 2392.

[87] J.A. Cronin, Phys. Rev. 161 (1967) 1483.

[88] A. Pich and E. de Rafael, Nucl. Phys. B358 (1991) 311.

[89] G. Altarelli and L. Maiani, Phys. Lett. 52B (1974) 351.

[90] M.K. Gaillard and B.W. Lee, Phys. Rev. Lett. 33 (1974) 108.

[91] M.A. Shifman, A.I. Vainshtein and V.I. Zakharov, Nucl. Phys. B120 (1977) 316.

[92] A.J. Buras, M. Jamin, M.E. Lautenbacher and P.H. Weisz, Nucl. Phys. B370 (1992) 69.

[93] J. Kambor, J. Missimer and D. Wyler, Phys. Lett. B261 (1991) 496.

[94] A.A. Bel'kov, G. Bohm, D. Ebert and A.V. Lanyov, Phys. Lett. B220 (1989) 459.

[95] W.A. Bardeen, A.J. Buras and J.-M. Gérard, Phys. Lett. B180 (1986) 133; Nucl. Phys. B293 (1987) 787; Phys. Lett. B211 (1988) 343;
W.A. Bardeen, in Ref. [8].

[96] M. Neubert and B. Stech, Phys. Lett. B231 (1989) 477; Phys. Rev. D44 (1991) 775;
B. Stech, Mod. Phys. Lett. A6 (1991) 3113; Nucl. Phys. B(Proc. Suppl.)23A (1991) 409.

[97] G. Ecker, A. Pich and E. de Rafael, Nucl. Phys. B303 (1988) 665.

[98] J. Kambor, J. Missimer and D. Wyler, Nucl. Phys. B346 (1990) 17.

[99] G. Ecker, Geometrical aspects of the non–leptonic weak interactions of mesons, in Proc. of the IX. Int. Conference on the Problems of Quantum Field Theory, Dubna, April 1990, Ed. M.K. Volkov (Dubna, 1990).

[100] G. Esposito-Farèse, Z. Phys. C50 (1991) 255.

[101] G. Ecker, J. Kambor and D. Wyler, Resonances in the weak chiral Lagrangian, preprint CERN-TH. 6610/92, Aug. 1992.

[102] C. Bruno and J. Prades, Rare kaon decays in the $1/N_C$ expansion, Marseille preprint CPT-92/P.2795, July 1992.

[103] J. Bijnens, G. Ecker and A. Pich, Phys. Lett. B286 (1992) 341.

[104] S. Fajfer and J.-M. Gérard, Z. Phys. C42 (1989) 425.

[105] H.-Y. Cheng, Phys. Rev. D42 (1990) 72.

[106] G. Ecker, A. Pich and E. de Rafael, Phys. Lett. B237 (1990) 481.

[107] G. Isidori and A. Pugliese, Chiral weak Lagrangian for vector mesons and $K \to 3\pi$ decay amplitudes, Univ. di Roma 'La Sapienza' preprint n. 849, Dec. 1991.

[108] C. Alliegro et al., Phys. Rev. Lett. 68 (1992) 278.

[109] T.J. Devlin and J.O. Dickey, Rev. Mod. Phys. 51 (1979) 237.

[110] J. Kambor, J.F. Donoghue, B.R. Holstein, J. Missimer and D. Wyler, Phys. Rev. Lett. 68 (1992) 1818.

[111] G. D'Ambrosio and D. Espriu, Phys. Lett. B175 (1986) 237.

[112] J.L. Goity, Z. Phys. C34 (1987) 341.

[113] H. Burkhardt et al. (NA31 Coll.), Phys. Lett. B199 (1987) 139.

[114] A. Cohen, private communication and paper in preparation.

[115] G. Ecker, A. Pich and E. de Rafael, Phys. Lett. B189 (1987) 363.

[116] L. Cappiello and G. D'Ambrosio, Nuovo Cim. 99A (1988) 153.

[117] G.D. Barr et al. (NA31 Coll.), A measurement of the decay $K_L \to \pi^0\gamma\gamma$, preprint CERN-PPE 92-110, April 1992; Phys. Lett. B242 (1990) 523.

[118] V. Papadimitriou et al. (E731 Coll.), Phys. Rev. D44 (1991) 573.

[119] L.M. Sehgal, Phys. Rev. D38 (1988) 808; ibid. D41 (1990) 161;
T. Morozumi and H. Iwasaki, Prog. Theor. Phys. 82 (1989) 371;
J. Flynn and L. Randall, Phys. Lett. B216 (1989) 221;
M.K. Volkov, To the problem of $K_L \to \pi^0\gamma\gamma$ decay, Dubna preprint JINR Rapid Comm. 6 [45] - 90.

[120] L. Cappiello, G. D'Ambrosio and M. Miragliuolo, Corrections to $K \to \pi\gamma\gamma$ from $K \to 3\pi$, Univ. di Napoli preprint DSFT-T-92/10, May 1992.

[121] A. Cohen, G. Ecker and A. Pich, in preparation.

[122] G. Ecker, H. Neufeld and A. Pich, Phys. Lett. B278 (1992) 337.

[123] H. Dykstra, J.M. Flynn and L. Randall, Phys. Lett. B270 (1991) 45.

[124] R. Funck and J. Kambor, The decays $K_{L,S} \rightarrow \pi^0\pi^0\gamma\gamma$ and $K_L \rightarrow \pi^0\pi^0 l^+ l^-$ in the effective chiral Lagrangian approach, TU München preprint TUM-T31-24/92 (1992).

[125] S.L. Adler and W.A. Bardeen, Phys. Rev. 182 (1969) 1517.

[126] R.J. Abrams et al., Phys. Rev. Lett. 29 (1972) 1118;
V.N. Bolotov et al., Sov. J. Nucl. Phys. 45 (1987) 1023.

[127] G. Ecker, H. Neufeld and A. Pich, in preparation.

CP- AND T- VIOLATIONS IN THE STANDARD MODEL

Jean-Marc Gérard [1]

Institut de Physique Théorique
Université Catholique de Louvain
Chemin du Cyclotron 2
B-1348 Louvain-la-Neuve (Belgium)

1. Time-reversal

The Newton's law

$$m\frac{d^2\vec{x}}{dt^2} = \vec{F}(\vec{x})$$

(1)

is quadratic in the time variable and, consequently, invariant under Time-reversal

$$t \to -t$$

(2)

if a non-dissipative force $\vec{F}$ acts on a particle of mass m. In other words, if $\vec{x}(t)$ is a trajectory, so is $\vec{x}(-t)$. Such a microscopic reversibility seems to contradict the macroscopic "Arrow of Time" we have to face everyday.

In 1897 [1], Planck claimed to have a solution to this apparent contradiction by noticing that the Maxwell equations for electromagnetism are linear in time. However, Boltzmann immediately replied by showing that these equations are in fact T-invariant if the magnetic field and the current are also reversed. Later on, he strongly argued that macroscopic irreversibility is not in contradiction with classical mechanics but is in fact due to specific boundary conditions. In particular, the tendancy to disorder (the second law of thermodynamics) stems form the way initial states are prepared. But the fascinating story of discrete symmetries in the fundamental interactions started with Boltzmann's key observation about Maxwell equations.

In 1964 [2], a microscopic CP-violation is observed in the $K° - \bar{K}°$ system. The validity of the CPT-theorem in the Standard Model implies then real microscopic T-violations in both weak and strong fundamental interactions.

---

[1]Chercheur qualifié du Fonds National de la Recherche Scientifique.

Quantitative Particle Physics, Edited by
M. Lévy *et al.*, Plenum Press, New York, 1993

Table 1. Transformation properties for the Maxwell equations.

	P	T	C
t	+	-	+
$\vec{x}$	-	+	+
ρ	+	+	-
$\vec{j}$	-	-	-
ϕ	+	+	-
$\vec{A}$	-	-	-
$\vec{E}$	-	+	-
$\vec{B}$	+	-	-

2. Maxwell Equations with C, P, and T

As mentioned above time-reversal invariance was first expressed by Boltzmann in the classical frame of electromagnetism where charge conjugation and parity can also be defined. The Maxwell equations (in Heaviside-Lorentz units with $c = 1$) in the vacuum

$$
\begin{aligned}
\vec{\nabla} \cdot \vec{E} &= \rho \\
\vec{\nabla} \cdot \vec{B} &= 0 \\
\vec{\nabla} \times \vec{E} &= -\partial \vec{B}/\partial t \\
\vec{\nabla} \times \vec{B} &= \vec{j} + \partial \vec{E}/\partial t ,
\end{aligned}
\tag{3}
$$

with

$$
\begin{aligned}
\vec{E} &= -\vec{\nabla}\phi - \partial \vec{A}/\partial t \\
\vec{B} &= \vec{\nabla} \times \vec{A} ,
\end{aligned}
\tag{4}
$$

are indeed invariant under

a) PARITY : $\qquad \vec{x} \xrightarrow{P} -\vec{x}$
b) TIME REVERSAL : $\qquad t \xrightarrow{T} -t$
c) CHARGE CONJUGATION : $\rho \xrightarrow{C} -\rho$

if the electromagnetic fields transform as in Table 1. Consequently, the corresponding covariant Lagrangian

$$
\mathcal{L}(A_\mu) = -\frac{1}{4}F_{\mu\nu}F^{\mu\nu} + j_\mu A^\mu,
\tag{5}
$$

with the photon field strength tensor

$$
F_{\mu\nu} = \partial_\mu A_\nu - \partial_\nu A_\mu,
\tag{6}
$$

is also invariant under these discrete symmetries. Defining the four-vector $x^\mu = (t, \vec{x})$ such that $x^\mu = g^{\mu\nu}x_\nu$ with the metric

$$
g^{\mu\nu} = \begin{pmatrix} 1 & & & \\ & -1 & & \\ & & -1 & \\ & & & -1 \end{pmatrix},
\tag{7}
$$

we obtain the transformation laws listed in Table 2.

Table 2. Transformation properties for the covariant Maxwell equations.

	P	T	C
x^{μ}	x_{μ}	$-x_{\mu}$	x^{μ}
j^{μ}	j_{μ}	j_{μ}	$-j^{\mu}$
A^{μ}	A_{μ}	A_{μ}	$-A^{\mu}$
$F^{\mu\nu}$	$F_{\mu\nu}$	$-F_{\mu\nu}$	$-F^{\mu\nu}$

In the presence of matter, the Lagrangian for Quantum Electro-Dynamics (QED) becomes

$$\mathcal{L}_{QED}(A_{\mu}, \Psi) = -\frac{1}{4}F_{\mu\nu}F^{\mu\nu} + i\bar{\Psi}\gamma^{\mu}[\partial_{\mu} - iQA_{\mu}]\Psi - \bar{\Psi}m\Psi \tag{8}$$

where Ψ is a fermionic operator with

$$\bar{\Psi} \equiv \Psi^{\dagger}\gamma^{0}, \tag{9}$$

and Q and m are the electric charge and the mass of the fermion Ψ, respectively. The Dirac matrices γ^{μ} satisfy the following relations

$$\gamma^{\mu}\gamma^{\nu} + \gamma^{\nu}\gamma^{\mu} = 2g^{\mu\nu} \tag{10}$$

and

$$\gamma^{0}\gamma^{\mu\dagger}\gamma^{0} = \gamma^{\mu}. \tag{11}$$

In the standard Pauli-Dirac representation, useful when discussing the non-relativistic limit of the Dirac equation, we have

$$\gamma^{0} = \begin{pmatrix} 1 & 0 \\ 0 & -1 \end{pmatrix}$$

$$\vec{\gamma} = \begin{pmatrix} 0 & \vec{\sigma} \\ -\vec{\sigma} & 0 \end{pmatrix} \tag{12}$$

where $\sigma_{i}(i = 1, 2, 3)$ are the usual Pauli matrices.

$$\sigma_{1} = \begin{pmatrix} 0 & 1 \\ 1 & 0 \end{pmatrix} \; ; \; \sigma_{2} = \begin{pmatrix} 0 & -i \\ i & 0 \end{pmatrix} \; ; \; \sigma_{3} = \begin{pmatrix} 1 & 0 \\ 0 & -1 \end{pmatrix} \tag{13}$$

Other representations are obtained through a unitary transformation

$$\gamma' = U \gamma U^{\dagger} \tag{14}$$

with the new matter fields $\Psi' = U\Psi$.

All experimental information on electromagnetic processes are consistent with the invariance under P, T and C. We therefore define the P, T and C transformation laws for matter fields Ψ such that $\mathcal{L}_{QED}(A_{\mu}, \Psi)$ defined in Eq.(8) remains invariant.

3. Fermion Fields

The Lagrangian density for electromagnetic interactions is invariant under Parity

$$P \; \mathcal{L}_{QED}(t, \vec{x}) \; P^{-1} = \mathcal{L}_{QED}(t, -\vec{x}) \tag{15}$$

if the anticommuting matter fields transform like

$$\Psi_P(t,\vec{x}) \;=\; P\Psi(t,\vec{x})P^{-1} \;=\; U_P\Psi(t,-\vec{x})$$
$$\bar{\Psi}_P(t,\vec{x}) \;=\; P\bar{\Psi}(t,\vec{x})P^{-1} \;=\; \bar{\Psi}(t,-\vec{x})U_P^{-1}, \qquad (16)$$

with

$$U_P U_P^\dagger = 1 , \qquad (17)$$

and

$$U_P \gamma^\mu U_P^{-1} = \gamma_\mu . \qquad (18)$$

In the Pauli-Dirac representation, U_P is proportional to γ^0. From Eq.(14), we conclude that U_P is given by

$$U_P = e^{i\alpha_P}\gamma^0 \qquad (19)$$

in any representation of the Dirac matrices.

While the Parity operator P is unitary :

$$\langle P\alpha \mid P\beta \rangle \;=\; \langle \alpha \mid \beta \rangle, \qquad (20)$$

the Time-reversal operator T is antiunitary :

$$\langle T\alpha \mid T\beta \rangle \;=\; \langle \alpha \mid \beta \rangle^* \;=\; \langle \beta \mid \alpha \rangle . \qquad (21)$$

This feature arises from the fundamental commutation relation in quantum mechanics

$$[\vec{x},\vec{p}] = i\hbar , \qquad (22)$$

which requires

$$T \, i \, T^{-1} \;=\; -i . \qquad (23)$$

In particular, a c-number transforms like

$$T \, c \, T^{-1} \;=\; c^* . \qquad (24)$$

It should be noted that T does not transform "bra" into "ket" states, but rather transform a "bra" ("ket") into its complex conjugate :

$$T \mid \alpha \rangle \;=\; U(K \mid \alpha \rangle) \;=\; U \mid \alpha \rangle^*$$
$$\langle \alpha \mid T^\dagger \;=\; (\langle \alpha \mid K)U^\dagger \;=\; \langle \alpha \mid {}^* U^\dagger \qquad (25)$$

with U, a unitary matrix. Treating the Dirac matrices as c-numbers, we obtain then

$$T \; \mathcal{L}_{QED}(t,\vec{x}) \; T^{-1} = \mathcal{L}_{QED}(-t,\vec{x}) \qquad (26)$$

if

$$\Psi_T(t,\vec{x}) \;=\; T\Psi(t,\vec{x})T^{-1} \;=\; U_T\Psi(-t,\vec{x})$$
$$\bar{\Psi}_T(t,\vec{x}) \;=\; T\bar{\Psi}(t,\vec{x})T^{-1} \;=\; \bar{\Psi}(-t,\vec{x})U_T^{-1} \qquad (27)$$

with

$$U_T U_T^\dagger = 1 \qquad (28)$$

and

$$U_T \gamma^\mu U_T^{-1} = +(\gamma^\mu)^t , \qquad (29)$$

t standing for transposed. In the Pauli-Dirac representation, we have

$$U_T = e^{i\alpha_T}\gamma^1\gamma^3 = -U_T^t \tag{30}$$

Finally, the QED Lagrangian is also invariant under Charge conjugation :

$$C \ \mathcal{L}_{QED}(t,\vec{x}) \ C^{-1} = \mathcal{L}_{QED}(t,\vec{x}) \tag{31}$$

if

$$\begin{aligned}
\Psi_C(t,\vec{x}) &= C\Psi(t,\vec{x})C^{-1} = U_C\bar{\Psi}^t(t,\vec{x}) \\
\bar{\Psi}_C(t,\vec{x}) &= C\bar{\Psi}(t,\vec{x})C^{-1} = -\Psi^t(t,\vec{x})U_C^{-1}
\end{aligned} \tag{32}$$

with

$$U_C U_C^\dagger = 1 \tag{33}$$

and

$$U_C^{-1}\gamma^\mu U_C = -(\gamma^\mu)^t \tag{34}$$

In the Pauli-Dirac representation, we have

$$U_C = e^{i\alpha_C}\gamma^0\gamma^2 = -U_C^t \tag{35}$$

and consequently,

$$U_C^t = -U_C \tag{36}$$

in any representation since $U_C' = UU_C U^t$.

4. Fermion Bilinears

Let us consider the 16 Dirac matrices which form a complete basis for Clifford algebra

$$\mathbf{1}, \ \gamma^5, \ \gamma^\mu, \ \gamma^\mu\gamma^5, \ \frac{i}{2}[\gamma^\mu,\gamma^\nu] = \sigma^{\mu\nu} \tag{37}$$

with

$$\gamma^5 = i\gamma^0\gamma^1\gamma^2\gamma^3 = (\gamma^5)^\dagger \tag{38}$$

such that the corresponding bilinear forms $(\bar{\psi},\psi)$ denoted according to their tensor character:

$$\begin{aligned}
s_{12}(x) &= \quad : \bar{\Psi}_1(x)\Psi_2(x) : \\
p_{12}(x) &= \quad : \bar{\Psi}_1(x)i\gamma^5\Psi_2(x) : \\
v_{12}^\mu(x) &= \quad : \bar{\Psi}_1(x)\gamma^\mu\Psi_2(x) : \\
a_{12}^\nu(x) &= \quad : \bar{\Psi}_1(x)\gamma^\mu\gamma^5\Psi_2(x) : \\
t_{12}^{\mu\nu}(x) &= \quad : \bar{\Psi}_1(x)\sigma^{\mu\nu}\Psi_2(x) :
\end{aligned} \tag{39}$$

are Hermitian. Keeping in mind that the fermion fields anticommute each other, and using the fact that :

$$\begin{aligned}
U_P \ \gamma^5 \ U_P^{-1} &= -\gamma^5 \\
U_T \ \gamma^5 \ U_T^{-1} &= (\gamma^5)^t \\
U_C^{-1} \ \gamma^5 \ U_C &= (\gamma^5)^t
\end{aligned} \tag{40}$$

we obtain Table 3.

Table 3. Action of the P,T and C operators on the bilinears

	$s_{12}(x^\rho)$	$p_{12}(x^\rho)$	$v^\mu_{12}(x^\rho)$	$a^\mu_{12}(x^\rho)$	$t^{\mu\nu}_{12}(x^\rho)$
P	$s_{12}(x_\rho)$	$-p_{12}(x_\rho)$	$v^{12}_\mu(x_\rho)$	$-a^{12}_\mu(x_\rho)$	$t^{12}_{\mu\nu}(x_\rho)$
T	$s_{12}(-x_\rho)$	$-p_{12}(-x_\rho)$	$v^{12}_\mu(-x_\rho)$	$a^{12}_\mu(-x_\rho)$	$-t^{12}_{\mu\nu}(-x_\rho)$
C	$s_{21}(x^\rho)$	$p_{21}(x^\rho)$	$-v^\mu_{21}(x^\rho)$	$a^\mu_{21}(x^\rho)$	$-t^{\mu\nu}_{21}(x^\rho)$

For illustration,

$$\frac{1}{2}\bar{\Psi}\sigma^{\mu\nu}F_{\mu\nu}\Psi \tag{41}$$

is invariant under P,T and C, while

$$\frac{1}{2}\bar{\Psi}i\gamma^5\sigma^{\mu\nu}F_{\mu\nu}\Psi \tag{42}$$

changes its sign under P and T (see Table 3). In the non-relativistic limit, these operators give respectively rise to the magnetic moment $\vec{\sigma}.\vec{B}$ and the electric dipole moment $\vec{\sigma}.\vec{E}$. Note that weak as well as strong interactions induce in principle an electric dipole moment as we will see later. In order to derive Table 3, we have assumed $\alpha^1_{C,P,T} = \alpha^2_{C,P,T}$. We will justify this phase convention in the framework of the Standard Model.

5. The Weak Gauge Interactions for quarks

The gauge sector of the standard electroweak Lagrangian [3]

$$\mathcal{L}_{gauge} = i(\bar{u}'_L, \bar{d}'_L)_j \; \gamma^\mu \; [\partial_\mu - ig\frac{\vec{\sigma}}{2}\vec{W}_\mu - i\frac{g'}{6}B_\mu] \begin{pmatrix} u'_L \\ d'_L \end{pmatrix}_j$$

$$+i(\bar{u}'_R)_j \; \gamma^\mu \; [\partial_\mu - i\frac{2g'}{3}B_\mu] \; (u'_R)_j$$

$$+i(\bar{d}'_R)_j \; \gamma^\mu \; [\partial_\mu + i\frac{g'}{3}B_\mu] \; (d'_R)_j \tag{43}$$

with j, the family index, u' and d', the weak eigenstates and

$$q_{R(L)} = \frac{1}{2}(1 \pm \gamma_5)q \,, \tag{44}$$

the right (left)-handed quark projections, is invariant under local $SU(2)_L \times U(1)$ transformations. The sector describing the couplings of quarks to the neutral gauge bosons:

$$Z_\mu \equiv W^3_\mu \cos\theta_W - B_\mu \sin\theta_W \quad \text{and} \quad A_\mu \equiv B_\mu \cos\theta_W + W^3_\mu \sin\theta_W \tag{45}$$

is manifestly invariant under the CP transformations :

$$\begin{aligned} B_\mu(t,\vec{x}) &\overset{CP}{\rightarrow} - B^\mu(t,-\vec{x}) \\ W^3_\mu(t,\vec{x}) &\overset{CP}{\rightarrow} - W^{3\mu}(t,-\vec{x}) \end{aligned} \tag{46}$$

if

$$(u'_L)_i \quad \overset{CP}{\rightarrow} \quad - (U^u_L)_{ij} \, U_C \, (u'_L)^*_j$$

$$(u'_R)_i \quad \overset{CP}{\rightarrow} \quad - (U^u_R)_{ij} \, U_C \, (u'_R)^*_j$$

$$(d'_L)_i \quad \overset{CP}{\rightarrow} \quad - (U^d_L)_{ij} \, U_C \, (d'_L)^*_j$$

$$(d'_R)_i \quad \overset{CP}{\rightarrow} \quad - (U^d_R)_{ij} \, U_C \, (d'_R)^*_j \tag{47}$$

with $U^{u(d)}_{L(R)}$, arbitrary unitary matrices acting in the weak flavor space.

On the other hand, the gauge interaction between the $W^\pm$ bosons

$$W^\pm_\mu = \frac{W^1_\mu \pm i W^2_\mu}{\sqrt{2}} \tag{48}$$

and the parity-violating charged current J^μ

$$J^\mu = \sum_{i=1}^{3} \bar{u}'_i \gamma^\mu (1 - \gamma^5) d'_i$$

$$\equiv 2 \bar{u}'_L \gamma^\mu d'_L \tag{49}$$

is given by

$$\mathcal{L}_{\text{charged current}} = \frac{g}{2\sqrt{2}} J^\mu W^+_\mu + h.c. \tag{50}$$

Under CP, $W^\pm_\mu$ transform like

$$W^\pm_\mu(t, \vec{x}) \quad \overset{CP}{\rightarrow} \quad - W^{\mu\mp}(t, -\vec{x}) \tag{51}$$

such that the gauge sector of the Lagrangian is also CP-invariant if we restrict the CP-transformations given in Eq. (47) :

$$U^u_L = U^d_L \equiv U_L \tag{52}$$

6. The Yukawa Interactions for quarks

The quarks with same electric charge mix among themselves through mass terms in the Yukawa sector :

$$\mathcal{L}_{\text{mass}} = \sum_{i,j=1}^{3} \{ \bar{u}'_i M^u_{ij} \left(\frac{1+\gamma^5}{2} \right) u'_j + \bar{u}'_i M^{u\dagger}_{ij} \left(\frac{1-\gamma^5}{2} \right) u'_j \} + (u \leftrightarrow d)$$

$$\equiv \bar{u}'_L M^u u'_R + \bar{u}'_R M^{u\dagger} u'_L + (u \leftrightarrow d) \tag{53}$$

The crucial observation is that gauge-invariance does not restrict the form of the quark mass matrices $M^{u,d}$. These matrices are arbitrary

$$M \neq M^t$$
$$M \neq M^\dagger$$
$$M \neq M^*$$

and imply therefore $C-, P-$ and $T-$ violations, respectively (see Table 3).

The polar decomposition theorem allows us to write the two 3×3 quark mass matrices as follows :

$$M^{u(d)} \;=\; H^{u(d)} \, X_R^{u(d)\dagger} \quad ; \quad H = H^\dagger$$
$$X_R X_R^\dagger = 1 \,. \tag{54}$$

By applying the following unitary transformations to the right-handed quark fields :

$$u_R' \to X_R^u \; u_R'$$
$$d_R' \to X_R^d \; d_R' \,, \tag{55}$$

the mass matrices become Hermitian. In this basis, the electroweak Lagrangian for quarks given by Eqs.(43) and (53) remains therefore invariant under the following (restricted) CP transformations

$$(u_{L,R}')_i \;\; \overset{CP}{\to} \;\; - \; U_{ij} U_C \; (u_{L,R}')_j^*$$
$$(d_{L,R}')_i \;\; \overset{CP}{\to} \;\; - \; U_{ij} U_C \; (d_{L,R}')_j^* \tag{56}$$

only if

$$U^\dagger H^u \; U \;=\; (H^u)^t$$
$$U^\dagger H^d \; U \;=\; (H^d)^t \,, \tag{57}$$

with U, an arbitrary unitary matrix. From Eq.(57), we conclude that CP-invariance requires

$$U^\dagger \, [H_u, H_d] \, U = -[H_u, H_d]^t. \tag{58}$$

Defining the Hermitian commutator

$$C \equiv i \, [H^u, H^d], \tag{59}$$

we obtain the non trivial relations

$$Tr \; C^{2n+1} \;= 0 \,, \tag{60}$$

with n, an integer. From the Cayley-Hamilton theorem applied to the Hermitian and traceless matrix C, we conclude that the condition of CP-invariance is automatically satisfied for two generations of quarks. On the other hand, for three generations, we get the unique constraint [4]

$$Tr \; C^3 \;=\; 3 \; det C \;=\; 0 \tag{61}$$

to have CP-conservation in the standard electroweak model.

The meaning of this condition can easily be understood in the physical basis defined by the quark mass eigenstates

$$(u_{L,R})_i \;=\; (V_u)_{ij} \; (u_{L,R}')_j$$
$$(d_{L,R})_i \;=\; (V_d)_{ij} \; (d_{L,R}')_j, \tag{62}$$

with

$$D_u \equiv \begin{pmatrix} m_u & & \\ & m_c & \\ & & m_t \end{pmatrix} = V_u \, H_u \, V_u^\dagger,$$

$$D_d \equiv \begin{pmatrix} m_d & & \\ & m_s & \\ & & m_b \end{pmatrix} = V_d \, H_d \, V_d^\dagger. \tag{63}$$

If $V_u \neq V_d$, the field redefinition given in Eq.(62) is obviously inconsistent with the CP transformations defined in Eq.(56) and we obtain the relation

$$\begin{aligned} detC &= -i \; det[D_u, V D_d V^\dagger] \\ &= 2(m_t - m_c)(m_t - m_u)(m_c - m_u)(m_b - m_s) \\ &\quad \times (m_b - m_d)(m_s - m_d) Im(V_{11} V_{21}^* V_{22} V_{12}^*) \end{aligned} \tag{64}$$

with V, the unitary matrix defined by

$$V \equiv V_u \; V_d^\dagger. \tag{65}$$

In conclusion, the Yukawa sector of the standard model with three generations of quarks produces CP-violation if and only if

a) $m_{u_i} \neq m_{u_j}$, $m_{d_i} \neq m_{d_j}$, $(i \neq j)$

b) $J \equiv Im(V_{11} \, V_{21}^* \, V_{22} \, V_{12}^*) \neq 0. \tag{66}$

7. The Cabibbo-Kobayashi-Maskawa Matrix

A more intuitive derivation is possible if one works from the beginning in the physical basis (see Eq.(62)) for the quark fields. Then, the up and down quark mass matrices given in Eq.(63) are diagonal, real and non degenerated and the Yukawa sector of the standard electroweak model is invariant under the following (generalized) CP-transformations

$$\begin{aligned} (u_{L,R})_j & \overset{CP}{\to} - e^{i\alpha_j} U_C \; (u_{L,R})_j^* \\ (d_{L,R})_j & \overset{CP}{\to} - e^{i\beta_j} U_C \; (d_{L,R})_j^* \end{aligned} \tag{67}$$

However, the charged current (c.c.) interaction reads now [5]

$$\mathcal{L}_{cc} = \frac{g}{\sqrt{2}}(\bar{u}_L V \gamma^\mu d_L W_\mu^+ + \bar{d}_L V^\dagger \gamma^\mu u_L W_\mu^-) \tag{68}$$

with V, the unitary matrix defined in Eq.(65).

For n generations of quarks, this $n \times n$ matrix V contains $n(n-1)/2$ Euler-type of angles, and consequently $n(n+1)/2$ phases. Using the phase freedom in the CP transformations given in Eq.(67),one can always remove $2n - 1$ of these phases, leaving the Yukawa sector invariant under the (restricted) <u>flavor-blind</u> CP transformations

$$(u_{L,R})_j \xrightarrow{CP} - U_C (u_{L,R})_j^*$$
$$(d_{L,R})_j \xrightarrow{CP} - U_C (d_{L,R})_j^* \tag{69}$$

This choice corresponds to the Kobayashi-Maskawa phase convention which we will adopt in this book. Within this convention, the action of the P, T and C operators on the bilinears is given in Table 3.

Under these CP transformations, we have

$$\bar{u}_L V \gamma^\mu d_L W_\mu^+ \xrightarrow{CP} \bar{d}_L V^t \gamma^\mu u_L W_\mu^- \neq \bar{d}_L V^\dagger \gamma^\mu u_L W_\mu^- \tag{70}$$

Therefore the $(n-1)(n-2)/2$ remaining phases of V are CP-violating since they imply

$$V \neq V^* \tag{71}$$

In particular, we recover the results (see Eq.(60)) that there is no possible CP-violation for two generations ($n = 2$) and a unique CP-violating phase determined by

$$\pm \ Im \ (V_{ij} \ V_{kj}^* \ V_{kl} \ V_{il}^*) \qquad \text{(no summation over } i, j, k, l) \tag{72}$$

for three generations ($n = 3$). Indeed, the unitarity of V :

$$\sum_j V_{ij} \ V_{kj}^* = \delta_{ik} \tag{73}$$

implies

$$\begin{aligned} Im \ V_{11} V_{21}^* V_{22} V_{12}^* &= -Im V_{23} V_{33}^* V_{32} V_{22}^* \\ &= +Im V_{11} V_{31}^* V_{33} V_{13}^* \ \ldots \end{aligned} \tag{74}$$

and ensures the quark field phase convention independence.

8. Parametrizations of the CKM Matrix

There are several parametrizations for the Cabibbo-Kobayashi-Maskawa (CKM) mixing matrix V. The standard one is given by [6]

$$V = \begin{pmatrix} V_{ud} & V_{us} & V_{ub} \\ V_{cd} & V_{cs} & V_{cb} \\ V_{td} & V_{ts} & V_{tb} \end{pmatrix}$$

$$= \begin{pmatrix} 1 & 0 & 0 \\ 0 & c_{23} & s_{23} \\ 0 & -s_{23} & c_{23} \end{pmatrix} \cdot \begin{pmatrix} c_{13} & 0 & s_{13}e^{-i\delta_{13}} \\ 0 & 1 & 0 \\ -s_{13}e^{i\delta_{13}} & 0 & c_{13} \end{pmatrix} \cdot \begin{pmatrix} c_{12} & s_{12} & 0 \\ -s_{12} & c_{12} & 0 \\ 0 & 0 & 1 \end{pmatrix}$$

$$= \begin{pmatrix} c_{12}c_{13} & s_{12}c_{13} & s_{13}e^{-i\delta_{13}} \\ -s_{12}c_{23} - c_{12}s_{23}s_{13}e^{i\delta_{13}} & c_{12}c_{23} - s_{12}s_{23}s_{13}e^{i\delta_{13}} & s_{23}c_{13} \\ s_{12}s_{23} - c_{12}c_{23}s_{13}e^{i\delta_{13}} & -c_{12}s_{23} - s_{12}c_{23}s_{13}e^{i\delta_{13}} & c_{23}c_{13} \end{pmatrix} \tag{75}$$

with three angles $\theta_{12}, \theta_{13}, \theta_{23}$ and one phase δ_{13}. The abreviations s_{ij} and c_{ij} denote $\sin \theta_{ij}$, and $\cos \theta_{ij}$ and we obtain

$$J = c_{12}\, c_{13}^2\, c_{23}\, s_{12}s_{13}s_{23}\, \sin \delta_{13} \tag{76}$$

If we define

$$\lambda = s_{12}$$
$$A = s_{23}/s_{12}^2$$
$$\rho = s_{13} \cos \delta_{13}/s_{12}s_{23}$$
$$\eta = s_{13} \sin \delta_{13}/s_{12}s_{23} \tag{77}$$

with $\lambda \sim 0.22$, the following approximate parametrization is sometime useful [7]

$$V = \begin{pmatrix} 1 - \lambda^2/2 & \lambda & A\lambda^3(\rho - i\eta) \\ -\lambda & 1 - \lambda^2/2 & A\lambda^2 \\ A\lambda^3(1 - \rho - i\eta) & -A\lambda^2 & 1 \end{pmatrix} + \mathcal{O}(\lambda^4) \tag{78}$$

and

$$J \simeq s_{12}^2\, s_{23}^2\, \eta. \tag{79}$$

9. Relations between Quark and Hadron Fields

Physical amplitudes involve hadron bound states. Within the Kobayashi-Maskawa phase convention, the quark field phase freedom has been used to remove unphysical phases in V. On the other hand, hadrons are formed via strong QCD interactions which are $C-, P-$ and $T-$ conserving. Consequently the quark bound states simply transform like the corresponding bilinears under P, T and C (see Table 3). For illustration, the strange pseudoscalars K° and $\bar{K}^\circ$ transform like $\bar{s}i\gamma_5 d$ and $\bar{d}i\gamma_5 s$, respectively:

$$\begin{aligned} P\,|K^\circ(\vec{p})\rangle &= -|K^\circ(-\vec{p})\rangle \\ T\,|K^\circ(\vec{p})\rangle &= -|K^\circ(-\vec{p})\rangle \\ C\,|K^\circ(\vec{p})\rangle &= |\bar{K}^\circ(\vec{p})\rangle\,. \end{aligned} \tag{80}$$

The generalization to the pseudoscalar nonet

$$\pi = \lambda_a \pi^a \quad (a = 0...8)$$

$$= \sqrt{2} \begin{pmatrix} \frac{\pi_3}{\sqrt{2}} + \frac{\eta_8}{\sqrt{6}} + \frac{\eta_0}{\sqrt{3}} & \pi^+ & K^+ \\ \pi^- & -\frac{\pi_3}{\sqrt{2}} + \frac{\eta_8}{\sqrt{6}} + \frac{\eta_0}{\sqrt{3}} & K^\circ \\ K^- & \bar{K}^\circ & -2\frac{\eta_8}{\sqrt{6}} + \frac{\eta_0}{\sqrt{3}} \end{pmatrix} \tag{81}$$

is straightforward

$$\begin{aligned} \pi(x^\rho) &\xrightarrow{P} -\pi(x_\rho) \\ \pi(x^\rho) &\xrightarrow{T} -\pi(-x_\rho) \\ \pi(x^\rho) &\xrightarrow{C} +\pi^t(x^\rho)\,. \end{aligned} \tag{82}$$

10. The CPT Theorem for transition amplitudes

We have seen that the action

$$I_{QED} = \int d^4x \mathcal{L}_{QED}(x) \tag{83}$$

for electromagnetic interactions with matter fields is invariant under P, T and C transformations. This holds true for strong interactions with quark fields. The <u>classical</u> QCD Lagrangian density has indeed a similar structure

$$\mathcal{L}_{QCD}(G_\mu, q) = -\frac{1}{4}G^a_{\mu\nu}G^{\mu\nu}_a + i\bar{q}\gamma^\mu[\partial_\mu - ig_s\frac{\lambda_a}{2}G^a_\mu]q - \bar{q}mq, \tag{84}$$

with $\lambda_a(a = 1,...8)$, the conventional Gell-Mann matrices. We obtain therefore the same transformation laws

$$
\begin{aligned}
P\ \mathcal{L}_{QCD}(t,\vec{x})\ P^{-1} &= \mathcal{L}_{QCD}(t,-\vec{x}) \\
T\ \mathcal{L}_{QCD}(t,\vec{x})\ T^{-1} &= \mathcal{L}_{QCD}(-t,\vec{x}) \\
C\ \mathcal{L}_{QCD}(t,\vec{x})\ C^{-1} &= \mathcal{L}_{QCD}(t,\vec{x})
\end{aligned} \tag{85}
$$

On the other hand, weak interactions break parity, time-reversal and charge-conjugation. We have seen that the CP (and T) invariance of the standard electroweak Lagrangian is only broken by the charged current (c.c.) gauge sector

$$\mathcal{L}_{c.c.} = \frac{g}{\sqrt{2}}\ (\bar{u}_L V\gamma^\mu d_L W^+_\mu + \ \bar{d}_L V^\dagger\gamma^\mu u_L W^-_\mu) \tag{86}$$

in the physical basis for quarks. The CP and T transformations on this sector of the electroweak Lagrangian are

$$(CP)\ \mathcal{L}_{c.c.}\ (CP)^{-1} = \frac{g}{\sqrt{2}}\ (\bar{d}_L V^\dagger\gamma^\mu u_L W^-_\mu + \ \bar{u}_L V^*\gamma^\mu d_L W^+_\mu) \tag{87}$$

and

$$T\ \mathcal{L}_{c.c.}\ T^{-1} = \frac{g}{\sqrt{2}}\ (\bar{u}_L V^*\gamma^\mu d_L W^-_\mu + \ \bar{d}_L V^\dagger\gamma^\mu u_L W^-_\mu) \tag{88}$$

can be interpreted as acting only on the CKM mixing matrix in the following way

$$
\begin{aligned}
V &\xrightarrow{CP} V^* \\
V &\xrightarrow{T} V^*.
\end{aligned} \tag{89}
$$

From this point of view, we conclude that the CP- and T- violating CKM mixing matrix is invariant under the combined CPT transformation. Consequently, the electroweak Lagrangian density also behaves as a neutral scalar field

$$\mathcal{L}_{EW}(x) \xrightarrow{CPT} \mathcal{L}_{EW}(-x) \tag{90}$$

under CPT. The action I corresponding to the three fundamental gauge interactions is therefore invariant under CPT.

In fact, the famous CPT theorem states that this holds true for any local quantum field theory [8]. The proof for a Lagrangian quantum field theory

$$\mathcal{L}\{\phi, A^\mu, \psi, \bar{\psi}\} \tag{91}$$

involving scalar (ϕ), vector (A^μ) and spinor (ψ) fields is based on Table 4.
In Table 4, the Lorentz indices of the bilinear form $\bar{\psi}\Gamma_{\mu_1....\mu_n}\psi$ are carried by either

Table 4. The CPT transformations for a Lagrangian quantum field theory.

	c	∂_μ	$\phi(x)$	$A^\mu(x)$	$\bar{\psi}\Gamma_{\mu_1\ldots\mu_n}\psi(x)$
CPT	c^*	$-\partial_\mu$	$\phi^\dagger(-x)$	$-A^\mu(-x)$	$(-1)^n\bar{\psi}\Gamma_{\mu_1\ldots\mu_n}\psi(-x)$

Dirac matrices (γ_μ) or derivatives (∂_μ). The induced $(-1)^n$ sign is derived from Table 3. Lorentz invariance requires $\mathcal{L}$ to be a scalar density transforming as

$$\mathcal{L}(x) \overset{CPT}{\rightarrow} \mathcal{L}^\dagger(-x) . \tag{92}$$

The Hermiticity condition leads then to the CPT theorem

$$\mathcal{L}(x) \overset{CPT}{\rightarrow} \mathcal{L}(-x) . \tag{93}$$

such that the action I is invariant.

The CPT theorem implies the existence of antiparticles $(\pi^-, \bar{K}^\circ, \bar{p}, ...)$ for charged fields $(\pi^+, K^\circ, p, ...)$:

$$|CPT\alpha\rangle \equiv CPT|\alpha\rangle \equiv |\bar{\alpha}\rangle \tag{94}$$

with the equality of masses :

$$
\begin{aligned}
m_\alpha = \langle\alpha|H_m|\alpha\rangle &= \langle\alpha|[(CPT)^{-1}(CPT)]H_m[(CPT)^{-1}(CPT)]|\alpha\rangle \\
&= [\langle CPT\alpha|H_m|CPT\alpha\rangle]^* \\
&= \langle\bar{\alpha}|H_m|\bar{\alpha}\rangle \\
&= m_{\bar{\alpha}} ,
\end{aligned}
\tag{95}
$$

and the equality of life-times :

$$
\begin{aligned}
\tau(\alpha) = [\Gamma(\alpha)]^{-1} &= \frac{1}{2\pi}\left[\sum_f \delta(m_\alpha - E_f)\langle\alpha|H|f\rangle\langle f|H|\alpha\rangle\right]^{-1} \\
&= \frac{1}{2\pi}\left[\sum_f \delta(m_{\bar{\alpha}} - E_{\bar{f}})\langle\bar{\alpha}|H|\bar{f}\rangle\langle\bar{f}|H|\bar{\alpha}\rangle\right]^{-1} \\
&= \frac{1}{2\pi}\left[\sum_f \delta(m_{\bar{\alpha}} - E_f)\langle\bar{\alpha}|H|f\rangle\langle f|H|\bar{\alpha}\rangle\right]^{-1} \\
&= \tau(\bar{\alpha})
\end{aligned}
\tag{96}
$$

The last equality is based on the completeness of the final states

$$\sum_f |CPTf\rangle\langle CPTf| = \sum_f |f\rangle\langle f|. \tag{97}$$

However the CPT-theorem has also an implication at the level of partial weak decays [9] if the subset of final states $|F_{out}\rangle$ considered is complete with respect to factorized strong interaction rescatterings :

$$|F_{out}\rangle = S_{strong}|F_{in}\rangle. \tag{98}$$

In the first order in the weak interactions, we have indeed the transition amplitude

$$
\begin{aligned}
\langle F_{out}|H_w|\alpha\rangle &= \langle F_{out}|T^{-1}TH_wT^{-1}T|\alpha\rangle \\
&= \langle TF_{out}|TH_wT^{-1}|T\alpha\rangle^* \\
&= \langle TF_{out}|(CP)^{-1}CPTH_w(CPT)^{-1}CP|T\alpha\rangle^* \\
&= \langle \bar{F}_{in}|H_w|\bar{\alpha}\rangle^*.
\end{aligned}
\tag{99}
$$

If strong rescatterings are factorizable, we obtain then

$$
\langle F_{out}|H_w|\alpha\rangle = S_{strong}\langle \bar{F}_{out}|H_w|\bar{\alpha}\rangle^*.
\tag{100}
$$

Using the unitarity of the S-matrix

$$
S^\dagger S = 1,
\tag{101}
$$

we conclude that the partial decay widths of a particule α and of its antiparticle are equal

$$
\Gamma(\alpha \to F) = \Gamma(\bar{\alpha} \to \bar{F})
\tag{102}
$$

if F stands for any subgroup of states couple to one another via strong final-state interactions. The subset F is therefore characterized by some "good" quantum number such as isospin or charm which are conserved in strong interactions. This subtelty due to the CPT theorem implies a reduction of CP-asymmetries in hadronic $K-$ and charmless $B-$ decays, respectively.

11. Necessary and Sufficient Conditions for CP Violation

One way to detect CP-violation is to look for specific decays which are strictly forbidden in the CP-invariance limit. A typical example is

$$
\eta \to \pi^+\pi^- .
\tag{103}
$$

The pseudoscalar η has indeed a odd CP-eigenvalue

$$
CP \,|\, \eta\rangle = -\,|\, \eta\rangle ,
\tag{104}
$$

while the final two-pion state with

$$
CP \,|\pi^\pm\rangle = -\,|\, \pi^\mp\rangle
\tag{105}
$$

is CP-even. A CP-odd Hamiltonian with non-vanishing transition matrix element is then necessary to induce this decay. The observation of a single event is therefore sufficient to establish a CP-violation. Other processes are the decay sequences of CP-even vector states

$$
\phi \to K^\circ\bar{K}^\circ(P-\text{wave}) \to (2\pi)(2\pi)
\tag{106}
$$

or

$$
\Upsilon(4s) \to B^\circ\bar{B}^\circ(P-\text{wave}) \to (\psi K_s)(\psi K_s) .
\tag{107}
$$

The transition via a CP-odd Hamiltonian H_- is however not sufficient to obtain a CP-asymmetry in decay widths. Let us indeed consider a CP-violating decay amplitude purely induced by H_-

$$
\begin{aligned}
A &= \langle f|H_-|i\rangle \\
&= \langle f_+|H_-|i_-\rangle + \langle f_-|H_-|i_+\rangle ,
\end{aligned}
\tag{108}
$$

with

$$CP|i_\pm\rangle = \pm|i_\pm\rangle$$
$$CP|f_\pm\rangle = \pm|f_\pm\rangle \,. \tag{109}$$

In that case, the CP-conjugate decay amplitude reads

$$\bar{A} = \langle \bar{f}|H_-|\bar{i}\rangle$$
$$= -A \tag{110}$$

such that the decay width

$$\Gamma = |\langle f|H_-|i\rangle|^2 \tag{111}$$

is identical to the CP-conjugate one

$$\bar{\Gamma} = |\langle \bar{f}|H_-|\bar{i}\rangle|^2 \,. \tag{112}$$

On the other hand, if both CP-even H_+ and CP-odd H_- Hamiltonians contribute to the process $i \to f$, we obtain the decay width asymmetry

$$\Gamma - \bar{\Gamma} = 4 Re(\langle f|H_+|i\rangle\langle f|H_-|i\rangle^*) \,. \tag{113}$$

In the absence of CP-conserving dynamical phase, this asymmetry is still vanishing. Let us indeed assume the following form for a physical amplitude

$$A = A_1 \, e^{\imath\delta_1} + A_2 \, e^{\imath\delta_2} \tag{114}$$

with $A_{1,2}$, two complex partial weak amplitudes with CP-conserving dynamical phases $\delta_{1,2}$. (In the Standard Model, the complexity of $A_{1,2}$ arises from the CKM factors.) Then, the amplitude for the CP-conjugate process reads

$$\bar{A} = A_1^* \, e^{\imath\delta_1} + A_2^* \, e^{\imath\delta_2} \neq A^* \,. \tag{115}$$

Defining the CP-asymmetry in decay widths,

$$a \equiv \frac{\Gamma - \bar{\Gamma}}{\Gamma + \bar{\Gamma}} = \frac{|A|^2 - |\bar{A}|^2}{|A|^2 + |\bar{A}|^2} \,, \tag{116}$$

we obtain

$$a = \frac{-2Im\,(A_1 A_2^*)\,\sin(\delta_1 - \delta_2)}{|A_1|^2 + |A_2|^2 + 2Re(A_1 A_2^*)\cos(\delta_1 - \delta_2)} \,. \tag{117}$$

A non-zero CP-asymmetry requires therefore at least two partial amplitudes with

- a relative CKM CP-violating phase

- a relative dynamical CP-conserving phase.

We know three ways to fulfil the second non-trivial requirement in K and B physics.

a) *Final state interaction phase shift in decays*

Let us consider for illustration the weak $K^\circ \to \pi^+\pi^-$ transition amplitude

$$A \equiv \langle (\pi^+\pi^-)_{out}|H_w|K^\circ\rangle \tag{118}$$

The CPT-invariance of the weak $\Delta S = 1$ Hamiltonian H_w implies (see Eq.(99))

$$A = \langle (\pi^+\pi^-)_{in}|H_w|\bar{K}^\circ\rangle^* \tag{119}$$

Strong elastic rescattering between the two charged pions conserves isospin:

$$
\begin{aligned}
|(\pi^+\pi^-)_{out}^I\rangle &= S|(\pi^+\pi^-)_{in}^I\rangle \\
&= e^{2i\delta_I}|(\pi^+\pi^-)_{in}^I\rangle
\end{aligned}
\tag{120}
$$

where δ_I is the phase shift between the initial and final state with isospin $I(=0,2)$. We find then

$$\langle (\pi^+\pi^-)_{out}^I|H_w|K^\circ\rangle = e^{2i\delta_I}\langle (\pi^+\pi^-)_{out}^I|H_w|\bar{K}^\circ\rangle^* \tag{121}$$

or

$$
\begin{aligned}
\langle (\pi^+\pi^-)_{out}^I|H_w|K^\circ\rangle &= e^{i\delta_I}A_I \\
\langle (\pi^+\pi^-)_{out}^I|H_w|\bar{K}^\circ\rangle &= e^{i\delta_I}A_I^* .
\end{aligned}
\tag{122}
$$

Notice that CPT-invariance requires equal partial decay widths for a definite isospin amplitude. This is due to the fact that the two-pion state cannot communicate via strong interactions with the three-pion state. The total $K^\circ \to \pi^+\pi^-$ amplitude is the superposition of the two isospin amplitudes

$$A = \sqrt{\frac{2}{3}}A_0 e^{i\delta_0} + \sqrt{\frac{1}{3}}A_2 e^{i\delta_2} \tag{123}$$

while its CP-conjugate reads

$$\bar{A} = \sqrt{\frac{2}{3}}A_0^* e^{i\delta_0} + \sqrt{\frac{1}{3}}A_2^* e^{i\delta_2} . \tag{124}$$

A CP-asymmetry arises from the interference of the two isospin amplitudes. In this example, the CP-conserving phase shift is induced by a "soft" rescattering at the hadron level. On the other hand, "hard" rescatterings among quarks can also occur for $B-$meson rare decays, the $B-$mass being above the $1GeV$ confining scale [10].

b) *Oscillation of unstable neutral mesons*
The transition amplitudes associated with a meson-antimeson oscillation contain a dispersive (M) and an absorptive (Γ) piece

$$
\begin{aligned}
A(M^\circ \to \bar{M}^\circ) &\equiv M_{12} - i\,\Gamma_{12}/2 \\
\bar{A}(\bar{M}^\circ \to M^\circ) &= M_{12}^* - i\,\Gamma_{12}^*/2 ,
\end{aligned}
\tag{125}
$$

such that we obtain a CP-asymmetry proportional to

$$Im\,(M_{12}\Gamma_{12}^*)\sin\pi/2 . \tag{126}$$

Notice that the dynamical phases in final state interactions and flavour-oscillations arise from the absorptive part induced by on-shell intermediate states. In a Feynman diagram language, we have physical cuts in penguin- and box-loop diagrams, respectively. The third way to get a relative dynamical phase is not of this nature.

c) *Time-dependent decay*

For a neutral meson M° oscillating before it decays, the time-dependence is given by the T-invariant Schrödinger evolution operator e^{-iHt}. Therefore, the CP (or T)-conjugate amplitude is again different from the complex-conjugate one. The relative dynamical phase is then given by the mass difference ΔM and we obtain a time-dependent asymmetry proportional to [11]

$$\sin \Delta Mt \qquad (127)$$

12. The Strong T- Violation

At the classical level, the QCD Lagrangian for the light quarks ($q = u, d, s$) coupled to the gluon fields $G_\mu^a (a = 1....8)$

$$
\begin{aligned}
\mathcal{L}_{QCD}(u,d,s) &= i \sum_q \bar{q}\gamma^\mu [\partial_\mu - ig_s \frac{\lambda_a}{2} G_\mu^a]q \\
&\equiv i \sum_q [\bar{q}_L \gamma^\mu D_\mu q_L + \bar{q}_R \gamma^\mu D_\mu q_R]
\end{aligned} \qquad (128)
$$

with

$$Tr\lambda_a\lambda_b = 2\delta_{ab} \qquad (129)$$

and

$$[\lambda_a, \lambda_b] = 2if_{abc}\lambda_c , \qquad (130)$$

is invariant under the chiral $U(3)_L \times U(3)_R$ symmetry

$$
\begin{aligned}
q_L &\rightarrow g_L \ q_L &, \ g_L \in U(3)_L \\
q_R &\rightarrow g_R \ q_R &, \ g_R \in U(3)_R
\end{aligned} \qquad (131)
$$

if one neglects their mass compared to the confining scale of one GeV. Below this scale, the $\bar{q}q$ neutral pairs condensate

$$\langle \bar{q}_L q_R \rangle = \langle \bar{q}_R q_L \rangle \neq 0 , \qquad (132)$$

such that the $U(3)_L \times U(3)_R$ global symmetry is broken into the vectorial $U(3)_V$ subgroup. One would therefore expect nine Goldstone bosons from this spontaneous symmetry breaking. The fact that only eight light pseudoscalars are observed (the so-called $U(1)_A$ problem [12]) is now understood [13] in terms of an anomaly appearing in the divergence of the flavor-singlet axial current

$$J_\mu^5 = \sum_q \bar{q}\gamma_\mu\gamma^5 q . \qquad (133)$$

At the quantum level, this hadronic current is not conserved anymore in the chiral limit ($m_q \rightarrow 0$). One finds indeed [14] the anomalous Ward identity

$$\partial^\mu J_\mu^5 = n_F \frac{g_s^2}{16\pi^2} G_a^{\alpha\beta} \tilde{G}_{\alpha\beta}^a , \qquad (134)$$

with $G^{\alpha\beta}$, the gluon field strength tensor,

$$G_a^{\alpha\beta} = \partial^\alpha G_a^\beta - \partial^\beta G_a^\alpha + g_s f_{abc} G_b^\alpha G_c^\beta , \qquad (135)$$

$\tilde{G}_{\alpha\beta} \equiv \varepsilon_{\alpha\beta\gamma\delta} G^{\gamma\delta}/2$, its dual and n_F, the number of quark flavors involved. Notice that the right-handed side of Eq.(134) can be written as a total derivative

$$\frac{g_s^2}{16\pi^2} G^{\alpha\beta} \tilde{G}_{\alpha\beta} = \partial^\mu K_\mu \ , \tag{136}$$

with

$$K_\mu = \frac{g_s^2}{16\pi^2} \varepsilon_{\mu\nu\rho\sigma} G_a^\nu (G_a^{\rho\sigma} - \frac{g_s}{3} f_{abc} G_b^\rho G_c^\sigma) \ . \tag{137}$$

The effective strong interaction Lagrangian describing the low-energy remnants of both the spontaneous and explicit breakings of the chiral $U(3)_L \times U(3)_R$ symmetry is given by [15]

$$\mathcal{L}_{eff} = \frac{f^2}{8} Tr \partial_\mu U \partial^\mu U^\dagger - \frac{i}{4} (\partial^\mu K_\mu) Tr(lnU - lnU^\dagger) + \frac{N}{2m_0^2 f^2} (\partial^\mu K_\mu)^2 \ . \tag{138}$$

The first term in Eq. (138) with

$$U = \exp i \frac{\sqrt{2}\pi}{f} \tag{139}$$

corresponds to the interaction of the Goldstone boson nonet

$$\pi = \sqrt{2} \begin{pmatrix} \frac{\pi_3}{\sqrt{2}} + \frac{\eta_8}{\sqrt{6}} + \frac{\eta_0}{\sqrt{3}} & \pi^+ & K^+ \\ \pi^- & -\frac{\pi_3}{\sqrt{2}} + \frac{\eta_8}{\sqrt{6}} + \frac{\eta_0}{\sqrt{3}} & K^\circ \\ K^- & \bar{K}^\circ & -2\frac{\eta_8}{\sqrt{6}} + \frac{\eta_0}{\sqrt{3}} \end{pmatrix} \tag{140}$$

The second term in Eq.(138) is introduced to reproduce the anomalous Ward identity given in Eq.(134). Let us indeed construct the flavor-singlet axial current. Flavor currents are obtained by promoting the $g_{L(R)}$ chiral transformations to local ones

$$q_{L,R} \rightarrow g_{L,R}(x) q_{L,R} \tag{141}$$

with left-handed (right-handed) external gauge fields $L(R)^\mu$ transforming in the following way

$$L(R)^\mu \rightarrow \frac{i}{g} g_{L,R}(x) \partial^\mu g_{L,R}^\dagger(x) + g_{L,R}(x) L(R)^\mu g_{L,R}^\dagger(x) \tag{142}$$

At the quark level, we just substitute the usual derivative acting on $q_{L,R}$ (see Eq.(128)) by a covariant one

$$\partial^\mu \rightarrow \partial^\mu - igL(R)^\mu \tag{143}$$

to obtain

$$\begin{aligned} \mathcal{L}(q) &\ni \ g \ \bar{q}_L^a \gamma_\mu L_{ab}^\mu q_L^b + g \ \bar{q}_R^a \gamma_\mu R_{ab}^\mu q_R^b \\ &\equiv \frac{g}{2} \left\{ J_\mu^L L^\mu + J_\mu^R R^\mu \right\} \end{aligned} \tag{144}$$

At the meson level, the corresponding left- and right-handed currents are obtained via the minimal substitutions

$$\partial^\mu U \rightarrow \partial^\mu U - igL^\mu U \tag{145}$$

and

$$\partial^\mu U \rightarrow \partial^\mu U + igU R^\mu \ , \tag{146}$$

respectively. The $\mathcal{O}(p)$ contribution to the hadronic currents is then derived by identifying

$$\mathcal{L}(\text{meson}) \ni \frac{igf^2}{4}Tr\{\partial_\mu UU^\dagger L^\mu + \partial_\mu U^\dagger U R^\mu\} \tag{147}$$

with Eq.(144). We obtain

$$
\begin{aligned}
(J_\mu^L)^{ab} &\equiv \bar{q}^a\gamma_\mu(1-\gamma^5)q^b \\
&= i\frac{f^2}{2}(\partial_\mu UU^\dagger)^{ba} + \mathcal{O}(p^3) \\
(J_\mu^R)^{ab} &\equiv \bar{q}^a\gamma_\mu(1+\gamma^5)q^b \\
&= i\frac{f^2}{2}(\partial_\mu U^\dagger U)^{ba} + \mathcal{O}(p^3)
\end{aligned} \tag{148}
$$

such that the flavor-singlet axial current is given by

$$
\begin{aligned}
J_\mu^5 &\equiv \frac{1}{2}Tr\,(J_\mu^R - J_\mu^L) \\
&= i\frac{f^2}{4}Tr\,\{\partial_\mu U^\dagger U - \partial_\mu UU^\dagger\}
\end{aligned} \tag{149}
$$

with

$$\partial^\mu J_\mu^5 = \frac{if^2}{4}Tr(\Box U^\dagger U - \Box UU^\dagger)\,. \tag{150}$$

Using then the equations of motion for the U and $U^\dagger$ fields

$$
\begin{aligned}
\Box U &= +\frac{2i}{f^2}(\partial^\mu K_\mu)U \\
\Box U^\dagger &= -\frac{2i}{f^2}(\partial^\mu K_\mu)U^\dagger\,.
\end{aligned} \tag{151}
$$

obtained from the variational principle applied to $U^\dagger$ and U, respectively, we recover the anomalous Ward identity given in Eq.(134).

The last kinetic term in Eq.(138) can be eliminated by varying $\mathcal{L}_{eff}$ with respect to the K_μ ghost field

$$\partial^\mu K_\mu = i\frac{f^2m_0^2}{4N}Tr(lnU - lnU^\dagger)\,. \tag{152}$$

The effective Lagrangian becomes then simply

$$\mathcal{L}_{eff} = \frac{f^2}{8}\left\{Tr\partial_\mu U\partial^\mu U^\dagger + \frac{m_0^2}{4N}[Tr(lnU - lnU^\dagger)]^2\right\} \tag{153}$$

in full agreement with the $1/N$-expansion approach [16], N being the number of colors. Expanding the exponential field U up to the first order in π, we obtain

$$\mathcal{L}_{eff} \ni \frac{1}{4}Tr\partial_\mu\pi\partial^\mu\pi - \frac{m_0^2}{2}\eta_0^2\,, \tag{154}$$

which is the free Lagrangian for **eight** Goldstone bosons and one massive pseudoscalar.

To reproduce the observed mass spectrum, we must introduce a small breaking of the flavor $SU(3)_V$. In the QCD Lagrangian, this symmetry is broken via the non-degenerate quark mass matrix

$$\mathcal{L}_{QCD}^{\text{mass}} = \bar{q}_L m q_R + h.c.\,, \tag{155}$$

with

$$m = \begin{pmatrix} m_u & & \\ & m_d & \\ & & m_s \end{pmatrix} \tag{156}$$

Similarly, at the $\mathcal{O}(p^2)$ we have one effective mass term

$$\mathcal{L}_{eff}^{mass} = \frac{f^2}{8} r Tr(mU^\dagger + h.c.) \tag{157}$$

leading to the spectrum

$$m_{K^\pm}^2 = \frac{r}{2}(m_s + m_u)$$
$$m_{K^0}^2 = \frac{r}{2}(m_s + m_d)$$
$$m_{\pi^\pm}^2 = \frac{r}{2}(m_d + m_u) \tag{158}$$

for the charged pseudoscalars. The neutral sector involves mixings. In the (π_3, η_8, η_0) basis, we have

$$m_{neutral}^2 = \frac{r}{2} \begin{pmatrix} (m_u + m_d) & \frac{1}{\sqrt{3}}(m_u - m_d) & \sqrt{\frac{2}{3}}(m_u - m_d) \\ \frac{1}{\sqrt{3}}(m_u - m_d) & \frac{1}{3}(m_u + m_d + 4m_s) & \frac{\sqrt{2}}{3}(m_u + m_d - 2m_s) \\ \sqrt{\frac{2}{3}}(m_u - m_d) & \frac{\sqrt{2}}{3}(m_u + m_d - 2m_s) & \frac{2}{3}(m_u + m_d + m_s) \end{pmatrix}$$

$$+ m_0^2 \begin{pmatrix} 0 & 0 & 0 \\ 0 & 0 & 0 \\ 0 & 0 & 1 \end{pmatrix} \tag{159}$$

In the isospin limit $m_u = m_d$, we recover the Gell-Mann-Okubo relation

$$m_{\eta_8}^2 = \frac{4}{3}m_K^2 - \frac{1}{3}m_\pi^2 , \tag{160}$$

and the physical η and η' eigenstates

$$\eta = \eta_8 \cos\theta_P - \eta_0 \sin\theta_P$$
$$\eta' = \eta_8 \sin\theta_P + \eta_0 \cos\theta_P \tag{161}$$

are obtained from the diagonalization of the two-by-two $\eta_8 - \eta_0$ submatrix, with

$$tg\, 2\theta_P = 2\sqrt{2}\left\{1 - \frac{3m_0^2}{2(m_K^2 - m_\pi^2)}\right\}^{-1} . \tag{162}$$

For $m_0 = 0$, we obtain the ideal mixing $\theta_P \sim 35°$ and, consequently, the mass relation $m_{\eta'} = m_\pi$ at the source of the $U(1)_A$ problem. Using the trace condition

$$m_0^2 = m_{\eta'}^2 + m_\eta^2 - 2m_K^2 , \tag{163}$$

one finds that the observed mass of the η' pseudoscalar requires in fact a large value

$$m_0 \sim 0.85 \; Gev . \tag{164}$$

From Eq.(162), we conclude that

$$\theta_P \sim -20° \tag{165}$$

in good agreement with experimental data.

Notice that the identification of $\mathcal{L}_{QCD}^{\text{mass}}$ with the mass terms in the effective Lagrangian gives a useful relation between the (pseudo)-scalar densities and the U field

$$\bar{q}_R^a q_L^b = -\frac{f^2 r}{8}(U)^{ba} + \mathcal{O}(p^2)$$

$$\bar{q}_L^a q_R^b = -\frac{f^2 r}{8}(U^\dagger)^{ba} + \mathcal{O}(p^2) \tag{166}$$

The P, T and C transformations defined for the pseudoscalar bound-states (in Eq.(82)) are consistent with

$$\begin{aligned} U(x^\rho) &\xrightarrow{P} U^\dagger(x_\rho) \\ U(x^\rho) &\xrightarrow{T} U^\dagger(-x_\rho) \\ U(x^\rho) &\xrightarrow{C} U^t(x^\rho) \end{aligned} \tag{167}$$

We have seen how the anomalous Ward identity provides a solution to the $U(1)_A$ problem. This resolution introduces however a new puzzle for the standard model of electroweak and strong interactions. We have indeed assumed a diagonal, real quark mass matrix m. However, this mass matrix arises from the Yukawa sector of the $SU(2)_L \times U(1)$ invariant Lagrangian. Its diagonalization requires in general a CHIRAL redefinition of the right-handed fields (see Eq.(55)). But the global phase of the mass matrix

$$\theta = \arg \det m \tag{168}$$

cannot be rotated away since the effective Lagrangian given in Eq.(153) is not invariant under a flavor-singlet axial transformation. The chiral transformation needed to get a real mass matrix

$$\begin{aligned} q_L &\rightarrow q_L \\ q_R &\rightarrow e^{-i\theta/n_F} q_R \end{aligned} \tag{169}$$

is indeed equivalent to

$$U \rightarrow e^{+i\theta/3} U \tag{170}$$

in the case of three flavors. This phase redefinition of U induces a new interaction in the effective Lagrangian

$$\delta\mathcal{L}_{eff}^\theta = i\frac{f^2 m_0^2 \theta}{8N} Tr(\ln U - \ln U^\dagger) . \tag{171}$$

Extending the argument for an arbitrary number of quark flavors, we conclude that the θ−angle

$$\theta = \arg \det M^u M^d \tag{172}$$

induced by the electroweak sector implies the following modification of the QCD Lagrangian

$$\begin{aligned} \delta\mathcal{L}_{QCD}^\theta &= \frac{\theta}{2} \partial^\mu K_\mu \\ &= \theta \frac{g_s^2}{32\pi^2} G^{\alpha\beta}\tilde{G}_{\alpha\beta} . \end{aligned} \tag{173}$$

The gauge interaction $G^{\alpha\beta}\tilde{G}_{\alpha\beta}$ is the QED analog of $\vec{E}.\vec{B}$ and violates therefore P and T (see Table 1). Due to non-perturbative "instanton" effects, this total derivative term leads to $P-$ and $T-$ violations in physical processes.

This is most easily seen at the effective level where the corresponding tadpole interaction derived from Eq.(171)

$$\delta\mathcal{L}_{QCD}^{\theta} = -\frac{\sqrt{3}}{2}\frac{fm_0^2}{N}\theta\ \eta_0 \tag{174}$$

provides the pseudoscalar η_0 a vacuum expectation value. A pseudoscalar field redefinition

$$U \to \begin{pmatrix} e^{-i\varphi_u} & & \\ & e^{-i\varphi_d} & \\ & & e^{-i\varphi_s} \end{pmatrix} U \tag{175}$$

is therefore necessary to obtain the correct $P-, T-$ and $C-$ invariant vacuum defined by

$$\langle 0|\pi|0\rangle = 0 . \tag{176}$$

This is implemented by imposing the extremal conditions

$$\frac{\partial\mathcal{L}_{eff}(\pi,\varphi)}{\partial\varphi_a}\Big|_{\pi=o} = o \tag{177}$$

on the modified effection Lagrangian. We obtain three relations

$$rm_a \sin\varphi_a = \frac{m_0^2}{N}(\theta - \sum_{i=1}^{3}\varphi_i) \text{ (no summation over } a) \tag{178}$$

which eliminate the terms linear in the pseudoscalar fields. The $\theta-$induced strong Lagrangian for the light pseudoscalars is then given by

$$\begin{aligned}\delta\mathcal{L}_{eff}^{\theta} &= -i\frac{f^2}{8}\frac{m_0^2}{N}(\theta - \sum_i\varphi_i)Tr(U - U^{\dagger})|_{\text{no linear terms in } \pi} \\ &= -\frac{1}{6\sqrt{2}}\frac{r}{f}m_a\sin\varphi_a\{Tr\pi^3 - \frac{1}{10f^2}Tr\pi^5....\} ,\end{aligned} \tag{179}$$

with

$$\begin{aligned} P\ \delta\mathcal{L}_{eff}^{\theta}(x^{\rho})\ P^{-1} &= -\delta\mathcal{L}_{eff}^{\theta}(x_{\rho}) \\ T\ \delta\mathcal{L}_{eff}^{\theta}(x^{\rho})\ T^{-1} &= -\delta\mathcal{L}_{eff}^{\theta}(-x_{\rho}) \\ C\ \delta\mathcal{L}_{eff}^{\theta}(x^{\rho})\ C^{-1} &= +\delta\mathcal{L}_{eff}^{\theta}(x^{\rho}) . \end{aligned} \tag{180}$$

Assuming φ_a small, we obtain

$$m_a\varphi_a \simeq \left\{r + \frac{m_0^2}{N}\left[\frac{1}{m_u} + \frac{1}{m_d} + \frac{1}{m_s}\right]\right\}^{-1}\frac{m_0^2}{N}\theta . \tag{181}$$

The observed mass hierarchy in the pseudoscalars

$$m_\pi^2 \ll m_K^2 \ll m_{\eta'}^2 \tag{182}$$

implies

$$rm_{u,d} \ll rm_s \ll m_0^2 \tag{183}$$

and allows to approximate the $P-$ and $T-$ violating effective Lagrangian in the following way

$$\delta \mathcal{L}^{\theta}_{eff} \simeq -\frac{1}{6\sqrt{2}f} r \frac{m_u m_d}{m_u + m_d} \theta \{Tr\pi^3 + ...\}$$

$$\simeq -2\sqrt{\frac{2}{3}} \frac{m_\pi^2}{f} \frac{m_u m_d}{(m_u + m_d)^2} \theta \{(\eta_8 + \sqrt{2}\eta_0)\pi^+\pi^- + ...\} . \qquad (184)$$

This cubic interaction implies the $P-$ and $T-$ violating process $\eta \rightarrow \pi^+\pi^-$ with a transition amplitude given by

$$A(\eta \rightarrow \pi^+\pi^-) = -2\sqrt{\frac{2}{3}} \frac{m_\pi^2}{f} \frac{m_u m_d}{(m_u + m_d)^2} (\cos\theta_P - \sqrt{2}\sin\theta_P)\theta . \qquad (185)$$

In conclusion, strong interactions violate P and T in the Standard Model for fundamental interactions even if the starting QCD Lagrangian is $P-$ and $T-$ invariant ($\theta_{QCD} = 0$). Nature tells us indeed that one cannot impose CP-invariance on the entire Lagrangian. Consequently, in this specific framework, the δ_{CKM} and θ CP-violating angles arise simultaneously from the Yukawa sector of the electroweak Lagrangian. They are however unrelated since the CKM phase is obtained from the diagonalization of the Hermitean mass matrices H^u, H^d and we know that

$$\arg \det(H^u H^d) = 0 \qquad (186)$$

The electric dipole moment of the neutron provides us with such a stringent constrain on θ [17] that the possibility to measure some effect in weak decay processes with our present technology is hopeless. Nevertheless attempts to understand the smallness of θ leads to extensions of the Standard Model which contain new sources for CP-violation. These new sources could compete with the CKM contributions for the $K-$ and $B-$ systems.

Acknowledgements

I would like to thank T. Nakada, K. Schubert and D. Wyler for useful comments.

References

[1] See A. Pais in "CP-violation in Particle Physics and Astrophysics", Ed. J. Tran Thanh Van (Edition Frontières 1990).

[2] J.H. Christenson, J.W. Cronin, V.L. Fitch and R. Turlay, Phys. Rev. Lett. 13 (1964) 138.

[3] S.L. Glashow, Nucl. Phys. 22 (1961) 579 ; S. Weinberg, Phys. Rev. Lett. 19 (1967) 1264 ; A. Salam, in Proc. 8 th Nobel Symposium, Aspenäsgarden, ed. N. Svartholm (Almqvist and Wiksell, Stockholm, 1968), p. 367.

[4] C. Jarlskog, Phys. Rev. Lett. 55 (1985) 1039 ; Z. Phys. C. 29 (1985) 491

[5] N. Cabibbo, Phys. Rev. Lett. 10 (1963) 531 ; M. Kobayashi and T. Maskawa, Progr. Theor. Phys. 49 (1973) 652.

[6] Review of Particle Properties, Particle Data Group, Phys. Lett. 239 B (1990) 1.

[7] L. Wolfenstein, Phys. Rev. Lett. 51 (1983) 1945.

[8] J. Schwinger, Phys. Rev. 82 (1951) 914 ; G. Lüders, Dansk. Math. Fys. Medd. 28, No. 5 (1954) 1 ; W. Pauli, in Niels Bohr and the development of Physics, ed. W. Pauli, Pergamon Press, New York (1955).

[9] A. Pais and S.B. Treiman, Phys. Rev. D12 (1975) 2744.

[10] M. Bander, D. Silverman and A. Soni, Phys. Rev. Lett. 43 (1979) 242 ; J.-M. Gérard and W.-S. Hou, Phys. Rev. D43 (1991) 2909.

[11] See for example I.I. Bigi, V.A. Khoze, N.G. Uraltsev and A.I. Sanda in "CP-violation", Ed. C. Jarlskog (World Scientific, Singapore 1989).

[12] S. Weinberg, Phys. Rev. D12 (1975) 3583.

[13] G. 't Hooft, Phys. Rev. Lett. 37 (1976) 8.

[14] J.S. Bell and R. Jackiw, Nuovo Cimento 60A (1969) 47; S.L. Adler, Phys. Rev. 177 (1969) 2426; S.L. Adler and W.A. Bardeen, Phys. Rev. 182 (1969) 1517.

[15] P. Di Vecchia and G. Veneziano, Nucl. Phys. B171 (1980) 253 ; C. Rosenzweig, J. Schechter and C.G. Trahern, Phys. Rev. D21 (1980) 3388.

[16] E. Witten, Ann. Phys. 128 (1980) 363.

[17] R.J. Crewther, P. Di Vecchia, G. Veneziano and E. Witten, Phys. Lett. 88B (1979) 123 ; 91B (1980) 487 (E).

HEAVY FLAVOR PHYSICS

Karl Berkelman

Newman Laboratory
Cornell University
Ithaca, NY 14853

1   INTRODUCTION

This is the written version of the lectures I gave at the 1992 Summer School on *Quantitative Particle Physics* at Cargèse, Corsica. Since this material had to be submitted for the proceedings before the lectures were given, it may not correspond exactly to what I said. Also, since the lectures were to be mainly pedagogical, there is no attempt to make this a definitive review of the latest data. I have not made a systematic survey of the references. The experimental results that are quoted are incomplete and in some cases outdated. They are included only for illustration. The lectures divide into three topics: (1) measurement of the CKM matrix, (2) CP violation in B decays, and (3) prospects for a CESR B Factory upgrade.

2   MEASUREMENT OF THE CKM MATRIX

2.1   The Cabibbo-Kobayashi-Maskawa Matrix

The pattern of fundamental fermions is now rather well established. They are grouped into three doublets of quarks and three doublets of leptons. Each quark doublet consists of a charge 2/3 quark and a charge $-1/3$ quark:

$$\begin{array}{ll} q = 2/3 \\ q = -1/3 \end{array} \qquad \begin{pmatrix} u \\ d \end{pmatrix}, \quad \begin{pmatrix} c \\ s \end{pmatrix}, \quad \begin{pmatrix} t \\ b \end{pmatrix}.$$

The symbols d, u, s, c, b, and t represent eigenstates of the strong interaction with conserved flavor quantum numbers. However, the mass eigenstates of the total Hamiltonian, including the weak interaction, are not the same as the flavor eigenstates, but are related to them through a unitary transformation of the charge $-1/3$ states.

When there were only two known quark doublets, this transformation was just a rotation in two-space, parametrized by θ_c, the Cabibbo angle [1]:

$$\begin{pmatrix} d' \\ s' \end{pmatrix} = \begin{pmatrix} \cos\theta_c & \sin\theta_c \\ -\sin\theta_c & \cos\theta_c \end{pmatrix} \begin{pmatrix} d \\ s \end{pmatrix},$$

Quantitative Particle Physics, Edited by
M. Lévy *et al.*, Plenum Press, New York, 1993

taking the flavor eigenstates d and s into the weak eigenstates d' and s'. Flavor changing weak decays take place through the coupling of the two members of a weak eigenstate doublet and a W boson. Thus in this two-doublet model we have for the decay amplitudes of the flavor eigenstates

$$\mathcal{A}(d \to uW) = \mathcal{A}(d' \to uW) \cos \theta_c,$$

$$\mathcal{A}(s \to uW) = \mathcal{A}(d' \to uW) \sin \theta_c.$$

The diagonal matrix elements govern decays within a flavor doublet ($d \to uW$ and $c \to sW$), while the off-diagonal elements are respnsible for the less favored decays between doublets ($s \to uW$ and $c \to dW$). One of the motivations for this formalism was that it automatically suppressed the useen flavor changing neutral current decays:

$$\mathcal{A}(s \to dZ) = \mathcal{A}(d' \to d'Z) \sin \theta_c \cos \theta_c + \mathcal{A}(s' \to s'Z) \cos \theta_c (- \sin \theta_c) = 0,$$

the "GIM mechanism" [2].

Even before charm was discovered Kobayashi and Maskawa [3] made the generalization to three quark doublets. In the three-space rotation the Cabibbo angle gets replaced by three Euler angles, and because this is quantum mechanics, there is the possibility of a complex phase as well. Rather than expressing the matrix in terms of the three angles and phase, we will use the more convenient parametrization by Wolfenstein [4]:

$$\begin{pmatrix} V_{ud} & V_{us} & V_{ub} \\ V_{cd} & V_{cs} & V_{cb} \\ V_{td} & V_{ts} & V_{tb} \end{pmatrix} = \begin{pmatrix} 1 - \lambda^2/2 & \lambda & A\lambda^3(\rho - i\eta) \\ -\lambda & 1 - \lambda^2/2 & A\lambda^2 \\ A\lambda^3(1 - \rho - i\eta) & -A\lambda^2 & 1 \end{pmatrix}.$$

In this approximate expression we have made the assumption that λ is small enough to justify keeping terms only up to λ^3.

The four CKM parameters λ, A, ρ, and η are basic constants of the Standard Model, on the same footing as the quark and lepton masses. We have to go beyond the Standard Model to find a relation between the CKM parameters and the quark masses or to understand why they have the values they do. To guide us in this, and to check the consistency of the three-doublet CKM version of the Standard Model, we have to measure the values of the four parameters, in redundant ways if possible.

2.2 Measurements with u, d, s, and c Quarks

To a very good approximation, tree-level weak decays within the first two doublets depend only on the one parameter λ, which is essentially the Cabibbo angle. The nuclear beta decay matrix element contains the factor V_{ud}. A comparison of beta decay lifetimes with the muon lifetime therefore measures $|V_{ud}|^2$. The result is

$$|V_{ud}| = 1 - \lambda^2/2 = 0.9739 \pm 0.0005,$$

which implies

$$\lambda = 0.227 \pm 0.002.$$

Rates for strangeness changing decays, such as the meson decays $K_L \to \pi^+ e^- \bar{\nu}_e$ and $K^+ \to \pi^0 e^+ \nu_e$ and the hyperon decays $\Lambda \to p e^- \bar{\nu}_e$, $\Sigma^- \to n e^- \bar{\nu}_e$, and $\Xi^- \to \Lambda(\text{ or } \Sigma^0) e^- \bar{\nu}_e$, are proportional to $|V_{us}|^2$ and yield the result

$$|V_{us}| = \lambda = 0.220 \pm 0.002.$$

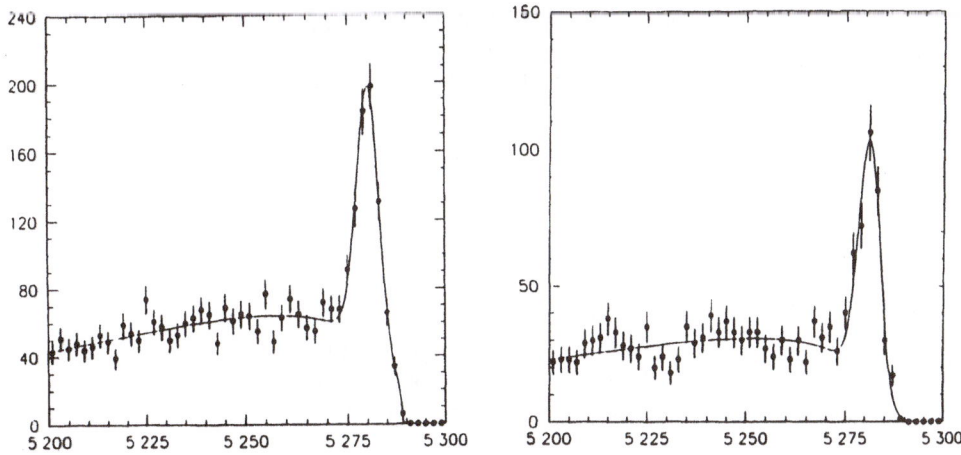

Figure 1. Reconstructed mass distributions of (left) $B^{\pm}$ and (right) $B^0/\overline{B}^0$ decay candidates, from CLEO [5].

Measurements with charm decays give less precise results. From the semileptonic decays of charm, $D^0 \to K^- e^+ \nu_e$ and $D^+ \to \overline{K}^0 e^+ \nu_e$, we get

$$|V_{cs}| = 1 - \lambda^2/2 = 0.96 \pm 0.08,$$

while from the analysis of neutrino production of dimuons, $\nu N \to \text{charm} \ldots \to \mu^+ \mu^-$ ..., we have

$$|V_{cd}| = \lambda = 0.205 \pm 0.014.$$

All measurements are reasonably consistent with the value

$$\lambda = 0.224 \pm 0.002.$$

2.3 Measurement of V_{cb}: B Decays to Charm

To measure any of the other three CKM matrix parameters, we need to involve quarks from the third doublet. In the absence of experimental information on the t quark, this means looking at b quark decays. Measured inclusive rates for B mesons to decay to D, D^*, D_s, ψ, ψ', and Λ_c are consistent within the 15% experimental accuracy with all b decays going through the bcW vertex. All of the reconstructed exclusive B decay modes (see Fig. 1) involve charm in the final state.

Ideally, to measure the matrix element V_{xb} for the coupling $b \to xW^-$, where x is the $q = +2/3$ quark (c or u) in either of the two lower mass doublets, one needs to measure the decay rate for $b \to xe^-\overline{\nu}_e$, which should be given by essentially the same formula as for the well known $\mu^- \to \nu_\mu e^- \overline{\nu}_e$, that is,

$$\Gamma(b \to xe^-\overline{\nu}_e) = \frac{G_F^2 m_b^5}{193\pi^3} |V_{xb}|^2 \Phi\left(\frac{m_x}{m_b}\right),$$

where G_F is the Fermi decay constant and Φ is a known phase space suppression factor which is 1 in the limit $m_x \ll m_b$. Since b quarks are not free, we use semileptonic decays

Table 1. Measurements of the b mean life [6].

Experiment	Picoseconds
JADE	$1.8^{+0.6}_{-0.5}$
Mark II	0.85 ± 0.27
TASSO	$1.56^{+0.45}_{-0.42}$
MAC	1.29 ± 0.29
DELCO	$1.17^{+0.32}_{-0.27}$
HRS	$1.02^{+0.41}_{-0.37}$
ALEPH	$1.29 \pm 0.06 \pm 0.10$
OPAL	$1.32^{+0.31}_{-0.25} \pm 0.15$
L3	$1.32 \pm 0.08 \pm 0.09$
Average	1.27 ± 0.07

of $\overline{B}$ mesons ($B^- = b\overline{u}$ or $\overline{B}^0 = b\overline{d}$) and assume that the light $\overline{u}$ or $\overline{d}$ is only a spectator in the weak interaction (see Fig. 4a).

The experimental measurement of a meson decay rate Γ involves two separate measurements, (a) the mean lifetime τ of the meson, and (b) the branching ratio $\mathcal{B}$ for an appropriate final state; that is, $\Gamma = \mathcal{B}/\tau$. The B lifetime has been measured at PEP, PETRA, and LEP [6], where B's are produced with enough velocity to move a measurable distance before decaying. Since none of the experiments is able to identify B decays, several techniques have been used to obtain an enriched sample of events. They rely either on the expectation that a b quark jet is wider than a jet containing only lighter quarks, or on the chance that the B decay will produce a high momentum lepton at a relatively large p_T with respect to the jet axis. Several measures of the decay length are also used: the impact parameter of the high p_T lepton with respect to the jet axis at the interaction point, the impact parameters of all of the tracks, the dipole moment of the extrapolated intersections with the jet axis, or the separation distance of reconstructed vertices. Monte Carlo simulations are used to estimate the effective B velocities, the effects of non-B background, and the relation between the measured distances and the decay distances. The agreement among the nine different measurements listed in Table 1 is rather good; the world average lifetime is

$$\tau = 1.27 \pm 0.07 \text{ ps}.$$

This is actually not a B^- or $\overline{B}^0$ lifetime, but a weighted average over the mix of B^-, $\overline{B}^0$, B_s, Λ_b, etc. produced in the experiments. If the spectator model dominated all b-hadron decays, it would not matter which quarks accompanied the b quark, but for now we have to assume that the measured lifetime is appropriate for the decay rates Γ that we are concerned with.

Suppose we consider the inclusive rate for B semileptonic decays. Then,

$$\Gamma = \frac{G_F^2 m_b^5}{193\pi^3} \left[|V_{cb}|^2 \Phi \left(\frac{m_c}{m_b} \right) + |V_{ub}|^2 \Phi \left(\frac{m_u}{m_b} \right) \right].$$

Since, as we shall see, $|V_{ub}| \approx 0.1|V_{cb}|$, and $\Phi(m_u/m_b) \approx 2\Phi(m_c/m_b)$, we can ignore the 2% contribution of V_{ub} in the rate. A measurement of the B inclusive semileptonic branching ratio, in combination with the lifetime measurement, will therefore yield a measurement of $|V_{cb}|$. The experiment is best performed near the electron-positron

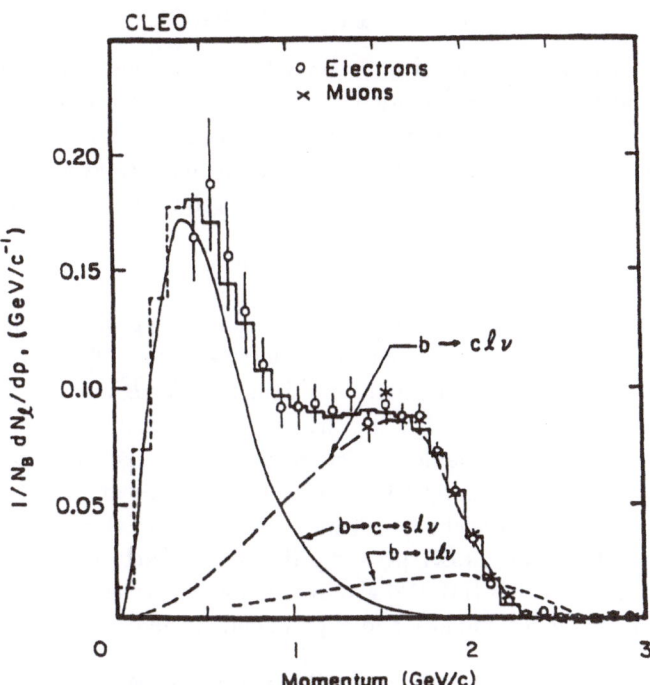

Figure 2. CLEO measurements of the inclusive electron and muon momentum spectra from decays of B mesons produced at the $\Upsilon(4S)$ resonance. The curves show the various contributions to the spectrum.

annihilation threshold where the $\Upsilon(4S)$ resonance gives an enhanced cross section. The CLEO [7] and ARGUS [8] experiments measure the inclusive lepton production rate at the resonance and subtract the continuum contribution measured below $B\overline{B}$ threshold:

$$e^+e^- \to \Upsilon(4S) \to B\overline{B} \to eX \text{ or } \mu X.$$

After they subtract the contribution of leptons from $B \to DX \to \ell X'$ decays (see Fig. 2), the leptons from $B \to \psi X \to \ell^+\ell^- X$, and the other particles misidentified as leptons, and account for detection efficiency and number of B's produced, the result is the semileptonic branching ratio averaged over the mix of charged and neutral B's produced in $\Upsilon(4S)$ decays. For the average of $\mathcal{B}(e\nu X)$ and $\mathcal{B}(\mu\nu X)$ they get

$$10.5 \pm 0.2 \pm 0.4\% \quad \text{(CLEO)},$$

$$10.2 \pm 0.5 \pm 0.2\% \quad \text{(ARGUS)}.$$

In addition to the quoted statistical and systematic errors there is a model dependence of about 2% coming from the uncertainty in the shape of the momentum spectrum (Fig. 2) assumed in correcting for the portion below the cut, the main source of uncertainty being the size of the contribution of multiparticle hadron final states such as $B \to D\pi\ell\nu$. Computing the decay rate $\Gamma = \mathcal{B}/\tau$ and solving for the CKM matrix element gives

$$|V_{cb}| = 0.047 \pm 0.005.$$

A subsample of the observed lepton events can be tagged as coming from neutral $\overline{B}^0$ decays, by identifying the accompanying B decay as $B^0 \to D^{*-}\pi^+$ or $B^0 \to D^{*-}\ell^+\nu$. This allows us to measure a semileptonic branching ratio for $\overline{B}^0$, which is $10.4\pm2.2^{+1.0}_{-1.1}\%$, consistent with the assumption that the charged and neutral B's have equal lifetimes.

If we can get a more reliable prediction for the rate and the shape of the lepton momentum spectrum in the simplest exclusive semileptonic mode, $B \to [D \text{ or } D^*]\ell\nu$, we can avoid some of the model dependence of the result for the inclusive semileptonic branching ratio. Based on CLEO [9] and ARGUS [10] exclusive data for D and D^* modes, five different models [11], [12] give results for $|V_{cb}|$ ranging from 0.039 to 0.042. Averaging over the modes, models, and experiments we get

$$|V_{cb}| = A\lambda^2 = 0.041 \pm 0.003 \pm 0.004.$$

This allows us to fix the Wolfenstein-CKM parameter,

$$A = 0.82 \pm 0.10.$$

Note that the coupling of the third generation to the second generation, V_{cb}, is much weaker than the coupling of the second to the first, V_{us} or V_{cd}. The small coupling suppresses the dominant b-to-c decay and makes it possible for rare processes — b-to-u decays, $B^0 - \overline{B}^0$ oscillations, higher order loop processes, and so on — to compete.

2.4 Measurement of V_{ub}: Charmless B Decays

For this matrix element we need to measure the rate for $b \to uW^-$. Again we use semileptonic B decays. To see these rare decays in the presence of the much more common decays to charm, we need an unambiguous, high-efficiency signature for the

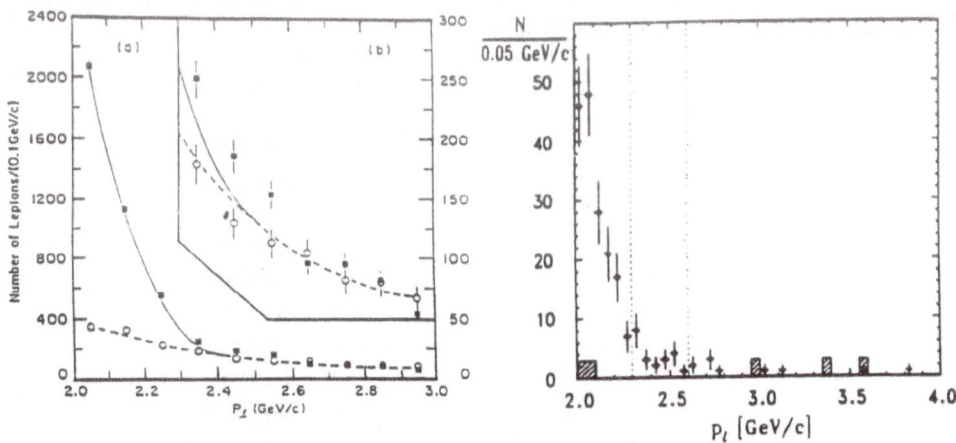

Figure 3. CLEO [13] (left) and ARGUS [14] (right) measured semileptonic B decay momentum spectra in the vicinity of the end point. The measured non-$B\overline{B}$ backgrounds are also shown.

Table 2. Results for $|V_{ub}/V_{cb}|^2$ in % for two experiments [13], [14] and four models [11], [12], [15].

Model	CLEO	ARGUS
Altarelli et al.	0.8 ± 0.2	1.0 ± 0.2
Isgur et al.	2.2 ± 0.6	3.2 ± 0.7
Wirbel, Stech, and Bauer	1.3 ± 0.4	1.4 ± 0.5
Körner and Schüler	0.9 ± 0.2	0.8 ± 0.2

charmless hadron final state in $B \to X_u \ell^- \overline{\nu}$. The endpoint of the momentum spectrum in the inclusive decay $B \to eX$ of a B at rest to an electron is

$$p_e^{max} = (M_B^2 - M_X^2)/2M_B.$$

For the dominant b-to-c decay $M_X > M_D$ and $p_e^{max} = 2.4$ GeV/c, while for the b-to-u decay M_X could be as low as m_π, and p_e^{max} is therefore higher, about 2.6 GeV/c. These maximum momentum figures hold whether the lepton is an electron or a muon, and include the boost due to the motion of the B in the decay of the $\Upsilon(4S)$ and the measurement error on the lepton momentum. We therefore look for evidence of the b-to-u charmless decay in the lepton momentum spectrum beyond the endpoint of the charm decay spectrum. The rate of such decays, normalized to the rate of charmed decays is then a measure of the ratio of CKM matrix elements:

$$\left|\frac{V_{ub}}{V_{cb}}\right|^2 = \frac{N(p_\ell > 2.4)}{N(p_{min} < p_\ell < 2.4)} \frac{f_c(p_\ell > p_{min})}{f_u(p_\ell > 2.4)}.$$

The fractions f_c and f_c of the lepton momentum spectra that are above the momentum cuts have to be obtained from theory.

The measurement is not easy. The inclusive lepton rate in the momentum range from 2.4 to 2.6 GeV/c is dominated by the non-$B\overline{B}$ background underlying the $\Upsilon(4S)$

Table 3. Branching ratio upper limits (90%) for charmless B decays (in 10^{-4}) [16], [17] compared with model predictions [18].

	Mode	Model	Data	Experiment
$\overline{B}^0 \rightarrow$	$\pi^+\pi^-$	0.2	< 0.26	CLEO
	$\pi^0\pi^0$		< 0.4	CLEO
	$\rho^\pm\pi^\mp$	0.6	< 5.2	ARGUS
	$a_1^\pm\pi^\mp$	0.6	< 5.7	CLEO
	$a_2^\pm\pi^\mp$	–	< 3.5	CLEO
	$\pi^+\pi^-\pi^0$	0.6	< 7.2	ARGUS
	$\pi^+\pi^+\pi^-\pi^-$	1.0	< 6.7	ARGUS
	$p\overline{p}$	–	< 0.4	CLEO
	$p\overline{p}\pi^+\pi^-$	–	< 2.9	CLEO
$B^- \rightarrow$	$\pi^+\pi^-\pi^-$	0.6	< 4.5	ARGUS
	$2(\pi^+\pi^-)\pi^-$	2	< 8.6	ARGUS
	$p\overline{p}\pi^-$	–	< 1.4	CLEO

resonance. This background must be measured accurately by running below the $B\overline{B}$ threshold. The statistical significance of the subtraction is enhanced by suppressing the background with a topology cut that rejects two-jet events and by fitting the background lepton momentum spectrum to a smooth curve before subtracting. One also has to avoid lepton tracks with mismeasured momenta, and subtract the contribution of misidentified hadrons, as well as leptons from ψ decays and from the $B \rightarrow DX \rightarrow \ell X'$ cascade. CLEO [13] and ARGUS [14] data (Fig. 3) both show an excess in the 2.4 to 2.6 GeV/c momentum range. CLEO confirms that the excess is due to b-to-u decays by showing that the topology of these events is essentially spherical, as expected for $B\overline{B}$ events but not for non-$B\overline{B}$ background. ARGUS sees the effect also in dilepton events where there is no non-$B\overline{B}$ background, and has reconstructed an event containing a $\overline{B}^0 \rightarrow \pi^+\mu^-\overline{\nu}_\mu$ decay and a $\overline{B}^0 \rightarrow D^{*+}\rho^-$ decay. The two experiments each measure $|V_{ub}/V_{cb}|^2$ with about a 20% relative error, but the accuracy of the result is limited by the models for f_u and f_c, which lead to a variation in $|V_{ub}/V_{cb}|^2$ of a factor of two (see Table 2). Averaging over experiments and models and using the previous result for $|V_{cb}|$, we get

$$|V_{ub}| = A\lambda^3|\rho - i\eta| = 0.005 \pm 0.001,$$

$$|\rho - i\eta| = 0.5 \pm 0.1.$$

Although no exclusive charmless hadronic decays of the B have yet been seen, the search goes on. Table 3 shows some of the measured branching ratio upper limits. The lowest ones are below 10^{-4}, while the theoretical predictions, with $|V_{ub}/V_{cb}| = 0.1$, are still lower. The ARGUS report of the observation of charmless decays $B^- \rightarrow p\overline{p}\pi^-$ and $\overline{B} \rightarrow p\overline{p}\pi^+\pi^-$ was not confirmed by CLEO and later ARGUS data. The analysis of the most recent CLEO data sample of two million produced B's now in progress should yield significantly lower limits and perhaps a few actual measurements.

2.5 Measurement of V_{td}: $B\overline{B}$ Mixing

Since the t-to-d decay is not yet accessible experimentally, we have to look for a loop process with a virtual t. In the box diagram (Figs. 4g and h) responsible for

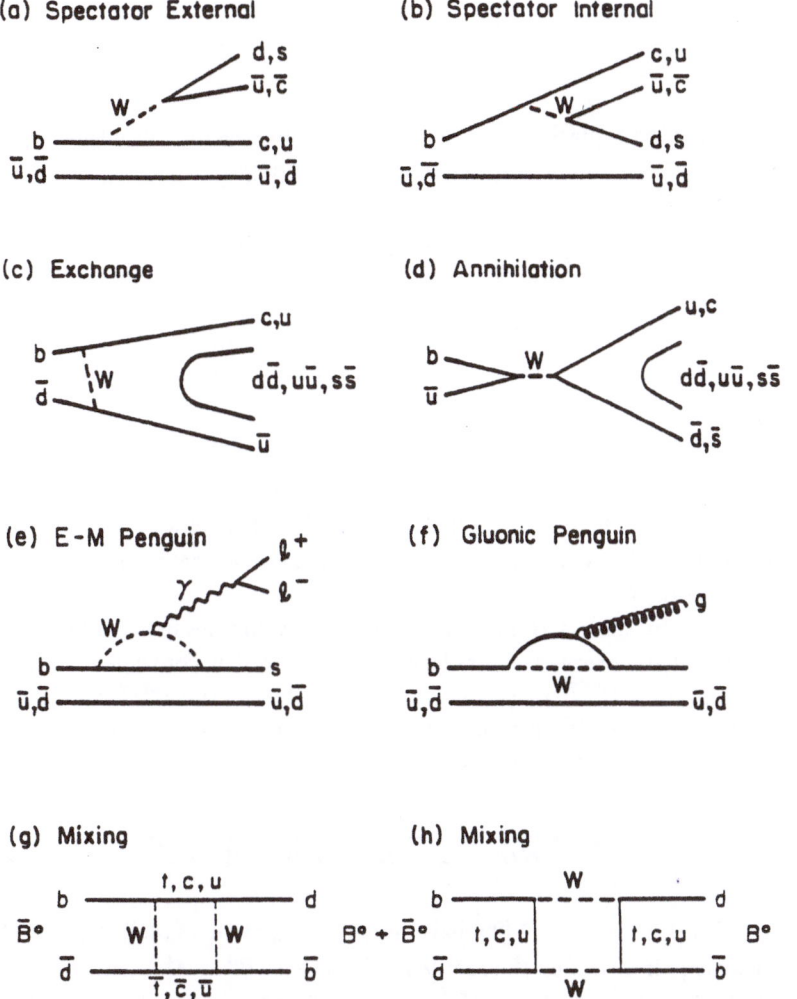

Figure 4. (a) through (f) Quark level diagrams for B meson decay. (g) and (h) Box diagrams for $\overline{B}^0 \to B^0$.

Table 4. Measurements of mixing fraction [19], [20], [26], [27].

Experiment	Beam	c.m. GeV	Measurement
ARGUS	e^+e^-	$\Upsilon(4S)$	$\chi_d = 0.18 \pm 0.05$
CLEO	e^+e^-	$\Upsilon(4S)$	$\chi_d = 0.141 \pm 0.027$
UA1	$\bar{p}p$	540	$\chi = 0.121 \pm 0.047$
ALEPH	e^+e^-	91	$\chi = 0.132^{+0.027}_{-0.026}$
L3	e^+e^-	91	$\chi = 0.178^{+0.049}_{-0.040}$

$B^0 - \overline{B}^0$ mixing the t quark intermediate state dominates the amplitude. The predicted ratio of the $B^0 \rightleftarrows \overline{B}^0$ oscillation frequency to the decay rate is

$$x_d = \frac{\Delta m}{\Gamma} = \frac{G_F^2}{6\pi^2} m_t^2 g(m_t^2/M_W^2) \tau_B f_B^2 B_B \eta_{qcd} |V_{tb} V_{td}^*|^2,$$

where

$$g(y) = \frac{1}{4} \left(1 + \frac{3 - 9y}{(y-1)^2} + \frac{6y^2 \ln y}{(y-1)^3} \right)$$

and thus,

$$g(m_t^2/M_W^2) \approx 0.60 \left(\frac{m_t}{140 \text{ GeV}} \right)^{-0.45}.$$

The QCD correction factor η_{qcd} is 0.85 and V_{tb} should be close to one if there are only three quark doublets. A measurement of x_d can therefore determine $|V_{td}|$, provided m_t and $f_B^2 B_B$ are known.

The short lifetime of B mesons and their low velocities in the $\Upsilon(4S) \to B\overline{B}$ decay make it difficult to observe the oscillation directly, so it is convenient to integrate over time. Starting from $B^0\overline{B}^0$ pairs, one can eventually observe $B^0 B^0$ and $\overline{B}^0\overline{B}^0$ pairs. In $e^+e^- \to B\overline{B}$ the two B's are in a p-wave state, odd under exchange, so that Bose statistics implies the mixing fraction

$$\chi_d = \frac{BB + \overline{BB}}{B\overline{B} + \overline{B}B + BB + \overline{BB}} = \frac{x_d^2}{2(1 + x_d^2)}.$$

ARGUS [19] and CLEO [20] have measured χ_d at the $\Upsilon(4S)$, tagging the identity of the evolved pair of B's by the semileptonic decays, $B^0 \to \ell^+ X$ and $\overline{B}^0 \to \ell^- X$. They subtract the background of non-$B\overline{B}$ events, hadrons misidentified as leptons, leptons from the $B \to c \to \ell$ cascade and from $B \to \psi X \to \ell^+\ell^- X$, and the contribution of $B^+ B^-$. The results for χ_d are shown in Table 4. Each of the two groups has confirmed its observation of mixing using a subset of the dilepton sample with lower background. ARGUS looks at $D^{*\pm} \ell^\pm \ell^\pm$ and CLEO looks at $K^\pm \ell^\pm \ell^\pm$. ARGUS has also reconstructed several mixed events. The average mixing fraction from ARGUS and CLEO is

$$\chi_d = 0.151 \pm 0.024,$$

which implies

$$x_d = 0.65 \pm 0.04.$$

This is not yet a measurement of $|V_{td}|$, since the mass of the t quark and the value of the decay constant $f_B^2 B_B$ are not known. Indirect experimental evidence points to a top mass in the range $m_t = 140 \pm 40$ GeV, while estimates of $f_B^2 B_B$ tend to cluster around $(130 \text{ MeV})^2$ in heavy quark theory and around $(220 \text{ MeV})^2$ in lattice calculations. However, we can write

$$|V_{td}| = (0.0130 \pm 0.0004) \left(\frac{m_t}{140 \text{ GeV}}\right)^{-0.78} \left(\frac{f_B \sqrt{B_B}}{130 \text{ MeV}}\right)^{-1}.$$

The fact that $|V_{td}|$ is approximately proportional to the fourth root of χ_d greatly reduces the contribution to the error from the mixing measurement.

This gives us another constraint on the two Wolfenstein parameters ρ and η in the CKM matrix:

$$|1 - \rho - i\eta| = (1.41 \pm 0.05) \left(\frac{m_t}{140 \text{ GeV}}\right)^{0.78} \left(\frac{f_B \sqrt{B_B}}{130 \text{ MeV}}\right).$$

Once m_t and $f_b \sqrt{B_B}$ are known and the theory used to extract $|V_{ub}|$ from the semileptonic decay spectrum is made more reliable, we will be able to combine the b-to-u data and the mixing data to fix ρ and η and thus complete our determination of the four parameters of the CKM matrix. However, if the Standard Model of three quark doublets is not correct, there may be more than four parameters involved, and we will have to measure the rest of the CKM matrix elements.

2.6 Measurement of V_{ts}: Loop Decays

Here again we have to look for a higher order process involving virtual t quarks. We can construct loop diagrams (see Figs. 4 e and f) for processes $b \to s\gamma$ or $b \to sg$, called "penguin" modes. As in the case of the box diagram, the heaviest quark dominates in the intermediate state. Although the theoretical predictions [22] for the inclusive radiative decays,

$$\mathcal{B}(b \to s\gamma) = 3.5 \times 10^{-4} \left|\frac{V_{ts}}{0.04}\right|^2 \left(\frac{m_t}{140 \text{ GeV}}\right)^{0.5},$$

are subject to large QCD corrections, they are considered reliable, within $\pm 10\%$ uncertainties coming from Λ_{qcd}. A very recent (and preliminary) CLEO measurement with inclusive photons [21] gives $\mathcal{B}(B \to s\gamma) = (3.5 \pm 2.1) \times 10^{-4}$. Together with the theoretical prediction this implies

$$|V_{ts}| = (0.040 \pm 0.009) \left(\frac{m_t}{140 \text{ GeV}}\right)^{-0.25}.$$

This is consistent with the expectation $|V_{ts}| \approx |V_{cb}|$ for any reasonable value of the top quark mass.

The prediction for the corresponding hadronic inclusive penguin rate,

$$\mathcal{B}(b \to sg, \, sgg, \, sq\bar{q}) = 0.017|V_{ts}/0.04|^2,$$

is probably less reliable even though it is rather insensitive to the top quark mass. Still more uncertain are the rates for individual exclusive modes. Nevertheless they are

Table 5. CLEO upper limits (90% confidence) on branching ratios, in 10^{-5}

Mode	Data	Prediction	Ref.
$K^+\pi^-$	< 6	7 - 10	[23],[24]
$K^{*0}\pi^0$	< 4	3	[24]
$K^+\pi^0$	< 4	6	[24]
$K^0\pi^+$	< 8	6	[24]

being searched for experimentally; Table 5 shows recently measured CLEO [21] limits for modes in which the limits are close to the predicted rates [24],[23]. Recently CLEO has seen evidence for $\overline{B}^0 \to K^-\pi^+$ at a preliminary rate level that is lower than the theoretical expectation.

Substituting a strange quark for the $\overline{d}$ quark in the B^0 meson gives us a B_s meson, which can also oscillate to its antiparticle. The amplitude is again given by the box diagram, the formula for $x_s = \Delta m_s/\Gamma_s$ having the same form as the above formula for x_d, with V_{ts} and $f_{B_s}^2 B_{B_s}$ substituted for V_{td} and $f_B^2 B_B$.

CLEO has taken data at the $\Upsilon(5S)$ resonance, which should be above $B_s\overline{B}_s$ threshold, but the statistical evidence for B_s is not yet compelling. Several LEP experiments [25] have seen D_s-lepton sign correlations indicative of B_s production in Z^0 decays. There are also high energy dilepton data that should be sensitive to $B_s\overline{B}_s$ mixing. The first evidence for $B\overline{B}$ mixing was actually reported by the UA1 collaboration [26], who saw an excess of like-sign muon pairs over the Monte Corlo calculation for the rate from mundane sources, such as pion, kaon, and charm decays and misidentification of other particles. The mixed fraction χ should have contributions from $B^0 - \overline{B}^0$ mixing and $B_s - \overline{B}_s$ mixing:

$$\chi = f_d\chi_d + f_s\chi_s.$$

The χ data from UA1 and the LEP experiments are shown in Table 4. The results are consistent with B^0 and B_s production fractions $f_d = 0.375$ and $f_s = 0.15$ and a value of χ_s near the limit 0.5, but the precision is not yet sufficient to enable one to extract a useful limit for x_s or for $|V_{ts}|$.

2.7 Summary of Measurements

Our present experimental knowledge of the elements of the Cabibbo-Kobayashi-Maskawa matrix is summarized below:

$$\begin{pmatrix} 0.9747 \pm 0.0011 & 0.220 \pm 0.002 & 0.005 \pm 0.001 \\ 0.21 \pm 0.03 & 1.1 \pm 0.2 & 0.041 \pm 0.005 \\ (0.013 \pm 0.0004)a^{-0.78}b^{-1} & (0.040 \pm 0.009)a^{-0.25} & ? \end{pmatrix},$$

where $a = m_t/(140 \text{ GeV})$ and $b = f_B\sqrt{B_B}/(130 \text{ MeV})$.

From these measurements we can derive values or constraints for the four Wolfenstein parameters:

$$\lambda = 0.224 \pm 0.002,$$

$$A = 0.82 \pm 0.10,$$

$$\sqrt{\rho^2 + \eta^2} = 0.5 \pm 0.1,$$

$$\sqrt{(1-\rho)^2 + \eta^2} = (1.41 \pm 0.05)a^{0.78}b.$$

Except in the case of the Cabibbo angle λ, the quoted error limits come from not from the experimental errors but from the uncertainties in the theory used to model the hadronic matrix elements in B decays. We can probably expect some gradual improvement in the theory as we accumulate more data on various B decay modes. There are good prospects for a flood of B decay data in the next few years from the upgraded CLEO detector operating at CESR with new record luminosities.

3 CP VIOLATION IN B DECAYS

3.1 The KM Description of CP Violation

The CP operation, which takes all particles into their mirror-image antiparticles, is not a symmetry of the universe, at least in our immediate vicinity, since we see a preponderance of protons, neutrons, and electrons and very few antiprotons, antineutrons, and positrons. It is one of the important tasks of physics to understand the source of the asymmetry.

In 1964 we learned that CP symmetry is violated in about 0.2% of the weak decays of neutral kaons [28]. This is almost certainly related to the eventual explanation for the particle-antiparticle abundance asymmetry. In 1973, before the discovery of the charmed quark, Kobayashi and Maskawa [3] realized that the most general 3×3 quark doublet rotation matrix for six quarks can contain an imaginary piece (the $i\eta$ in the Wolfenstein parametrization) that can give rise to CP violation at the level observed in kaon decay.

If the Cabibbo-Kobayashi-Maskawa matrix is unitary, the relation

$$V_{td}V_{tb}^* + V_{cd}V_{cb}^* + V_{ud}V_{ub}^* = 0$$

defines a triangle in the complex plane (see Fig. 5a). The triangle can be replotted in ρ versus η by normalizing the dimensions (i.e., dividing by $A\lambda^3$) so that the base lies along $(\rho, \eta) = (0,0)$ to $(1,0)$ (Fig. 5b). The upper two sides are then defined by $V_{ub}^*/(A\lambda^3) = \rho + i\eta$, measured in charmless B decays, and $V_{td}/(A\lambda^3) = 1 - i\rho - i\eta$, which can be determined from $B^0 - \overline{B}^0$ mixing once the top mass is known. CP violation should occur if the altitude or area of this unitarity triangle is nonzero.

In particular, the ϵ that parametrizes the admixture of the "wrong" CP state in the mass (i.e., decay) eigenstates of the neutral kaons depends on the CKM matrix parameters, especially the η parameter:

$$|\epsilon| = (2.26 \pm 0.02) \times 10^{-3} \quad [29],$$

$$\approx A^2 \eta B_K \left(0.0036 a^{0.21} + 4.8 A^2 \lambda^4 (1-\rho) a^{1.55}\right) \quad [30],$$

where $a = m_t/(140 \text{ GeV})$ and $B_K \approx 0.8 \pm 0.2$ [31] is a theoretically estimated K meson structure parameter. Figure 5c shows the constraints on the location of the apex of the unitarity triangle given by the measurements of $|V_{ub}|$, $B - \overline{B}$ mixing, and $|\epsilon|$ for various assumed values of m_t.

The ratio ϵ'/ϵ, characterising the direct CP violation in the two-pion decay, is also proportional to η. However, since the situation in both experiment and theory is unsettled, this does not yet provide a useful constraint.

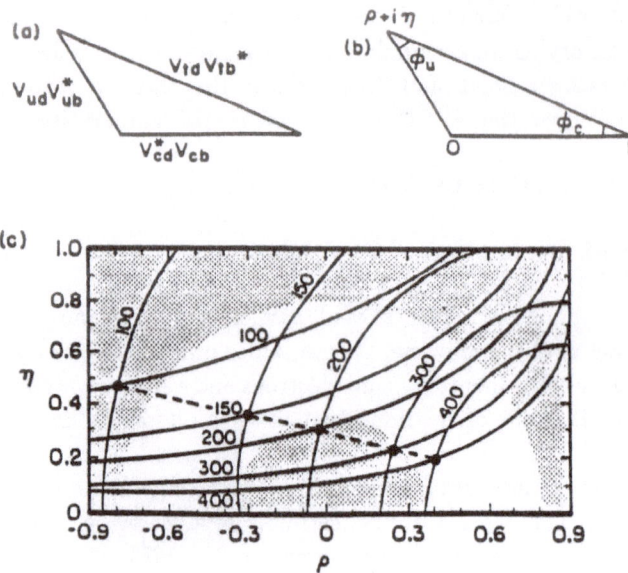

Figure 5. (a) The unitarity triangle. (b) The same triangle with normalized dimensions, plotted in the space of the Wolfenstein parameters ρ and η. (c) Constraints on the (ρ, η) of the upper vertex of the unitarity triangle. The arcs centered at (1,0) are from $B\overline{B}$ mixing, and the lsoping curves are from ϵ. The shaded regions are excluded by the measurements of $|V_{ub}|$ in charmless semileptonic B decay. The numbers are values of m_t in GeV, and the dashed line connects (ρ, η) values implied by the combination of mixing and ϵ for various m_t.

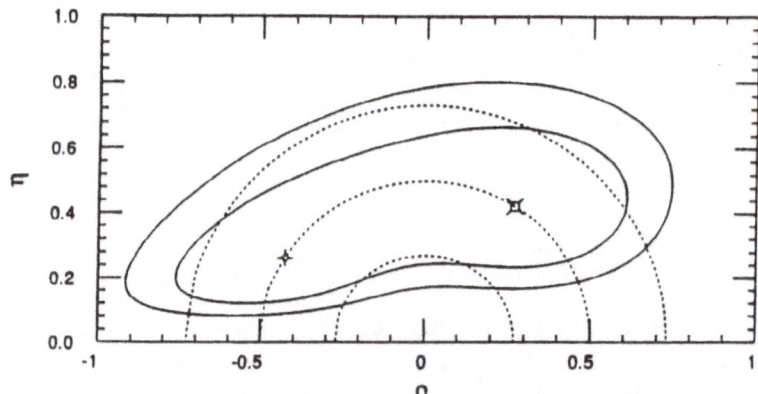

Figure 6. Contours of 68% and 90% confidence levels (inner and outer solid curves) for the (ρ, η) location of the apex of the normalized unitarity triangle, obtained in ref. [32].

One can put together the available experimental information from B and K decays, the experimental constraints on the t quark mass from the direct searches and from the measurement of m_W/m_Z, and theoretical estimates for $f_B\sqrt{B_B}$ and B_K to plot a combined χ^2 probability for the location of the apex of the unitarity triangle in ρ versus η. A recent survey [32] is shown in Fig. 6. In spite of the fact that the measurements of b-to-u decays, $B - \overline{B}$ mixing, and ϵ are becoming rather accurate, the allowed values of ρ and η still cover a wide range. The main problems are the uncertainties in m_t and in $f_B\sqrt{B_B}$. Nevertheless, there is overall consistency with the Kobayashi-Maskawa picture of CP violation; that is, the fact that CP violation in kaon decay requires $\eta \neq 0$ is consistent with the most likely values for the lengths of the sides of the unitarity triangle, obtained from measurements on B decays.

A crucial test of this picture is the observation of CP violation in B decay. Since its partner t quark is heavier, the b quark can decay only through mixing of quark families, and the resulting rate suppression allows rare process to compete. The situation is similar to the case of the decay of the strange quark, but in the case of the b quark the suppression of the favored decays is greater ($V_{cb} \sim \lambda^2$ versus $V_{us} \sim \lambda$) and since there are more lighter quark species to decay to, there are more opportunities for interference effects to make CP asymmetries. In the next sections I will discuss some of the more experimentally accessible CP violation effects in B decays, and derive the requirements that have to be satisfied for a facility that can carry out such experiments.

3.2 CP Violation in Propagation

We first consider this case, familiar from the neutral kaon system. If CP is not an invariant, then the mass eigenstates of the neutral B's, B_+ and B_- (analogous to the K_S and K_L), are not CP eigenstates, but instead:

$$|B_\pm\rangle = \frac{2}{1+\epsilon_B^2}\left((1+\epsilon_B)|B^0\rangle \pm (1-\epsilon_B)|\overline{B}^0\rangle\right).$$

Thus, when B^0 oscillates to $\overline{B}^0$ and vice versa, the rate for $B^0 \to \overline{B}^0$ will not equal the rate for $\overline{B}^0 \to B^0$. The B situation is different from the K situation, because there is no dominant B decay mode, the final CP eigenstates have very low branching ratios, and the two mass eigenstates have approximately the same lifetime. As a consequence, the Standard Model predicts [33], [34] a much smaller effect than in the kaon system: $\epsilon_B < 2 \times 10^{-4}$.

Although this form of CP violation happens as the B mesons propagate, oscillating to their antiparticles, the only way we can detect it is to observe the B's decaying. For example, we can distinguish B^0 from $\overline{B}^0$ by the semileptonic decay:

$$B^0 \to \ell^+ X \quad \text{and} \quad \overline{B}^0 \to \ell^- X.$$

Starting with $B^0\overline{B}^0$ pairs, produced for instance in $\Upsilon(4S)$ decays, the measurable CP violating asymmetry is

$$A = \frac{\ell^+\ell^+ - \ell^-\ell^-}{\ell^+\ell^+ + \ell^-\ell^-} \approx 4\,\mathcal{R}e\,\epsilon_B < 10^{-3}.$$

The number of produced $B\overline{B}$ pairs required to see an asymmetry A of s standard deviations significance is

$$N(B\overline{B}) = \frac{2s^2}{B^2A^2\epsilon^2\chi_d} \sim \frac{2(4)^2}{(0.2)^2(10^{-3})^2(0.6)^2(0.2)} \sim 10^{10},$$

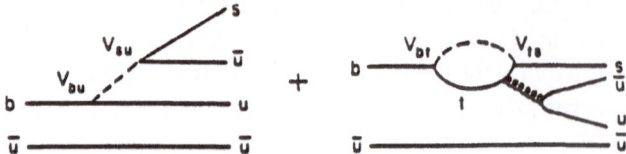

Figure 7. Interfering diagrams for $B^- \to K^-\pi^0$: (a) spectator and (b) penguin.

where $\mathcal{B}$ is the semileptonic branching ratio and ϵ is the lepton detection efficiency. The prospect of seeing a 4 standard deviation effect looks pretty hopeless if the Standard Model is correct. Nevertheless, we are keeping an eye on this asymmetry as the data accumulate, just in case there are surprises.

3.3 CP Violation in Charged B Decays

I will start the discussion of this case by considering a particular example, the decay $B^\mp \to K^\mp\pi^0$. The decay $B^- \to K^-\pi^0$ can proceed through either the spectator diagram, with $b \to u + s\bar{u}$ (see Fig. 7a), or through the penguin loop process with $b \to s + u\bar{u}$ (Fig. 7b). If we represent the two amplitudes for the B^- decay by

$$\mathcal{A} = ae^{i\alpha}V_{ub}V_{us}^* \quad \text{(spectator)}$$

$$\mathcal{B} = be^{i\beta}V_{tb}V_{ts}^* \quad \text{(penguin)},$$

the corresponding amplitudes for the CP conjugate B^+ decay will be

$$\overline{\mathcal{A}} = ae^{i\alpha}V_{ub}^*V_{us}$$

$$\overline{\mathcal{B}} = be^{i\beta}V_{tb}^*V_{ts}.$$

The CP violating rate difference with then be

$$\Gamma - \overline{\Gamma} = |\mathcal{A} + \mathcal{B}|^2 - |\overline{\mathcal{A}} + \overline{\mathcal{B}}|^2 = 4ab\sin(\alpha - \beta)\mathcal{I}m(V_{ub}V_{us}^*V_{tb}^*V_{ts}).$$

The violation will be nonzero only if (a) there are two interfering amplitudes of comparable magnitude, (b) they have different strong interaction phases, and (c) they involve different weak phases with small quark mixing matrix elements. This is characteristic of CP violation signals in other modes as well.

Starting with a sample of equal numbers of B^+ and B^-, one looks for an asymmetry in the occurance of $K^+\pi^0$ and $K^-\pi^0$ final states. In terms of the Wolfenstein parameters, the asymmetry is

$$A = \frac{\Gamma - \overline{\Gamma}}{\Gamma + \overline{\Gamma}} = \frac{2ab\eta\sin(\alpha - \beta)}{a^2\lambda^2(\rho^2 + \eta^2) + b^2\lambda^{-2} + 2ab\rho\cos(\alpha - \beta)}.$$

Predictions of the asymmetry [33],[35] are not very reliable, because of the uncertainties in the parameters a, b, α, and β, but it might be as high as 10% either in this mode or in some similar mode, like $(K$ or $K^*) + (\pi, \eta, \rho, \omega,$ or $\phi)$ in various charge combinations. The asymmetry can be much more impressive in nonstandard models; in the charged Higgs model [36] it might be as large as 70%.

Other examples involve a spectator diagram interfering with annihilation, for instance $B^{\pm} \to \pi^{\pm}\rho^0$ or $D^{*\pm}D^0$. The interference can involve different decay paths, such as $B^- \to (D^0 K^- \text{ or } \overline{D}^0 K^-) \to K_S^0 \pi^0 K^-$. It is not necessary that it be a charged B decay mode, as long as there is a quantum number in the final state (such as strangeness) that identifies with sufficiently high probability whether the decaying meson was a B^0 or a $\overline{B}^0$; for instance, $B^0 \to K^+\pi^-$ versus $\overline{B}^0 \to K^-\pi^+$. It is also possible to see CP violating asymmetries in partial decay rates by observing momentum and/or spin correlations [37].

These searches can be carried out in any experiment in which B and $\overline{B}$ are produced in equal numbers. One simply compares the total numbers of observed B and $\overline{B}$ decays to the chosen mode. Tagging and decay length measurements are not needed. Many of the favorable modes are charmless modes, which do not suffer from the low detection efficiencies typical of charmed final states.

The total number of produced $B\overline{B}$ pairs required to get an asymmetry A with significance s standard deviations is $N(B\overline{B}) = s^2/\mathcal{B}A^2\epsilon$. Provided that there is at least one mode with branching ratio times efficiency $\mathcal{B}\epsilon > 10^{-5}$ and asymmetry $A > 10\%$, then 1.6×10^8 events would be sufficient to establish CP violation to 4 standard deviations.

Because of the simplicity of the measurement and the likelihood that some mode will have an appreciable $\mathcal{B}A^2$ product, this case offers the best chance of being the first experimental indication of CP violation in B decay. Although our present ignorance of the hadronic matrix elements involved in the description of these decays would prevent us from extracting quantitative information on the weak parameters (such as η) from measured asymmetries, there is hope that by the time we have observed an asymmetry we will have accumulated enough data on the various branching ratios to advance considerably our theoretical understanding.

3.4 CP Violation by Interference through $B^0 - \overline{B}^0$ Mixing

To avoid the small and poorly known interference effects that make it difficult to understand CP violation from charged decay asymmetries, we can let the rather large $B^0 - \overline{B}^0$ oscillation amplitude provide the alternate decay route needed for interference. A non-flavor-specific final state f^0 like $\pi^+\pi^-$ can be reached by either B^0 or $\overline{B}^0$ decay. The interference between $B^0 \to f^0$ and $B^0 \to \overline{B}^0 \to f^0$ can be of opposite sign from the interference between $\overline{B}^0 \to \overline{f}^0$ and $\overline{B}^0 \to B^0 \to \overline{f}^0$.

To observe the CP violating asymmetry in the decay rates we have to know for each decay whether the produced meson was a B^0 or a $\overline{B}^0$, but we cannot tell from the final states, since f^0 and $\overline{f}^0$ are common to both. We therefore have to rely on the fact that B mesons are produced in pairs of opposite flavor. In order to tag the flavor of one we observe the other decaying into a flavor-specific final state, for example

$$B^0 \to \ell^+ X \quad \text{vs.} \quad \overline{B}^0 \to \ell^- X$$

or
$$B^0 \to K^+ X \ (\overline{b} \to \overline{c} \to \overline{s}) \quad \text{vs.} \quad \overline{B}^0 \to K^- X \ (b \to c \to s).$$

Since the tag meson can also oscillate to its antiparticle in a correlated way, we have to consider the joint time evolution of the $B^0\overline{B}^0$ pair, which depends on its charge conjugation state C. For the simpler case of a CP eigenstate $f^0 = \overline{f}^0$ the B^0 tagged and $\overline{B}^0$ tagged differential decay rates are given for $C = \pm 1$ by

$$\mathcal{R} = \mathcal{R}(B^0\overline{B}^0 \to B^0 f) = K e^{-T_f} e^{-T_t}[1 + \sin 2\phi \, \sin x_d(T_t \pm T_f)],$$

$$\overline{\mathcal{R}} = \mathcal{R}(B^0\overline{B}^0 \to \overline{B}^0 f) = Ke^{-T_f}e^{-T_t}[1 - \sin 2\phi \, \sin x_d(T_t \pm T_f)].$$

Here K is a constant depending on the branching ratios for the CP eigenstate mode and for the tag mode, T_f and T_t are the two decay times in units of the mean life τ, and $x_d = \Delta m/\Gamma \approx 0.6$ is the mixing parameter. The angle ϕ contains the weak interaction parameters. In the case of decays involving charm, such as ψK_S or D^+D^- it is in fact ϕ_c the lower-right interior angle in the unitarity triangle; in the charmless decays, like $\pi^+\pi^-$, it is ϕ_u, the upper angle (see Fig. 5b). Note that ϕ_c and ϕ_u, which can be directly measured, have no dependence on the strong dynamics.

If the $B\overline{B}$ pair is produced in the $C = +1$ state, we can integrate the differential rates over decay times T_f and T_t and obtain an overall rate asymmetry

$$A = d(x_d) \, \sin 2\phi.$$

The factor d is called the dilution factor and in this case has the value

$$d_{+1}(x_d) = \frac{2x_d}{(1 + x_d^2)^2} \approx 0.64.$$

If $C = -1$, the corresponding asymmetry vanishes identically: $d_{-1} = 0$. To recover a CP violation measurement in the $C = -1$ case one can either measure the differential rates $\mathcal{R}$ and $\overline{\mathcal{R}}$ as functions of $T_f - T_t$ and fit for $\sin 2\phi$ or form a "rectified" overall asymmetry A_{rect} by counting the events with $T_f - T_t < 0$ with opposite sign from the events with $T_f - T_t < 0$. The fitting procedure allows a measurement of $\sin 2\phi$ with an effective dilution factor (assuming no degradation due to experimental resolution in the time measurements),

$$d_{-1}^{fit}(x_d) = 1.2\frac{x_d}{1 + x_d^2} \approx 0.55.$$

Because of the C of the virtual photon, the reaction $e^+e^- \to B\overline{B}$ always produces a $B\overline{B}$ in the $C = -1$ state. One can produce a $B\overline{B}$ pair in a $C = +1$ state by first producing a B^* which decays radiatively to B:

$$e^+e^- \to B^*\overline{B} \text{ (or } B\overline{B}^*) \to B\overline{B}\gamma.$$

The cross section above the B^* threshold is a factor of 7 smaller than at the $\Upsilon(4S)$ resonance [38], where only $B\overline{B}$ is produced. So to carry out the measurement of the asymmetry without having to observe the decay times one has has to increase the integrated luminosity by a factor of about $7 \times d_{-1}^{fit}/d_{+1} \approx 6$. If the increased luminosity is available, however, one can make this measurement at a conventional symmetric e^+e^- collider.

To take advantage of the larger cross section at the $\Upsilon(4S)$, however, one needs to be able to observe the time difference $T_f - T_d$ between the decay of the CP violating mode and of the tag. This has to be done by observing the separation distance between vertices formed by extrapolating measured tracks from the detector into the beam pipe. The mean decay distance is

$$\langle \ell \rangle = \gamma\beta c\tau = \gamma\beta \times (380 \ \mu\text{m}),$$

which at the energy of the $\Upsilon(4S)$ is

$$\langle \ell \rangle = \sqrt{(M_{4S}/2m_B)^2 - 1} \times c\tau = 23 \ \mu\text{m}.$$

This is not a large enough distance to expect to measure accurately. The problem is that the B's are produced almost at rest in the frame of the $\Upsilon(4S)$. The solution is to collide electrons and positrons of different energies so that the $\Upsilon(4S)$ is not at rest in the laboratory and the produced B's are boosted. The product of the two beam energies is constrained by the requirement that the total energy in the center of mass be the 10.58 GeV mass of the $\Upsilon(4S)$, and the difference determines the boost:

$$E_1 E_2 = (M_{4S}/2)^2, \qquad \gamma\beta = (E_1 - E_2)/M_{4S}.$$

A combination like $E_1 = 8$ GeV and $E_2 = 3.5$ GeV gives $\langle \ell \rangle = 160\ \mu$m, which is large enough to be measurable with a high resolution tracking detector at a distance of 2 to 3 cm from the beam line.

Because the measurement of CP violation in neutral B decays to CP eigenstates like ψK_S and $\pi^+\pi^-$ offers the advantage of a direct, unambiguous test of the Kobayashi-Maskawa picture of CP violation, it is the favored choice for experimental study. The goal then is to build an asymmetric e^+e^- collider to run at a center of mass energy of 10.58 GeV with luminosity sufficient to allow a measurement in a reasonable time, say several years, of the time-dependent CP violating asymmetry in one or several charm modes such as ψK_S and in one or several charmless modes like $\pi^+\pi^-$. One will then have direct measurements of two of the angles in the unitarity triangle. If the measurements of the angles and sides are consistent, then we will have confirmed standard picture of CP violation and will have measured the imaginary quark-mixing parameter $i\eta$ that completes our description of the weak interactions of the quarks. Our task now is to estimate the luminosity required in such a collider.

3.5 Required Luminosity

The rectified asymmetry observed in the experiment can be written as

$$A_{obs} = r_{sig} d \sin 2\phi = \frac{(N_+ - N_-) - (\overline{N}_+ - \overline{N}_-)}{(N_+ - N_-) + (\overline{N}_+ - \overline{N}_-)},$$

where $r_{sig} = signal/background$, $N_+ (\overline{N}_+)$ is the number of $B^0 (\overline{B}^0)$ tagged decays observed with $T_f - T_t > 0$, and $N_- (\overline{N}_-)$ is the number of $B^0 (\overline{B}^0)$ tagged decays observed with $T_f - T_t < 0$. Assuming that the asymmetry is not large and that the accuracy is statistics limited, the r.m.s. experimental error in A is $1/\sqrt{N}$, where N is the total number of observed tagged decays:

$$N = 2\sigma L \mathcal{B}_{obs} \epsilon \epsilon_{tag}.$$

Here $\sigma = 0.5$ nb is the $B^0 \overline{B}^0$ production cross section, L is the integrated luminosity, $\mathcal{B}_{obs} = \mathcal{B}/r_{sig}$ is the observed branching ratio and ϵ is the detection efficiency for the CP violating decay, and ϵ_{tag} is the effective tagging efficiency, which can be written in terms of the probabilities for observing a right tag and a wrong tag,

$$\epsilon_{tag} = (r + w)\left(1 - \frac{2w}{r + w}\right)^2.$$

Then the integrated luminosity required for a measurement of $\sin 2\phi$ to an accuracy $\delta \sin 2\phi$ is given by

$$L = \frac{1}{2\sigma \mathcal{B} r_{sig} \epsilon \epsilon_{tag} d^2 (\delta \sin 2\phi)^2}.$$

Table 6. Factors in the calculation of required luminosity

Factor	ψK_S	$\pi^+\pi^-$
2σ	1.1 nb	1.1 nb
$\mathcal{B}$	4×10^{-4}	2×10^{-5}
r_{sig}	> 0.98	0.85, 1.00
ϵ	$0.30 \times \mathcal{B}_\psi$	0.63, 0.69
ϵ_{tag}	0.40	0.22+0.18
d	0.49	0.49
$\delta \sin 2\phi$	0.1	0.1
N	435	496
L	59 fb^{-1}	78 fb^{-1}

Table 7. Tagging efficiencies

Method	$r + w$	$w/(r + w)$	ϵ_{tag}
$e^\pm, \mu^\pm,\ p > 1.4$ GeV/c	0.10	0.04	0.09
$e^\pm, \mu^\pm,\ p = 0.6 - 1.4$ GeV/c	0.10	0.20	0.04
$e^\pm,\ p < 0.6$ GeV/c	0.07	0.16	0.03
$K^\pm$	0.38	0.14	0.20
$D + \pi$'s	0.09	0.01	0.09
All tags (minus overlaps)			0.40

We will estimate the required L for the modes most favorable for measuring $\sin 2\phi_c$ and $\sin 2\phi_u$ to an accuracy of ± 0.1. Table 6 shows the assumed values of the various factors in the calculation. Most of them are taken from CLEO data or based on extrapolations of CLEO experience. The branching ratio $\mathcal{B}$ for the $\pi^+\pi^-$ mode is not known but is taken from theory [18] and the CLEO measurement of $|V_{ub}|$.

The tagging efficiency ϵ_{tag} deserves more comment. High momentum leptons, say with $p > 1.4$ GeV/c, indicate the flavor of the B through the semileptonic decay:

$$B^0 \to \ell^+ X, \qquad \overline{B}^0 \to \ell^- X.$$

The wrong tags come from the decay of secondary charm:

$$B^0 \to \overline{D}x \to \ell^- Xx. \quad \text{and} \quad \text{conjugate.}$$

At low momentum, say $p < 0.6$ GeV/c, the secondary charm leptons predominate and can be used as a tag in the opposite sense, although in this momentum range muons are not reliably identifiable. Even in the intermediate momentum range one can use the lepton information in correlation with a measurement of the missing (neutrino) momentum. Charged kaons from the $b \to c \to s$ chain tag the initial flavor, but other processes ($W \to c\bar{s}$, $g \to s\bar{s}$, etc.) contribute to the wrong tags. In events where one decaying B has already been reconstructed, the efficiency for identifying a D or D^* in the rest of the event is enhanced enough for this to be a useful source of tags. In all, about 40% of the CP violating B decays can be tagged as being accompanied by a B^0 or a $\overline{B}^0$. Note that in the $\pi^+\pi^-$ measurement the efficiency ϵ and the signal purity r_{sig} depend on the type of tag.

Table 8. Expected experimental resolution in decay length

CP violating mode	tag	$\Delta z, \mu$m FWHM
ψK_S	high-p leptons	159
	low-p leptons	246
	$K^\pm$	191
	$D + \pi$'s	153-208
$\pi^+\pi^-$	high-p leptons	151

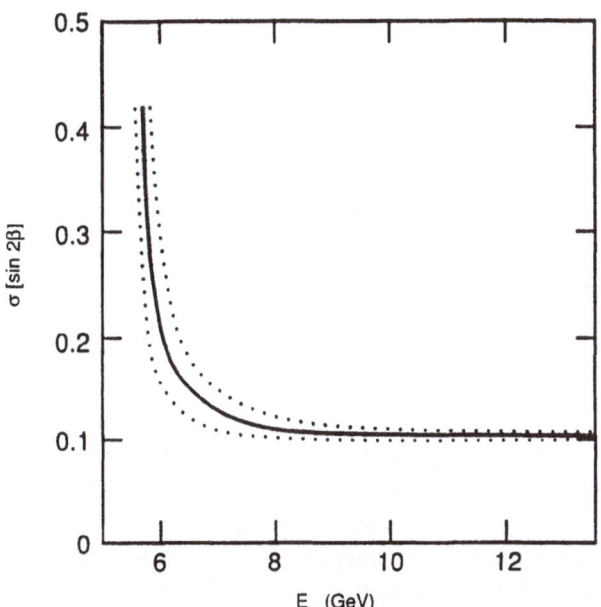

Figure 8. Accuracy in the measurement of $\sin 2\phi_c$ versus the higher of the two beam energies [39]

Table 9. Modes useful for measuring ϕ_c

Mode	Branching ratio $\mathcal{B}$, 10^{-4}	Efficiency ϵ	$\mathcal{B}\epsilon$, 10^{-6}
ψK_S	4	0.29	16
$\psi' K_S$	< 3		< 6
$\psi K^{*0} \to \psi K_S \pi^0$	4	0.15	8
χK_S or χK^{*0}	?		?
$D^+ D^-$	3.7	0.0059	2.2
$D^{*+} D^-$, $D^+ D^{*-}$	15	0.0036	5.4
$D^{*+} D^{*-}$	16	0.0028	4.5
$D^0 \pi^0$, or ρ, ω $(D^0 \to CP$ eigenstate)	2	0.007	1.6
$D^{*0} \pi^0$, or ρ, ω $(D^{*0} \to D^0 \to CP$ eigenstate)	2	0.007	1.6

Table 10. Modes useful for measuring ϕ_u

Mode	Relative sensitivity $\mathcal{S}$	Branching ratio $\mathcal{B}$, 10^{-4}	Efficiency ϵ	$\mathcal{S}\mathcal{B}\epsilon$, 10^{-6}
$\pi^+ \pi^-$	1	0.2	0.69	14
$\rho^\pm \pi^\mp$	1/2	0.7	0.40	14
$a_1^\pm \pi^\mp$	1/5	0.6	0.25	2
$\rho^+ \rho^-$	1/9	0.2	0.25	1

The dilution factor $d_{-1}^{ft}(x_d)$, which would be 0.55 in the case of an ideal detector, gets degraded to 0.49 by the experimental resolution in the measurement of $T_f - T_t$. This has been studied by Monte Carlo simulation based on conservative assumptions about the performance of the precision tracking detector close to the beam pipe. Table 8 summarizes the expected resolution widths Δz. The full-width half-maximum resolutions are to be compared with the tag-f separation distance for maximum CP asymmetry,

$$\Delta_m z = 0.58 \gamma \beta c\tau / x_d \approx 140 \ \mu\text{m},$$

where we have used the $\gamma\beta$ appropriate for $8 + 3.5$ GeV colliding energies. Actually, the accuracy in the measurement of $\sin 2\phi$ is not sensitive to the machine energy difference provided the high energy ring has at least 7 GeV (see Fig. 8).

Our conclusion is that $N \sim 500$ tagged decays are needed to determine a $\sin 2\phi$ value to ± 0.1. To accomplish this for both ψK_S and $\pi^+ \pi^-$ modes requires an integrated luminosity of $L \sim 80$ fb^{-1}. If the long term average luminosity is about one-third of the peak luminosity, as it is now at CESR, then this integrated luminosity can be achieved in 32 months of running with a peak luminosity of $\mathcal{L} = 3 \times 10^{33}$ cm^{-2}sec^{-1}. We therefore set this peak luminosity as the goal for our e^+e^- B Factory.

We have concentrated our attention on the ψK_S and $\pi^+ \pi^-$ modes for measuring the angles ϕ_c and ϕ_c in the unitarity triangle. Tables 9 and 10 show some of the other modes that can be used. Those that involve several interfering diagrams and those that are not CP eigenstates may involve a more complicated analysis [40],[37],[41],[44], but they should be usable nevertheless. If the standard picture of CP violation is correct, data from the additional modes can be combined with the data from ψK_S or $\pi^+\pi^-$ to improve the accuracy of the determination of ϕ_c or ϕ_u. Various nonstandard models [36],[42],[43] predict that there will be no direct CP violation in B decay or that the measured lengths and angles will not be consistent with a closed triangle or that modes that give the same asymmetry in the Kobayashi-Maskawa picture will actually have different asymmetries. The extra modes would then give independent information that could lead to a deeper understanding of the theory.

So far we have not mentioned the lower-left angle of the unitarity triangle. With more luminosity it is possible to measure this angle, using the interference of $B \to D^0 X$ with $\overline{B}^0 \to \overline{D}^0 X$ when the D decays to a final state that is accessible to both D^0 and $\overline{D}^0$ [40],[44].

4   B FACTORY AT CESR

4.1   Introduction

As we have seen in the previous lectures, the understanding of CP violation depends on observing it in B decays. In particular, unambiguous measurements of the relevant weak interaction parameters require the observation of the asymmetry in the time evolution of of $B^0\overline{B}^0$ pairs produced at the $\Upsilon(4S)$ resonance decaying into final states such as $\psi K + \ell^+ X$ and $\psi K + \ell^- X$. The required experimental facility, called a B Factory, must have (a) peak luminosity of the order of 3×10^{33} cm^{-2}s^{-1}, so that the asymmetries can be measured in a few years of running, (b) asymmetric electron and positron energies, in order to boost the $\Upsilon(4S)$ in the laboratory so that the decay distances can be observed, (c) a precision vertex detection system capable of resolving decay lenths of the order of 100μm.

The Cornell Electron-Positron Storage Ring (CESR) has been operating at the b-quark threshold since 1979. It has been upgraded from its original single bunch configuration to the present scheme with seven bunches in each beam, colliding at one interaction point and separated elsewhere by electrostatic fields. The focusing at the interaction point has also been strengthened. The CESR peak luminosity at the $\Upsilon(4S)$ is now 2.5×10^{32} cm^{-2}s^{-1}, which is six times the design luminosity for that energy and a world's record for storage rings.

During the last 13 years the CESR experimental program has concentrated on B mesons. The B was in fact discovered at CESR, as was the $\Upsilon(4S)$ resonance, the prime decay mode $B \to \psi K$ for CP violation measurements, and most of what we now know about B mesons. In 1989 the original CLEO detector was replaced by a more powerful CLEO-II detector with unexcelled energy and angle resolution for both charged and neutral products of B decays. Given our experience with B physics and high luminosities and the considerable investment already made in the CESR machine and the CLEO-II detector, it was natural then for us to start planning in 1988 for an upgrade that would make it possible to study CP violation. A formal proposal was submitted in Feb., 1991.

Dictated by the goals outlined above, there are three major features that form the basis for the CESR-B design.

(1) The two beams are stored in separate rings with different energies, 8 and 3.5 GeV.

(2) The beams cross at a small angle.

(3) The gain in luminosity relative to CESR is achieved by storing many more bunches, with individual bunch parameters similar to those in CESR.

4.2 Design Concept

(a) Asymmetric Energies

The mean decay length of a B meson, produced almost at rest in the frame of the $\Upsilon(4S)$, is boosted in the lab to

$$\langle \ell \rangle = \beta\gamma c\tau = \left(\frac{E_1 - E_2}{M_{4S}}\right) \times (380\mu\text{m}),$$

where E_1 and E_2 are the two ring energies. They are also constrained by the condition that the c.m. energy be that of the $\Upsilon(4S)$ resonance:

$$E_1 E_2 = (M_{4S}/2)^2 = (5.29 \text{ GeV})^2.$$

The energy asymmetry should be sufficient to allow measurements of the decay times in the region where the CP asymmetries are expected, but not so large as to complicate the design of the detector or to make the high energy ring too expensive. To find the most cost effective compromise one has to know the time scale of the decay asymmetry, the resolution of the innermost tracking detector, and its distance from the beam line. The maximum asymmetry occurs when the $c\Delta t$ difference between the decay of interest (say $\rightarrow \psi K$) and the tag ($\ell^{\pm} X$, for instance) is $0.52c\tau/x_d \approx 330\mu\text{m}$. A silicon microstrip detector [45] with intrinsic r.m.s. resolution $30\mu\text{m}$ or better, located at a distance of 2 to 3 cm, can achieve extrapolated vertex accuracies of the order of $70\mu\text{m}$ [46]. This suggests that a boost factor $\beta\gamma \geq 0.3$ should be adequate, that is, $E_1 \geq 7.1$ GeV for the higher energy ring. The conclusion is confirmed by careful Monte Carlo simulation [46], [39], [47]. Figure 8 shows how the accuracy of the measurement of the CP asymmetry depends on the the choice of E_1. Evidently, any value above about 7 GeV is sufficient. To provide some margin without incurring too much additional cost, we choose 8 GeV for the high energy ring. This then implies 3.5 GeV for the other ring. It is a fortunate choice, since the present CESR ring was designed for optimum operation at 8 GeV per beam, before the b quark was discovered.

Granted that the measurement of CP violation is greatly facilitated by asymmetric beam energies, is it really possible to operate a high luminosity collider in such a configuration? We know from experience in many storage rings that the stability of the beams is very sensitive to any imbalance, for example in beam currents. Computer simulations of the beam-beam interaction [48], however, indicate that the energy imbalance will not lead to any sacrifice in luminosity, provided that certain parameters are matched between the two beams, such as the product of energy and current $E \times I$, and the bunch sizes σ_x, σ_y, and σ_z and the β functions at the interaction point. The two rings should also have the same circumference. We have worked these conditions into our design for CESR-B. Although the ISR and HERA are very different kinds of machines, the fact that they have already operated with energy asymmetries gives us some confidence.

Nevertheless, we should be prepared for the possibility that it will be difficult to reach the required luminosity with unequal beam energies. We have designed CESR-B so that it can be easily modified to run with equal energies. The measurement of

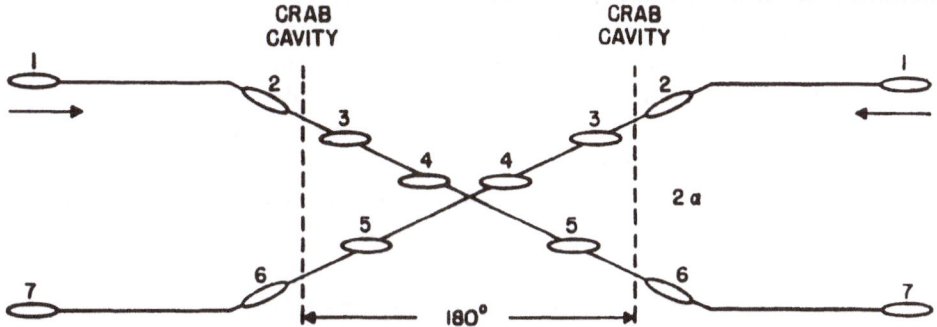

Figure 9. The crab compensation scheme. The ovals show the beam bunches (top view) at successive times as they pass through the interaction region.

CP violation then would have to depend on the observation of decay rate asymmetries in charged B decays at the $\Upsilon(4S)$ or neutral B decays above the threshold for $e^+e^- \to B\overline{B}^*$. Although the former measurement is somewhat compromised by theoretical uncertainties, and the latter suffers about a factor of six in effective $B\overline{B}$ production rate, they nevertheless provide an alternative to asymmetric beam energies.

(b) Crossing Angle

All existing storage rings operate with head-on collisions; that is, in the interaction region the two beams travel on the same orbits. When the orbits cross at an angle α (see Fig. 9) that is larger than the bunch aspect ratio σ_x/σ_z one loses in geometric overlap of the colliding bunches and the luminosity is decreased. More important, however, is the fact that the beam-beam interaction, which now includes transverse forces that depend on the longitudinal coordinate in the bunch, can excite synchrobetatron resonances that destabilize the beam. Such an effect is believed to be responsible for limiting the luminosity in the original configuration of the DORIS rings, the only previous case where angle crossing was tried [49].

Using a "crab compensation" scheme [50], one can cross the beam orbits at an angle and still have head-on bunch collisions. One applies opposite transverse kicks to the head and tail of the bunch at one-quarter betatron wavelength before it reaches the crossing point so that when the bunch arrives it has rotated by the amount of the crossing angle (see Fig. 9). If the opposing bunch has also been rotated, the two will superpose at the interaction point. In fact, in the frame of reference that moves sideways at velocity $c\sin\alpha$ the two bunches collide head-on with zero crossing angle. Lorentz covariance requires that the beam-beam interaction will then be the same as if the orbits intersected in the lab with no crossing angle. Of course, one has to remove the transverse motion of each bunch with a compensating kick as the bunch leaves the interaction region downstream.

The problem is then to produce the transverse kicks. One way, the method we have planned for CESR-B, is to provide four rf cavities operating in a transverse horizontal deflection mode, phased so that each cavity exerts zero net impulse on the center of each bunch. The technology of such cavities is actually not new; they are rather similar in concept to the rf separators used in secondary beam lines.

We have to insure that the kicks can be made to the required accuracy to keep the collisions head-on and to cancel the net transverse oscillation. In CESR we have recently measured, without crab compensation, the tune shift and luminosity for beams colliding at an angle $\alpha \approx 2$ mr (the largest crossing angle we can make with electrostatically separated beams in a single ring), and we have found no degredation compared with head on collisions. That means that the $\alpha = 12$ mr crossing can tolerate an error in crab compensation of at least 2 mr.

There are several important advantages of angle crossing compared with head-on collisions. First, it is the easiest and most natural way to separate the two beams away from the interaction point. It minimizes the number of magnet elements that must be designed to accommodate two beams of different energies. The geometry of the crossing is independent of the energy ratio of the two beams, which can then be tuned to optimize performance or even set equal. Head-on beams would have to be separated by either magnetic or electrostatic fields, which would generate intense synchrotron radiation background close to the sensitive elements of the detector. The only disadvantage of the angle crossing option is the necessity of installing crabbing cavities.

(c) High Luminosity

In the high beam current limit the luminosity in $\text{cm}^{-2}\text{s}^{-1}$ can be written in the form

$$\mathcal{L} = 2.17 \times 10^{34} \frac{(1+r)\xi E I}{\beta_y^*},$$

where $r = \sigma_y/\sigma_x$ is the transverse aspect ratio of the beams at the crossing, E is the beam energy in GeV, ξ is the beam-beam tune shift parameter, and β_y^* is the vertical focus depth of field parameter in cm. The parameters can be evaluated for either beam.

For most storage rings $r \approx 0$. Round beams ($r = 1$) could in principle provide more luminosity, but so far no one has come up with a satisfactory design of interaction region optics that accomplishes this without a heavy penalty in synchrotron radiation background. So we use $r \approx 0$, as in CESR.

The tune shift ξ parametrizes the focusing effect of one beam on the other. The nonlinear part of this beam-beam interaction excites destabilizing resonances and sets a limit on the beam density. Empirically, high luminosity storage rings perform best at $\xi \approx 0.03$, although higher values have been reached in special circumstances. We therefore take that as a design figure for CESR-B.

We would like to minimize β_y^* by focusing the beam as strongly as possible at the interaction point. This involves very strong quadrupole magnets placed as close as possible to the beam crossing. Although it is difficult to achieve low β_y^* values under the simultaneous constraints of physically realizable field gradients, space interference with the detector, the geometry of the crossing beams, and the requirement of low synchrotron radiation background, it is nevertheless possible to find solutions in the $\beta_y^* \sim 1$ cm range. The real limitation, however, comes from the requirement that the bunch length σ_z not be greater than the depth of focus β_y^*. Otherwise, much of the bunch overlap occurs where the beams are not tightly focussed, the so-called "hourglass" effect. It is very difficult to make bunches shorter than the $\sigma_z = 1.5$ cm now reached in CESR. We therefore assume $\beta_y^* = 1.5$ cm in CESR-B.

The beam current I is the product $i_b n_b$ of the current per bunch and the number of bunches. The single bunch current i_b is limited by wake field effects; that is, at the current limit the electromagnetic field left behind by the head of the bunch blows up the tail of the bunch. CESR with 2.3×10^{11} particles per bunch is already near this limit. Since there is no practical cure for the single bunch instability, we cannot expect to use higher bunch currents in CESR-B than already achieved in CESR.

Table 11. Parameter lists

	CESR-B	CESR ('96)	CESR ('92)	PEP-2
Operating energies, GeV	8, 3.5	5.3	5.3	9, 3.1
Crossing angle, mr	±12	±2	0	0
Luminosity, 10^{33} cm^{-2}s^{-1}	3	1.7	0.25	3
Circumference, m	765	768	768	2199
Bunches/beam	164	∼ 40	7	1746
Bunch separation, m	4.66	∼ 5, 60	110	1.26
Particles/bunch, 10^{11}	0.84, 1.92	∼ 2	2.3	0.41, 0.59
β_y^*, cm	1.5	1.4	1.8	3.0, 1.5
Beam-beam tune shift	0.03	0.03	0.03-0.04	0.03
RF system, MW	sc, 6.25	sc	nc	nc, 13.13
Detector	CLEO II+	CLEO II+	CLEO II	new
Upgrade cost, M\$	106	10	-	167+det.

The only remaining possibility for making more than an order of magnitude increase in CESR luminosity is to increase greatly the number of circulating bunches, now seven. Since we have to have separate rings to accommodate the two beam energies, the bunches of one beam can interfere with those of the other only where they cross. The longitudinal separation of the bunches (4.66 m) must be large enough so that at the point of the first near miss, close to the crossing point, the two beam orbits have diverged sufficiently (a few cm) to allow the bunches to pass without perturbing each other. The luminosity goal can be achieved with $n_b = 164$ bunches in each ring. A more serious difficulty is the multibunch wake field effect. The electromagnetic wake of one bunch can disturb the following bunches and limit the total current. The wake fields can be minimized by avoiding resonant structures that allow them to build up. Provided the buildup times for these resonant modes is not too short compared with the natural damping time constants, they can be controlled by feedback applied to individual bunches; when you detect that the centroid of a bunch is moving from its nominal position, you can push it back with a pulsed field.

The parameter list for CESR-B is shown in Fig. 10 (for more details see the design report [51]). Some of the important parameters are compared in Table 11 for CESR-B, CESR, and the SLAC proposal [39]. The very large increase in the number of circulating bunches, a factor of 23.4 in the CESR upgrade, implies very large beam currents, of the order of Amperes. This leads to a number of engineering challenges that have to be met in the construction of the B Factory.

4.3 Engineering Challenges

A relativistic electron in a circular orbit emits synchrotron radiation. The radiated power is given by

$$P = \frac{2}{3} r_e m_e c^3 \frac{\gamma^4}{\rho^2},$$

where $r_e = 2.818 \times 10^{-15}$ m, $\gamma = E_e/m_e c^2$, and ρ is the radius of curvature. For the 0.87 Ampere 8 GeV beam this amounts to 4.16 MW. The intense xray radiation raises a number of technical problems.

PARAMETER	HER	LER
E_{beam} [GeV]	8.00	3.50
Luminosity [10^33/cm^2/sec]	3.00	
ξ_v/β^* (1+r) [/m] (lum. coef.)	2.04	
n_b (number of bunches)	164	
r (aspect ratio)	0.020	
N [x10^11] (e/bunch)	0.84	1.92
I_{tot} [Amps] (current in one beam)	0.87	1.98
ξ_v (tune shift parameter)	0.03	
$2\pi R$ [m] (circumference)	764.84	
ρ_{arc} [m] (bending radius in arc)	87.90	20.00
σ^*_H [mm]	0.37	
σ^*_v [μm]	7.5	
β^*_H [m] (focusing param at ip)	1.00	
β^*_v [cm]	1.50	
θ_c [mr] (crossing half-angle)	12.00	
σ_L [cm] (bunch length)	1.00	
ε_H [x10^-7 m] (emittance)	1.35	
α_p [10^-2] (momentum compaction)	0.84	1.10
D^*_H [m] (dispersion at ip)	0.00	
Q_s (synchrotron tune)	0.085	
Q_B (betatron tune)	12.70	11.70
σ_E/E [10^-4] (energy spread)	8.16	6.42
U_o [MeV] (s.r. loss/turn)	4.80	0.73
P_{sr} [MW] (per beam)	4.16	1.44
V_c [MV] (accel cavity voltage)	32.67	11.59
U_c [MV] (crab voltage)	1.83	0.80
β_{crab} [m]	25.00	
k [V/pc] (loss parameter)	5.63	3.79
P_{BCM} [kW] (per beam)	65.53	230.47
n_c (number of s.c. cavities/ring)	12.00	4.00
N_k (number of 0.8 MW klystrons)	6.00	2.00
P_{rf} [MW] (available rf power)	4.80	2.40
λ_{rf} [cm] (rf wavelength)	60.00	
τ_E [ms] (energy damping time)	4.26	12.32
$\tau_{x,y}$ [ms] (betatron damping time)	8.51	24.65

Figure 10. Parameter list for CESR-B.

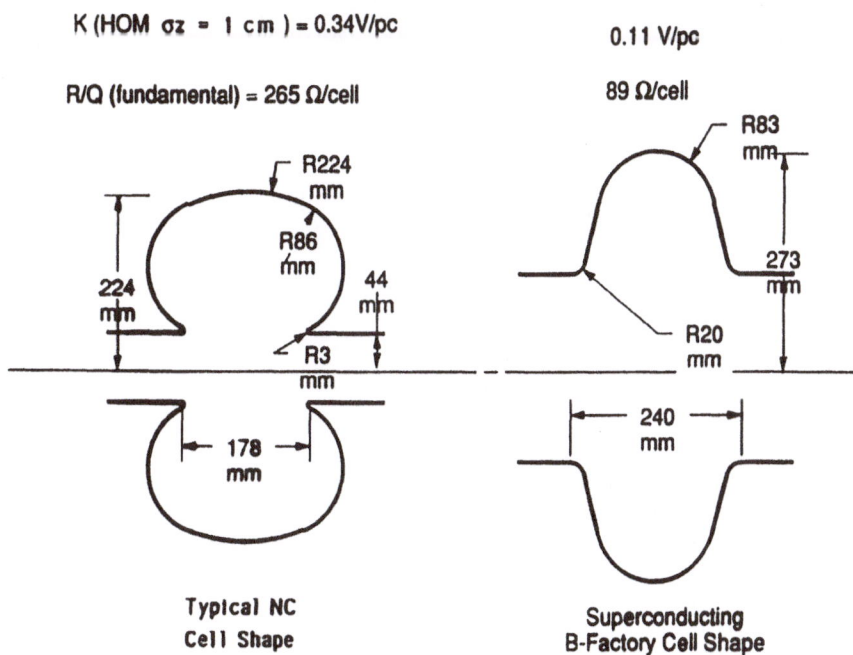

K (HOM σz = 1 cm) = 0.34V/pc   0.11 V/pc

R/Q (fundamental) = 265 Ω/cell   89 Ω/cell

**Typical NC
Cell Shape**

**Superconducting
B-Factory Cell Shape**

Figure 11. Comparison of rf cavity shapes: normal conducting nose-cone structure (left) and superconducting large-pipe structure (right).

(1) The energy has to be put back into the beam by a system of rf accelerating cavities, while maintaining minimum impedence $Z \sim V_{wake}/I_{beam}$ for resonating wake fields excited by the beam.

(2) The xrays must be absorbed and the heat carried away.

(3) The gases desorbed by the bombardment of the beam pipe walls must be pumped out fast enough to maintain an adequate vacuum, for long beam lifetime and low experimental background.

(4) The radiation must not be allowed to strike the detector elements, even on first or second bounce.

I will discuss our solutions to these problems under the headings of rf, vacuum, and shielding for the detector.

(a) RF System

The rf system not only has to replace the power radiated by the beam, but it has to shorten the bunches so that the condition $\sigma_z \sim \beta_y^*$ can be met. We need an accelerating voltage of 33 MV per turn in the 8 GeV ring, considerably more than that required to compensate for synchrotron radiation, in order to provide enough phase focussing to make a bunch length $\sigma_z = 1$ cm.

Providing high V_{acc} in rf cavities exacts an extra price in power dissipated in the cavity walls. Although in existing e^+e^- storage rings practically all of the rf power goes into the beam, it is not true for a B Factory with short bunches. The dissipated power P_{diss} is related to V_{acc} through a "shunt impedance" factor, which is the product of a parameter R/Q that is determined by the cavity-and-mode geometry and the Q of the

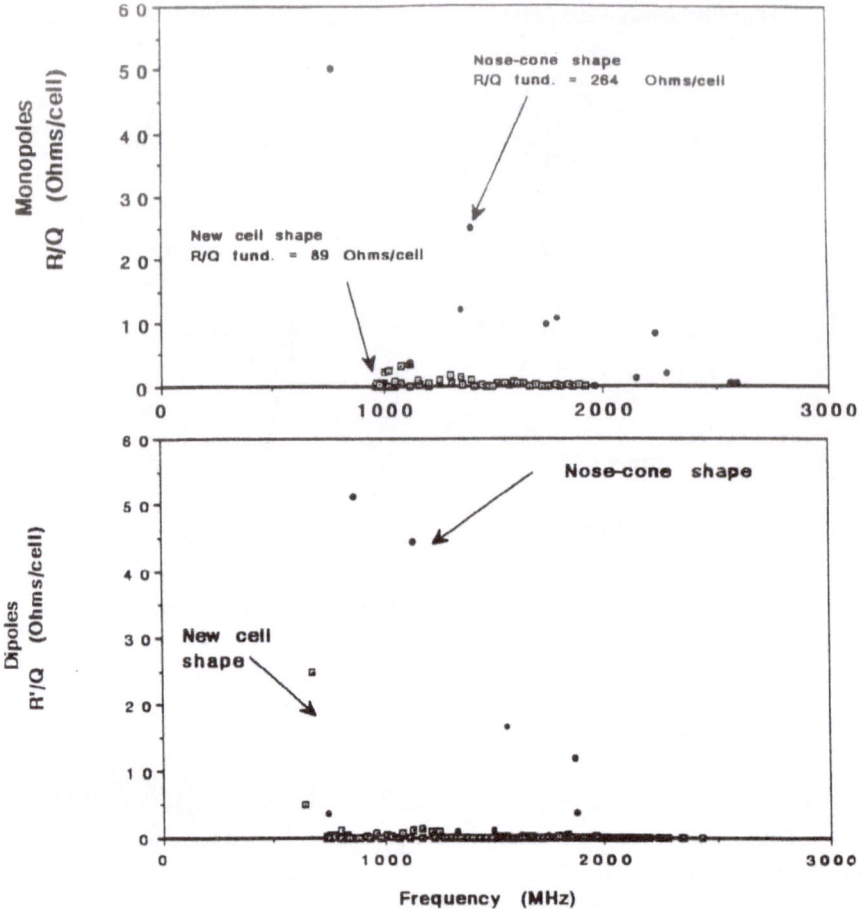

Figure 12. R/Q values for higher modes of CESR-B (squares) and CESR (black circles) rf cavities.

cavity, which is determined mainly by the surface conductivity:

$$P_{diss} = \frac{V_{acc}^2}{(R/Q)Q}.$$

To minimize the power for a given voltage, we want to maximize the shunt impedence $(R/Q)_0 Q_0$ for the fundamental accelerating mode of the cavity.

Higher frequency cavity modes, excited by the passage of the beam bunches, contribute to the multibunch instability. For any particular mode n the cavity contribution to the resonant wake will be proportional to the shunt impedence for that mode,

$$V_{wake,n} = I_{beam,n}(R/Q)_n Q,$$

as will be the corresponding power loss,

$$P_{diss,n} = I_{beam,n}^2 (R/Q)_n Q.$$

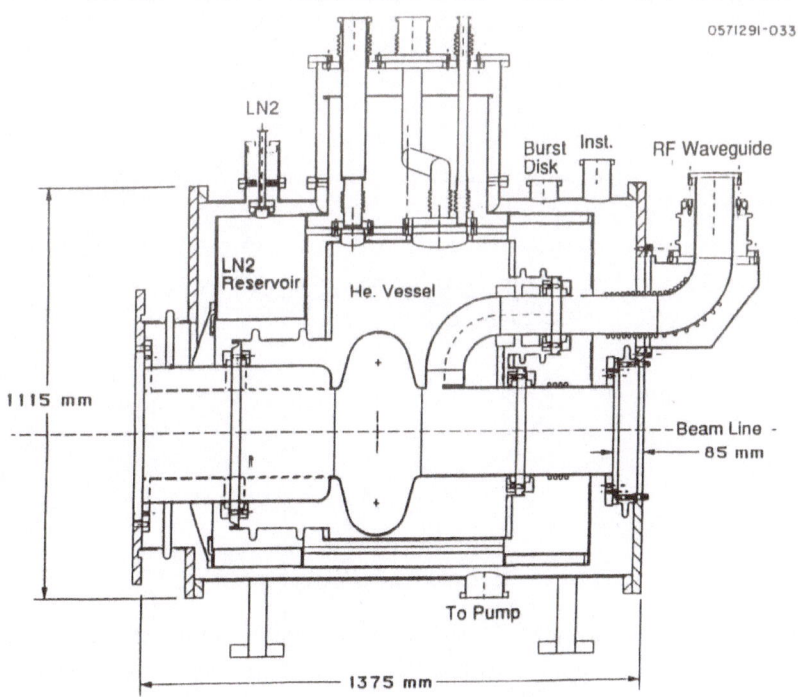

Figure 13. The CESR-B superconducting rf cavity in its cryostat.

Table 12. Comparison of CESR and CESR-B rf cavities

Mode	CESR, Cu		CESR-B, Nb	
	R/Q	Q	R/Q	Q
Fundamental (500 MHz)	264 Ω	30000	89 Ω	10^9
Worst transverse HOM	50 Ω	700	25 Ω	< 200
Typical transverse HOM	5-20 Ω	700	< 3 Ω	< 75
Worst longitudinal HOM	50 Ω	700	3.5 Ω	70

So the problem is to maximize the shunt impedence for the fundamental mode while minimizing it for all higher order modes.

If the cavity is made of copper, $Q \approx 30000$, and $(R/Q)_0$ is maximized by using a reentrant "nose-cone" geometry, as in the present CESR cavities (Fig. 11), for which $(R/Q)_0 Q = 8$ MΩ per cell. The $(R/Q)_n$ for the higher modes has to be brought down to much lower values in a B Factory (say 100 Ω per cell) than would be tolerable in a machine such as CESR with much less beam current. Since this is difficult with the nose-cone geometry, we would end up having to depend on very strong feedback suppression of the multibunch instabilities, a very risky solution.

Instead we propose to use superconducting niobium for the cavity walls. With $Q \approx 10^9$ it is not necessary to optimize the geometry of the cavity for a large $(R/Q)_0$. Instead, a more open geometry with a large beam pipe radius (Fig. 11) allows the higher frequency modes to propagate away and be absorbed outside the cavity, resulting in higher mode shunt impedences that are typically well below 100 Ω (see Table 12 and

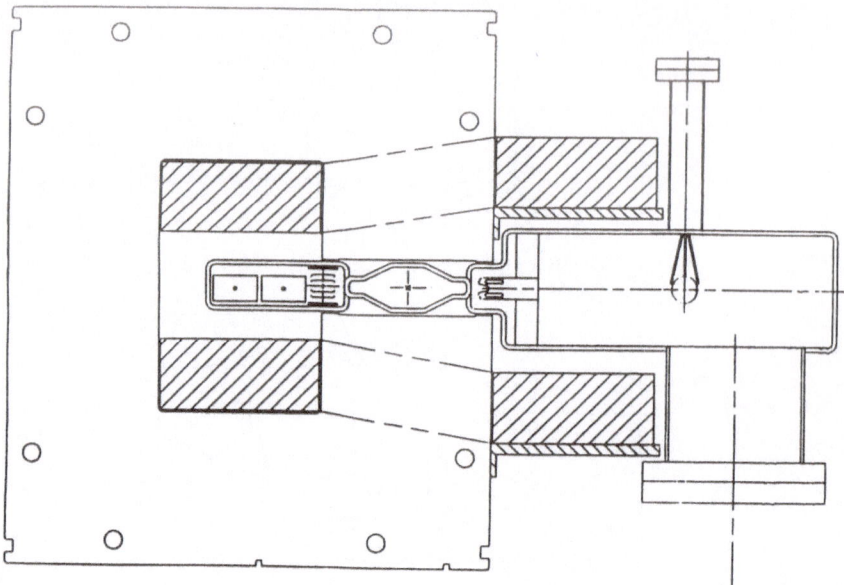

Figure 14. Cross section view of vacuum chamber and pumps in a high-field bending magnet of the CESR-B high energy ring. On the left of the beam pipe are non-evaporable getter pumping modules; on the right is a titanum sublimation pump

Fig. 12). The cavity we have designed (we are now testing the prototype) has only one resonant mode, the fundamental accelerating mode at 500 MHz. The higher modes are very strongly suppressed and only two, the lowest transverse modes, require a modest amount of feedback to stabilize the multibunch motion. Figure 13 shows a diagram of the cavity in its cryostat.

There is a practical limitation in the amount of input microwave power that can be fed into a cavity through a vacuum window. It is about 0.5 MW. That determines the number of cavities needed around the ring. Superconducting cavities have negligible wall dissipation and higher order mode power loss compared to copper cavities, for which only about half the power would go into the beam. A superconducting cell can therefore develop a much higher V_{acc} for the same 0.5 MW input power. The contribution to the higher mode wake field impedence is therefore further minimized by requiring fewer cells. A total of 16 cells, 4 of of them in the low energy ring, is more than enough for CESR-B. In addition 4 single-cell cavities are required for the crab compensation. These would also be superconducting and would operate at similar field strength values, but with a geometry optimized for the transverse mode.

Superconducting cavities are not a novelty in accelerators. They are being used at Tristan, LEP, and HERA. The CESR superconducting rf group has 17 years experience in superconducting cavity research and development, and performed the first tests of such cavities in an accelerator and in a storage ring.

(b) Vacuum

The beam lifetime is determined mainly by beam-gas scattering. For a 3(5) hour lifetime for the 8(3.5) GeV beam we need an average residual gas pressure of 5×10^{-9} Torr. In the vicinity of the interaction region we would like to do a factor of 5 better, in

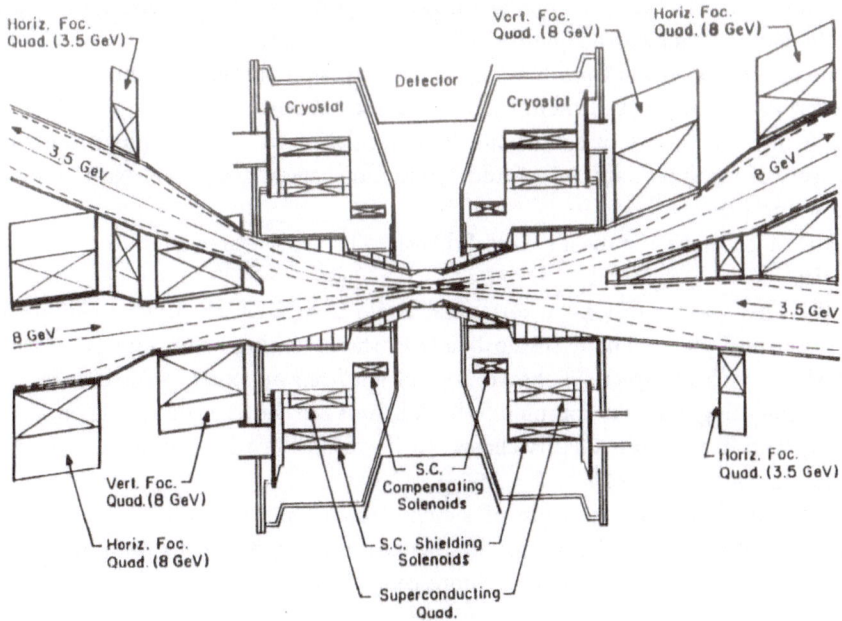

Figure 15. Plan view of the layout of the intersection region. The vertical scale of the picture is expanded by ×10.

order to reduce the background in the detector from scattered electrons and radiated photons. It is not trivial to maintain such a pressure against the desorption of gas from the chamber walls bombarded by the very intense synchrotron radiation.

The solution involves a number of innovations: (a) replacing the CESR aluminum vacuum chamber by a copper chamber, to better handle the radiation and beam image current heat loads, (b) absorbing the radiation in a separate chamber connected to the beam chamber by a narrow opening (see Fig. 14), (c) using distributed pumping by titanium sublimation and nonevaporable getters spread along the magnet arcs, and (d) using auxilliary discrete sputter ion pumps for nongetterable gases. The vacuum system represents a major part of the CESR-B upgrade project, 12% of the total cost.

(c) Shielding the Experiment

The CP violation measurements depend on locating decay vertices with an accuracy better than 100 μm, which in turn requires a high resolution tracking detector, say a silicon strip device, surrounding a thin beryllium beam pipe at a distance of 2 to 3 cm from the beam line. It is not trivial to operate such a detector successfully so close to circulating Ampere beams. One has to examine carefully the sources of background, find appropriate shielding strategies, calculate the detector dosage for the proposed interaction region geometry, and evaluate the detector damage, the currents, and the background counting rates.

There are two background sources: (1) synchrotron radiation hitting the interaction region beam pipe, either directly or after one or more scatterings off nearby surfaces, and (2) secondary high energy electrons and photons from beam-gas Coulomb scattering and bremsstrahlung hitting the detector. The synchrotron radiation is in the xray region, and is very intense especially near the beam line. The beam pipe must be shadowed from all sources, including secondary and tertiary scattering surfaces, and from beam tails as well as the core. The beam-gas background is more diffuse, more difficult to shield, and depends on the quality of the vacuum. We have written very detailed simulations of the background production and propagation through the interaction region geometry, and have checked them by reproducing the measured backgrounds observed in CESR under a variety of conditions. The simulation has been applied to many candidate sets of machine parameters and interaction region geometry. Most designs fail; very few survive.

Firure 15 shows the most successful design. It incorporates a number of important strategies.

(1) The beams cross at a horizontal angle $\alpha = \pm 12$ mr. Both rings are in the same horizontal plane. There are no bends in the incoming beam lines, and therefore no nearby sources of direct synchrotron radiation from on-axis beam particles.

(2) Focusing magnets common to both beams are centered on the incoming beam.

(3) The beam emittance is chosen to be rather small to minimize synchrotron radiation in the quadrupoles.

(4) The shape and location of synchrotron radiation masking is optimized to prevent radiation hitting the interaction region beam pipe, and to minimize the effect of tipscattering, backscattering, and fluorescence from the masks.

(5) The inner surface of the 1 mm thick beryllium beam pipe is coated with a 100 μm layer of aluminum and a 25 μm layer of copper to absorb xrays at glancing angles.

(6) Extra pumping is to be installed near the interaction region to reduce the gas pressure to one-fifth of the average in the arcs.

As a result the predicted radiation dose for the silicon detector is less than an acceptable 10 krad/year, and the dose to an argon/ethane wire chamber cell is less than 0.02 C/cm-yr, well below the acceptable level of 0.1 C/cm-yr. The occupancy in the silicon detector inner layer adds up to 2.2% of the channels, and 98% of typical $B\overline{B}$ events will be free of any spurious track segments in silicon.

4.4 Status of the Proposal

In February, 1991 a proposal to upgrade CESR to a B Factory was submitted to the National Science Foundation. The upgrade would make maximum use of the the existing facility: the laboratory, the 760 m circumference accelerator tunnel, the 8 GeV magnet ring, the linac-plus-synchrotron injection system (it has already performed at the current per bunch and total current levels required), and the CLEO-II detector. The major items of new construction would be the following: the 3.5 GeV magnet ring, a new copper vacuum chamber for both rings, a superconducting rf system, focusing magnets for the interaction region, and some additional building space.

At the same time SLAC proposed to the DOE to upgrade the PEP machine to a B Factory [39]. CESR-B and PEP-2 parameters are compared in Table 11. The KEK lab in Japan has also proposed a B Factory [52]. In July, 1991 the CESR-B cost estimate was reviewed by a panel of experts named by the NSF. The NSF and DOE at first decided to have a joint technical review of both proposals but then in January, 1992, decided to postpone further consideration of a US B Factory until it was clear that either one could be funded. In April, 1992, an advisory subpanel of HEPAP [53] concluded that if inflation adjusted DOE funds for high energy physics (exclusive of SSC construction) continued level for the next several years then there would be enough

Table 13. CESR Upgrade schedule

Phase	Date	Improvements	$\mathcal{L}_{pk}$ goal $\text{cm}^{-2}\text{s}^{-1}$
I	-1993	single i.r., new Cu rf cavities	4.5×10^{32}
II	1994-5	± 2.5 mr crossing angle, 27 bunches/beam	6×10^{32}
III	1996	SC rf cavities, 40 bunches/beam	1.7×10^{33}
IV	1996-9	2nd ring, ± 12 mr crossing angle, SC i.r. quads	3×10^{33}

to begin construction of PEP-2 in 1996. Also, the NSF announced [54] that if funds for large projects continued level for the next several years then there would be enough to begin construction of the second ring for CESR-B in 1996.

Meanwhile we are carrying out an active R&D program with the following objectives:

(a) to finish the detailed engineering design of the B Factory components – superconducting cavities, vacuum chamber and pumps, feedback circuitry, magnets for the low energy ring and interaction region,

(b) to prototype these components and in most cases test them in the present CESR ring,

(c) to make all the modifications to the CESR ring that are required for it to serve as the 8 GeV B Factory ring and that are consistent with continued running as a single ring collider,

(d) to get operational experience with high beam currents and superconducting rf cavities,

(e) to raise the luminosity of CESR in the present single ring configuration to as high a level as possible.

The key to increasing the single ring CESR luminosity is a scheme to collide the beams at a small crossing angle, allowing each of the present seven bunches to be replaced by a train of several closely spaced bunches. As the beam current increases more new components, such as rf cavities, are required to sustain the increased synchrotron radiation power. We expect to proceed on this combined program of luminosity upgrading and B Factory preparation with a schedule (see Table 13) that culminates with the beginning of construction of the second ring in 1996.

The present CLEO detector is also being upgraded. The data readout system has recently been speeded up so that CLEO can now write data to tape at the rate of 30 Hz. Further improvements may double that speed. Construction of a three layer silicon strip vertex detector system has begun. This will be fitted around a new 2 cm radius beryllium beam pipe. Research is continuing on schemes to increase the particle identification capabilities of the detector. A new main drift chamber will be built to accommodate the more restrictive geometry of the B Factory interaction region. This will be an opportunity to improve further the momentum resolution and reduce the amount of material in the path of the particles. With these improvements the CLEO detector will be the ideal instrument for the investigation of rare B decays in the next few years and later for the study of CP violation in B decays.

References

[1] N. Cabibbo, *Phys. Rev. Lett.* 10: 531 (1963).

[2] S.L. Glashow, J. Iliopoulos, and L. Maiani, *Phys. Rev.* D2: 1285 (1970).

[3] M. Kobayashi and T. Maskawa, *Prog. Theor. Phys.* 35: 252 (1977).

[4] L. Wolfenstein, *Phys. Rev. Lett.* 51: 1945 (1984).

[5] D. Bortoletto et al. (CLEO), *Phys. Rev.* D45: 21 (1992).

[6] Particle Data Group, *Phys. Lett.* B239: 1 (1990); D. Decamp et al. (ALEPH), CERN preprint PPE/90-116 (1990); G. Alexander et al. (OPAL), CERN preprint PPE/91-92 (1991); B. Adeva et al. (L3), L3 preprint 32 (1991).

[7] S. Henderson et al. (CLEO), preprint CLNS 91/1101, 1991, submitted to *Phys. Rev.* D.

[8] H. Albrecht et al. (ARGUS), *Phys. Lett.* B249: 359 (1990).

[9] D. Bortoletto et al. (CLEO), *Phys. Rev. Lett.* 63: 1667 (1989).

[10] H. Albrecht et al. (ARGUS), *Phys. Lett.* 219: 121 (1989); 197: 452 (1987); 229: 175 (1989).

[11] N. Isgur, D. Skora, B. Grinstein, M.B. Wise, *Phys. Rev.* D39: 799 (1989).

[12] M. Wirbel, B. Stech, M. Bauer, *Z. Phys.* C29: 637 (1985); J. Körner and J.G. Schüler, *Z. Phys.* C38: 511 (1988); W. Jaus, *Phys. Rev.* D41: 3394 (1990).

[13] R. Fulton et al. (CLEO), *Phys. Rev. Lett.* 64: 16 (1990).

[14] H. Albrecht et al. (ARGUS), *Phys. Lett.* B234: 409 (1990).

[15] G. Altarelli, N. Cabibbo, G. Corbo, L. Maiani, *Nucl. Phys.* B327: 353 (1989); C. Ramirez, J.F. Donoghue, G. Burdman, *Phys. Rev.* D41: 1496 (1990).

[16] D. Bortoletto et al. (CLEO), *Phys. Rev. Lett* 62: 2436 (1989); CLEO private communication.

[17] H. Albrecht et al. (ARGUS), "International Conference on High Energy Physics", Singapore, Aug., 1990.

[18] M. Bauer, B. Stech, and M. Wirbel, *Z. Phys.* C34: 103 (1987).

[19] H. Albrecht et al. (ARGUS), *Phys. Lett.* B192: 245 (1987); "International Conference on High Energy Physics", Singapore, Aug., 1990.

[20] M. Artuso et al. (CLEO), *Phys. Rev. Lett.* 62: 2233 (1989); A. Bean et al. (CLEO), *Phys. Rev. Lett.* 58: 183 (1987); CLEO private communication.

[21] K. Lingel (CLEO), "Rencontres de Moriond", March, 1992.

[22] R. Grigjanis, H. Navelet, M. Sutherland, and P. O'Donnell, *Phys. Lett.* B213: 355 (1988); A. Ali, C. Greub, preprint DESY 90-102 (1990); N.G. Despande and J. Trampetic, *Phys. Rev. Lett.* 61: 2583 (1988); N.G. Deshpande and P. O'Donnell, "Proceedings of the Workshop towards Establishing a b Factory", Syracuse, Sept., 1989, p. 1.35 and 1.66.

[23] M.B. Gavela et al., *Phys. Lett.* B154: 425 (1985).

[24] L.-L. Chau and H.Y. Cheng, *Phys. Rev. Lett.* 53: 1037 (1984); *Phys. Lett.* B165: 429 (1986).

[25] ALEPH, "Rencontres de Moriond", March, 1992; P. Roudeau (DELPHI), CERN Particle Physics Seminar, April 7, 1992.

[26] C. Albajar et al. (UA1), *Phys. Lett.* B186: 247 (1987).

[27] W. Bartl et al. (JADE), *Phys. Lett.* B146: 437 (1984); H.R. Band et al. (MAC), *Phys. Lett.* B200: 221 (1988); D. Decamp et al. (ALEPH), CERN preprint PPE/90-194; B. Adeva et al. (L3), L3 preprint 20, Nov 2, 1990.

[28] J.H. Christenson, J.W. Cronin, V.L. Fitch, and R. Turlay, *Phys. Rev. Lett.* 13: 138 (1964).

[29] Particle Data Group, *Phys. Lett.* B239: 1 (1990).

[30] J.L. Rosner, A.I. Sanda, and M.P. Schmidt, "Proceedings of the Workshop on High-Sensitivity Beauty Physics at Fermilab", Fermilab, Nov., 1987, page 165; C. Hamza-oui, J.L. Rosner, and A.I. Sanda, *ibid.*, page 215.

[31] C.S. Kim, J.L. Rosner, and C.-P. Yuan, *Phys. Rev.* D42: 96 (1990).

[32] J.L. Rosner, University of Chicago preprint EFI 92-02, Jan., 1992.

[33] Much of the theory of CP violation in B decays was first developed by I. Bigi and A. Sanda. Their work is summarized in I. Bigi, V.A. Khoze, N.G. Uraltsev, and A. Sanda in "CP Violation", ed. C. Jarlskog, World Scientific, Singapore, 1988, and preprint SLAC-PUB-4476 (1987). These include references to earlier work.

[34] J. Hagelin, *Nucl. Phys.* B193: 123 (1981); L.M. Sehgal and M. Wanninger, *Phys. Rev.* D42: 2324 (1990).

[35] H. Simma, G. Eilam, and D. Wyler, ETH Zurich preprint.

[36] W. Weinberg, *Phys. Rev. Lett.* 37: 657 (1976).

[37] G. Valencia, *Phys. Rev.* D39: 3339 (1989).

[38] D.S. Akerib et al. (CLEO), *Phys. Rev. Lett.* 67: 1692 (1991).

[39] "An Asymmetric B Factory Based on PEP", SLAC-372, Feb., 1991.

[40] M. Gronau, *Phys. Rev. Lett.* 63: 1451 (1989); M. Gronau and D. London, *Phys. Rev. Lett.* 65: 3381 (1990).

[41] B. Kayser, M. Kuroda, R.D. Peccei, and A.I. Sanda, *Phys. Lett.* B237: 508 (1990).

[42] Y. Nir et al., *Nucl. Phys.* B345: 301 (1990); *Phys. Rev.* D42: 1473 amd 1477 (1990).

[43] J.-M. Gérard and T. Nakada, *Phys. Lett.* B261: 474 (1991); B. Winstein, University of Chicago preprint, EFI 91-54.

[44] I. Dunietz, preprints CERN-TH-6161/91 and -6239/91.

[45] M. Ogg et al., "Detector for a B Factory", Cornell report CLNS 91-1047, Feb., 1991.

[46] K. Lingel et al., "Physics Rationale for a B Factory", Cornell report CLNS 91-1043, Feb., 1991.

[47] "Proposal for an Electron Positron Collider for Heavy Flavour Particle Physics and Synchrotron Radiation", Paul Scherrer Institute report PSI-PR-88-09, July, 1989; T. Nakada et al., "Feasibility Study for a B-Meson Factory in the CERN-ISR Tunnel", CERN 90-02.

[48] S. Krishnagopal and R. Siemann, Beam energy inequality in the beam-beam inter-action, Cornell preprint CLNS89/967 (1989).

[49] A. Piwinski, Satellite resonances due to the beam-beam interaction, *IEEE Transactions on Nuclear Science*, NS-24, 1408 (1977).

[50] K. Oide, K. Yokoya, *Phys. Rev.* A40: 315 (1989), based on a suggestion by Palmer for linear colliders. See also A. Piwinski, "Simulations and Tolerances for Crab Crossing", DESY-HERA 90-04 (1990).

[51] K. Berkelman et al., *CESR-B Conceptual Design for a B Factory Based on CESR*, Cornell report CLNS 91-1050, Feb., 1991.

[52] *Task Force Report on Asymmetric B-Factory at KEK*, KEK report, Feb., 1990.

[53] M. Witherell et al., *1992 HEPAP Subpanel on the U.S. Program of High Energy Physics Research*, U.S. Department of Energy, April, 1992.

[54] D. Sanchez, invited talk at the Washington meeting of the American Physical Society, April, 1992.

PHYSICS AT HERA

Günter Wolf

Deutsches Elektronen Synchrotron DESY
2 Hamburg 52, Germany

INTRODUCTION

On May 31st this year, the two HERA experiments H1 and ZEUS observed for the first time electron - proton collisions in their central detectors. This feat was the culmination of eight years of prototyping, construction and commissioning of the machine [1] and the detectors [2,3].

In parallel with the construction of the machine and detectors, the horizon of HERA physics has been substantially expanded as documented in several workshops. The most recent and complete proceedings are those from 1987 (ref. 4) and 1991 (ref. 5).

This report focusses on the results obtained by H1[6] and ZEUS[7] in the first running period in July [8] (partly after this meeting) .

THE HERA COLLIDER

Layout

The layout of HERA is shown in fig.1. Two separate magnet systems guide the e and p beams around the 6.3 km long ring. DESY and PETRA

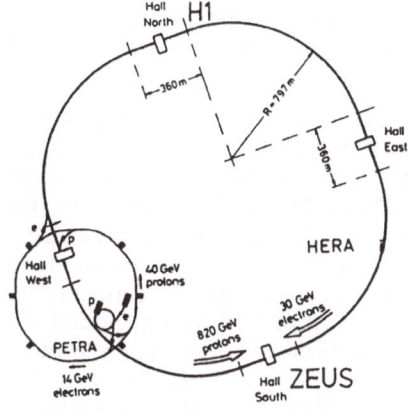

Fig. 1. Layout of HERA

Quantitative Particle Physics, Edited by
M. Lévy *et al.*, Plenum Press, New York, 1993

Table 1. HERA design parameters

	electron ring	proton ring
circumference		6336 m
energy	30 GeV	820 GeV
e - p c.m. energy		314 GeV
magnetic bending field	0.164 T	4.682 T
bending radius of dipole	610 m	584 m
circulating current	60 mA	160 mA
number of bunch buckets	220	220
number of bunches	210	210
time between beam crossings		96 ns
luminosity		$1.5 \cdot 10^{31}$ cm^{-2}s^{-1}
polarization time at E_e = 30 GeV	25 min	

serve as injectors. There are four interaction regions, two of which are occupied by H1 and ZEUS. A third region has been earmarked for the proposed experiment HERMES[9]. Table 1 shows some of the parameters of the collider.

Commissioning Phase

Construction of HERA started in April 1984. In August of 1987 the 6.3 km long tunnel was completed. A year later the electron ring made of normal conducting magnets was put into operation. The proton ring was closed in fall of 1990. Most of the proton magnets are superconducting and are operated at 4.4 K providing a bending field of 4.68 T. The first protons were stored in April 1991 at the injection energy of 40 GeV. Having learned how to accelerate protons and to operate proton and electron beams simultaneously the machine crew began with collision studies. Luminosity was observed for the first time on October 19, 1991 by colliding one electron bunch of 12 GeV with one proton bunch of 480 GeV. The measured luminosity was about $L = 10^{26}$ cm^{-2}s^{-1}. In the following weeks the electron energy, the number of bunches and the beam currents were increased. After reaching $L = 2.10^{28}$ cm^{-2}s^{-1} with 10 bunches per beam at beam energies E_e = 26.6 GeV and E_p = 480 GeV, HERA operations were interrupted and the experiments H1 and ZEUS were installed in the interaction regions during a 4 months shut down.

In April of this year HERA resumed operation. By mid-April electrons were stored at 26.6. GeV; one month later protons were accelerated to the nominal energy of 820 GeV and on May 31st the two experiments registered the first collisions between the two beams. The H1 and ZEUS detectors were brought on-line concurrently with the beam tests of HERA: no special running period was provided for running-in or tuning of the detectors. Data taking started at the end of June with 10 bunch operation, maximum beam currents of 1 - 2 mA and maximum luminosities around 6.10^{-28}cm^{-2}s^{-1}. In a first one month long period data were recorded for integrated luminosities of 2 - 3 nb^{-1}. Physics results from this running period were reported a week after data taking has finished[6,7].

After a shut down and machine tests a second data taking period has begun in mid-September.

Performance

The construction of HERA presented the machine builders with many challenges. Some of these will be briefly discussed.

Superconducting magnets.   HERA has been the first accelerator where the superconducting magnets have been built in industry. The design of the dipole magnets has started with that from the FNAL magnet, changing in the course of the development from a warm to a cold iron yoke. Production procedures for dipoles were developed at DESY. Following the construction of prototype magnets at DESY and in industry, the 422 dipole magnets were built by Ansaldo (Italy) and ABB (Germany). The quadrupoles were developed and prototypes were constructed by Saclay. The 224 quadrupoles were built by KWU + Noell (Germany) and Alsthom (France). The dipoles and quadrupoles delivered exceed the required operating currents and fields by more than 30%.

Persistent sextupole currents. Protons are injected into HERA from PETRA at an energy of 40 GeV which is a factor of 20 lower than the operating energy of 820 GeV. At 40 GeV persistent currents producing a sextupole component in the dipole magnets present a major disturbance. However, the persistent current sextupole varies little from magnet to magnet and is well reproducible. It is compensated by correction coils wound directly on the dipole and quadrupole beam pipes. Flux creep in the superconductor and other effects cause the magnetization currents to decay approximately logarithmically in time. This drift is also compensated using the correction coils. The required strength of the correction elements at injection and during acceleration is determined by measuring continuously the dipole and sextupole fields in two reference magnets, powered in series with the ring magnets.

Superconducting cavities. The power lost by the electron beam due to synchrotron radiation amounts to 7.6 MW at 30 GeV and 60 mA beam current. Up to ~ 28 GeV the energy loss is compensated by 88 normal conducting cavities recuperated from PETRA. For higher energies eight 2 x 4 cell superconducting cavities operating at 500 MHz were developed and built in industry. Without beam they provide an accelerating gradient of 5 MV/m.

Proton beam stability. The lifetime of the proton beam in single beam operation is well above 100 h. In colliding beam mode lifetimes in excess of 50 h have been obtained regularly. While the transverse size of the electron beam is limited by synchrotron radiation, no such mechanism is operative for protons. The longitudinal bunch length can be limited to ~ 10 cm (rms) by operating the 52 MHz and 208 MHz RF systems concurrently. For the first two data taking periods this was not done routinely. Typical bunch lengths were 12 - 50 cm. Since the bunch length of the electron beam is much smaller (~ 8 mm rms) the interaction volume was half of the proton bunch length, i.e. 6 - 25 cm (rms).

In operations with 10 consecutive bunches no bunch - bunch interactions were observed for either the electron or the proton beam. With increasing number of bunches bunch - bunch interactions are expected to occur. Feed back systems which had been tested in PETRA were installed in HERA to counteract these collective effects.

Luminosity. The design value for the specific luminosity (luminosity per bunch and mA bunch currents) is 4.4 10^{29} cm^{-2}s^{-1} mA^{-2}. This value was achieved in the operation with 10 on 10 bunches and beam currents of up to 3 mA. The maximum luminosity observed to date was 3.10^{29} cm^{-2}s^{-1}. The average luminosity collected by the experiments per day has increased by roughly a factor of 7 from the first to the second run period (see fig. 2). The major limitation of the luminosity at present is a limit on the electron current of about ~ 7 mA whose cause is not yet understood and which is under active study.

ZEUS INTEGRATED LUMINOSITY

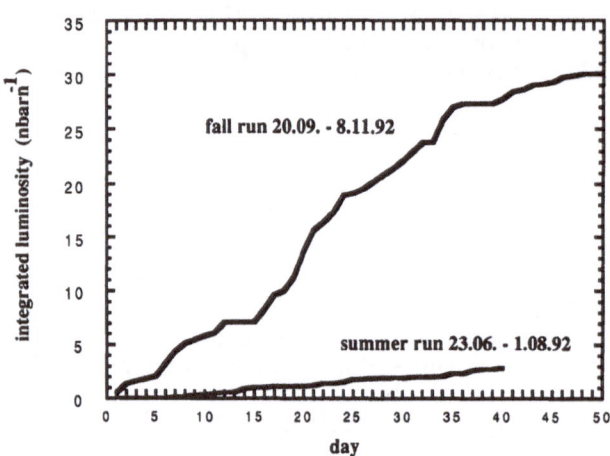

Fig. 2.   Luminosity per day collected by ZEUS during the first and second running period (from U. Schneekloth).

Polarization. After coasting for some time the electrons become polarized with spins being antiparallel to the direction of the bending field as a result of the Sokholov - Ternov effect [10]. The build - up time for the polarization is determined by the synchrotron radiation and is given by $P(t) = P_0$ (1- exp $(-t/t_p)$) with $P_0 = 92\%$ and $t_p = 98$ r^2 R E^{-5}, t_p in s, r bending radius in m, R average radius in m and E electron beam energy in GeV. This prediction holds for a perfect machine. Already small imperfections e.g. in the magnet lattice may produce depolarization effects and make the depolarization time shorter than the build-up time thereby destroying the polarization. Only a few shifts were devoted sofar to the study of beam polarization. An electron energy near 26.67 GeV was chosen. Beam polarization in HERA was observed for the first time in fall of 1991 at the level of P = 5 - 9%. After realignment of magnets P increased to 18% by April 1992. After the summer shutdown and with some tuning suggested by tracking programs a polarization of 58±5% was measured[11] (see fig. 3).

For particle physics, instead of transverse polarization electrons of definite helicity (left - or right handed) are needed. This can be achieved with the help of a pair of spin rotators installed at the interaction regions. A pair of spin rotators was designed and built[12] and is available for installation and testing in HERA.

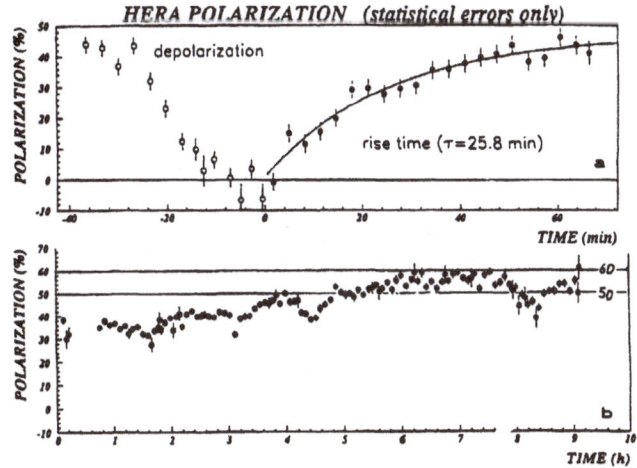

Fig. 3.   a) The transverse electron polarization at F_e = 26.7 GeVas a function of time.
           b) Beam polarization during a long fi¹'.

THE EXPERIMENTS

Detector Challenges

HERA produces a large variety of reactions with widely differing energy flows. This feature together with the desire for detecting and identifying the constituents such as electron, photon, quarks and gluons which participate in these reactions places different requirements on the detector. The large momentum imbalance between incident electrons and protons and the nature of space - like processes send most particles into a narrow cone around the proton direction. The observation of deep inelastic (DIS) neutral current (NC) scattering, e p -> e X (fig. 4a), is fairly straightforward. It produces a high energy electron whose transverse momentum is balanced by the current jet. The remants from the breakup of the proton escape mostly unseen down the beam pipe. The variables x and Q^2 which describe the process (see below) can be determined from the energy and angle of either the electron or the current jet. This requires a precise electromagnetic and hadronic calorimeter with the calibration known at the 1 - 2% level.

In contrast, in charged current (CC) scattering, e p -> ν X (fig. 4b), only the current jet can be observed. The idendification of such events is based on the observation of missing transverse momentum carried away by the neutrino. It requires a hermetic calorimeter which covers the full solid angle such that e.g. photons, neutrons or K^0_2 cannot escape undetected. The variables x and Q^2 are measured from the current jet.

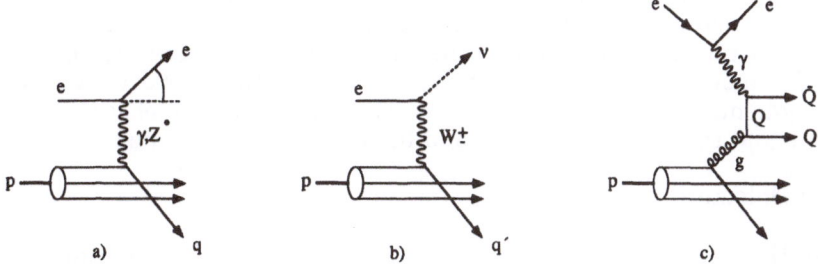

Fig. 4.   Diagrams for NC and CC scattering and for photon - gluon fusion

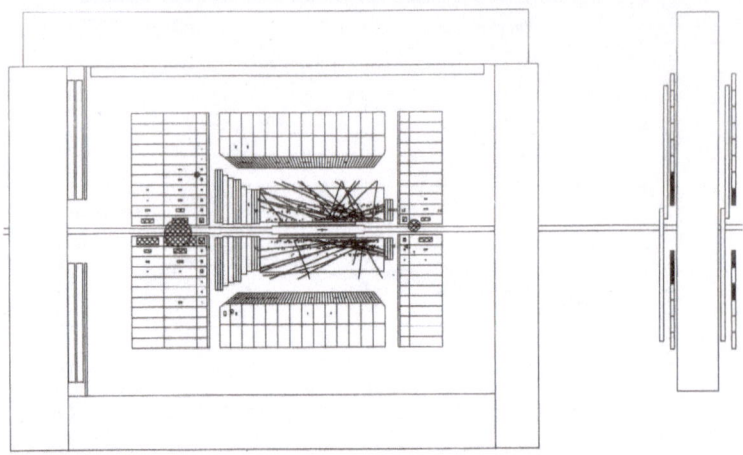

Fig. 5.   A background event produced by a proton interaction upstream (to the right) of the ZEUS detector

The observation of processes at low Q^2, in particular from scattering on soft partons or from production of pairs of charm or bottom mesons (fig. 4c) is more difficult. The energy deposited in the calorimeter often is only a few GeV. Additional information from tracking detectors which surround the interaction point is necessary for their identification.

Background presents another challenge. The number of events from e - p interactions is tiny (10^{-3} - 10^{-5}) compared to the background events produced for instance by beam protons on the beam pipe wall or in the residual gas. What is worse, this type of background deposits a large amount of energy in the detector. A typical background event is shown in fig. 5 where 225 GeV are observed in the calorimeter and many tracks in the tracking detector. At proton design current the background rate is expected to be around 10 - 100 kHz. The detector must be able to discriminate quickly - within a few microseconds - against background events although both beams cross each other every 96 ns. The high background rates combined with the short bunch crossing interval forced the HERA experiments to develop novel concepts of electronic readout and triggering, concepts which are suitable also for detectors at the next generation of pp colliders, SSC and LHC.

The H1 and ZEUS detectors are driven by their choice of calorimeter. ZEUS uses a compensating uranium - scintillator calorimeter which provides the best possible energy resolution for hadrons. Compensation means that electromagnetic particles (electrons, photons) and hadrons of the same energy yield the same pulse height, e/h = 1. The radioactivity of the (depleted) uranium provides an extremely stable calibration signal, the mean life time of ^{238}U being $6.5 \cdot 10^9$ years. The H1 calorimeter uses liquid argon for readout which promises a very stable and precise energy calibration and allows a high transverse and longitudinal segmentation. The calorimeter is noncompensating, e/h = 1.1 - 1.2. However, due to the high segmentation and using software weighting with the observed shower profile equal signals for electrons and hadrons can be obatined.

The H1 Detector[2]

The H1 detector is displayed in fig. 6. The liquid - argon calorimeter (LA) covers the angular region $4^0 < \Theta < 155^0$ (the foward direction, $\Theta = 0^0$, is

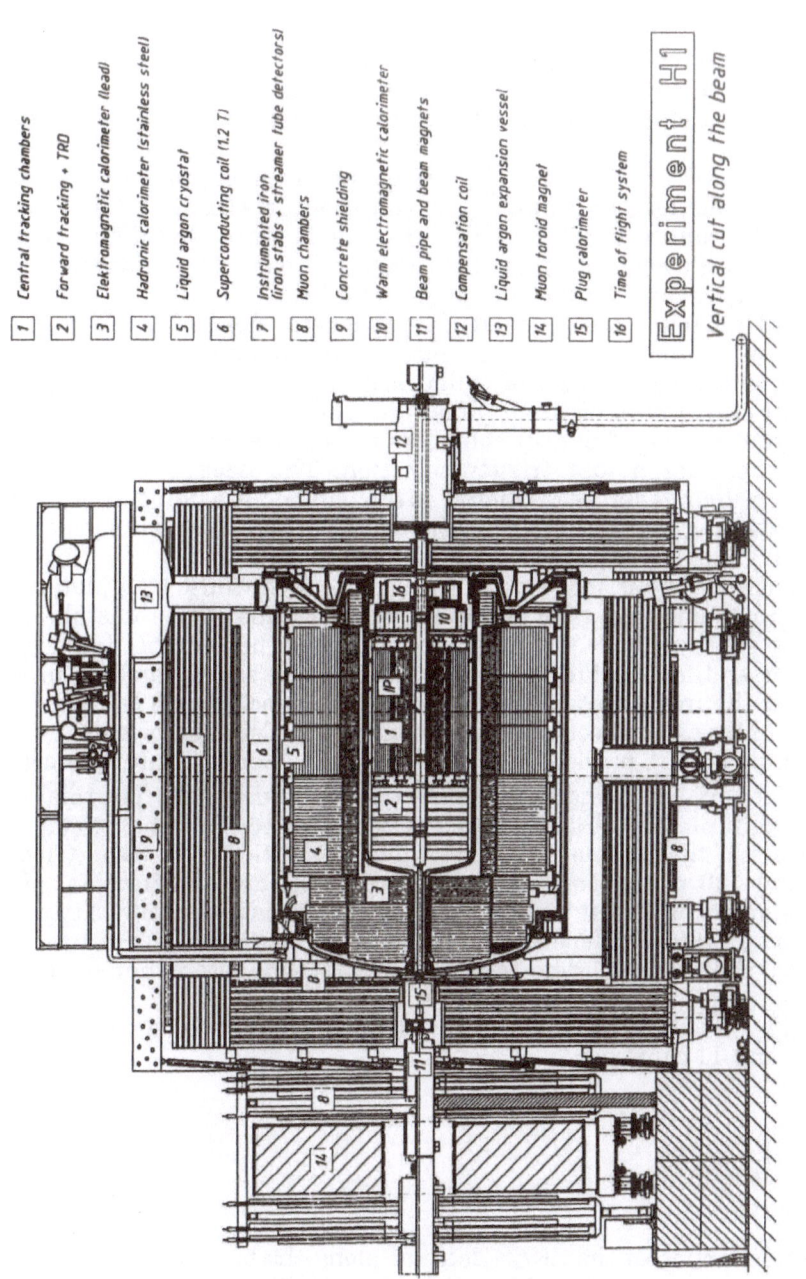

1 Central tracking chambers
2 Forward tracking + TRD
3 Elektromagnetic calorimeter (lead)
4 Hadronic calorimeter (stainless steel)
5 Liquid argon cryostat
6 Superconducting coil (1,2 T)
7 Instrumented iron
  (iron slabs + streamer tube detectors)
8 Muon chambers
9 Concrete shielding
10 Warm electromagnetic calorimeter
11 Beam pipe and beam magnets
12 Compensation coil
13 Liquid argon expansion vessel
14 Muon toroid magnet
15 Plug calorimeter
16 Time of flight system

Experiment H1

Vertical cut along the beam

0 1 2 3 4m

Fig. 6.   Schematic view of the H1 detector.

217

given by the direction of the proton beam). The calorimeter is longitudinally subdivided into an electromagnetic section with lead plates and a hadronic section with stainless steel plates as absorbers. The total depth varies between 8 and 4.5 absorption lengths. The calorimeter was optimized for a precise measurement and identification of electrons and for a stable energy calibration for electrons and hadrons. The energy resolution σ/E for electrons is $12\%/\sqrt{E} \oplus 1\%$ ($\oplus$ means quadratic addition) and $45\%/\sqrt{E} \oplus 1\%$ for hadrons (after weighting) as measured with test beams (fig. 7). The calorimeter has been in operation since April 1991 and has fullfilled all expectations. Since then the charge collection has changed by less than 0.2% and the number of dead or problematic readout channels is below 0.1%.

In the backward direction the calorimeter is supplemented by an electromagnetic lead - scintillator calorimeter (BEMC) followed by time-of-flight counters (TOF-VETO). In the forward direction the plug calorimeter with copper plates and silicon diode readout extends the energy measurement for hadrons down to angles of 0.7^{0}.

Charged particles are tracked in a magnetic field of 1.2 T which is produced by a superconducting solenoid that surrounds the calorimeter. The tracking system consists of cylindrical jet - and z - drift chambers in the central region, and of three radial and three planar drift chambers in the forward direction. The drift chambers are interleaved with proportional wire chambers for a fast trigger selection. The backward direction is covered by a four layer proportional wire chamber providing space points up to a scattering angle of 175^{0}. In forward direction a transition radiation detector (TRD) enhances the electron identification.

The magnet yoke is made of 10 layers of 7.5 cm thick iron plates. The gaps are instrumented with limited streamer tube (LST) chambers for measuring energy which has not been fully absorbed in the liquid argon calorimeter and for tracking of muons. Large area LST chambers in front and behind the iron yoke and an iron toroid magnet plus 6 layers of drift chambers in forward direction complete the muon detection system.

The luminosity is measured by observing the bremsstrahlung process e p -> e p γ at very small angles to the electron beam direction. The final state electron and photon are detected in coincidence in electromagnetic calorimeters of the luminosity detector LUMI positioned at 33 m (electron tagger) and 100 m (photon tagger) upstream (in proton direction) of the central detector (fig. 8). At nominal luminosity the rate of luminosity events is between 50 - 100 kHz depending on the selection criteria.

Figure 9 shows the H1 detector on its way into the interaction region.

The H1 collaboration at present consists of 320 physicists from 32 institutes and 11 countries.

The ZEUS detector[3]

A cross section of the ZEUS detector along the beams is shown in fig. 10. The main component is the uranium - scintillator calorimeter (CAL) subdivided mechanically into the forward (FCAL), barrel (BCAL) and rear (RCAL) calorimeters. The CAL covers polar angles from 2.6^{0} to 176.1^{0} and 99.7% of the total solid angle. It consists of a total of 80 modules. Every module is made of up to 180 layers of 3.3 mm thick depleted uranium

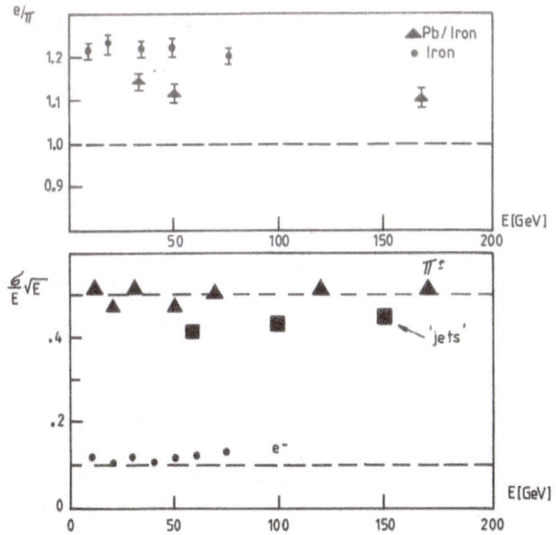

Fig. 7.   Test beam results for the H1 calorimeter. Top: e/π ratio, bottom: energy resolution for electrons and hadrons (after software weighting).

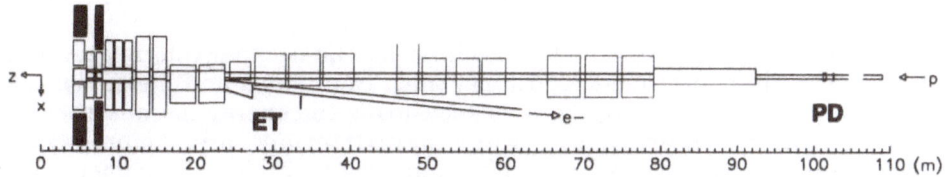

Fig. 8.   Layout of the H1 luminosity detector.

Fig. 9   The H1 detector on its way to the beam position.

plates plus 2.6 mm thick scintillator plates. Wave - length shifter bars transport the light to photomultipliers. The modules are subdivided longitudinally into an electromagnetic and two (one) hadronic sections in FCAL, BCAL (RCAL) presenting a total depth of 7 to 4 absorption lengths. The scintillator plates form 5 x 20 cm^2 (10 x 20 cm^2) cells in the electromagnetic section and 20 x 20 cm^2 in the hadronic sections of FCAL, BCAL (RCAL).

The calibration of the photomultipliers is being monitored with the signal (UNO) from the radioactivity of the uranium to a precision of < 0.2%. The pulse heights of electrons and hadrons (fig. 11a) are equal to within 3%, i.e. e/h = 1.0 ± 0.03 (fig. 11b), for momenta above 3 GeV/c. The energy resolution as measured in the test beam is for electrons σ/E = 18%/√E (E in GeV) and for hadrons 35%/√E (fig. 11c). The calorimeter yields also an accurate time measurement. The time resolution of a calorimeter cell is 1.5/√E ⊕ 0.5 ns or < 1 ns above 3 GeV.

In the course of an upgrade program, the transverse segmentation of the forward and rear parts of the calorimeter is being increased by inserting a plane of 3 x 3 cm^2 silicon diodes after the first 3 radiation lengths.

Charged particles coming from the interaction point are detected in the tracking system. It consists of a vertex detector, a cylindrical jet - type drift chamber and planar drift chambers in the forward and backward directions. In the forward direction planar transition radiation chambers are used for enhanced electron identification. The tracking detectors are surrounded by a thin - walled solenoid which produces a magnetic field of up to 1.8 T.

The iron yoke serves as the absorber for the backing calorimeter and as a muon filter. It is made of 7.5 cm thick iron plates and instrumented with proportional tube chambers for measuring the energy not absorbed in the uranium calorimeter. For identification and momentum measurement of muons the yoke is magnetized to 1.6 T with copper coils. Large area LST chambers measure the position and direction of muons in front and behind the iron yoke. In the forward direction a spectrometer of two iron toroids and drift - and LST chambers identifies muons and measures their momenta up to 100 - 150 GeV/c.

For luminosity measurement the same reaction and a setup similar to that of H1 is used.

Very forward scattered protons (transverse momenta < 1 GeV/c) are measured in the leading proton spectrometer which uses the proton ring magnets for momentum analysis and detects the scattered protons in 6 stations with silicon strip detectors mounted very close to the beam at distances between 26 and 96 m. The stations have been installed, the detectors not yet.

Particles produced by the proton beam upstream of the detector are detected in the VETOWALL. For monitoring the time structure and other properties of the two beams a ring counter C5 has proven to be invaluable. It is made of two lead - scintillator layers and mounted on the beam pipe behind RCAL. C5 registers the halo particles accompanying both beams.

The central detector of ZEUS is shown in fig. 12.

The ZEUS experiment is performed by a joint effort of 450 physicists from 11 countries and more than 50 institutes.

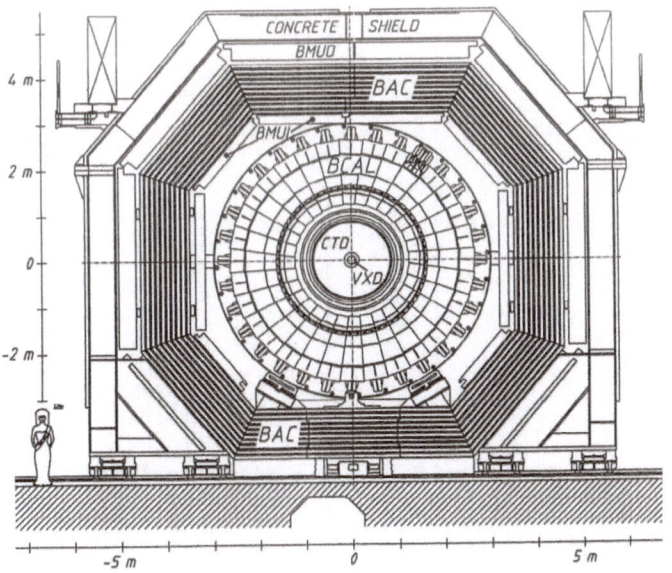

Fig. 10 a. Cross section of the ZEUS detector transverse to the beams

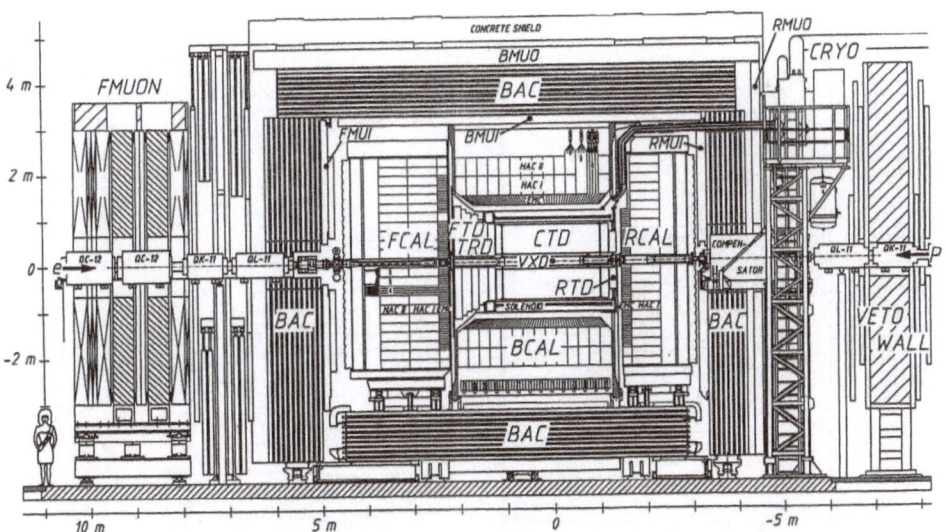

Fig. 10 b. Cross section of the ZEUS detector along the beams

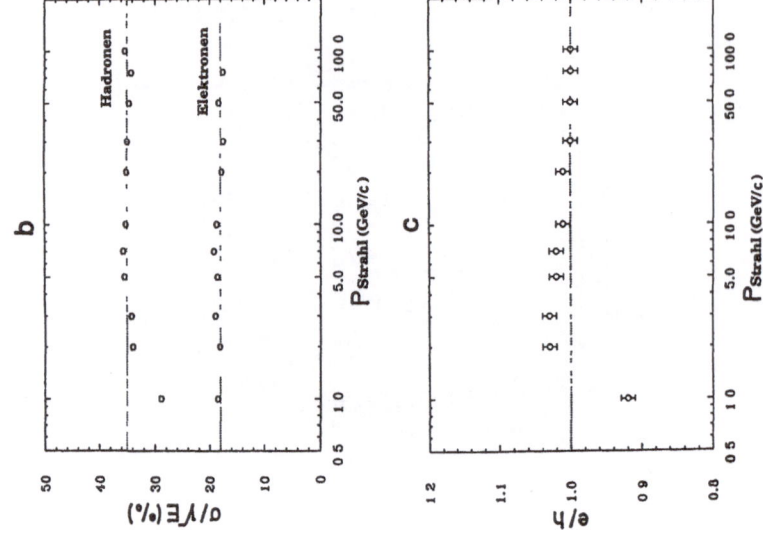

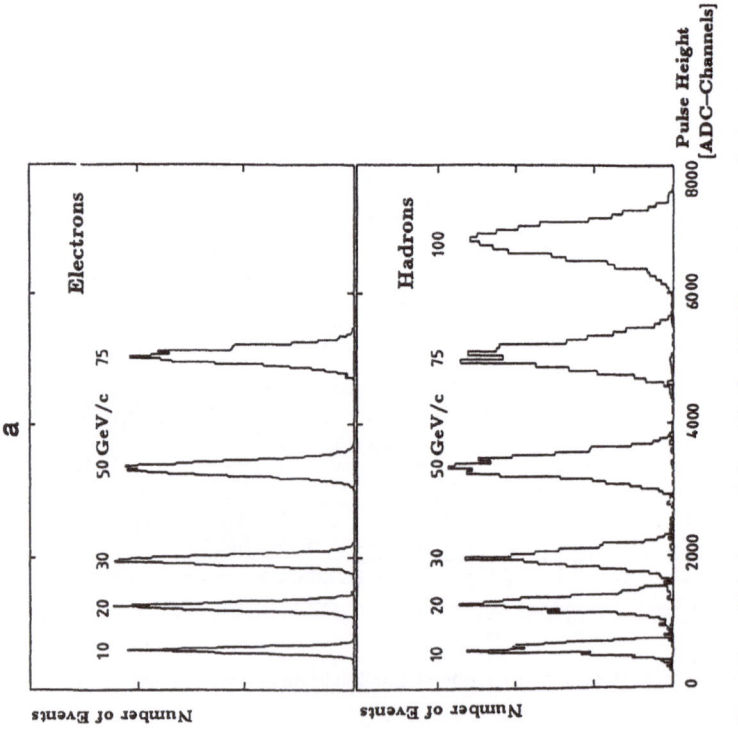

Fig. 11. Measurements with the ZEUS prototype calorimeter:
(a) Pulse height distributiuons for electrons and hadrons;
(b) Energy resolution for electrons and hadrons;
(c) e/h ratio as a function of momentum

Fig. 12. View of the ZEUS detector: in the foreground the uranium - scintillator calorimeter

Trigger Selection

The selection of interesting events during data acquisition proceeds in four (three) trigger steps for H1 (ZEUS). The selection at trigger level one is made after 2.4µs (H1) and 4.4 µs (ZEUS), respectively. Up to this point information from 200 000 to 300 000 electronic channels are stored dead time free in analog or digital pipelines for 25 (H1) and 46 (ZEUS) consecutive beam crossings, respectively. Global information from various components like the calorimeter energy sums obtained by summing over specific regions of the calorimeter are stored in trigger pipelines. In case an interesting event is detected, the signal pipelines are stopped and the data for the bunch crossing(s) in question are digitized. The digitized data are used on the next trigger level(s) for a more restrictive event selection.

The final event selection is done in computer farms. At this point the complete digitized information for the event is available and a first reconstruction of the event is performed. The filter farms consist of a large number of fast processors with computing power of ~ 300 (H1) to 1000 MIPS (ZEUS), each processor processing one event at a time. The rate of accepted events varies in both experiments between about 3 and 7 Hz; typical event sizes are 60 kByte (H1) to 140 kByte (ZEUS). The accepted events are reconstructed off-line in processor farms with sufficient computing power to have the reconstructed events available for analysis within a few hours after data taking.

RUNNING CONDITIONS

The typical bunch configuration of HERA during the first weeks of data taking is sketched in fig.13: 10 consecutive proton bunches (1 - 10) and 9 electron bunches (1 - 9) are filled. The 10th proton bunch is unpaired; it has no electron partner and is used for studying proton beam induced background. Similarly, the electron bunch 19 is unpaired and is used to measure electron beam induced background. Typical beam currents were 1 - 2 mA.

The luminosity was measured by detecting the process e p -> e'p γ as mentioned before. Figure 14 shows the scatter plot of the energies E_e', E_γ for the final state electron and photon as determined by the luminosity detector. There is a well isolated band of events for which the sum of the two energies is equal to the energy of the electron beam, $E_e' + E_\gamma = E_e$, as expected for the luminosity reaction. However, bremsstrahlung of the electron beam on the residual gas in the beam pipe, e A -> e' A' γ, satisfies the same condition. The subtraction of this background was done by measuring the gas bremsstrahlung with the unpaired electron bunch and scaling with the currents of the unpaired electron bunch and the total electron beam (see fig. 15). The achieved precision of the luminosity measurement is at present 10 - 15%.

The total integrated luminosity collected in the first running period was about 3 nb^{-1} for each of the two experiments, 50 - 75% of which was used for physics analysis.

Suppression of background events at the trigger stage to a manageable rate was not a particularly difficult task, mostly because the beam currents were only a few percent of their nominal values. By far the most copious background was produced by proton interactions upstream of the detector (see fig. 5). The strategies for suppressing unwanted background and selecting electron - proton collisions were different for the two experiments.

ZEUS Data Taking

In the ZEUS experiment, the trigger selection was made on the basis of

-    energies and arrival times measured in the calorimeter,
-    energy detected in the electron calorimeter of LUMI,
-    veto signals from the C5 counter or the VETOWALL.

The calorimeter time information has turned out to be a powerful handle for rejecting proton beam background. This is illustrated in fig. 16. In an ep collision particles are emitted from the interaction point, IP, and arrive at the calorimeter cells at times t = 0 while a proton interaction upstream of the detector such as shown in fig. 5 deposits energy in the RCAL about 10 ns *earlier*. The 10 ns difference corresponds to twice the distance between RCAL and the IP. Of course, in FCAL also the proton induced background arrives at t = 0. The measured distribution of FCAL (t_{FCAL}) versus RCAL (t_{RCAL}) times is shown in fig. 16c for events with more than 1 GeV deposited in a calorimeter cell in both FCAL and RCAL. The ep events with t_{FCAL} ~ t_{RCAL} ~ 0 are well separated from the background which clusters around t_{RCAL} = -10 ns, t_{FCAL} - t_{RCAL} = 10 ns. Note, there are about 1000 times more background than ep events.

Samples of event pictures are shown in fig. 17. The first event stems from neutral current scattering at Q^2 ≈ 2550 GeV^2, x ≈ 0.07, with an electron seen in BCAL and a high energy jet in FCAL (fig. 17a). The jet near the proton beam is presumably produced by the proton remnants. The high energy jet and the electron are back-to-back in the transverse plane and balance transverse momentum as expected for an NC event. The interaction point is marked by tracks detected in the cylindrical drift

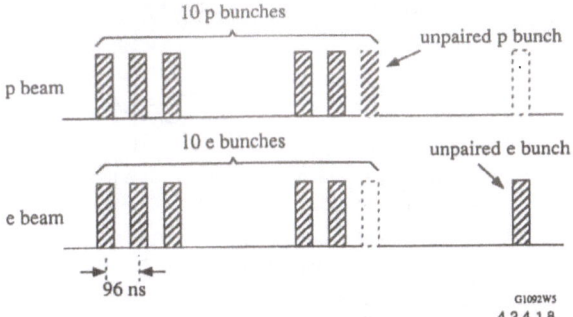

Fig. 13 Sketch of the HERA bunch configuration

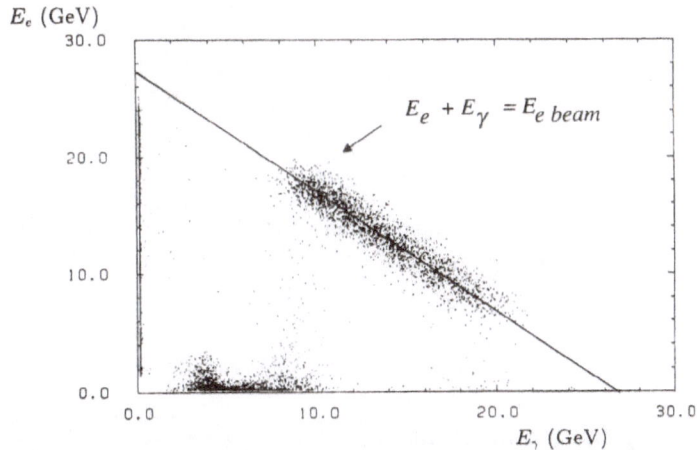

Fig. 14. Distribution of the electron energy versus the photon energy measured in the ZEUS luminosity detector

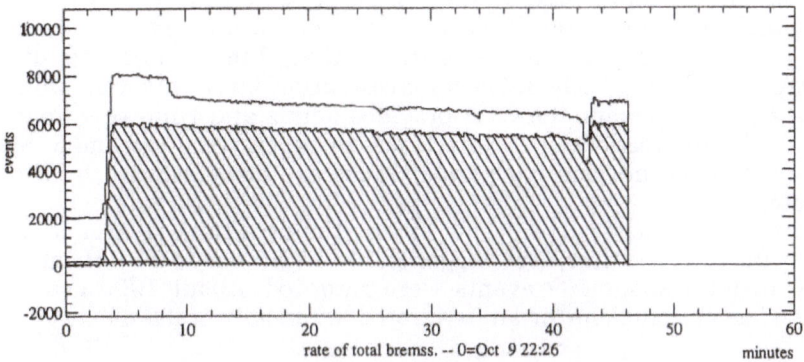

Fig. 15. The total bremsstrahlung rate (open curve) and the luminosity after subtraction of beam gas bremsstrahlung (shaded curve) as a function of time (from ZEUS)

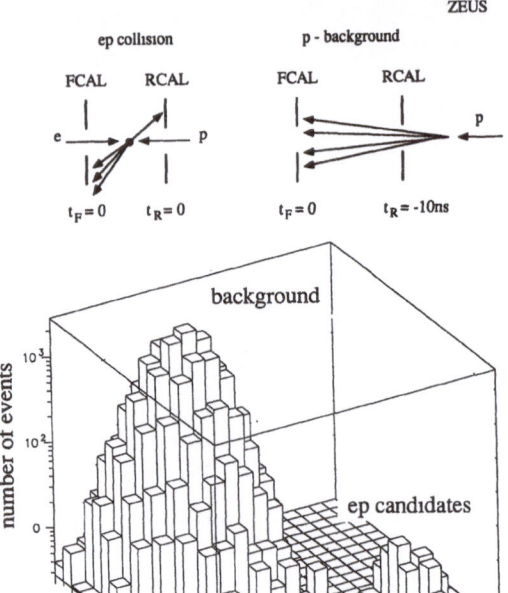

Fig. 16.   Distribution of the signal time measured by ZEUS in the RCAL (t_R) versus the difference t_F - t_R between signal times seen in FCAL and RCAL

chamber (CTD). The second event (fig. 17b) shows a low Q^2, low x event (Q^2 = 6 GeV^2, x = 0.0004). The electron is produced very close to the beam and is only seen in RCAL. The third event (fig. 17c) is due to quasireal photoproduction: it has energy deposition and charged tracks in the central detector plus an electron of 10.5 GeV in the LUMI electron calorimeter. The c.m. energy of the photon - proton system for this event is 230 GeV; the equivalent proton energy for a stationary proton would be 28 TeV.

The event pictures were produced by including every calorimeter cell with an energy more than 60 (100) MeV in the electromagnetic (hadronic) section. The calorimeter is seen to be very clean. The information on charged particle tracking is still limited due to missing digitizing electronics which is scheduled for installation later this year and during 1993. For the moment the CTD provides only z and r-phi coordinates from the z-by-timing readout of 16 wire layers in superlayers 1, 3 and 5. Since the start of the second running period tracks are also recorded by the vertex detector.

The rate of background events hitting the detector was around a kHz. It was reduced by the first level trigger to 10 - 15 Hz and to 3 - 5 Hz at the third level. In total about 10^6 events were recorded. About 1000 events were selected as ep collisions for an integrated luminosity of 2.5 nb^{-1}.

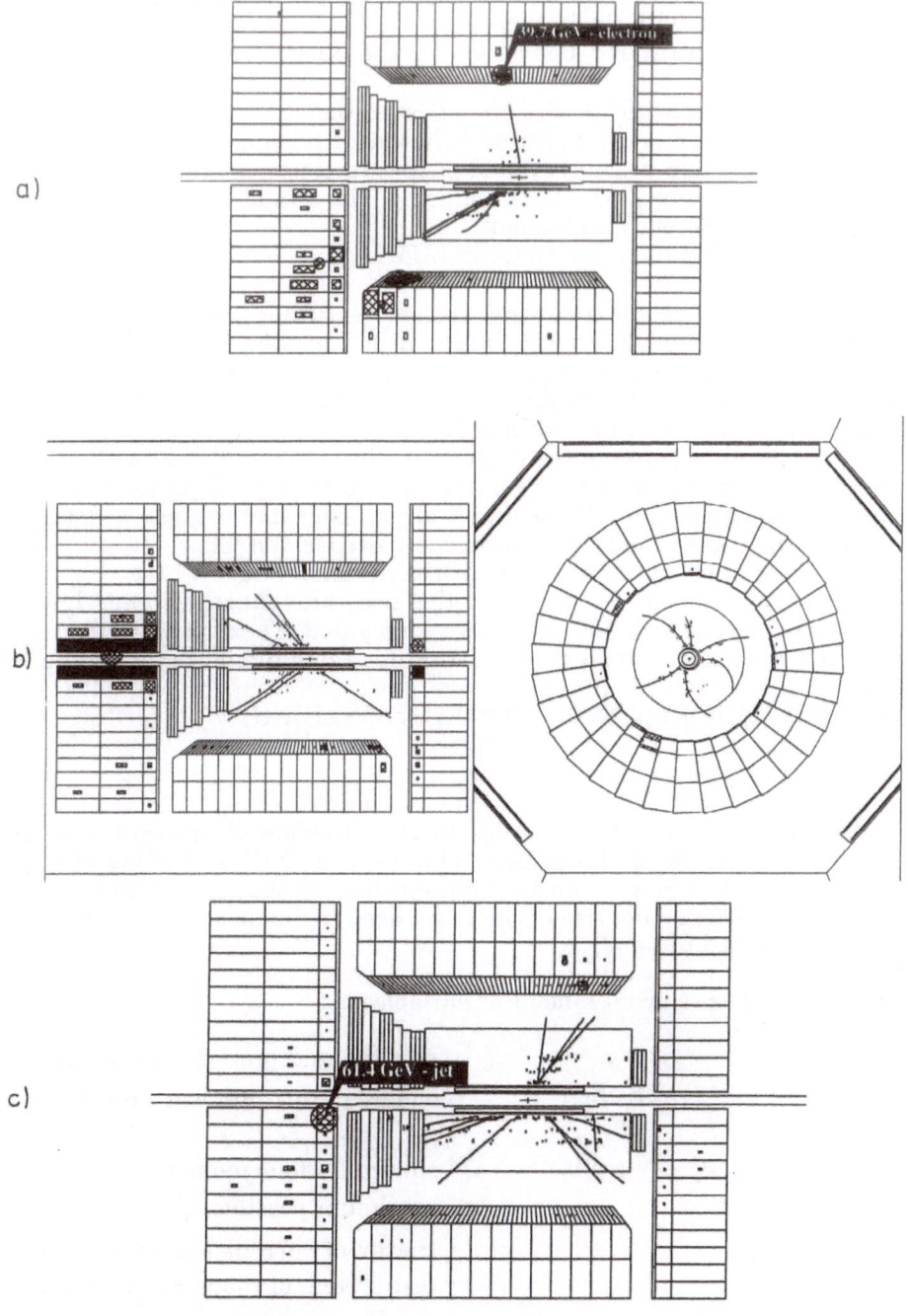

a)

b)

c)

Fig. 17. Event pictures observed by ZEUS:
a) Deep inelastic NC scattering at $Q^2 \approx 2550$ GeV^2, x ≈ 0.07
b) Deep inelastic NC scattering at $Q^2 \approx 6$ GeV^2, x ≈ 0.0004
c) Photoproduction at a total c.m. energy of 230 GeV

H1 Data Taking

The suppression of background at the trigger stage was accomplished in the H1 experiment with the help of

- the scintillator hodoscope in the backward direction (TOF - Veto),
- the proportional wire chambers of the central tracker requiring at least one ray pointing to the vertex region,
- the liquid-argon (LA) and backward electromagnetic calorimeters (BEMC) requiring clusters of > 8 (4) GeV in the first (second) component.
- the electron calorimeter of LUMI.

The readout of the tracking system is rather complete and well understood. A spectacular event produced by a proton interaction on the rest gas in the beam pipe is shown in fig. 18. It has 21 protons identified via dE/dx in the central jet chamber. The dE/dx distribution for background events shows well isolated bands of π, K, p and d (fig. 19).

The overall response of the detector is illustrated in fig. 20 with events from NC scattering and photoproduction. The high longitudinal and transverse segmentation of the liquid argon calorimeter gives a detailed account of the energy deposition for single particles and jets.

The rate of background events in the detector was ~ 3 kHz; this was reduced to 20 Hz and 3 - 5 Hz at the first and fourth trigger levels, respectively. A total of 5.10^6 events were recorded. The event sample used for the physics analysis corresponds to an integrated luminosity of 1.5 nb^{-1}.

DEEP INELASTIC ELCTRON PROTON SCATTERING

Physics Introduction

The incoming electron couples to the electroweak current j which probes the structure of the proton. The neutral (NC) and charged (CC) components of the current can be distinguished by the observation of the final state electron or neutrino. The basic deep inelastic scattering process (DIS) is illustrated in fig. 21.

Kinematics. The relevant kinematic variables are

E_e, E_p	electron and proton beam energies
$s = (e + p)^2 = 4\,E_e\,E_p$	square of the total c.m. energy
$q^2 = (e - e')^2$	
$\quad = -2\,E_e\,E_e{}'\,(1 + \cos\theta_{e'}) = -Q^2$	square of four momentum transfer
$Q^2{}_{max} = s$	maximum possible Q^2 value
$\nu = q \cdot p / m_p$	energy of current j as measured in rest system of the incoming proton
$\nu_{max} = s/(2m_p)$	maximum energy transfer
$y = (q \cdot p)/(q \cdot e) = \nu/\nu_{max}$	fraction of energy transfer
$x = Q^2/(2\,q \cdot p) = Q^2/(2\,m_p\nu)$ $\quad = Q^2/(y\,s)$	Bjorken scaling variable and fraction of the proton momentum carried by the struck parton
$\triangle = \hbar/Q$	smallest size of objects that can be resolved in the proton

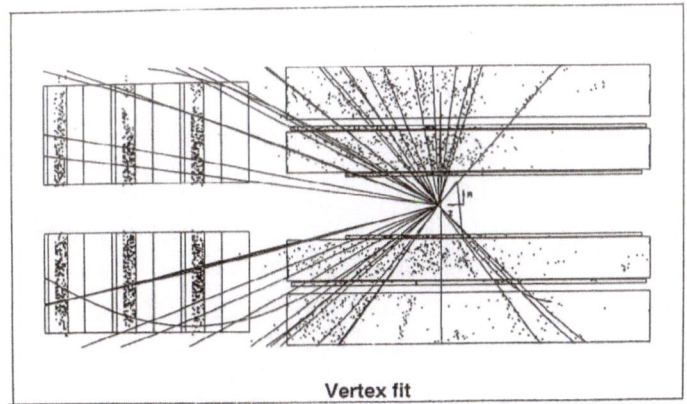

Vertex fit

Fig. 18. Example of a proton - gas interaction in the H1 detector with 21 final
state protons identified by dE/dx

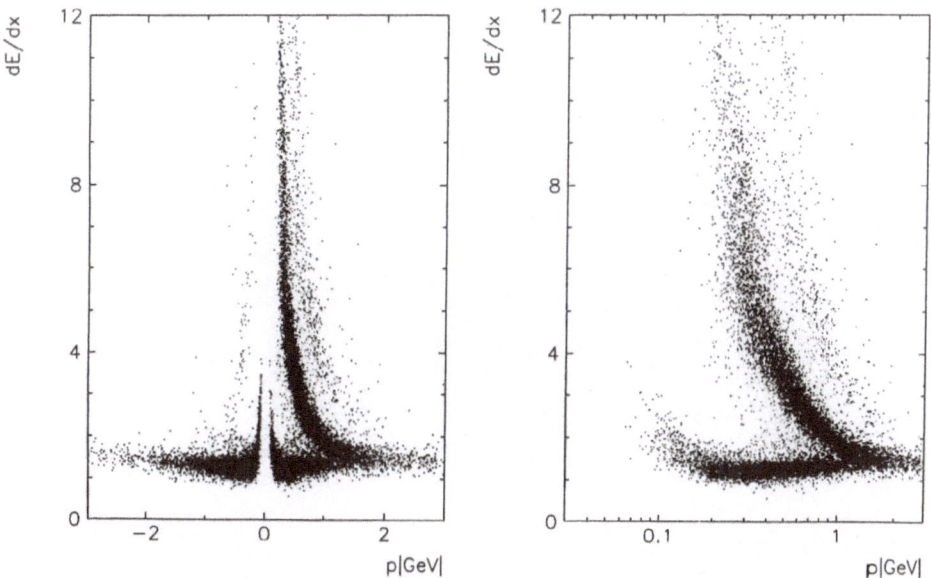

Fig. 19. dE/dx spectra observed by the central tracker of H1 for negative and positive
particles

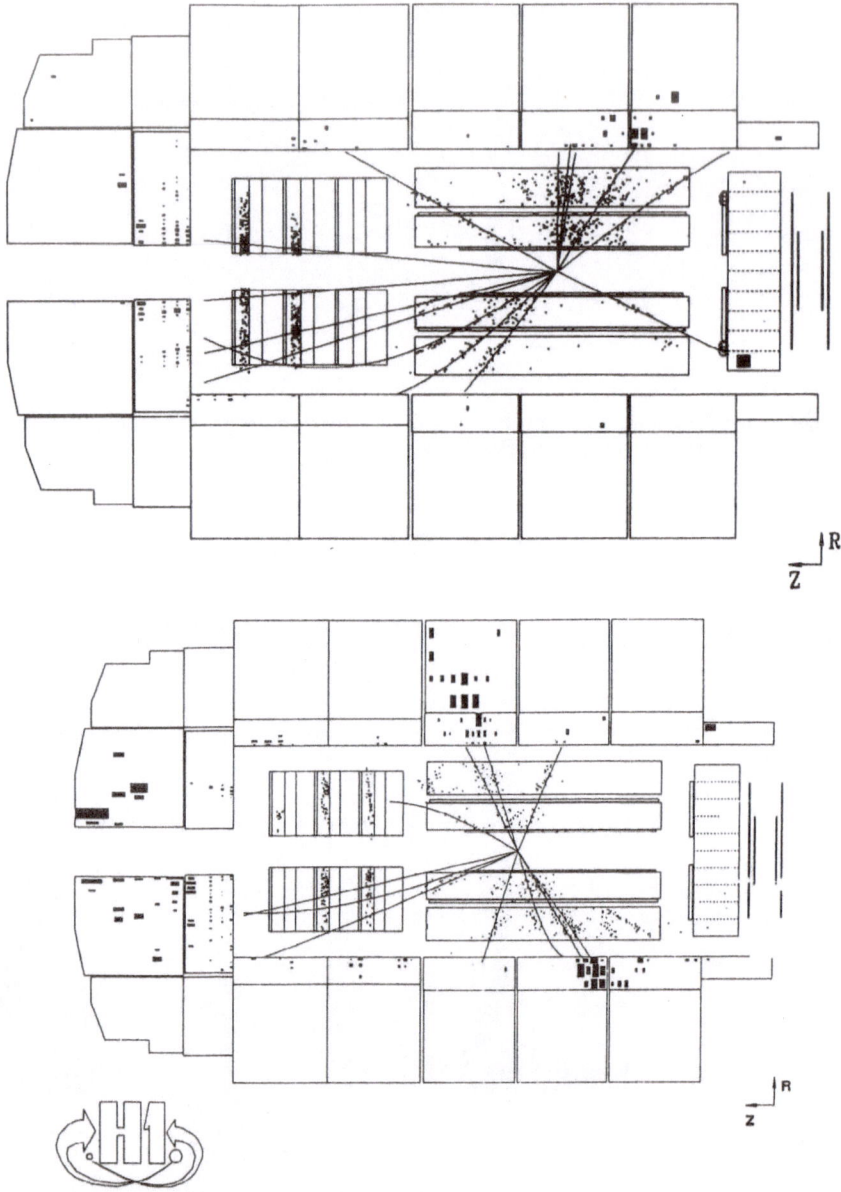

Fig. 20. Event pictures observed by H1:
top: Deep inelastic NC scattering at $Q^2 = 103$ GeV^2, x = 0.004
bottom: Photoproduction with two jets in the final state

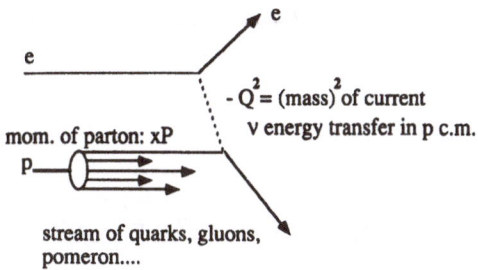

Fig. 21.   Diagram for ep scattering

Table 2.   Kinematic region accessible at HERA (E_e = 30 GeV, E_p = 820GeV) and in previous experiments

	HERA	pre - HERA
s (GeV^2)	10^5	10^3
maximum attainable Q^2 (GeV^2)	40 000	400
$\triangle$ (cm^{-1})	$1 . 10^{-16}$	$1 . 10^{-15}$
ν_{max} (GeV)	52 000	500
minimum x at Q^2 = 10 GeV^2	$1 . 10^{-4}$	$1 . 10^{-2}$

where $\theta_{e'}$ is the lepton scattering angle measured w.r.t. the incoming proton direction and the electron and proton masses, m_e, m_p, have been neglected.

Table 2 compares the kinematic range accessible at HERA and in previous lepton - nucleon scattering experiment. The maximum energy transfer is increased by a factor of ~ 100: HERA is equivalent to a fixed target experiment with an incident electron beam of 52 TeV. The Q^2 domain over which lepton nucleon scattering can be measured is also increased by two orders of magnitude. Since the typical Q values in DIS are much larger than the proton mass the electron interacts with one of the partons (quarks, gluons, ..) rather than with the proton as a whole: HERA is in reality an electron - quark (or gluon) collider.

For the analysis of ep interactions at HERA an understanding of the event kinematics is helpful. It is complicated by the fact that the c.m. system is not at rest. The large momentum excess of proton over electron beam pushes most final state particles into the proton direction. The correlations between energy and angle of electron and current jet (ignoring gluon emission) are shown in fig. 22 in the x - Q^2 plane.

Cross sections.   The cross sections for NC and CC scattering are related to structure functions F_i of the proton:

NC, e p -> e X:

$$\frac{d^2\sigma(\gamma+Z^0)}{dx\,dy} = \frac{4\pi\alpha^2}{s\,x^2\,y^2}$$

$$\cdot\,[(1 - y + y^2/2)\,F_2(x,Q^2) - y^2/2\,F_L(x,Q^2) \pm (y^2/2 - y)\cdot xF_3(x,Q^2)]$$

The upper (lower) sign applies to e^- (e^+) p scattering. For $Q^2 > 1$ GeV^2 the contribution of the longitudinal structure function F_L is small (Callan - Gross relation[13]). The F_3 contribution arises from Z^0 exchange and is significant only when Q is comparable to the Z mass.

CC, e p -> ν X:

$$\frac{d^2\sigma(W)}{dx\,dy} = \frac{G_F^2\,s}{2\pi}\,\frac{1}{(1+Q^2/M_W^2)^2}$$

$$\cdot\,[(1 - y + y^2/2)\cdot F_2(x,Q^2) \pm (y^2/2 - y)\cdot xF_3(x,Q^2)]$$

where G_F is the Fermi coupling constant, $G_F = 1.02 \cdot 10^{-5} / m_p^2$. As before, the upper (lower) sign applies to e^- (e^+)p scattering. The structure functions F_i from NC and CC scattering are in principle independent. They can be related via the quark parton model, however.

The structure functions measured in previous experiments have been extrapolated by QCD evolution[14] to the HERA regime (see fig. 23). The number of events expected from NC and CC scattering are shown in fig.24 for 500 pb^{-1}. The large NC rates at low Q^2 stem from photon exchange. At $Q^2 > M_Z^2$ contributions from Z - exchange become equally important. The requirement of a minimum of 100 events leads to a maximum Q^2 value of about 35 000 GeV^2 up to which NC measurements are feasible. The event rate for CC scattering at low Q^2 is much smaller. However, for $Q^2 > M_Z^2$ the CC cross section exceeds that for NC scattering (see fig. 25). The practical Q^2 limit for CC studies is around 40 000 GeV^2.

Small x physics. Small x physics is a new and exciting field of lepton - nucleon scattering pioneered by Gribov, Levin and Ryskin[16]. The possibility of accessing this region at HERA has stimulated an intensive discussion[17]. Since $x = Q^2/(2\,m_p\,\nu)$, small x - values are attained for fixed Q^2 by making the energy transfer ν large. At HERA, for $Q^2 = 10$ GeV^2, x - values as small as 10^{-4} can be reached which is a factor of 100 smaller than in previous experiments (see table 2). The NC cross section is favorably large in this regime as shown in fig. 26: For instance, the nominal yearly luminosity of 100 pb^{-1} should yield 10^6 events with $10^{-4} < x < 10^{-3}$, $10 < Q^2 < 20$ GeV^2.

Consider scattering (fig. 27a) at small x but not too small Q^2 such that α_s is small, e.g. $Q^2 > Q_0^2 = 10$ GeV^2. As x -> 0 the numbers of gluons and sea quarks of the proton are expected to grow beyond any limits (Regge picture), e.g. the number of gluons with momentum fractions x, x+dx should behave as $G(x) \sim x^{-3/2}$. Since the transverse size of the partons is

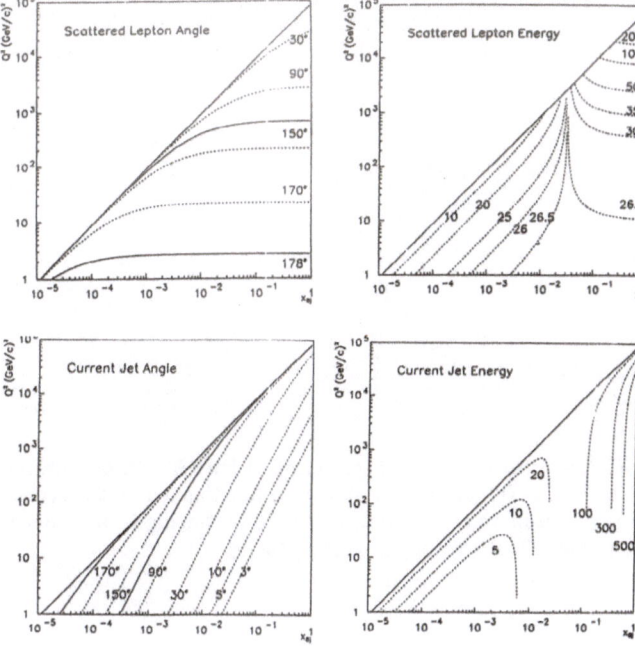

Fig. 22. The x - Q^2 dependence of the angle and energy of the scattered electron and the current jet (ignoring gluon radiation) for beam energies of E_e x E_p = 30 GeV x 820 GeV; the angles are measured w.r.t. the proton beam direction

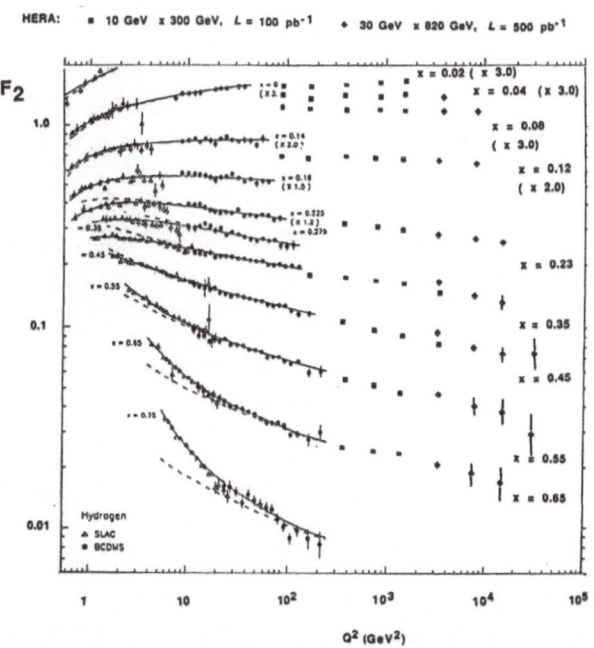

Fig. 23. The structure function F_2 for protons as mesasured by SLAC and BCDMS and as expected from HERA experiments at E_e x E_p = 30 GeV x 820 GeV with L = 500 nb^{-1}; the predictions for the HERA data have been taken from ref. 15

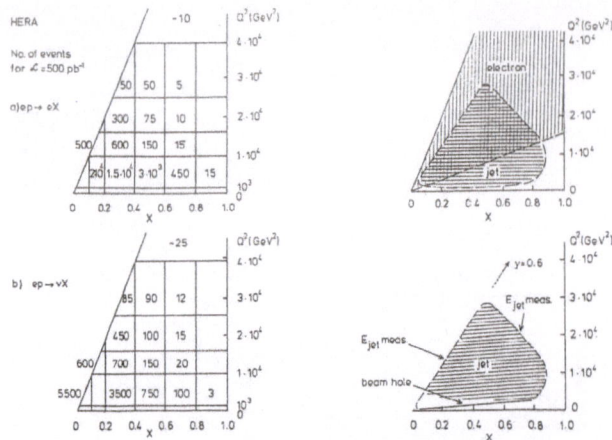

Fig. 24. Event rates for NC and CC scattering at HERA with $L = 500$ pb^{-1} calculated with LUND - LEPTO and EHLQ structure functions, and, the regions where x and Q^2 can be well measured at HERA for NC scattering from either the electron or the jet and for CC scattering only from the jet

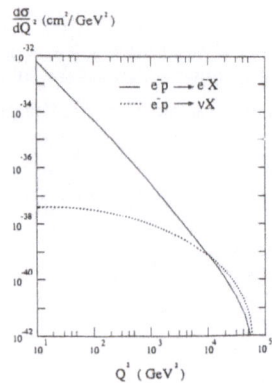

Fig. 25. The cross sections for NC and CC scattering as a function of Q^2

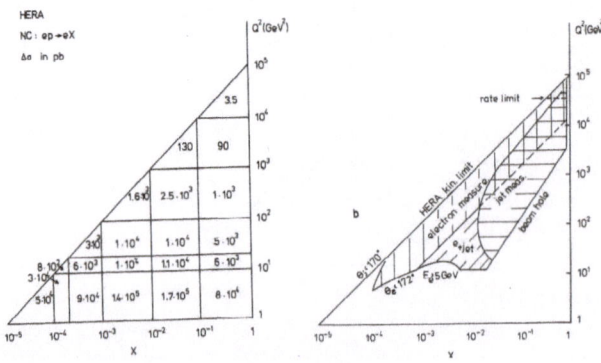

Fig. 26. a) The NC cross section at HERA for E$_e$ x E$_p$ = 30 GeV x 820 GeV calculated with LUND-LEPTO and EHLQ structure functions
b) The region where x and Q^2 can be well measured

234

fixed $(1/Q)$ and since the partons are confined to the proton and their number grows as $x \to 0$, there must be an $x = x_{crit}$ below which partons begin to overlap (fig. 28). This must lead to saturation of the structure functions as $x \to 0$ (fig. 29). A possible mechanism of parton overlap is depicted in fig. 27b : two ladders start from two different partons and begin to interact.

Two estimates for x_{crit} are given in fig. 30 as a function of Q^2 (from ref. 18). The first is characterized by a radius of 5 GeV^{-1} (proton radius) and assumes the parton distribution in the proton to be uniform. In this case the saturation region is barely within reach at HERA. For $Q^2 = 4$ GeV^2 x_{crit} $= 10^{-4}$. However, the low - x partons may concentrate around the valence quarks and form hot spots[19]. Assuming a hot spot radius of 2 GeV^{-1} observation at HERA looks very promising (fig. 30). The amount of saturation one may expect e.g. for the gluon structure function is shown in fig. 31 for the two models.

The smallest - x data for $Q^2 > 5$ GeV^2 that were presented before HERA are shown in fig. 32. They show recent measurements of F_2 by NMC[20] in μp scattering for x - values between 0.008 and 0.5. It is remarkable that the prediction for F_2 obtained by fitting previous data from higher x - values[21] fails to fit the NMC data: the NMC data indicate a faster rise of F_2 as x approaches zero. Inclusion of the new NMC data in the structure function fits has resulted in the predictions[22] D_0 and D_- shown in fig. 33. The two sets differ in the assumption on whether the gluon structure function is constant or diverges as x goes to zero: $xG(x,Q^2) \sim$ constant $(\sim x^{-0.5})$ for D_0 (D_-). While the two sets give identical results for x > 0.01 they make markedly different predictions for F_2 for $x < 10^{-3}$: at x = 10^{-4} the difference is a factor of three. Also indicated in fig. 33 is an estimate of the effects of parton saturation: they are small for a uniform proton but large in the hot spot model for $x = 10^{-4}$.

Deviations from the standard Altarelli - Parisi (GLAP) evolution at very small values of x are also expected for a "technical" reason. In the GLAP evolution for each additional factor of α_s only terms $\sim (\log Q^2) (\log 1/x)$ are kept while $(\log 1/x)$ terms are neglected. This approximation has been avoided in the Lipatov evolution[23].

Experimentally accessible x - Q^2 region.

Standard x - Q^2 region: The x - Q^2 region accessible to experiments depends on the structure of the events and on the detector. For NC events, the values of x and Q^2 can be determined from the energy and direction of *either* the scattered electron or the current jet. For CC events, where the scattered lepton is a neutrino, x and Q^2 can be measured only with the current jet. Figure 24b shows for nominal beam energies and standard x and Q^2 values the regions over which x and Q^2 can be measured well from the electron and the jet parameters, respectively. The main limitations stem from the precision with which the electron and jet energies can be measured, and from the size of the beam holes (see below). For NC scattering, structure function measurements should be feasible for basically the full range of x and Q^2. In the case of CC scattering precise measurements will be difficult for y > 0.6 and below y $\approx$ 0.02. The well measurable region can be extended by operating HERA at smaller beam energies.

Small x - region: The major limitation for NC studies at very small x and low Q^2 values comes from the cutouts provided in the forward and rear

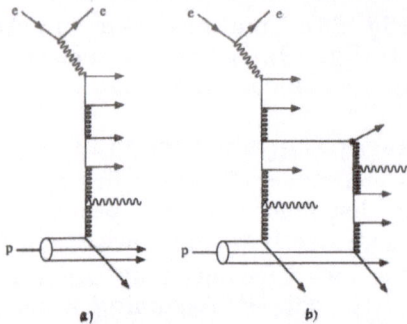

Fig. 27. Diagrams for low - x scattering: a) with a single ladder; b) with two ladders started from two different incoming partons and where the two ladders interact

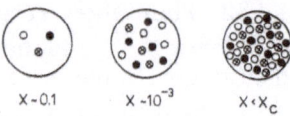

Fig. 28. The parton density in the nucleon for different values of x

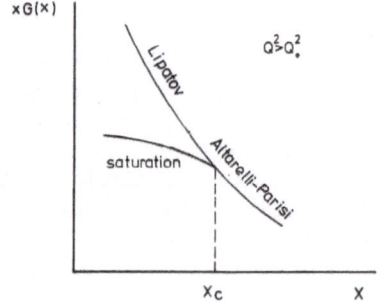

Fig. 29. Qualitative behavior of the gluon structure function at very small x values

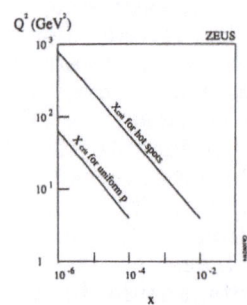

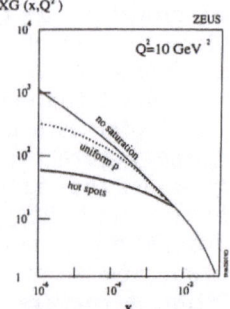

Fig 30. Model predictions for the x - Q^2 behavior of x_{crit} for a uniformly populated proton (lower curve) and for the hot spot model (upper curve); from ref. 18

Fig. 31 Model predictions for the x behavior of the gluon function at $Q^2 = 10$ GeV^2 assuming no saturation or saturation for a uniformly populated proton and for the hotspot model; from ref. 18

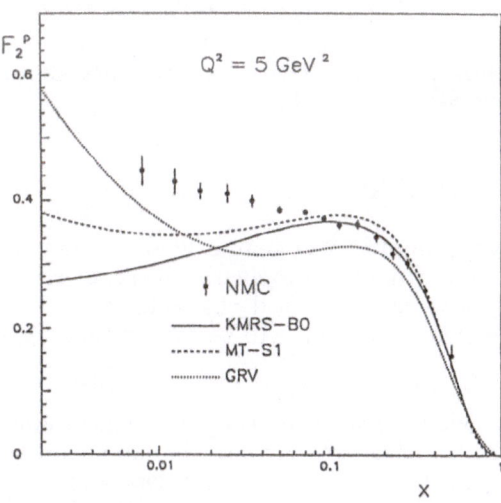

Fig. 32. The structure function F_2 as measured in μp scattering by NMC[20], together with predictions obtained from fits to previous DIS data[21]

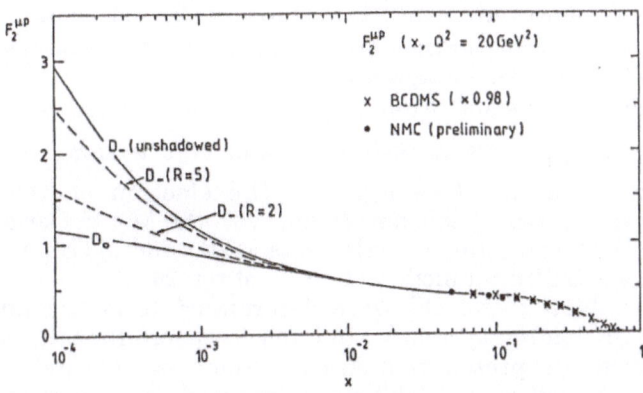

Fig. 33. Prediction for F_2 at very small x - values obtained from a fit [22] to the new data from NMC and other experiments assuming as x -> 0 : x G(x) ~ $x^{-0.5}$ (D_) and x G(x) ~ constant (D$_0$)

calorimeters for beam passage. Typical cross sections of these beam holes are 20 x 20 cm^2 (ZEUS). The effective hole in the acceptance is somewhat larger since a reliable energy measurement requires that the point at which electron or jet enter the calorimeter to be some distance away from the cutout. Figure 26b shows an educated guess for the well measurable region. It follows from the requirements

$$\theta_{current\ jet} < 172^0,\ \ \theta_{electron} < 172^0,\ \ E_{electron} > 5\ GeV$$

and from the beam hole. The HERA experiments should be well suited for the region $x > 10^{-4}$, $Q^2 > 10\ GeV^2$.

Results from H1 and ZEUS

For both experiments the prime goal in this first analysis was to establish that events from deep inelastic NC scattering can be isolated from background and have the expected characteristics and cross section. The description of the event selection will be rather sketchy; details can be found in refs. 6,7.

ZEUS analysis. Neutral current scattering was studied with an integrated luminosity of 2.2 nb^{-1}. The ZEUS analysis has been based primarily on the calorimeter information requiring energy deposition, correct timing and an identified electron with an energy > 5 GeV. By visual scanning most remaining background events from cosmic rays, beam - wall or beam - gas scattering and beam associated muons were removed (total of 8% of the events). The total energy and longitudinal momentum as measured in the calorimeter were limited to the range 37.2 GeV < E_{tot} - P_L < 60 GeV. For complete solid angle coverage energy - momentum conservation requires E_{tot} - P_L = 2 E_e = 57.2 GeV for NC events. The NC events are clearly visible in the E_{tot} - P_L distribution which is shown in fig. 34 before the cut. The distribution of the event vertex along the beam (Z - direction) shows the NC sample to be free of beam background (fig. 35).

The transverse momentum vectors of the electron, p_{T_e} , and of the hadron system, p_{T_h}, are back-to-back and of the same magnitude as required for NC events (see fig. 36). The electron energy and angle distributions are in reasonable agreement with the Monte Carlo predictions calculated with the structure function sets MTB1 and MTB2 (constant and diverging gluon structure function at x = 0) of ref. 24.

The variables x and Q^2 were determined from the angles of the electron and the hadron system[27]. In order to ensure that the current jet is separated from the proton remnants to some extent and its direction measurable a y_h - cut of y_h > 0.02 was introduced. Here y_h is the y variable determined from the total energy and longitudinal momentum of the hadron system (Jacquet - Blondel method[26]), $y_h = \Sigma\ (E^i - p_L^{\ i})\ /\ (2\ E_e)$.

H1 analysis. NC candidates were selected from data corresponding to an integrated luminosity of 1.5 nb^{-1} requiring energy in the liquid argon or BEMC calorimeters, rejecting out-of-time events with the time-of-flight information (TOF-VETO) , requiring - for low x - a track coming from the origin and an identified electron. The energy spectrum in the backward calorimeter BEMC is shown in fig. 37 for successive selection steps. The top distribution was obtained after requiring a cluster of > 3.5 GeV in the

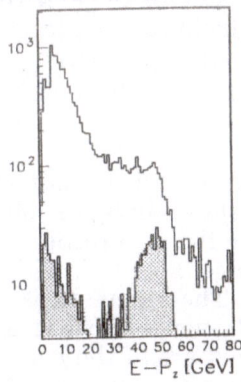

Fig. 34. The distribution of $E_{tot} - P_L$ as measured by ZEUS. The candidates for NC scattering show up as a bump near $2 \cdot E_e = 53.2$ GeV.

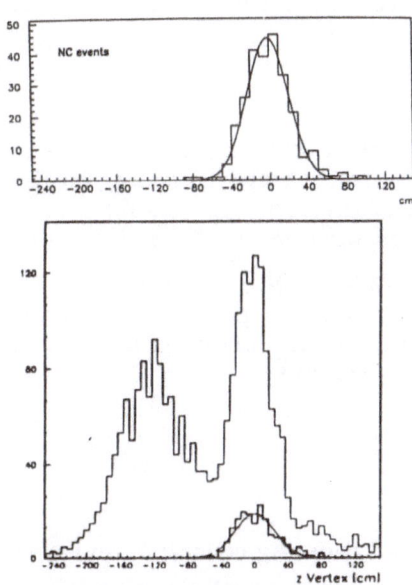

Fig. 35.

Distribution of the vertex coordinate along the beam for ZEUS events from NC scattering (top) and beam - gas interactions (bottom)

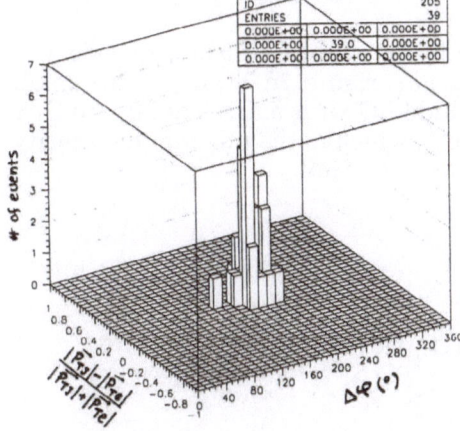

Fig. 36.

Distribution of the nomalized difference of transverse momenta and azimuth angles of electron and jet for NC events from ZEUS with a scattered electron of $E_{e'} > 10$ GeV, $\theta_e < 172^0$ and a current jet of $E_{jet} > 15$ GeV

calorimeters. Imposing the timing cut and the electron selection reduces the rate of candidates by a factor of 10^5. Most of the events below $E_{e'} = 14$ GeV arise from photoproduction (fig. 38); at higher electron energies the sample of 186 events is almost purely from NC scattering. The transverse momentum in these events is balanced.

H1 and ZEUS results. The event distributions in the x - Q^2 plane measured by the two experiments are shown in fig 39. Most events lie in the Q^2 region between 10 and 100 GeV^2. In the x - direction the events populate mainly the region of very small x - values, $x < 10^{-2}$, with a sizeable number of events between $x = 10^{-4}$ and 10^{-3}. The kinematical region over which the HERA experiments have detected events exceeds by far the region covered by previous experiments (dashed area): the accessible region is extended in x direction by two orders of magnitude from $x = 10^{-2}$ (previously) to 10^{-4} (now) at $Q^2 \approx 10 \, GeV^2$.

The observed x and Q^2 distributions within their present errors agree with the expectations from popular structure functions (figs. 40, 41). Also the average cross sections for $8 \cdot 10^{-4} < x < 8 \cdot 10^{-3}$ from ZEUS (fig. 42) are in agreement with the predictions calculated from the structure function sets of refs. 22, 24, 27.

Leptoquarks

Leptoquarks (LQ) are expected in many models. Production of leptoquarks in e - p collisions can proceed in several ways depending on the couplings and the leptoquark mass. The most favorable process for HERA is s - channel production(fig. 43): the electron with energy E_e and a quark with energy $x \cdot E_p$ form a leptoquark of mass squared $M_{LQ}^2 = 4 \, x \, E_e \, E_p = xs$. In the x - Q^2 distribution leptoquark production will populate a narrow band with fixed x. For narrow leptoquarks, the production cross section and the branching ratio into electron + X are given by[28]

$$\sigma_{LQ} = \pi / (4s) \cdot (g_L^2 + g_R^2) \cdot u(x = M_{LQ}^2/s)$$
$$BR(LQ \to e + u) = (g_L^2 + g_R^2) / (2 \, g_L^2 + g_R^2)$$

where g_L and g_R are the left and right handed coupling constants and $u(x)$ is the u-quark density in the proton. The coupling strengths are unknown. For $g = 0.3$ the coupling would have the same strength as the electromagnetic coupling, $g^2/4\pi = \alpha$.

The x - Q^2 distributions measured by H1 and ZEUS show no evidence for a leptoquark signal which is not surprising in view of the small luminosity of ~ 1 - $2 \, nb^{-1}$. The data exclude LQ with masses of 30 - 50 GeV for $g = 0.3$ (Refs. 6,7,29). A run with $1 \, pb^{-1}$, which is the nominal luminosity per day, will extend the mass range up to ~ 200 GeV.

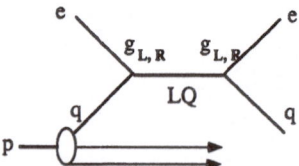

Fig. 43.   Diagram for leptoquark production

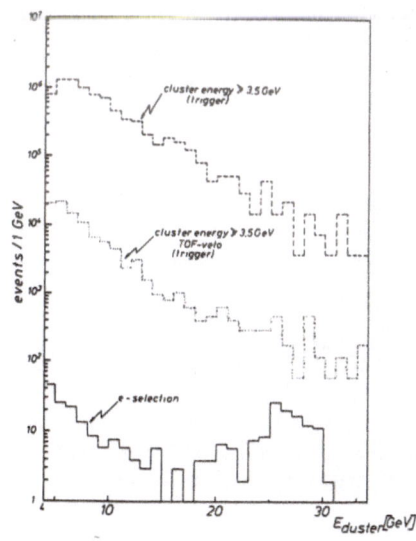

Fig. 37. Event rates for low Q^2 DIS candidates from H1 versus cluster energy at different levels of selection.

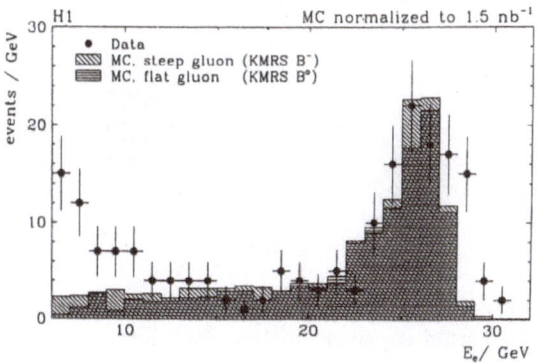

Fig. 38. Electron energy distribution for NC candidate events of H1 compared with predictions for two sets of structure functions from ref. 24

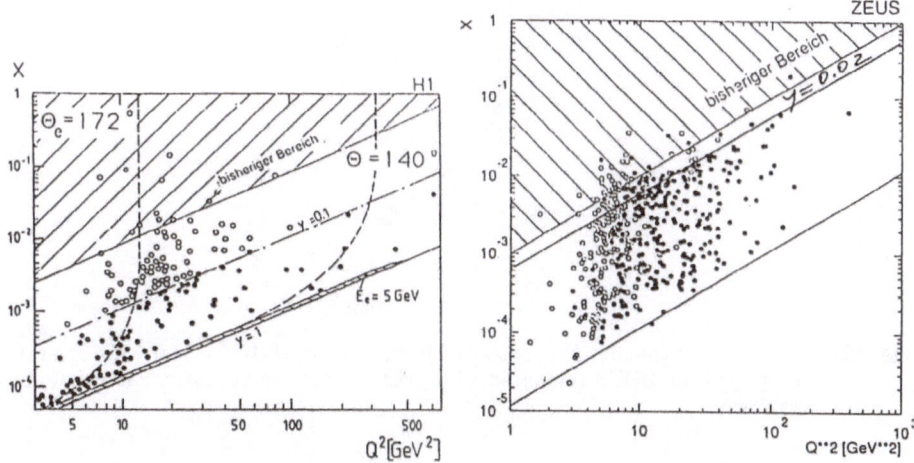

Fig. 39. The x - Q^2 distribution of NC events from H1 (top) and ZEUS (bottom). The kinematic region covered by previous experiments has been dashed

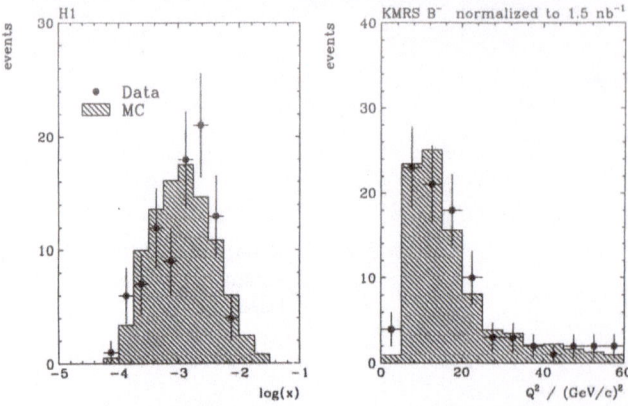

Fig. 40.   Distributions of x and Q^2 for NC events from H1 in the range $0.05 < y < 0.55$ compared to an absolute prediction due to a set of structure functions from ref. 22

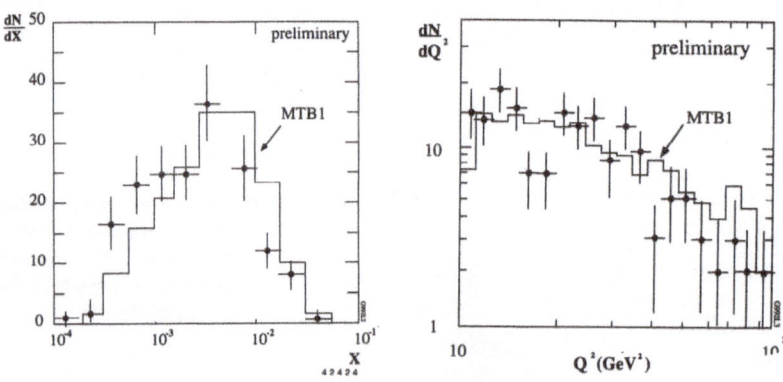

Fig. 41.   Distributions of x and Q^2 for NC events from ZEUS for $y > 0.02$ compared to the normalized prediction made with a set of structure functions from ref. 24

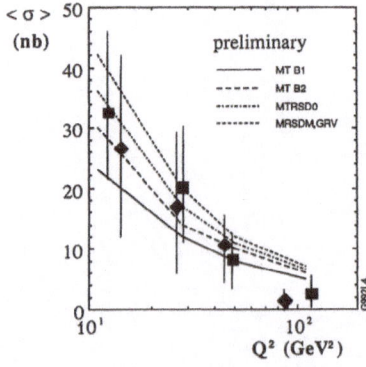

Fig. 42.   The cross sections for NC scattering in the x range $8 \cdot 10^{-4} - 8 \cdot 10^{-3}$ for several Q^2 intervals from ZEUS compared with predictions from various structure function sets[22,24,27]

TOTAL PHOTOPRODUCTION CROSS SECTION

Physics Introduction

For nonafficionados a measurement of the total photoproduction cross section, $\sigma_{tot}(\gamma p)$, at high energies may seem rather dull. A closer look at the underlying physics, however, shows that the subject has very interesting aspects. It has recently been reviewed by G. Schuler[30] and the following discussion draws heavily on his report.

Next to $p\bar{p}$ interactions, γp scattering is the only other hadronic or hadron - like reaction for which the total cross section can currently be measured at c.m. energies of several hundreds of GeV. In the VDM picture the photon couples to vector mesons V (ρ, ω, ϕ ..) which in turn interact with the proton. As a result $\sigma_{tot}(\gamma p)$ is related to the cross sections for Vp, and via the quark - parton model, for $\pi^{\pm}p$ and $K^{\pm}p$ scattering. Like the purely hadronic reactions, $\sigma_{tot}(\gamma p)$ in this model has three pieces coming from pomeron exchange, nonpomeron exchange and from partonic scattering in hadron - hadron interactions (fig. 44a-c). The behavior of $\sigma_{tot}(\gamma p)$ at high energies is therefore expected to be very similar to that of $\sigma_{tot}(p\bar{p})$ (although the hard scattering part of Vp is presumably larger than for pp due to a harder x - distribution of quarks).

The photon, in addition to its hadronic features, however, possesses a property which makes it distinctly different from hadrons: it couples directly to quarks and the coupling is pointlike. This leads to additional hard scattering processes which become prominent at high energies and which are not present in hadron - hadron interactions. They are represented by two types of diagrams: the first ("direct photon process") results from photon - gluon fusion into a quark - antiquark pair (fig. 45a) and from photon scattering off a quark in the proton under the emission of a gluon ("QCD Compton process"), see fig. 45b. But the quarks coupling to the photon can also emit gluons (e.g. fig. 45c), and either a quark or a gluon may participate in the hard scattering(fig. 45d,e). Together with the hard scattering of the hadronic (VDM) photon the hard scattering due to the anomalous quark and gluon content of the photon constitute the "resolved photon processes". They are summed in the photon structure functions F^{γ} which describe the quark and gluon content of the photon. The hard photon processes - direct or resolved - give rise to quark and gluon jets (sometimes called "minijets") with large transverse momenta. Their cross sections have been calculated and found to depend critically on the minimum momentum transfer p_{Tmin} down to which the integration is performed. As will be seen below, at HERA energies hard scattering originating from photon constituents is clearly visible.

The behavior of $\sigma_{tot}(\gamma p)$ at high energies is closely linked to the density of soft or low - x partons in the proton. The relation between the proton structure function F_2 and the total transvers and longitudinal photon proton cross sections is given by

$$F_2(x, Q^2) = \frac{(1-x)\,\nu^2\,Q^2}{4\pi^2\,\alpha\,(Q^2 + \nu^2)}\,(\sigma_{trans} + \sigma_{long})$$

where ν is the photon energy for a stationary proton. As $Q^2 \to 0$: $\sigma_{long} \ll \sigma_{trans} = \sigma_{tot}(\gamma p)$. For small $x \ll 1$ and high energies, $\nu^2 \gg Q^2$, the result for $Q^2 \to 0$ is:

$$F_2(x, Q^2) \approx \frac{Q^2}{4\pi^2\alpha}\,\sigma_{tot}(\gamma p).$$

Under the assumption that for very small x and Q^2 F_2 is of the form

$$F_2(x, Q^2) \sim x^{-\gamma}\,(Q^2)^\delta = (Q^2)^{-\gamma + \delta}\,(2m_p\,\nu)^\gamma \approx (Q^2)^{-\gamma + \delta}\,W^{2\gamma}$$

where W is the total γp c.m. energy. Continuity for F_2 is preserved as $Q^2 \to 0$ if $F_2 \sim Q^2$ and therefore

$$\sigma_{tot}(\gamma p) \sim W^{2\gamma}.$$

The energy dependence of the total cross section at high energies is driven by the parton density of the proton at low x.

The definition for F_2 used here is the standard one, given in above. A "microscopic" treatment might separate F_2 into a term describing the "proton proper" (whatever this is) and a term due to resolved photons. In this case F_2 is expected to be dominated by the x distribution of the gluon in the resolved photon structure function[30].

The following remark on this point is due to G. Schuler. A "microscopic" treatment might start from a dispersion relation for $F_2(x,Q^2)$,

$$F_2(x, Q^2) = \frac{Q^2}{4\pi}\int_{4m_\pi^2}^{\infty} dk^2\,\frac{\rho(x, k)}{(k^2 + Q^2)^2}$$

and split the k^2 - integration into a low and a high mass contribution. The low mass integral corresponds to the (low mass) VDM contribution while the high mass part describes the direct photon contribution which is related to $F_2(x, Q^2)$ for the proton as measured in DIS. In other words, the low mass part is associated with the internal structure of the photon, while the high mass part is connected with the proton structure. Yet there is no unique distinction between the photon structure, the proton structure, and the properties of the interaction.

Observations of "diffractive" air showers[31] and an excess of muons in very energetic cosmic air showers[32] have suggested $\sigma_{tot}(\gamma p)$ to rise much faster than $\sigma_{tot}(p\bar{p})$ at beam energies above ~ 100 TeV. This rise has been attributed to the semi-hard scattering arising from the resolved photon contributions. Since gluon initiated processes dominate, the rise is driven by the x-dependence of the gluon distribution for photons and protons[33].

H1 and ZEUS Results

Photoproduction at HERA is part of neutral current scattering, ep -> eX, with the exchanged photon being almost real, $Q^2 \approx 0$. For events of this type the electron is scattered in the direction of the electron beam. It can be tagged in the electron calorimeter of the luminosity monitor (fig. 8) and its energy, E_e', measured. This determines the energy k of the exchanged photon in the HERA system, $k = E_e - E_e'$, and the photon proton c.m. energy W, $W^2 = 4 k E_p$. For a minimum energy of the scattered electron of 4 GeV c.m. energies up to 300 GeV can be measured at HERA which is equivalent to a photon of energy $\nu = W^2/(2m_p) \approx 47$ TeV striking a stationary proton.

Since the standard DIS trigger of ZEUS had an acceptance of only 4% for photoproduction events a special run was performed with reduced energy thresholds of the calorimeter trigger. Data were taken over a period of 7 hours of running yielding a total luminosity of 233 μb^{-1} and 53k events. The main requirements for photoproduction events were an electron tagged in LUMI (scattering angle < 6 mrad), a minimum energy of 1.1 GeV in the rear calorimeter (RCAL) and proper calorimeter timing. In total 212 events satisfied these conditions of which 182 were accepted after a visual scan. The event vertex was reconstructed from the tracking information for 72% of the events. The vertex position peaks close to the origin in contrast to the uniform distribution characteristic of beam gas events (fig. 46). For the final sample of 97 events the energy of the scattered electron was required to lie in the range 10 - 16 GeV (W = 186 - 233 GeV) resulting in maximum and average Q^2 values of 0.02 and 0.0006 GeV^2, respectively. Hence the contribution from longitudinal photons can safely be neglected and the measured process is scattering of (almost) real photons on protons.

The main difficulty in the extraction of $\sigma_{tot}(\gamma p)$ lies in the fact that the trigger acceptance A depends strongly on the type of process. It is low for the "elastic" reaction $\gamma p \to \rho p$ (fig. 44a) and sensitive to the details of the model, A = 10 - 27%. For these events the proton disappears mostly undetected in the forward direction and the $\pi^{\pm}$ from ρ decay escape often unseen in the backward direction. For inelastic diffractive processes (fig. 44b) A = 40 - 50% while for low - p_t reactions (fig. 44c) A = 70%; for hard processes such as direct and resolved photon contributions A = 80%. The final acceptance calculation was made assuming 20% from elastic plus diffractive processes and 80% from low p_t processes. The resulting predictions for the energy spectra seen in the three sections of the calorimeter give a reasonable description of the data (fig. 47). One point is particularly noteworthy: the data show 18 events with zero energy deposition in the FCAL. These events are candidates for the elastic and inelastic diffractive processes. The Monte Carlo predicts 16 events, in good agreement with the data.

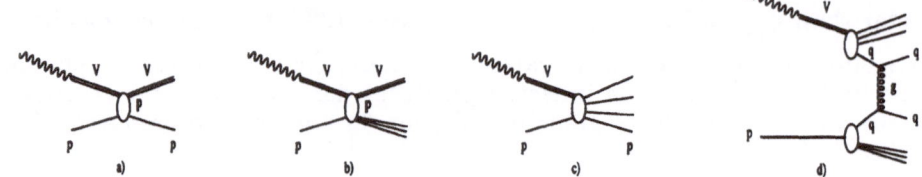

Fig. 44. Diagrams for photoproduction via VDM of vectormesons V by elastic (a) and inelastic diffractive scattering (b), for photoproduction via VDM by nondiffractive processes (c) and hard scattering (d)

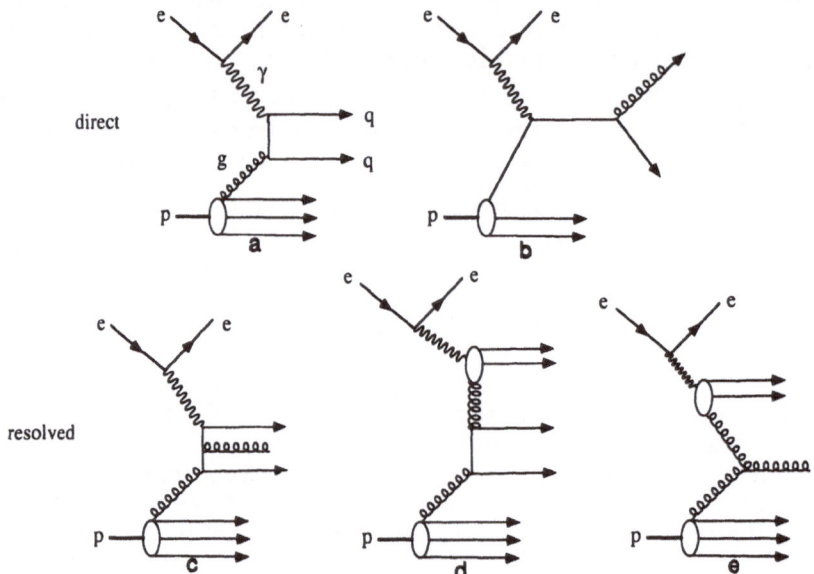

Fig. 45. Diagrams for direct (a,b) and resolved photon processes (c -e)

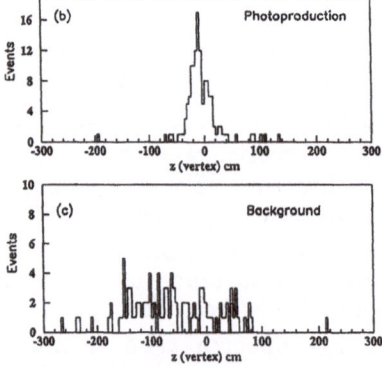

Fig. 46. Distribution of the vertex coordinate along the beam for ZEUS events from photoproduction (top) and beam - gas interactions (bottom)

The photoproduction candidates in the H1 experiment were obtained by requiring a tagged electron ($E_e' = 9 - 19$ GeV), a charged track pointing to the vertex region and at least one reconstructed hit in the backward proportional chamber. Most of the remaining background was due to proton gas interactions. It was statistically subtracted using the energy spectrum of the tagged electron which is substantially different for the signal and background events (fig. 48). This resulted in a sample of 602 ± 61 events corresponding to a luminosity of 1 μb^{-1}.

The acceptance was calculated assuming 28% elastic plus inelastic diffractive events and either 72 % low p_t events or 52% low p_t and 20% hard scattering events.

The total cross section values obtained by the two experiments are

$$\sigma_{tot}(\gamma p) = \begin{array}{l} 150 \pm 15 \,(\text{stat}) \pm 19 \,(\text{syst}) \,\mu b \quad <W> = 192 \text{ GeV, H1, preliminary[6]} \\ 154 \pm 16 \,(\text{stat}) \pm 32 \,(\text{syst}) \,\mu b \quad <W> = 210 \text{ GeV, ZEUS[7,34]} \end{array}$$

Radiative corrections were not included; they are estimated to reduce the cross section by 2.5 % (H1) and 1% (ZEUS). The systematic errors reflect the uncertainties in the determination of the acceptance and the luminosity.

Figure 49 shows $\sigma_{tot}(\gamma p)$ as a function of W above the resonance region (W> 1.75 GeV) as measured in previous experiments up to W = 18 GeV [35] and by H1 and ZEUS. No dramatic rise is observed between 18 and 200 GeV. The two solid curves labelled DL[36] and ALLM[37], are Regge-type analyses which used the lower energy photoproduction measurements, together with proton structure function data, to predict the high energy behavior. The other four curves are based on the assumption that the total cross section is a sum of a soft part plus the contributions from the direct and resolved photons[38]. They depend critically on the choice of photon structure function F^γ and on the parameter $p_{T min}$ which is the lower integration limit for the hard processes. Here, p_T is the transverse momentum cutoff in the c.m. system of the hard subprocess. The dashed - dotted lines use the parametrization of F^γ from ref. 39 with $p_{T min} = 2.0$ GeV/c for the lower (1.4 GeV/c for the upper) line. The dashed (dotted) line uses F^γ from ref. 40 and $p_{T min} = 2.0$ (1.4 GeV/c). A critical appraisal of the different calculations can be found in ref. 41. The measurements favor the lower theoretical predictions. A strong rise of the cross section that might have been expected from direct and resolved photons is postponed to higher energies, if at all.

HARD SCATTERING IN PHOTOPRODUCTION

The presence of direct and resolved photon processes suggests a substantial amount of hard scattering in photoproduction. H1 [6,42] and ZEUS[7,43] analyzed the transverse momentum behavior of events from photoproduction and ep scattering at low Q^2.

H1 analysis

The study was made with two event samples recorded for a total luminosity of 0.9 nb^{-1}. The selection criteria for sample I were:

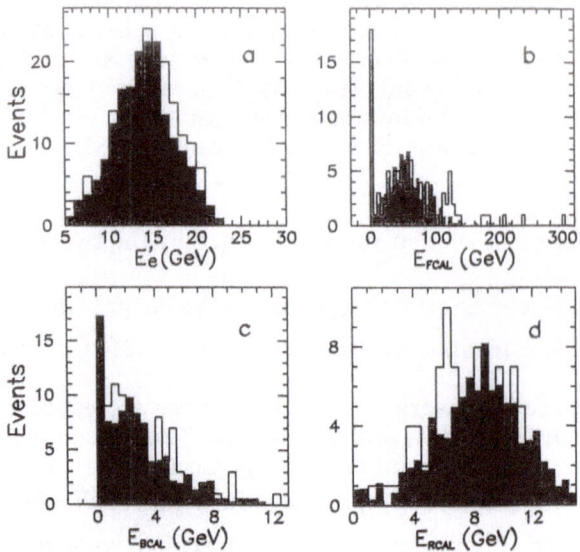

Fig. 47. Photoproduction data from ZEUS:
(a)The energy distribution of the scattered electron as measured by LUMI;
(b) - (d) The energy deposited in FCAL, BCAL and RCAL for photoproduction events with electron energies between 10 and 16 GeV; the shaded distributions are the predictions from the Monte Carlo

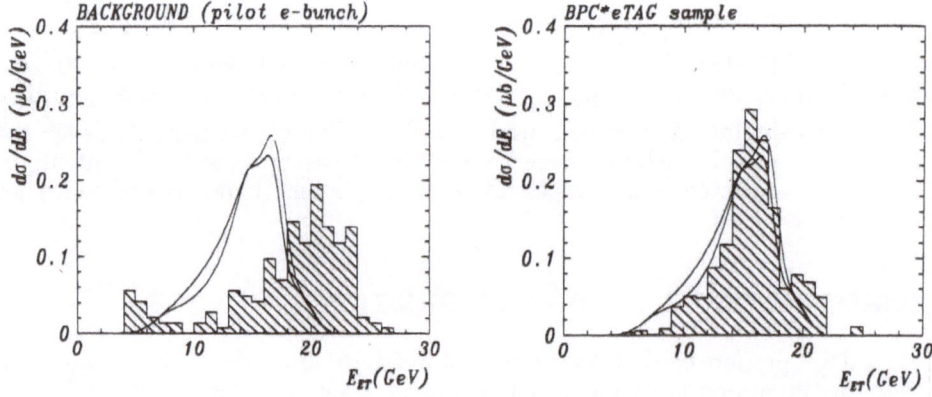

Fig. 48. Energy spectrum from H1 of tagged electrons for background events from the unpaired electron bunch (left) and for tagged events (right), compared with the tagger acceptance (curves)

- the scattered electron detected in LUMI
- a track from the interaction region
- $0.35 < y_e < 0.7$ and $y_h > 0.2$

    where y is the scaled photon energy $y = \nu/\nu_{max}$

    calculated from the scattered electron $y_e = 1 - E_e'/E_e(1 + \cos \theta e')/2$

    from the hadrons meas'd in calorimeter $y_h = \Sigma (E^i - p_L^i) / (2 E_e)$

- energy > 5 GeV in the forward region ($\theta < 25^o$).

A total of 330 events satisfied the requirements. The background from beam gas interactions was estimated to be $< 9\%$. Tagging of the scattered electron in LUMI ensures the events to come from photoproduction. The range for the c.m. energy W is 175 - 250 GeV.

The transverse momentum distribution is shown in fig. 50. For $p_t^2 < 1$ GeV^2 a steeply falling distribution is observed which is well described by a gaussian with an rms of ~ 350 MeV/c. At higher p_T^2 the data show a second component with a long tail of high transverse momenta indicating the presence of hard scattering. A jet search was performed with the calorimeter requiring jets of more than 3 GeV in cones of $\Delta R = (\Delta \eta^2 + \Delta \phi^2)^{1/2} < 1$ where ϕ is the azimuthal angle. For the 19 events with two or more jets the c.m. energy of the two most energetic jets is on average 12.5 GeV.

Since the photon energy is known, $E_\gamma = E_e - E_e'$, the momenta of the two jets are constrained. At the parton level, equating the jets with partons of equal transverse energy E_t and polar angles θ_1 and θ_2, and neglecting initial state radiation:

$$2 E_\gamma/E_t = \tan \theta_1/2 + \tan \theta_2/2$$

Figure 51 shows the distribution of θ_1 versus θ_2 for the data and the predictions for events from direct plus resolved photons and direct photons alone. There is a clear distinction predicted for direct and resolved processes: the direct process yields rarely events where both jet angles are below 100°: only 0.9 events should be found in this region. Instead, 9 of the 19 measured events lie in this region in agreement with the predictions for the direct and resolved processes.

The indication that the data require a resolved photon contribution has been further substantiated by a second sample (II) for which an electron tagged in LUMI was not demanded; thus, these events are not necessarily from photoproduction. However, events with a substantial Q^2 were removed by requiring that there be no electron with $E > 10$ GeV in $140^o < \theta < 176^o$. A total of 51 event were selected in this way.

Figure 52 shows the energy flow per event as a function of the polar angle θ for events where the two most energetic jet cones have $\theta < 100^o$. In this case the direct process predicts no energy flow beyond $\theta \sim 130^o$ in marked contrast with the data which show a sizeable energy flow up to 180°. The prediction for the direct *plus* resolved photon processes is in good agreement with the data.

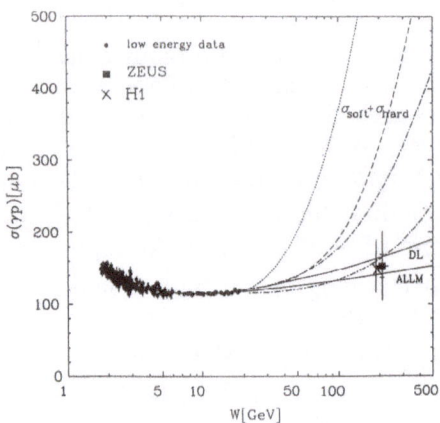

Fig. 49. The total γp cross section as a function of the c.m. energy W as measured below 18 GeV and by H1 and ZEUS near W = 200 GeV. The lower solid curve is the prediction of the ALLM[37] parametrization and the higher solid curve is that of DL[36]. The dotted (dashed) line uses the LAC1 parametrization[40] for the photon with p_{Tmin} = 1.4 GeV/c (2 GeV/c). The dash-dotted lines use the DG parametrization for the photon[39] with p_{Tmin} = 1.4 GeV/c (upper line) and p_{Tmin} = 2 GeV/c (lower).

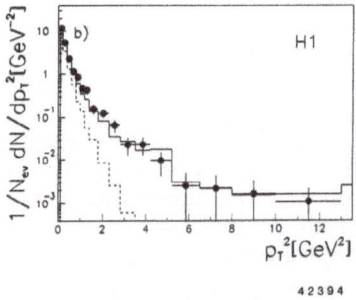

42394

Fig. 50. Distribution of $p_T 2$ for charged tracks as observed by H1 from ep data compared with 'soft' (dashed line) and 'soft + QCD' Monte Carlo simulation (full line)

250

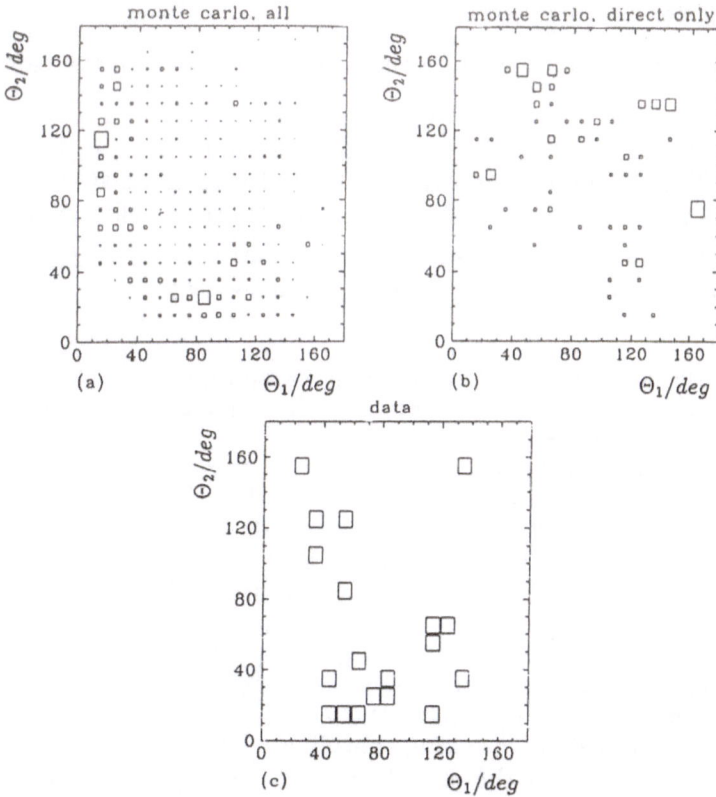

Fig. 51. Correlation of the two highest p_T jets in polar angle for H1 events with an electron tag: (a) Monte Carlo predictions for resolved and direct processes, (b) for direct processes only, (c) correlation for the 19 events with a tagged electron.

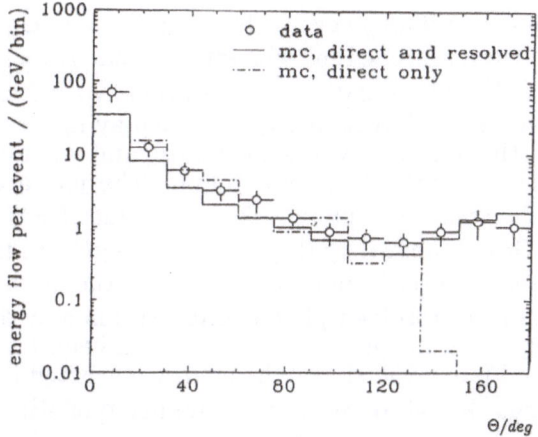

Fig. 52. Energy flow per event versus polar angle as measured by H1; the open points represent the data; the full line is taken from the Monte Carlo simulation for resolved and direct photons; the dotted line gives the prediction for the direct processes alone

ZEUS analysis

The events were selected by requiring in addition to the calorimeter trigger

- proper timing observed by the calorimeter,
- more than 10 GeV in FCAL and > 2.5 GeV in RCAL,
  or: > 20 GeV in total energy, > 10 GeV in total transverse energy and $\sum (E^i - p_L{}^i) / (2 E_e) > 12$ GeV when summed over all calorimeter cells,
- no scattered electron observed in the calorimeter with $y_e < 0.7$. This cut ensures that the accepted events have $Q^2 < 4$ GeV^2.

The cuts were satisfied by 576 events from a total luminosity of 2.2 nb^{-1}. The contamination from beam gas events is < 3% and < 2% from deep inelastic events with $Q^2 > 4$ GeV^2.

The data are consistent with originating from photoproduction. This is shown by comparing the full sample with the 96 events for which the scattered electron was detected in LUMI with $5 < E_e' < 22$ GeV. These events have $Q^2 < 0.02$ GeV^2. The ratio of tagged to total number of events agrees well with the Monte Carlo prediction of 20% for pure photoproduction events. The distribution of the γp c.m. energy W is shown in fig. 53a. For the full sample W is in the range from 100 to 295 GeV; for the tagged electron events it is narrower, $160 < W < 260$ GeV, reflecting the restricted electron energy range. Deep inelastic events would have a completely different shape, peaking sharply at small W.

The distributions of the total transverse energy E_T and the missing transverse momentum p_{Tmis} are shown in figs. 53b,c. The overall p_T is well balanced as expected for photoproduction, the average p_{Tmis} being 1.5 GeV/c. The E_T distribution exhibits, by contrast, a tail that extends beyond 10 GeV: 391 events have $E_T > 10$ GeV. The tagged electron events show the same behavior (dashed histogram). The tail is much larger than that expected from soft γp interactions which extends to a maximum of 10 GeV (dashed-dotted curve in fig. 53b). A comparison of the cross sections shows that 20 % of all photoproduction events have an $E_T > 10$ GeV.

The events were searched for jets with jet cone $\Delta R < 1$, jet transverse momentum > 4 GeV and $\eta < 2$ ($\theta > 15^0$): 41 events were found with two jets. One of these events is shown in fig. 54a displaying two clear jets. The distributions of the jet transverse momenta and azimuthal angles are shown in fig. 55b,c for the two - jet sample. The jets are predominantly back-to-back in the transverse plane. The E_T values for the two - jet events are all above 16 GeV as shown by the shaded histogram in fig. 53b.

As discussed before, a major fraction of the high E_T events is expected to come from resolved photon interactions and therefore to show substantial energy flow in the backward (electron beam) direction from the photon remnants. Figure 55d shows the energy deposited in the RCAL (rear calorimeter) versus the minimum pseudorapidity η of either of the two jets. Note, jets in the forward hemisphere have $\eta > 0$, and in the backward hemisphere $\eta < 0$. If direct photon interactions were the sole origin of these events, sizeable energy in RCAL would be expected for events with the minimum $\eta_{min} < -1$, falling essentially to zero as the jets become more

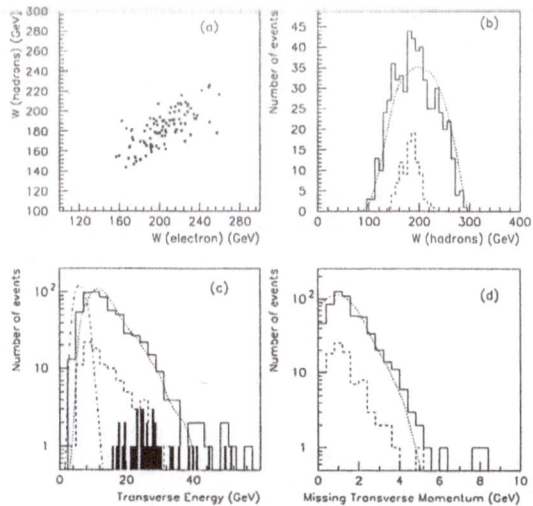

Fig. 53. Hard scattering in photoproduction as analyzed by ZEUS:

(a) Distribution of the γp c.m. energy W for all events (solid histogram) and for events with a tagged electron (dashed histrogram) ; the curves show the Monte Carlo expectations;

(b) Total transverse energy distribution for all events (solid histogram) and for events with a tagged electron (dashed histogram), and for events with a twojet structure (black histogram); the dashed curve shows the expectations from the Monte Carlo simulation including direct and resolved photon contributions with p_{Tmin} = 1.5 GeV/c ; the dashed-dotted curve shows the transverse energy distribution for soft γp interactions

(c) Missing transverse momentum distribution for all events (solid histogram) and for events with a tagged electron (dashed histogram). The dotted curve shows the Monte Carlo expectations.

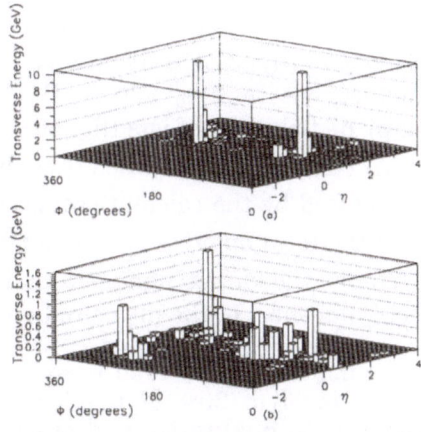

Fig. 54. Examples for events from hard scattering in photoproduction as observed by ZEUS in η - ϕ space:

(a) An event with a two - jet structure;

(b) An event showing a two - jet structure plus additional energy in the direction of the incident electron associated with the photon remnant distant from the RCAL region ($\eta_{min} > 0$). This trend is indeed observed in fig. 55d. However, in addition the data show events with as much as 4 - 12 GeV detected in RCAL, even when both jets are far from the RCAL itself, the nearest jet being as much as three units of rapidity away. These events are interpreted as originating from the resolved photon process, where the two partons go forward and the photon remnants go approximately in the direction of the incident electron, which is close to the direction of the photon. One of the resolved photon candidates is shown in fig. 54b. It has two jets in the forward direction and a cluster of energy going backward.

Discussion of the Results

The findings of the two experiments are consistent. Photoproduction at c.m. energies of 100 - 300 GeV exhibits an excess of events with large transverse momenta. For a fraction of these events large transverse momentum jets have been observed which is evidence for the presence of hard scattering. In events with two jets the jets tend to be back-to-back in the transverse plane. Although the statistics is limited, there is a clear indication that in some of the events where both jets go forward, in addition a substantial amount of energy is emitted in the backward direction, close to the direction of the incident photon. This event topology is expected for resolved photon processes.

Hard scattering in photoproduction was predicted to arise from the direct and resolved photon processes. H1 and ZEUS find that the observed gross features as well as the cross sections can be accounted for by these processes. They were calculated together with the soft contributions by Monte Carlo in full detail, using the PYTHIA generator[44] (H1) as well as PYTHIA and HERWIG[45] (ZEUS) with the photon structure functions of ref. 39, the proton structure function MTB1 of ref. 24, including initial and final state parton showers, fragmentation into hadrons in the Lund scheme[46], and describing the detector performance within the framework of GEANT[47].

The predictions calculated for a p_{Tmin} cutoff of 1.5 GeV/c give a good account of the observed data. This is shown by H1 by the solid curves for the p_T^2 distribution (fig. 50) and the energy flow in two - jet events (fig. 52). The corresponding comparison by ZEUS is shown for the E_T and p_{Tmis} distributions (dashed curves in figs. 53b,c). H1 has made an absolute comparison with the p_T distribution of jets (fig. 56) and observes agreement within a factor 2 for $p_T > 10$ GeV. ZEUS finds that the cross section for high E_T events is well reproduced by the calculation: fig. 55a compares the data with the prediction for the total ep cross section for E_T values larger than a given E_T^0 energy. If only the direct photon contribution is included the predicted cross section falls short of the measured one by at least an order of magnitude.

In conclusion H1 and ZEUS observed hard scattering in high energy photoproduction at the level of 20% of the total photoproduction cross section. The data are in good qualitative and quantitative agreement with theory which predicted the existence of hard scattering as a result of direct and resolved photon processes. The observed event topologies give an independent indication for the presence of resolved photon contributions.

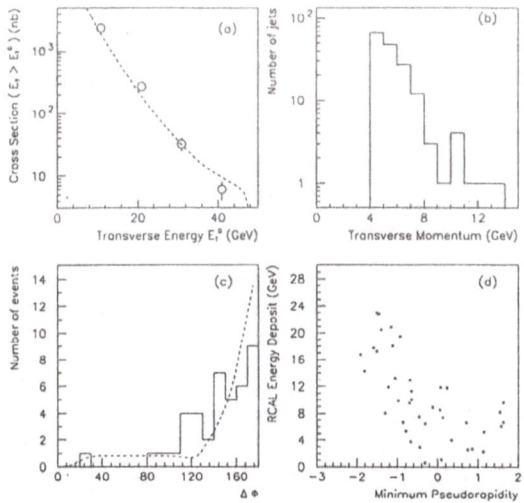

Fig. 55. Hard scattering in photoproduction as analyzed by ZEUS:
(a) Total ep cross section for transverse energies $E_T > E_t^0$. The curve is the cross section predicted by HERWIG with $p_{T_{min}}$ = 1.5 GeV/c, including direct and resolved photon contributions.
(b) Transverse momentum distribution of the identified jets;
(c) Distribution of the difference in azimuth between jets for events with two jets; the dotted curve shows the Monte Carlo prediction;
(d) Correlation between energy deposited in the RCAL versus the minimum rapidity of the two jets in the two-jet sample.

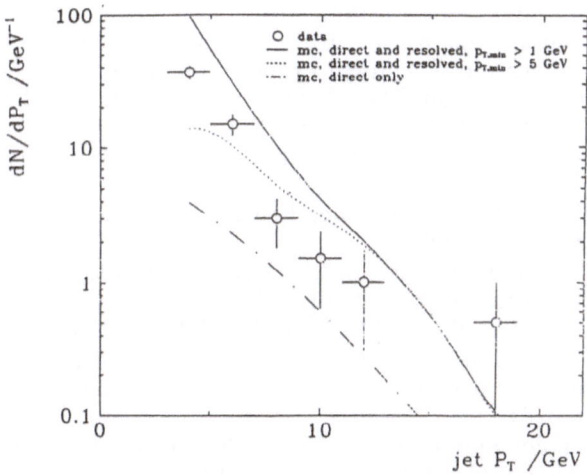

Fig. 56. Hard scattering in photoproduction as analyzed by H1: p_T distribution of jets with p_T > 3 GeV. The solid line is the prediction of PYTHIA for $p_{T_{min}}$ = 1 GeV/c, the dashed dotted line corresponds to $p_{T_{min}}$ = 5 GeV/c. The dashed - dotted line is the prediction if resolved photon contributions are excluded ($p_{T_{min}}$ = 1 GeV/c).

CONCLUDING REMARKS

HERA has joined the family of active high energy colliders and the experiments H1 and ZEUS sofar have encountered no problem to enter a new territory of electron proton collisions. The data which led to the results discussed above were obtained in the first month of data taking and were analyzed quasi instantaneously. Although the luminosity collected in this first period was small, the experiments presented already exciting results. More excitement can be expected from the second run period which offers 10 - 20 more luminosity. One of the topics which will benefit from the higher luminsity is deep inelastic neutral scattering where each factor of ten increase in luminosity will enlarge the accessible Q2 range by another order of magnitude.

Acknowledgements

I am grateful to F. Barreiro, H. Kowalski, E. Lohrmann and F. Sciulli for a critical reading of the manuscript, and to G. Schuler for many comments on the current theoretical picture of photoproduction. S. Gharavi has been helpful with the preparation of the drawings.

REFERENCES

In the following "HERA Workshop 1991" stands for Physics at HERA, Proc . Workshop 1991, ed. by W. Buchmüller and G. Ingelman, April 1992.

1. HERA, A Proposal for a Large Electron - Proton Colliding Beam Facility at DESY, DESY HERA 81 -10 (1981);
B.H. Wiik, Electron - Proton Colliding Beams, The Physics Programme and the Machine, Proc. 10th SLAC Summer Institute, ed. A. Mosher, 1982, p. 233, and Proc. XXVI Int. Conf. High Energy Physics, Dallas, 1992;
G. - A. Voss, Proc. First Euro. Acc. Conf., Rome, 1988, p. 7.
2. H1 Collaboration, Technical Proposal for the H1 Detector (1986).
3. ZEUS Collaboration, The ZEUS Detector, Technical Proposal (1986), and: The ZEUS Detector, Status Report 1989.
4. Proc. HERA Workshop 1987, ed. by R. Peccei, August 1988.
5. HERA Workshop 1991
6. H1 collaboration, F. Eisele, Proc. XXVI Int. Conf. High Energy Physics, Dallas, 1992.
7. ZEUS collaboration, B. Löhr, Proc. XXVI Int. Conf. High Energy Physics, Dallas, 1992; I. Gialas, ibid.
8. F. Eisele and G. Wolf, Physikalische Blätter 48 : 787 (1992).
9. HERMES Collaboration, A proposal to measure the spin - dependent structure functions of the neutron and the proton at HERA (1990)
10. A.A. Sokolov and M. Ternov, Sov. Phys. Doklady 8 :1203 (1964).
11. D.B. Barber et al., DESY Report 92 - 136 (1992).
12. J. Buon and K. Steffen, DESY report 85 - 128 (1985).
13. C.G. Callan and D.G. Gross, Phys. Rev. Lett. 22 : 156 (1969).
14. V.N. Gribov and L.N. Lipatov, Sov. J. Nucl. Phys. 15 : 438, 675 (1972);
G. Altarelli and G. Parisi, Nucl. Phys. 126 : 297 (1977).
15. J. Blumlein et al., Akad. Wissenschaften, Berlin-Zeuthen, Report 88 - 01 (1988)

16. L.V. Gribov, E.M. Levin, M.G. Ryskin, Phys. Rep. 100 : 1(1983).
17. see e.g.J. Bartels and J. Feltesse, HERA Workshop 1991, Vol 1, p. 131, and references given there.
18. V.T. Kim and M.G. Ryskin, DESY Report 91 - 064 (1991).
19. A.H. Mueller, Proc. Small x - Workshop at DESY, 1990, ed. by A. Ali and J. Bartels, Nucl. Phys. 18C : 125 (1991).
20. NMC Collaboration, P. Amaudruz et al., CERN-PPE/92-124 (1992), submitted to Phys. Lett.
21. J. Kwiecinski et al., Phys. Rev. D42 : 3645 (1992);
    J.G. Morfin and W.K. Tung, Z. Phys. C52 : 13 (1992);
    M. Glück, E. Reya, and A. Vogt, Z. Phys.C48 : 471 (1990).
22. A.D. Martin, R.G. Roberts and W.J. Stirling, Durham University preprint DTP - 92 -16 (1992).
23. L.N. Lipatov, Sov. J. Nucl. Phys. 23 : 338 (1976); E.A. Kuraev, L.N. Lipatov and V.S. Fadin, Sov. Phys. JETP 45 : 199 (1977); Y.Y. Balitskii and L.N. Lipatov, Sov. J. Nucl. Phys. 28 : 822 (1978).
24. J.G. Morfin and W.K. Tung, Z. Phys. C52 : 13 (1992).
25. S. Bentvelsen, J. Engelen and P. Kooijman, HERA Workshop 1991, Vol. 1. p.23
26. F. Jacquet and A. Blondel, Proc. Study of an ep Facility in Europe, ed. U. Amaldi, 79/48 (1979), p. 391
27. M. Glück, E. Reya, and A. Vogt, Z. Phys.C52 :127 (1992).
28. see F. Schrempp, HERA Workshop 1991, Vol. 2, p. 1034, and references given there.
29. ZEUS Collaboration, I. Gialas, Proc. XXVI Int. Conf. High Energy Physics, Dallas, 1992.
30. G. Schuler, HERA Workshop 1991, Vol. 1, p. 131.
31. Chacaltaya and Pamir collaborations, contribution to VI Int. Symp. on V. H. E. Cosmic Ray Interactions, ICRR-Rpt-216-90-9, 1990.
32. G. Yodh, Nucl. Phys. B (Proc. Suppl.) 12 : 277 (1990).
33. M. Drees, F. Halzen, and K. Hikasa, Phys. Rev. D39 : 1310 (1989);
    T.K. Gaisser et al., Phys. Lett. 243 : 444 (1990);
    R. Ghandi et al., Phys. Rev. D42 : 263 (1990).
34. ZEUS Collaboration, M. Derrick et al., Phys. Lett. B293 : 465 (1992).
35. S. I. Alekhin et al., CERN - HERA 87 - 01 (1987) and references given there.
36. A. Donnachie and P.V. Landshoff, Nucl. Phys. B244 : 322 (1984).
37. H. Abramowicz et al., Phys. Lett. B269 : 465 (1991).
38. M. Drees and K. Grassie, Z. Phys. C28 : 451 (1985);
    R.S. Fletcher etal., Phys. Rev. D45 : 377 (1992);
    R. Ghandi and I. Sarcevic, Phys. Rev. D44 : R10 :
    J.R. Forshaw and J.K. Storrow, Phys. Lett. B268 : 116 (1991);
    G. Schuler and J. Terron, HERA Workshop 1991, Vol. 1, p. 599.
39. M. Drees and K. Grassie, Z. Phys. C28 : 451 (1985).
40. H. Abramowicz, K. Charchula and A. Levy, Phys. Lett. B269 : 458 (1991).
41. G. Schuler and J. Terron, HERA Workshop 1991, Vol. 1, p. 599.
42. H1 Collaboration, T. Ahmed et al., DESY Report 92 - 142 (1992).
43. ZEUS Collaboration, M. Derrick et al., DESY Report 92 - 138 (1992), submitted to Phys. Lett.
44. T. Sjöstrand, HERA Workshop 1991, Vol. 3, p1405.
45. G. Marchesini et al. Comp. Phys. Comm., 67 : 465 (1992).
46. T. Sjöstrand, Comp. Phys. Comm., 39 : 347 (1986); T. Sjöstrand and M. Bengtsson, Comp. Phys. Comm., 43 : 367 (1987).
47. GEANT program manual, CERN program library (1992).

PHYSICS WITH HADRON COLLIDERS

Melvyn J. Shochet

Enrico Fermi Institute and Department of Physics
University of Chicago
Chicago, IL 60637
USA

INTRODUCTION

There are many physics issues addressed by hadron collider experiments. Studies of the production of jets, direct photons, and large P_T W and Z bosons test next to leading order QCD calculations. Precision measurements of the mass and lifetime of the W provide stringent constraints on the Standard Model of Electroweak interactions. The need for extensions to the Standard Model is probed with the search for new massive objects, such as higher mass intermediate vector bosons, supersymmetric particles, or indirectly by looking for quark compositeness.

Although there is much to present on each of these topics, the present lectures will concentrate on yet another focus of hadron collider experiments, heavy flavor physics. The first lecture will be on the search for the top quark. The second lecture will address b physics, both production and decay.

THE SEARCH FOR THE TOP QUARK

This is purely an experimental question of the strategy for searching for this as yet unobserved heavy fermion. I will present in detail the methods employed by the most sensitive experiment, CDF [1].

Introduction

Within the context of the Standard Model, the top quark must exist. The b quark must have a partner since it has a measured weak isospin of 1/2 [2]. Moreover, anomaly cancellation requires that for each generation

$$N_c \sum_{\text{quarks}} Q_i + \sum_{\text{leptons}} Q_i = 0$$

This fails for the third generation unless the top quark exists.

Quantitative Particle Physics, Edited by
M. Lévy *et al.*, Plenum Press, New York, 1993

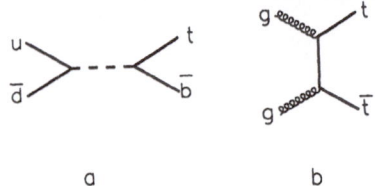

Figure 1. Top quark production via (a) W decay and (b) gluon fusion.

Table 1. Lower limits on the top quark mass

e^+e^-:	Limit:	Method:
PETRA/PEP	$15 - 22\ GeV/c^2$	R, event shape
Tristan	$30\ GeV/c^2$	R, event shape
SLC	$41\ GeV/c^2$	Z^0 width, event shape
LEP	$45\ GeV/c^2$	Z^0 width, event shape
$\bar{p}p$:		
Tevatron and $Sp\bar{p}S$	$48\ GeV/c^2$	W width
UA1	$60\ GeV/c^2$	isolated leptons
UA2	$69\ GeV/c^2$	isolated leptons
CDF	$91\ GeV/c^2$	isolated leptons

The search for the top quark has been underway since the bottom quark was discovered at Fermilab in 1977. The initial guess for the top mass was $15\ GeV/c^2$ based on the apparent geometric progression of the quark masses: M_s ($0.5\ GeV/c^2$), M_c ($1.5\ GeV/c^2$), M_b ($5\ GeV/c^2$). Since that time there has been a steady increase in the experimental lower limit on the mass as shown in Table 1 [3]. At $\bar{p}p$ colliders, two methods have been employed to search for the top quark. The W lifetime is sensitive to an open W decay channel into $t\bar{b}$, independent of how the top quark subsequently decays. The direct lepton search requires the standard charged weak current decay, but is sensitive to any top mass, not just $M_{top} < M_W$ as with the W lifetime method. Here we will consider only the direct top quark search.

In $\bar{p}p$ collisions, there are two major top quark production mechanisms as shown in Figure 1, from W decay and through gluon fusion. The W decay diagram only contributes significantly if $M_{top} < M_W - M_b \approx 75\ GeV/c^2$. At $\sqrt{s} = 1.8\ TeV$, $t\bar{t}$ production through gluon fusion dominates for all M_{top} (Fig 2) [4].

The decay of the top quark in the minimal Standard Model occurs via the charged weak current, $t \rightarrow Wb$, with the W real or virtual depending on the top quark mass.

$$gg \quad \rightarrow \quad t\bar{t} \rightarrow W^{(*)}b + W^{(*)}\bar{b}$$

Each W decays with a branching ratio of $1/9$ into each generation of leptons, and a branching ratio of $3/9$ (due to color) into $u\bar{d}$ or $c\bar{s}$. The all hadronic final state has the largest combined branching ratio ($4/9$), but the $t\bar{t}$ signal would be overwhelmed by QCD production of multiple quark and gluon jets.

Thus in order to observe a signal above background, at least one W must be required to decay into leptons. We will first consider the case where both W bosons decay into leptons, one into $e\nu$ and the other into $\mu\nu$. Then we will look at the final state in which

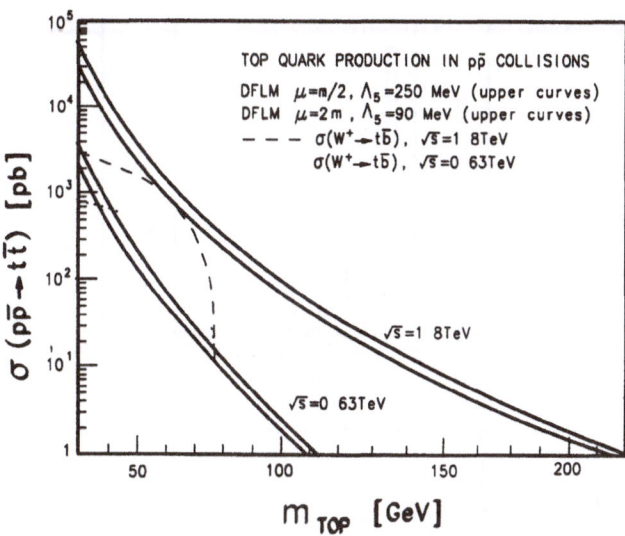

Figure 2. Expected top quark production cross section in $\bar{p}p$ collisions at both CERN and Fermilab energies.

one W decays into $e\nu$ with the other W decaying into quarks. Finally we will consider extensions to both of these searches.

$t\bar{t} \to e\mu X$

The decay chain

$$t\bar{t} \quad \to \quad WbW\bar{b} \to e\nu b\mu\nu\bar{b}$$

provides the final state with the lowest background. Unlike the single lepton modes, QCD production of W + jets doesn't contribute. And unlike the decay into two electrons or two muons, there is no background from the production of γ^*, Z^0, J/ψ, or Υ. The major background

$$gg \to b\bar{b} \to ce\nu\bar{c}\mu\nu$$

produces relatively low P_T leptons. Another possible background, W pair production ($q\bar{q} \to W^+W^- \to e\nu\mu\nu$) doesn't have a cross section competitive with $t\bar{t}$ for $M_{top} \lesssim 150\ GeV/c^2$. Possible background from $Z \to \tau\tau \to e\nu\nu\mu\nu\nu$ can be easily removed as we shall see.

It is important to note that large P_T charged leptons provide a good signature because they can be cleanly separated from the much more abundant charged hadrons. Figures 3 and 4 show the electron and muon selection variables for $W \to l\nu$ events. The hadron background is rather flat in these variables. The detection efficiency for high P_T electrons or muons incident on the active part of the detector is 75-95% depending on the criteria used.

There are a number of variables that are useful in separating signal from background.

- E_T of the electron and P_T of the muon. Top decay produces large P_T leptons, while $b\bar{b}$ background produces leptons with much lower P_T (Fig. 5).

- The missing E_T in the event. Events from $t\bar{t}$ production have large missing E_T due to the two large E_T neutrinos. Bottom events, on the other hand, have small

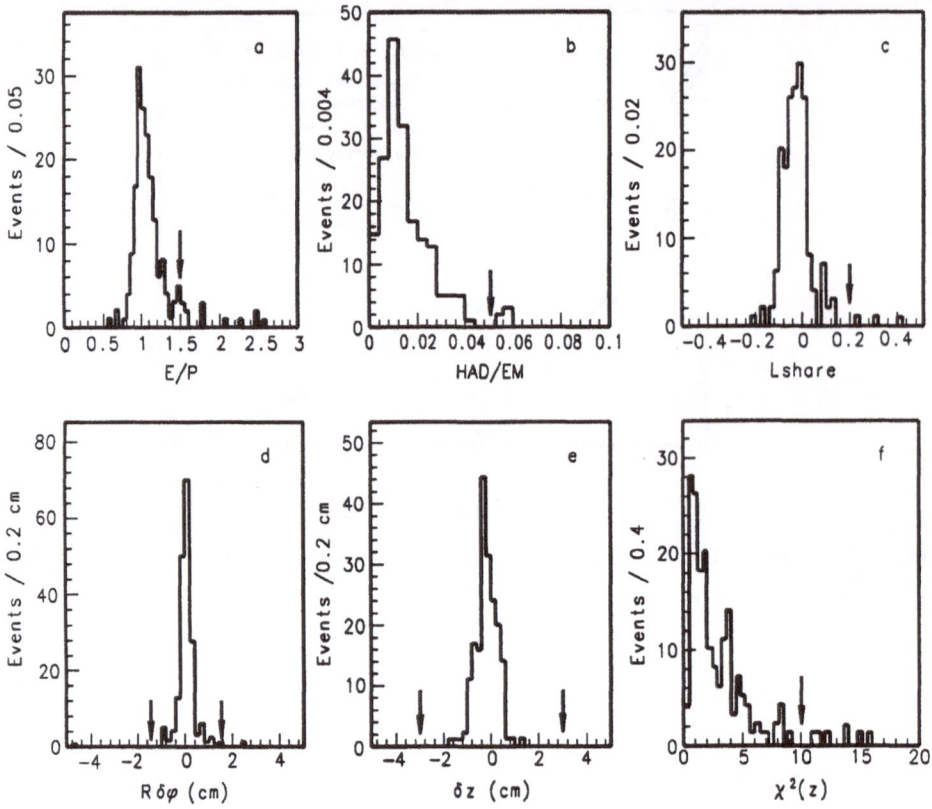

Figure 3. Distributions from $W \rightarrow e\nu$ events of variables used to select electrons. The arrows show the location of the usual cuts. (a) The ratio of the calorimeter energy to the track momentum. (b) The ratio of energies deposited in the hadronic and electromagnetic calorimeters. (c) A variable that describes the transverse size of the calorimeter shower. (d) Matching between the extrapolated track and the shower centroid in the azimuthal direction. (e) Matching between the extrapolated track and the shower centroid in the beam direction. (f) The chisquare for the comparison of the transverse shower shape with that measured in an electron test beam.

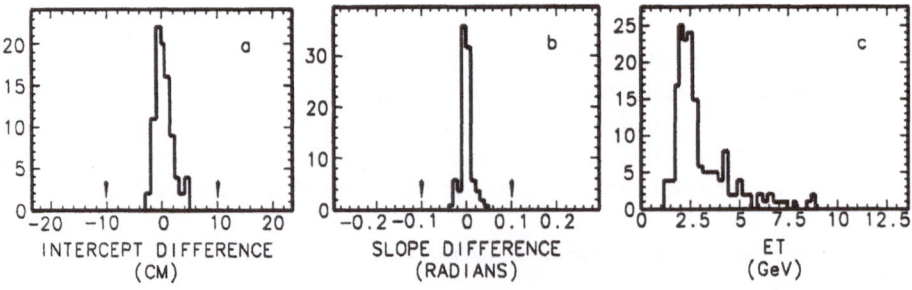

Figure 4. Distributions from $W \rightarrow \mu\nu$ events of variables used to select muons. (a) The difference between the extrapolated track from the central tracking chamber and the location of the track stub in the muon chamber. (b) The difference in slope between the extrapolated track and the muon chamber track stub. (c) The total E_T in the calorimeter cell through which the muon passes.

262

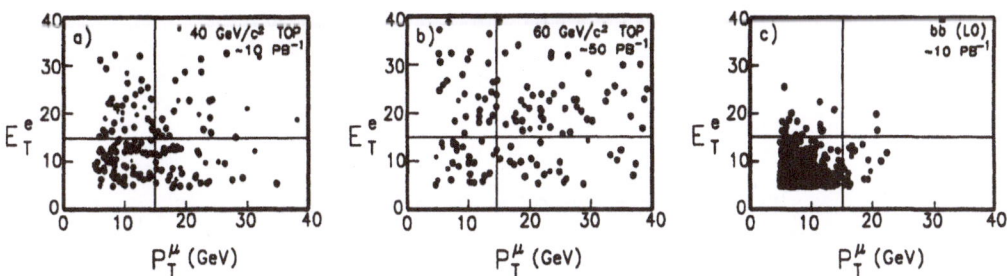

Figure 5. Electron E_T versus muon P_T for Monte Carlo simulations of (a) 40 GeV/c^2 top, (b) 60 GeV/c^2 top, and (c) leading order $b\bar{b}$ production. The location of the cuts that will be applied to the data are shown.

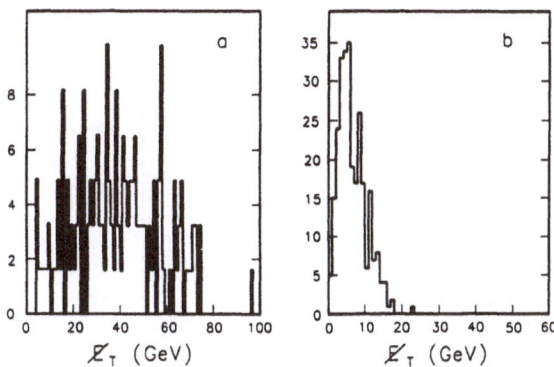

Figure 6. Missing E_T distributions from (a) 80 GeV/c^2 top and (b) $b\bar{b}$ Monte Carlo samples. Note the difference in the missing E_T scales for the two figures.

missing E_T because the requirements of large E_T^e and P_T^μ select the region of the b decay Dalitz plot where P_T^ν is small (Fig. 6).

- The azimuthal angular separation, $\Delta\phi^{e\mu}$, between the electron and the muon. Top events would produce a broad $\Delta\phi^{e\mu}$ distribution because of the large mass of the decaying mesons. Bottom production is characterized by peaks near 0° and 180° (Fig. 7). Z decay, $Z \to \tau\tau \to e\nu\mu\nu$, produces a peak at 180° because of the low τ mass. Figure 8 shows these expected distributions.

- Lepton isolation. The isolation of an electron can be characterized by

$$I = \frac{E_T(\text{cone, R} = 0.7) - E_T^e}{E_T^e}$$

where $E_T(\text{cone, R} = 0.7)$ is the transverse energy deposited in the calorimeter within a cone of radius 0.7 in $\eta - \phi$ space centered on the electron. For top decay $(t \to Wb \to e\nu b)$, the large top mass results in a large separation between the e and b and thus an isolated electron. In bottom decay $(b \to e\nu c)$, the electron is much closer to the charm quark and thus less isolated (Fig. 9).

The CDF $e\mu$ data were selected solely on the basis of E_T^e and P_T^μ. Figure 10 shows why the requirement on both variables was > 15 GeV. Only one $b\bar{b}$ background event was expected above that value for the integrated luminosity collected by CDF (4.1 pb^{-1}). Figure 11 shows the CDF data. The bulk of the data looks like $b\bar{b}$ production

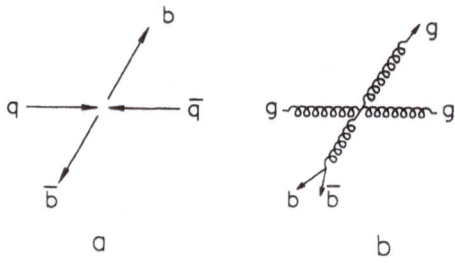

Figure 7. (a) The leading order diagram for b production produces b quarks 180° apart in azimuth. (b) The gluon splitting next to leading order diagram produces b quarks very close in azimuth.

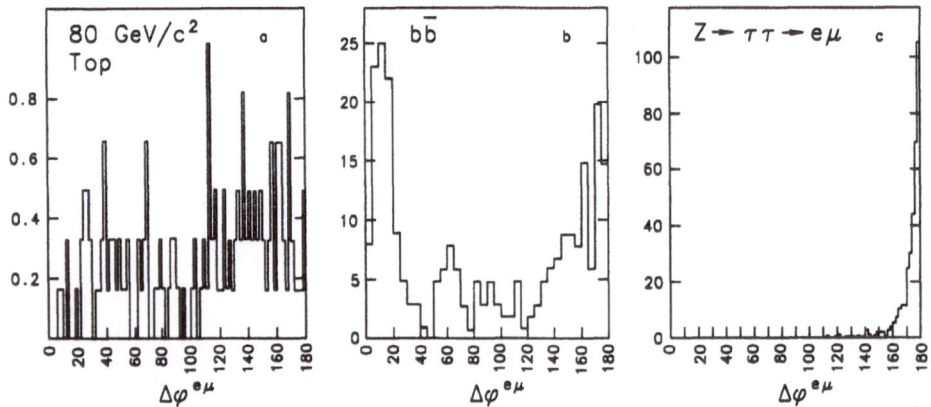

Figure 8. The expected $e\mu$ azimuthal separation for (a) an 80 GeV/c^2 top quark, (b) $b\bar{b}$ production, and (c) $Z \to \tau\tau \to e\mu\nu\nu$.

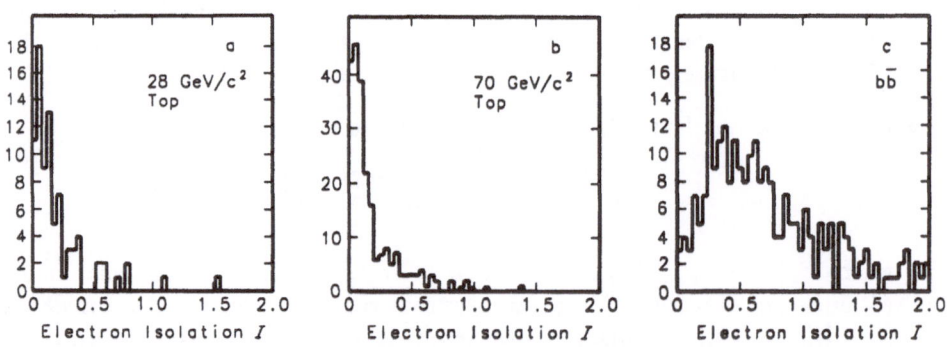

Figure 9. Electron isolation in Monte Carlo samples of (a) 28 GeV/c^2 top, (b) 70 GeV/c^2 top, and (c) $b\bar{b}$ events.

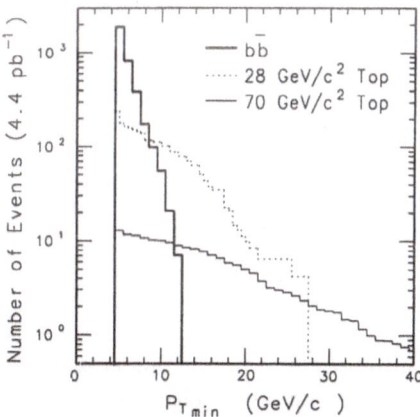

Figure 10. Monte Carlo predictions for the event rates as a function of the minimum lepton P_T accepted.

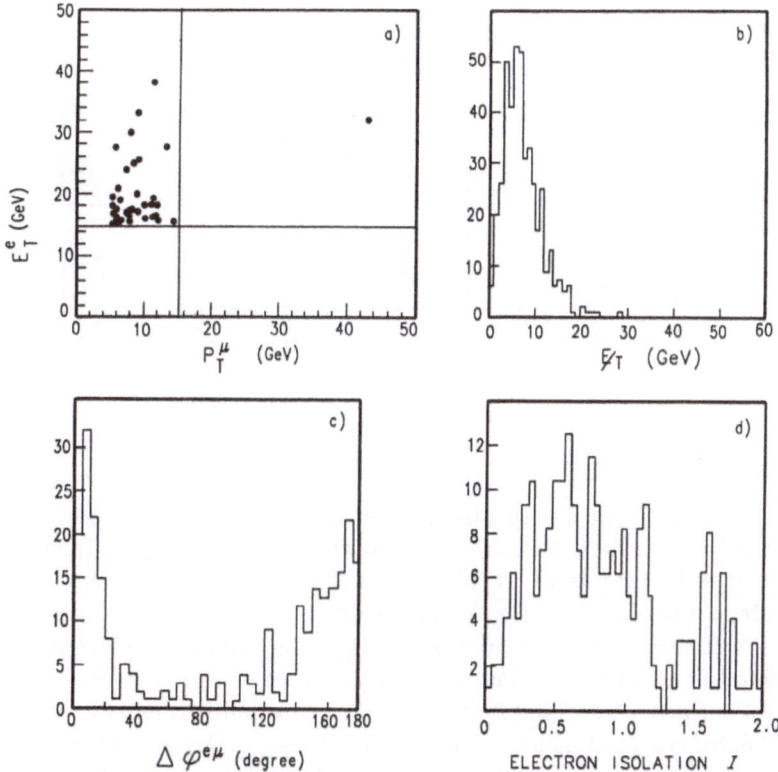

Figure 11. CDF data from the $e\mu$ top quark search. (a) The electron E_T versus the muon P_T . The lines show the location of the cuts in these variables. (b) The missing E_T , (c) dilepton azimuthal separation, and (d) electron isolation for the data with looser lepton transverse momenta requirements.

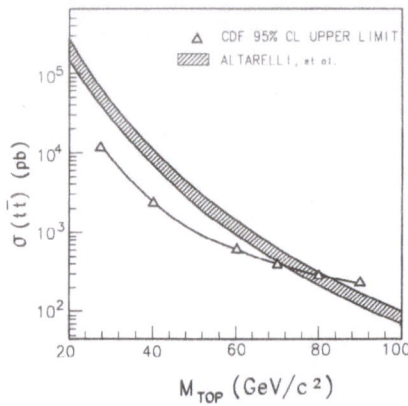

Figure 12. The upper limit on the $t\bar{t}$ production cross section from the CDF $e\mu$ search. Also shown is the next to leading order calculation of the cross section.

Table 2. Characteristics of the top candidate event. Calorimeter E_T is used in the P_T column for the electron and jet clusters.

	Charge	P_T [GeV/c]	η	ϕ [degrees]
Central Electron	+	31.7	-0.8	132
Central Muon	−	42.5	-0.8	269
Forward Muon	+	9.9	-2.0	98
Jet 1		14	1.1	341
Jet 2		5	-2.8	88

(compare Figure 11b,c,d with Figures 6, 8, and 9). However there is one event with very large E_T^e and P_T^μ. The characteristics of this event are given in Table 2. The event could be from $t\bar{t}$ decay, but it could just as well be a background event. With only one candidate, positive identification is impossible.

CDF calculated the upper limit on the $t\bar{t}$ production cross section using the calculated detection efficiency and Poisson statistics based on one observed event. Including the event without performing a background subtraction is conservative since it raises the calculated cross section upper limit. Also included in the calculation are the systematic uncertainties from lepton identification efficiency, the calculated P_T distribution for $t\bar{t}$ production, the top quark fragmentation function, and the experiment's integrated luminosity. The 95% confidence level upper limit on the cross section as a function of the top mass is shown in Figure 12 along with the next to leading order theoretical prediction [4]. The mass limit is taken where the experimental upper limit crosses the lower end of the theoretical prediction. From this, CDF concluded that the top mass is greater than 72 GeV/c^2 at the 95% confidence level.

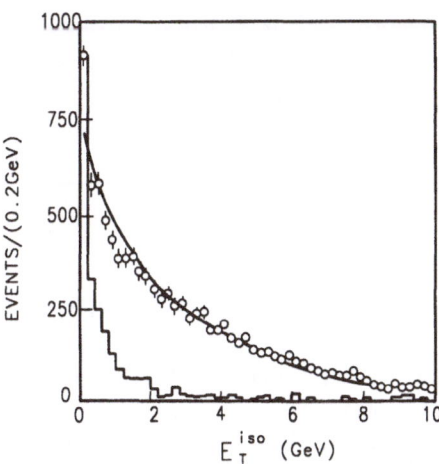

Figure 13. Isolation for electrons with $E_T < 20$ GeV (circles), a sample that should be largely b decay. The solid curve is a $b\bar{b}$ Monte Carlo prediction, and the histogram is a 75 GeV/c^2 top prediction. The excess data in the first bin is due to residual W and Drell Yan events in the sample.

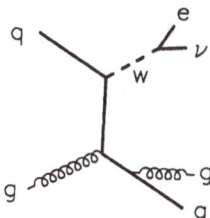

Figure 14. Diagram producing W + 2 jets.

$t\bar{t} \rightarrow e\nu+$ jets

The final state containing a single electron plus jets

$$gg \rightarrow \bar{t}t \rightarrow W\bar{b}Wb \rightarrow e\nu\bar{b}q\bar{q}b$$

has a combined branching ratio 6 times larger than that for the $e\mu$ final state. There are however experimental difficulties that complicate this search. For $M_{\text{top}} \lesssim 120\ GeV/c^2$, the probability of detecting all four quark jets is small because the b quarks have low energy and consequently don't appear jetlike in the detector. This forced CDF to search for events with an electron, missing E_T, and at least two jets of observed $E_T > 10$ GeV.

There are two major sources of background. The production of b quarks

$$gg \rightarrow \bar{b}b \rightarrow e\nu\bar{c}q\bar{q}c$$

produces low E_T electrons and neutrinos. Moreover this background can be reduced by requiring the electron to be isolated (Fig. 13). In this analysis, isolation is defined as the E_T in the calorimeter cells surrounding the cell hit by the electron. The more serious background is due to the QCD production of W + jets (Fig. 14). This background cannot be removed by simple cuts because the event characteristics are so similar to

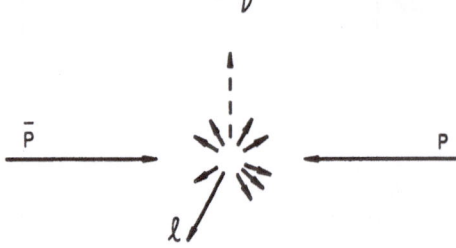

Figure 15. W decay into $l\nu$ in a $\bar{p}p$ collision.

that of a top quark signal. Rather a statistical method is employed to separate signal from background in the final sample.

The variable used for this separation is the $e\nu$ transverse mass. The usual invariant mass cannot be calculated due to the presence of the neutrino. Because it does not interact in the detector, the neutrino's momentum can only be inferred using conservation of momentum. Figure 15 shows a $\bar{p}p$ collision producing hadronic debris plus a W that decays into $l\nu$. Since the initial $\bar{p}p$ state has zero net momentum, the final state must also have no net momentum.

$$\sum \vec{P}_f = 0$$
$$\vec{P}_l + \vec{P}_\nu + \sum_{\text{hadrons}} \vec{P} = 0$$
$$\vec{P}_\nu = -\vec{P}_l - \sum_{\text{hadrons}} \vec{P}$$

In principle this can be measured well. In practice, however, only the transverse components of $\vec{P}_\nu$ can be calculated, because in hadron collisions significant longitudinal momentum can be carried by particles going undetected down the beam pipe. Thus the transverse mass is used, the three dimensional analog of the four dimensional invariant mass:

$$M_T = \sqrt{2P_T^l P_T^\nu (1 - \cos \Delta\phi^{l\nu})}$$

where $\Delta\phi^{l\nu}$ is the azimuthal separation between the leptons. The transverse mass has its maximum at M_W when the electron and neutrino have no longitudinal momentum. The transverse mass distribution is similar in shape to a Jacobian peak, with most of the information on the W mass coming from the location of the upper edge of the distribution.

The data in the $e + \geq 2$ jet sample are shown in Figure 16a. The concentration of events at low missing E_T and electron E_T near the trigger threshold is due to the $b\bar{b}$ background. The solid line in the figure represents a cut designed to remove most of this background. For very high top quark mass ($> 65 \ GeV/c^2$), a tighter cut (dashed lines) is used to further reduce background. Figure 16b and c show the expected distributions for a 70 GeV/c^2 top quark and the W + 2 jet background. The top signal is concentrated at lower E_T^e and missing E_T than the W background because the top quark decays to a virtual W when $M_{top} < M_W + M_b$. Thus the invariant mass of the final state $e\nu$ is less than the W mass, and the transverse momenta of the e and ν are smaller than they would be for the decay of an on-shell W. This translates into an $e\nu$ transverse mass distribution that is softer than for the W + jet background.

Figure 17a shows the data along with the expected shapes for signal and background. The data is consistent with pure background. This conclusion depends on an accurate

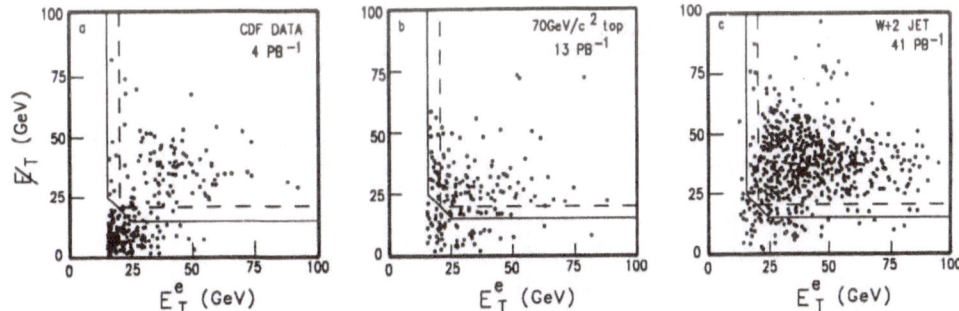

Figure 16. Electron E_T versus missing E_T for (a) CDF $e +$ ≥ 2 jet sample, (b) 70 GeV/c^2 top Monte Carlo, and (c) W + 2 jet Monte Carlo. The solid (dashed) lines correspond to the loose (tight) cuts that are applied to the data. The trigger requirement, $E_T^e > 15$ GeV, has been applied.

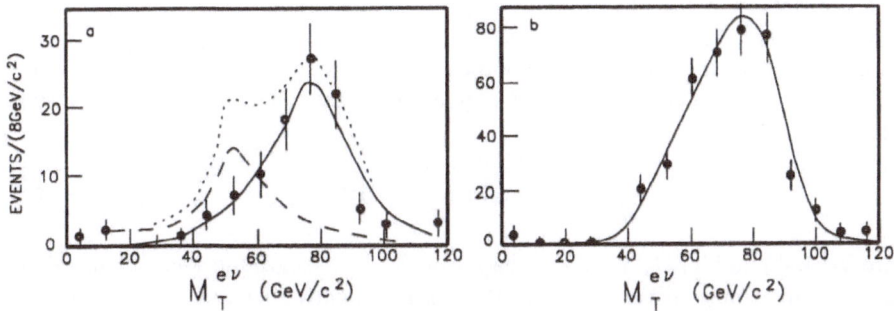

Figure 17. (a) The CDF $e\nu$ transverse mass distribution for $e + \nu + \geq 2$ jet data. The solid curve is the Monte Carlo W + 2 jet shape. The expected distribution for a 70 GeV/c^2 top is shown by the dashed curve. The dotted curve is the sum of the other two curves. (b) CDF $e + \nu + 1$ jet data compared to the Monte Carlo expectation for QCD W + 1 jet production.

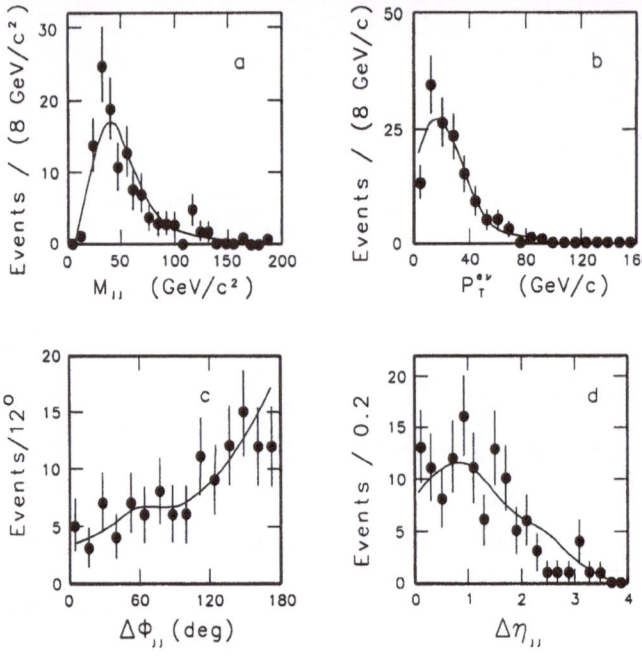

Figure 18. The (a) dijet invariant mass, (b) P_T of the $e\nu$ system, (c) two jet azimuthal separation, and (d) two jet rapidity separation for CDF $e + \nu + \geq 2$ jet data. The curves are from a W + 2 jet Monte Carlo.

simulation of the background transverse mass distribution. The simulation can be checked with a similar data sample in which the top quark contribution would be very small. Such a sample, $e + \nu + 1$ jet, is shown in Figure 17b. The agreement between the data and the background simulation is excellent. The assumption that the $e + \nu + \geq 2$ jet data sample is entirely QCD W + jet background can be checked by looking at a number of other variables. Figure 18 shows the 2 jet invariant mass, the transverse momentum of the $e\nu$ system, and the azimuthal and rapidity separation between the two jets. In each case the agreement between the data and background simulation is excellent.

To obtain the top contribution for a given top mass, CDF fits the transverse mass spectrum to

$$\frac{dN}{dM_T^{e\nu}} = \alpha T(M_T^{e\nu}) + \beta W(M_T^{e\nu})$$

where $W(M_T^{e\nu})$ and $T(M_T^{e\nu})$ are the shapes of the W background and top signal transverse mass distributions respectively. W and T are normalized so that $\alpha = \beta = 1$ for the QCD predicted cross sections. The results of the fit are α and β along with their uncertainties. Table 3 gives these results for different assumptions for the top quark mass.

The data are consistent with the QCD W + jet prediction alone, which has an overall theoretical normalization uncertainty of 30 – 35%. The results of the fit are combined with the systematic uncertainties to obtain the upper limit on the $t\bar{t}$ production cross section. The major systematic sources are the detector jet energy scale and integrated luminosity, along with the effects of the underlying event, initial state gluon radiation, and top quark fragmentation. The 95% confidence level upper limit on the cross section is shown in Figure 19. At the 95% confidence level, the top quark mass must be $> 77\ GeV/c^2$.

Table 3. Results of the transverse mass fits to the $e + \nu + \geq 2$ jet data along with the statistical fit uncertainties.

M_{top} [GeV/c^2]	α	β	χ^2 ($N_{df} = 10$)
40	0.07 ± 0.05	1.27 ± 0.14	9.7
50	0.06 ± 0.05	1.29 ± 0.14	10.4
60	0.11 ± 0.08	1.26 ± 0.15	10.4
70	$0.00^{+0.12}_{-0.00}$	1.28 ± 0.13	9.4
75	$0.00^{+0.18}_{-0.00}$	1.28 ± 0.13	9.4
80	$0.00^{+0.27}_{-0.00}$	1.28 ± 0.13	9.4

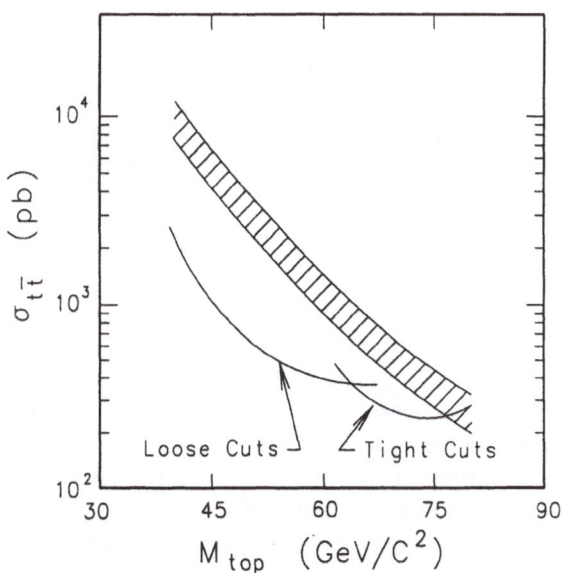

Figure 19. The CDF upper limit on the top production cross section from the $e + \nu + \geq$ 2 jet data sample, along with the next to leading order theoretical prediction.

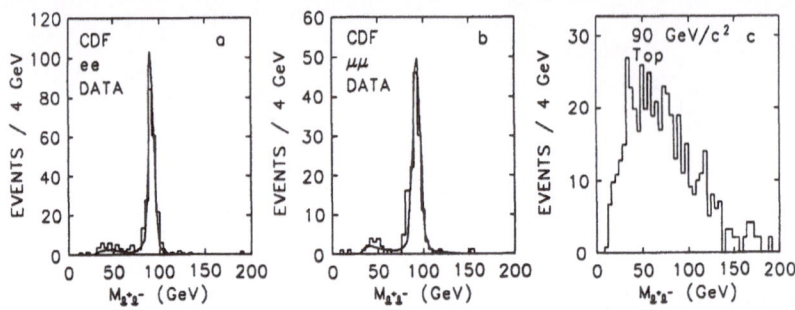

Figure 20. Dilepton invariant mass for (a) CDF ee data, (b) $\mu\mu$ data, and (c) a 90 GeV/c^2 top simulation.

Extended Dilepton Search

Although the $e\mu$ final state is the cleanest dilepton channel in which to search for the top quark, the ee and $\mu\mu$ channels also can be used. In extending the dilepton search, we increase the acceptance for the $gg \rightarrow b\bar{b} \rightarrow l^+l^-$ and $Z \rightarrow \tau\tau \rightarrow l^+l^-$ backgrounds. More important, however, is a new major source of background, γ^* and $Z \rightarrow l^+l^-$. Much of the Z background can be removed by cutting out the dilepton invariant mass range $75 < M_{l^+l^-} < 105 \ GeV/c^2$ (Fig. 20). In addition, we have to make use of the other discriminants mentioned earlier, missing E_T ($\not{E}_T$) and $\Delta\phi^{l^+l^-}$ (Fig. 21). The requirements are

$$\not{E}_T > 20 \ GeV$$

$$20° \leq \Delta\phi^{l^+l^-} \leq 160°$$

As shown in Figure 22, there are no additional events in the signal region for the extended dilepton top search. The combined top mass limit from the dilepton searches, $e\mu$, ee, and $\mu\mu$, is $M_{top} > 84 \ GeV/c^2$ at the 95% confidence level (Fig. 23).

Extended Single Lepton Search

The technique employed in the $e + \nu + $ jets search cannot be used for high mass top, since if $M_{top} > M_W + M_b$, the W from $t \rightarrow Wb$ is onshell. In this case, the $e\nu$ transverse mass distributions for signal and background are identical. Thus another discriminant is needed. CDF chose to look for a b quark in the event. Top events have two b quarks in each event ($\bar{t}t \rightarrow W\bar{b}Wb \rightarrow e\nu\bar{b}q\bar{q}b$ or $\mu\nu\bar{b}q\bar{q}b$), whereas the QCD produced W + jets background rarely contains b quarks. Here the b quark is tagged through its semileptonic decay into a muon; $b \rightarrow \mu\nu c$ occurs with a 10% branching ratio. The single high P_T lepton sample (e or μ) was searched for the presence of an additional muon with $P_T < 15 \ GeV/c$. The upper limit on P_T^μ was placed both because muons from b typically have low energy and to avoid double counting with the dilepton $e\mu$ and $\mu\mu$ searches. A low P_T muon candidate also had to be outside the cones of radius 0.6 (in $\eta - \phi$ space) centered on the two leading jets. In top decay, these jets would be from the hadronic decay of a W; the b quarks are not usually near these jets. This cut has the advantage of greatly reducing fake muon candidates. Hadrons can fake muons either by penetrating the absorber iron or by decay in flight before entering the iron. The large hadron multiplicity in jets make the fake muon rate near a jet core rather large.

Figure 24 shows the distance between the low energy muon in an event and the nearest of the two high P_T jets. There are no events with $R > 0.6$. The CDF top

272

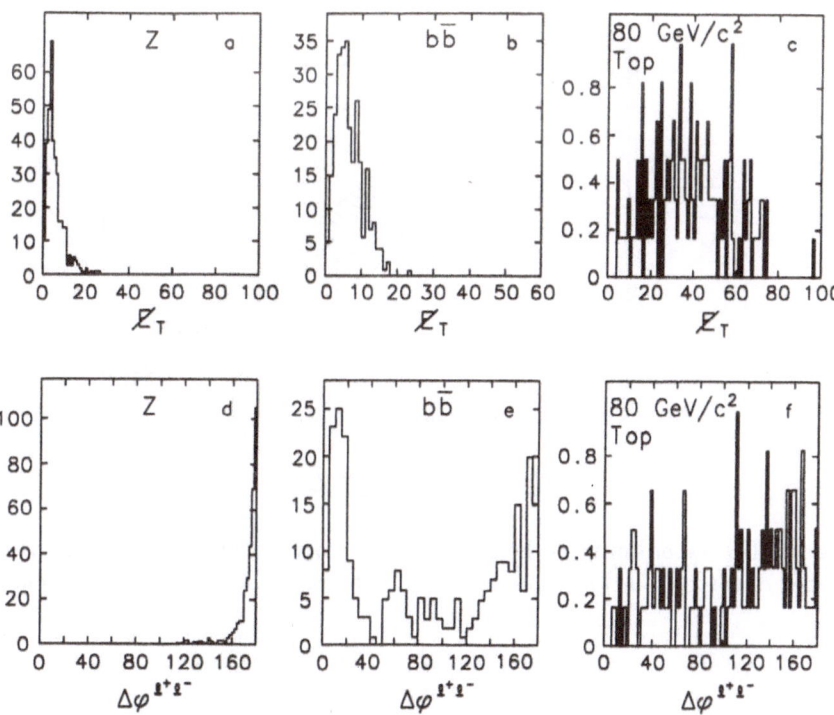

Figure 21. The missing E_T distributions expected for (a) Z, (b) $b\bar{b}$, and (c) 80 GeV/c^2 top. The $\Delta\phi$ distributions from simulated samples of (d) Z, (e) $b\bar{b}$, and (f) 80 GeV/c^2 top.

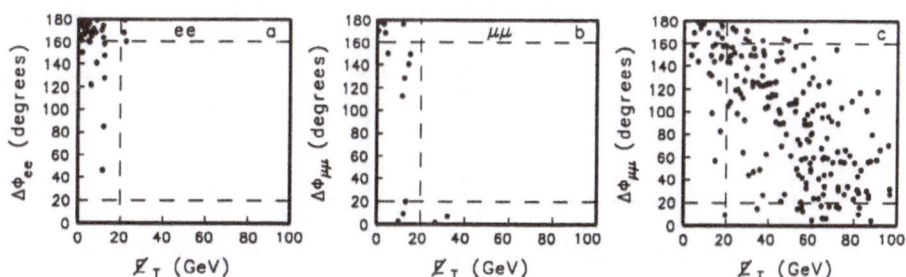

Figure 22. The CDF (a) ee and (b) $\mu\mu$ data, and (c) simulated 90 GeV/c^2 top production. The dashed lines show the cuts applied to the data.

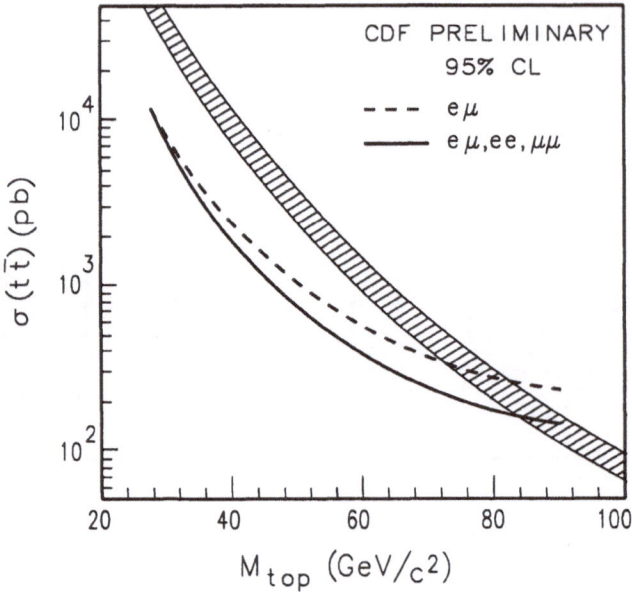

Figure 23. The upper limits on the $t\bar{t}$ production cross section from the $e\mu$ (dashed) and combined dilepton (solid) searches along with the range of the theoretical prediction.

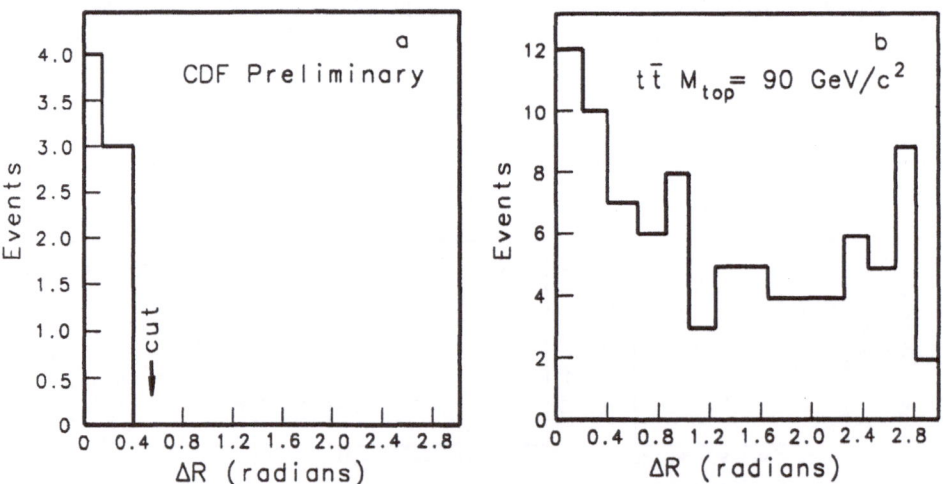

Figure 24. The distance in $\eta - \phi$ space between a low energy muon candidate and the closest of the two high P_T jets in (a) the e or $\mu + \nu +$ jets data sample, (b) a 90 GeV/c^2 $t\bar{t}$ Monte Carlo sample.

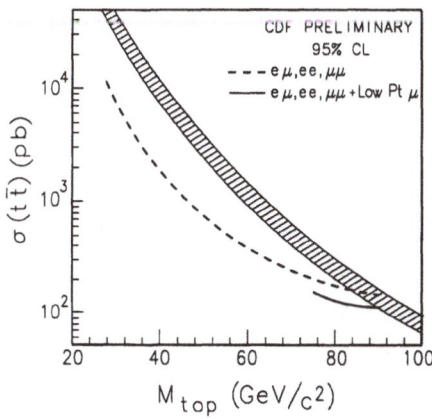

Figure 25: The CDF upper limit on the $t\bar{t}$ production cross section from the combined searches.

mass limit from the combined extended dilepton and extended single lepton searches is (Fig. 25)

$$M_{top} > 91\ GeV/c^2 \qquad @\ 95\%\ CL$$

Prospects for the Future

As we have seen, the current lower limit on the top quark mass is 91 GeV/c^2 at the 95% confidence level. The indirect evidence based on the consistency of the minimal Standard Model (Z decay, M_W, ν scattering, etc.) provides an upper limit, $M_{top} < 200\ GeV/c^2$. However more than a 3σ discrepancy with the Standard Model prediction would be required before the Standard Model would be abandoned. Thus to test the Standard Model, the top mass range up to 250–300 GeV/c^2 must be explored.

During the next few years, Fermilab will replace the Main Ring, the high energy injector for the Tevatron, with a new superconducting accelerator, the Main Injector. With it, the Tevatron luminosity should increase by approximately a factor of 50 over that achieved during the last data run. For a two year run with the Main Injector, the integrated luminosity should be $1 fb^{-1}$. This will provide greatly increased sensitivity for a high mass top quark.

Detection of heavy top can occur in the $l\nu + 4jet$ and $ll\nu\nu + 2jet$ final states, where l includes electrons and muons. The branching ratio is 30% for the former and 5% for the latter, while the detection efficiency is approximately 30% for the single lepton mode and 25% for the dilepton mode. These efficiencies might in fact be larger, because there is no need to cut hard on the lepton identification variables. The dominant background is from W and Z decay into electrons or muons, so cutting hard reduces signal and background similarly. Table 4 gives, as a function of the top quark mass, the number of $t\bar{t}$ events produced and the number that should be detected in each mode for a 1 fb^{-1} data sample.

The number of signal events of course is not the only, or perhaps even the major consideration. The size of the background is also of crucial importance. For the single

Table 4. Number of $t\bar{t}$ events that should be produced and detected for a 1 fb^{-1} data sample as a function of M_{top}.

M_{top} [GeV/c²]	$N_{t\bar{t}}^{produced}$	$N_{l+4\,jet}^{detected}$	$N_{ll+2\,jet}^{detected}$
100	80,000	<7200	1000
140	15,000	1350	200
180	3300	300	40
220	1000	90	12
260	350	30	5
300	120	10	2

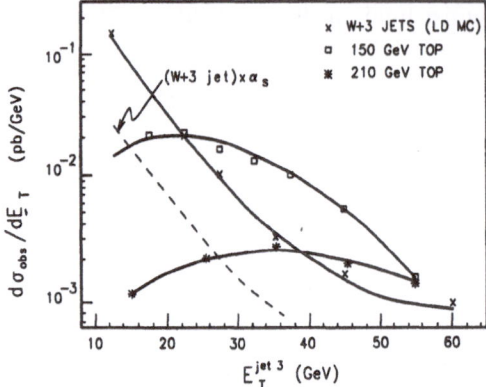

Figure 26. The E_T spectrum for the third highest E_T jet in single charged lepton events from 150 and 210 GeV/c^2 top, a W + 3 jet Monte Carlo, and a W + 4 jet estimate taken to be α_s times the W + 3 jet spectrum.

charged lepton final states, the dominant source of background is QCD W + 4 jet production. CDF does not as yet have a large sample of W + 4 jet events. Consequently we have to rely on Monte Carlo simulations. At the time these estimates were made, the W + 4 jet calculation was not yet available. We used the W + 3 jet calculation and multiplied the cross section by α_s to approximate the effect of requiring an additional jet. This is consistent with the CDF cross section ratio (W + 0 jet)/(W + 1 jet)/(W + 2 jet)/(W + 3 jet). Figure 26 shows the E_T spectrum of the third highest E_T jet for 150 and 210 GeV/c^2 top as well as for the background. The background jet E_T spectrum is rapidly falling, in contrast to the top decay spectrum which becomes harder as M_{top} increases. By selecting a jet E_T threshold that increases with M_{top}, a satisfactory signal to noise ratio can be maintained over a large M_{top} range extending to over 200 GeV/c^2.

If this proves not to be sufficient, a significant improvement in the signal to noise ratio can be obtained by identifying one or both of the b jets in the event. Low energy leptons from semileptonic b decay can be used to tag b quarks, but a 10% branching ratio penalty must be paid. A higher b tagging efficiency can be obtained by observing the secondary vertex from b decay. The CDF silicon vertex detector should have a 10–15 μ impact parameter resolution, to be compared to $c\tau \simeq 300\mu$ for B mesons. It is estimated that at least one b jet can be identified in approximately 50% of heavy top events.

For the dilepton final states, there are two major sources of background to high mass top. The QCD production of Z + 2 jets followed by the decay $Z \rightarrow ee$, $Z \rightarrow \mu\mu$,

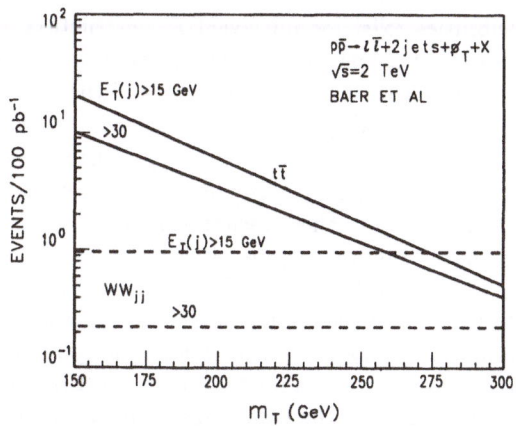

Figure 27. The dilepton event rate for a top signal and the WW background as a function of the top quark mass. Curves are shown for two different jet E_T thresholds.

or $Z \to \tau\tau \to ll\nu\nu\nu\nu$ can be easily removed using the dilepton invariant mass, $\not{E}_T$, and $\Delta\phi^{ll}$. The more difficult background is vector boson pair production, $p\bar{p} \to WW + 2$ jets $\to ll\nu\nu + 2$ jets. Again, by choosing a jet E_T threshold that increases with M_{top}, a good signal to noise ratio can be maintained over the accessible M_{top} range (Fig. 27) [5].

If these background estimates prove accurate, there should be a significant number of detected $t\bar{t}$ events (≥ 25 $l + \nu + 4$ jet events and ≥ 5 $ll\nu\nu + 2$ jet events per detector) with good signal to noise up to $M_{top} = 260 - 270$ GeV/c^2. Approximately 10 single lepton and a few dilepton events are expected per detector at $M_{top} = 300$ GeV/c^2. Thus the entire range allowed in the Standard Model would be covered.

If a signal appears, there are a number of ways that its identity as a top quark can be tested. The number of events with 0, 1, or 2 identified secondary vertices should be consistent with two b jets per $t\bar{t}$ event. The secondary vertex detection efficiency can be measured with the inclusive lepton data sample, which is mostly from b decay. One can also look at the ratio of the numbers of single lepton and dilepton events. This should be consistent with two W bosons per event. There can be additional confirmation of the presence of two W bosons using, for example, the $l\nu$ transverse mass and the dijet invariant mass. Finally, one can see if the production cross section is consistent with the QCD prediction. The theoretical cross section uncertainty is approximately $\pm$ 20-30%.

Another important issue is the accuracy with which the top quark mass can be measured. For the single charged lepton mode, the longitudinal momentum of the neutrino can generally be obtained using the measured $\not{E}_T$ and the W mass constraint for the charged lepton plus neutrino. There is a quadratic ambiguity, but choosing the solution with the smaller neutrino longitudinal momentum is correct most of the time. One can then in principle calculate the masses of both the t and $\bar{t}$. For one, use the invariant mass of the charged lepton, the neutrino, and the jet that came from the b quark associated with the W that decayed into $l\nu$. For the other, use the invariant mass of the three other jets in the event ($t \to Wb \to q\bar{q}b$). Unfortunately there is a serious problem of jet combinatorics. How do you know which jet to associate with the $l\nu$? And since many events will have more than four jets (e.g. from initial state gluon radiation), which three come from a common top quark? The combinatoric problem can be greatly reduced if one or both of the b jets can be identified, by observing either the secondary vertex or a semi-leptonic decay.

A number of possibilities exist for full event reconstruction. One can perform a standard 2-C fit, since $P_{\nu z}$ is unknown and there are three constraints, M_{W1}, M_{W2}, and $M_t = M_{\bar{t}}$. Likelihood techniques are also being developed that resolve the combinatoric ambiguity and include the effect of jet energy and $\not{E}_T$ resolutions by using an overall event likelihood as a function of top quark mass. The likelihood can include the difference between the fit and measured jet energies, the proton structure functions, the t and W decay configurations, etc. These techniques work well for Monte Carlo signal samples, giving a top mass resolution as low as 5-7 GeV/c^2. However questions still remain about the effect of the backgrounds. Here again, the identification of b jets can help, since this should greatly reduce the size of the background.

For the double leptonic decays, full reconstruction is not possible due to the presence of two neutrinos in the event. However a number of mass estimators have been studied. The simplest and one of the best is the transverse momentum of the b jet. It appears that a top mass resolution of 10 GeV/c^2 is possible in this mode.

Finally, comparing the event rate with the calculated cross section provides an estimate of M_{top} with an uncertainty of $\leq 10\%$. This of course assumes that the branching ratio for $t \to Wb$ is 100%.

Once the top quark is discovered and its mass is measured, a precision test of the Standard Model is possible by comparing the top mass and the W mass. A brief summary of the prospect for measuring the W mass is thus in order.

With a 1 fb^{-1} data sample, more than 10^6 $W \to l\nu$ events and 10^5 $Z \to ll$ events will be detected. The very large Z sample is critical since it is used to study and measure many of the sources of systematic uncertainty in the W mass: calorimeter energy scale, detector resolution, P_T of the W, effect of electron energy leakage on the measured P_T of the ν, background, and the mass fitting procedure.

The statistical uncertainty in M_W should be $\lesssim 30$ MeV/c^2. The dominant systematic uncertainty may well be the imprecise knowledge of the structure functions, which affects the W rapidity distribution. However the measurement of the W charge asymmetry will give the needed u to d ratio for the relevant range of x and q^2. If no unexpected new sources of systematic uncertainty arise, it is possible that the W mass can be measured to $\sim \pm 50$ MeV/c^2.

Such a measurement, coupled with the measurement of M_{top}, provides a powerful test of the Standard Model at the level of electroweak radiative corrections (Fig. 28). If the result disagrees with the Standard Model, it is obviously extremely important. On the other hand, even if it is consistent with the Standard Model, it can provide information about the Higgs mass.

B PHYSICS AT HADRON COLLIDERS

Heavy flavors provide a window on many important physics issues. The production process, $p\bar{p} \to b\bar{b}X$, is a testbed for QCD calculations since higher order diagrams make a large contribution here and there are a large number of scales in the problem [6]

$$\sqrt{s} \gg P_T \gg M_b \gg \Lambda_{\text{QCD}}$$

Moreover, the future of electroweak studies using the b system depends on the value of the total cross section (how many b quarks can be produced) and the differential cross sections (how efficiently can the second b in an event be tagged). With sufficient numbers of b quarks, significant electroweak studies can be carried out. These include $B\bar{B}$ mixing and rare B decays to obtain information about CKM matrix elements,

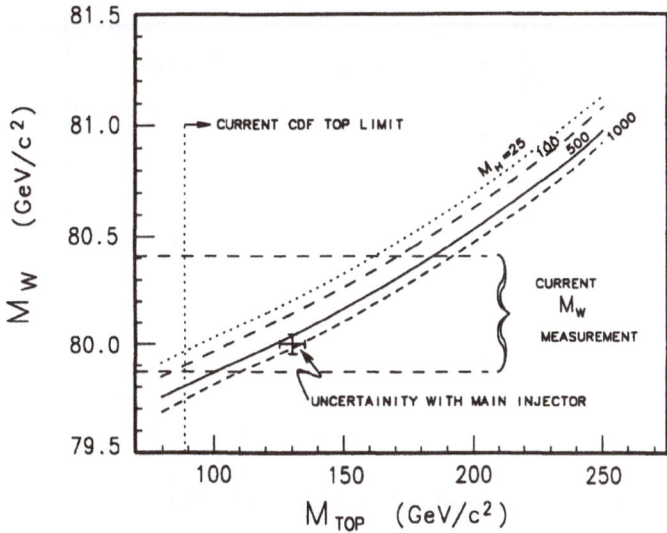

Figure 28. The Standard Model relation between the W and top masses for different values of the Higgs mass. The data point shows the precision possible with a 1 fb^{-1} data sample.

the search for forbidden decays to investigate extensions to the Standard Model, and hopefully CP violation in B decay where CP asymmetries may be large.

b Production

For b production via leading order QCD diagrams (Fig. 29a,b,c), the b and $\bar{b}$ have equal and opposite transverse momenta. Valence $q\bar{q}$ annihilation (Fig. 29c) dominates when $2M/\sqrt{s} \gtrsim 0.1$. At Fermilab Collider energies, this condition is satisfied for heavy top quark production but not for b production. When $2M_b/\sqrt{s} \ll 1$ (it is ~ 0.005 at Fermilab), the two-gluon initial state dominates, and higher order diagrams (Fig. 29d,e) can give a larger contribution than the leading order diagrams. This is due to the large gluon density at small x, the increased color factor at a 3-gluon vertex, and the cross section enhancement for diagrams containing t-channel vector exchange.

The dominant higher order diagrams are gluon splitting (Fig. 29d) and flavor excitation (Fig. 29e), which essentially is initial state gluon splitting. Understanding these

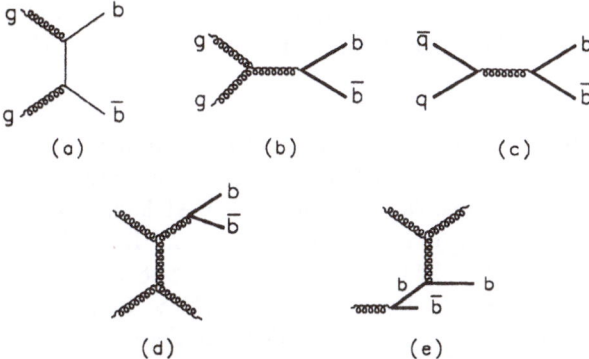

Figure 29. (a,b,c) Leading order b production diagrams. (d,e) Next to leading order diagrams.

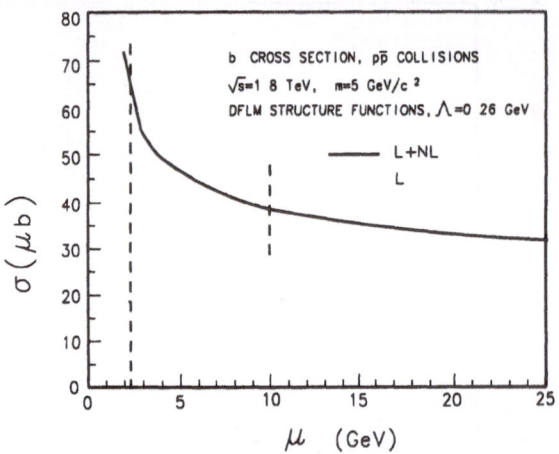

Figure 30. The dependence of the calculated b production cross section on the renormalization scale for leading order (dashed curve) and leading plus next to leading order (solid curve) diagrams.

higher order production mechanisms is important for at least two reasons. First there is the theoretical interest in understanding higher order QCD processes. Figure 30 shows the dependence of the cross section on the renormalization scale [7]. Note that contrary to the usual expectation for well behaved perturbation expansions, the dependence is stronger when the next to leading order diagrams are included. This may be due to the large next to leading order contribution and the resulting need to include yet higher order diagrams in the calculation. Second, there is the implication for flavor tagging the second b in $b\bar{b}$ events, since the P_T and rapidity correlations between the b and $\bar{b}$ are quite different in the leading order and the various next to leading order diagrams. The prospect for measuring $B\bar{B}$ mixing and studying CP violation at hadron colliders thus depends on understanding the b production mechanisms.

The major experimental challenge in doing b physics is separating b events from the much more copious light quark background. Since the largest b branching ratio into final states without a neutrino is only a few percent, most of the studies so far have concentrated on inclusive final states. We will first consider inclusive lepton samples where the major challenges are separating moderate P_T electrons and muons from misidentified hadrons, and determining the charm, W, Z, and γ^* contributions to the data samples. Then we will look at the analyses of data samples containing $J/\psi \rightarrow \mu^+\mu^-$ where b production must be separated from other sources of J/ψ such as $\chi \rightarrow J/\psi + \gamma$. Finally, we will consider the recent reconstruction of exclusive final states in b decay.

Inclusive Lepton Channels. The study of b production at hadron colliders was initially carried out by the UA1 collaboration [8]. Their primary b physics data is the inclusive muon sample, chosen because the thick UA1 hadron absorber allows muon detection in and near hadron jets. Unfortunately π and K decay in the jets produces a large background. Of their 20,000 events with $P_T^\mu > 6$ GeV/c, approximately 70% are background. This fraction drops to 35% for $P_T^\mu > 10$ GeV/c. A UA1 focus is the $10 < P_T^\mu < 15$ GeV/c range where the decay background is manageable and the contribution from resonances (W, Z, γ^*, J/ψ, Υ) is small ($\sim 6\%$). To separate $b\bar{b}$ from $c\bar{c}$, they define the variable $P_T^{rel} \equiv P^\mu sin\theta_{rel}$ where θ_{rel} is the angle between the muon and the nearest jet. The larger b mass results in a larger P_T^{rel}. Figure 31 shows the

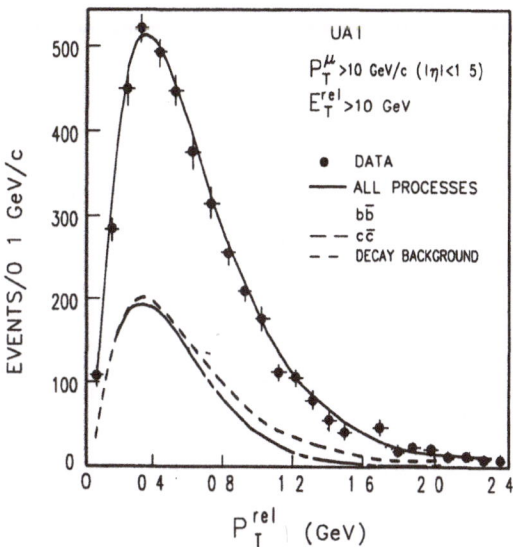

Figure 31. P_T^{rel} for the UA1 single muon data sample. The solid line is the sum of $b\bar{b}$, $c\bar{c}$, and decay background contributions.

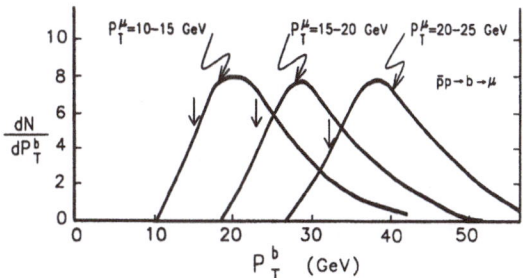

Figure 32. The b quark spectrum for different bins of muon P_T, as predicted by the UA1 Monte Carlo.

UA1 data fit to a sum of $b\bar{b}$, $c\bar{c}$, and π/K decay contributions [9]. The UA1 result on the fraction of $b\bar{b}$ is

$$\frac{N_{b\bar{b}}}{N_{b\bar{b}} + N_{c\bar{c}}} = 0.76 \pm 0.12$$

in agreement with predictions.

A Monte Carlo simulation is used to convert the measured muon differential cross section, $d\sigma/dP_T^\mu$, into the b quark production cross section, $d\sigma/dP_T^b$. The simulation uses b jet fragmentation, semileptonic B branching ratios, and B decay kinematics as measured in e^+e^- collisions. Figure 32 shows the relationship between the observed muon P_T and the parent b P_T . The resulting b cross section, integrated over the central three units of rapidity and over P_T above the P_T^{min} plotted, is shown in Figure 33. The results agree well with the next to leading order QCD calculation.

CDF used its electron data sample to study b production [10]. The advantage is relatively low background; misidentified electrons and unidentified gamma conversions constitute $\sim 30\%$ for $P_T^e > 7$ GeV/c. The disadvantage in using electrons is the difficulty in identifying electrons within jets. However this mostly affects the charm contribution

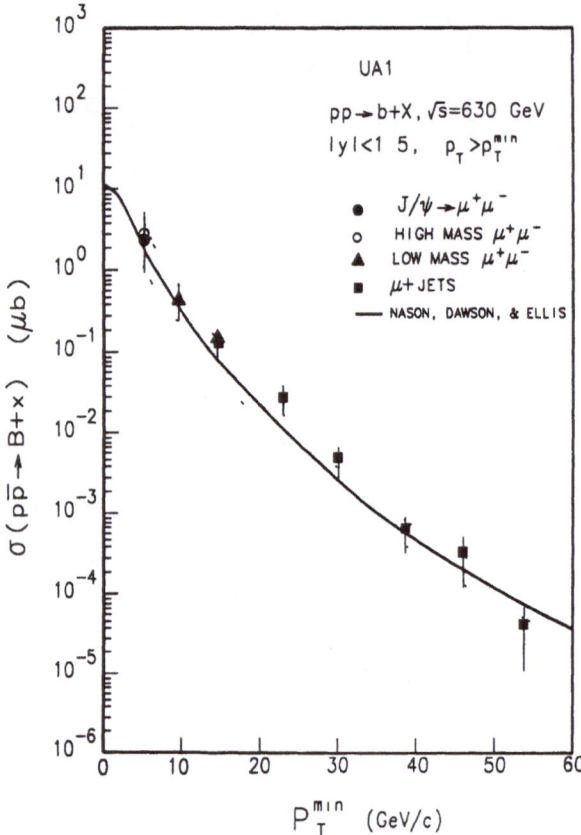

Figure 33. The UA1 *b* quark production cross section integrated over $|y| < 1.5$ and $P_T > P_T^{min}$. The data points come from the single muon analysis as well as dimuon analyses. The curves show the range of the next to leading order calculation.

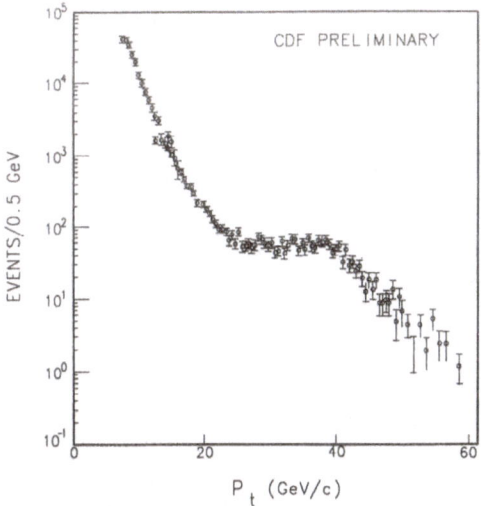

Figure 34. The CDF electron P_T spectrum.

rather than the b signal. CDF selects its electron sample with $P_T > 7$ GeV/c based on the transverse and longitudinal shower shape, the agreement between the track momentum and the calorimeter energy, and position matching of the extrapolated track and cluster centroid. In addition, identified gamma conversions are removed. Figure 34 shows the electron P_T spectrum. The shoulder above 25 GeV/c is due to W and Z decay. W bosons are easily removed by requiring that there be small missing E_T in the event; events are removed as Z contamination if the electron and another high P_T track have an invariant mass near M_Z. The electron spectrum after W and Z removal is shown in Figure 35. The shape agrees well with that predicted by ISAJET plus the CDF detector simulation. Note that charm is expected to contribute 3-15% depending on the electron P_T .

As for UA1, the relation between the P_T^e and P_T^b spectra is obtained from a Monte Carlo study that incorporates the results from e^+e^- colliders. The CDF cross section is shown in Figure 36. The data lie somewhat above the upper end of the theoretical prediction.

Independent evidence that these inclusive lepton events are indeed from b decay comes from CDF [11]. Their high resolution tracking chamber allows them to search for resonances near the electron. Since B meson semileptonic decay usually produces a D meson in the final state, identifying a D near the electron would confirm that the electron was produced by B decay. Figure 37a shows the B decay diagram. Note that the K and the e have the same sign electric charge. CDF looked for $D \rightarrow K\pi$ in a cone (R=1.0) around the electron. Figure 38 shows the D^0 peak when the e and K have the same sign, but no peak when they have the opposite sign. The number of events expected in the D^0 peak is 67 ± 20; 75 ± 17 are observed. Figure 39 shows that, for events in the D^0 peak, the $eK\pi$ invariant mass does not exceed M_B, as required if these are decay products of a single B meson.

CDF also looked for the charge correlation between the electron and a K from the decay of a K^*. The quark decay chain $b \rightarrow c \rightarrow s$ translates into the meson decay chain $\overline{B} \rightarrow e^- DX \rightarrow e^- \overline{K^*}X \rightarrow e^- K^- \pi^+ X$. Thus the electron and kaon must have the same sign charge. This is to be contrasted with an electron from $c\bar{c}$ production and decay, $c\bar{c} \rightarrow q\bar{q}se^-\overline{\nu}\bar{s}$. Here the s and $\bar{s}$ have equal probability to fragment into a K^*. Thus

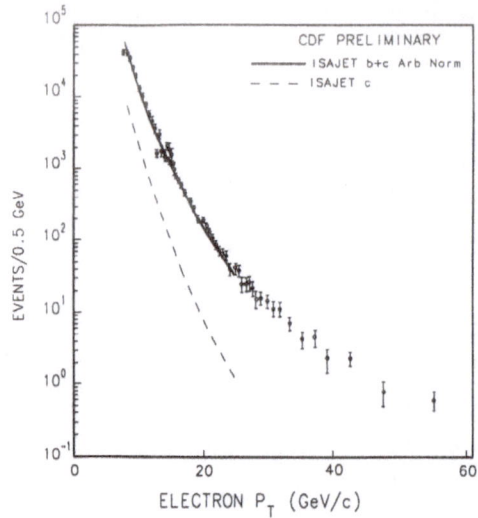

Figure 35. The CDF electron P_T spectrum after removing W and Z events. The curves show the predicted shape for c and $b + c$ production.

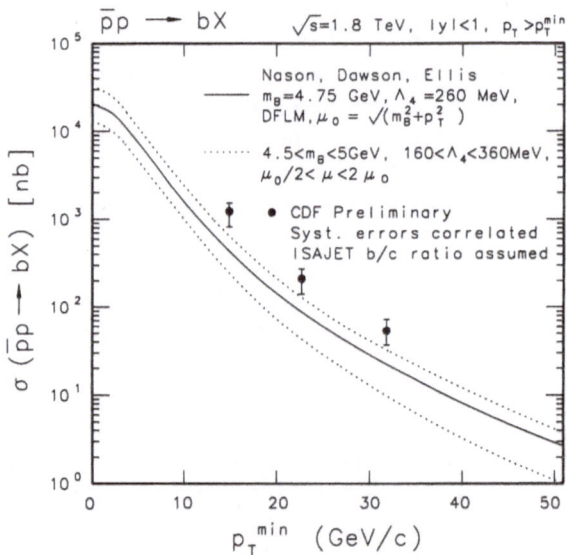

Figure 36. The CDF b production cross section integrated over $|y| < 1$ and $P_T^b > P_T^{min}$.

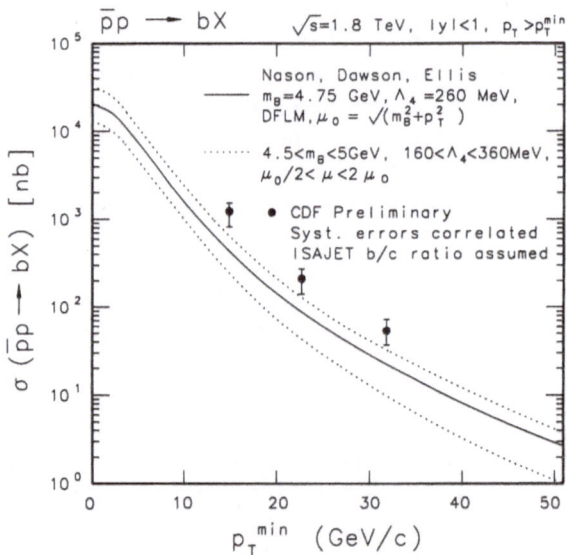

Figure 37. The decay diagram for (a) $B^- \to D^0 e^- \overline{\nu} \to K^- \pi^+ e^- \overline{\nu}$, (b) $\overline{B}_s \to e^- \overline{\nu} \phi X$.

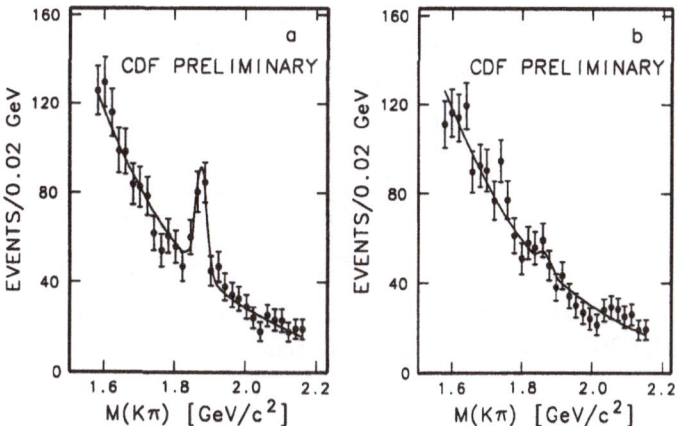

Figure 38. The CDF $K\pi$ invariant mass when (a) the K and e have the same sign charge, (b) when the K and e have opposite sign charge.

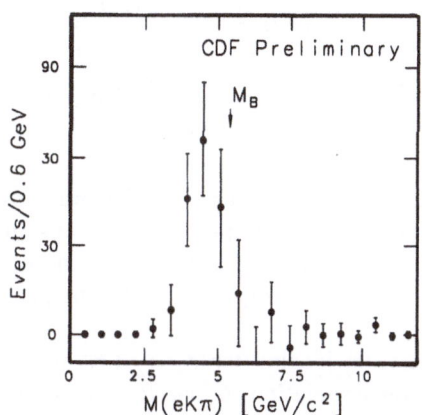

Figure 39. The CDF $eK\pi$ invariant mass.

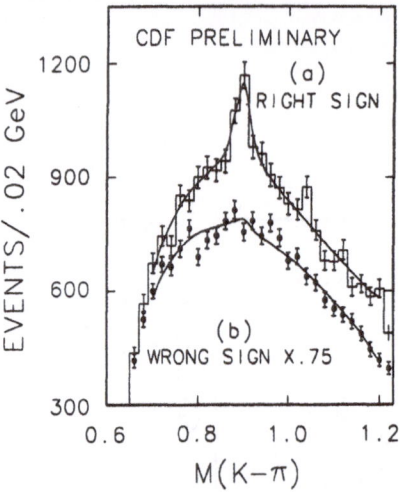

Figure 40. The CDF invariant mass spectrum for $K\pi$ pairs found near an electron in events in which the e and K have (a) the same sign charge, (b) the opposite sign charge.

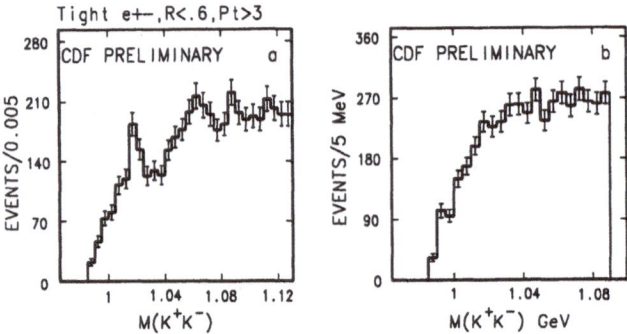

Figure 41. The CDF invariant mass spectrum for K^+K^- pairs found (a) near electrons and (b) near electrons in a photon conversion sample.

one would expect approximately equal numbers of same and opposite sign eK pairs. Figure 40 shows the $K\pi$ invariant mass spectrum for the same sign and opposite sign eK events. As expected for a data sample that is rich in b quarks, a K^* peak of the correct magnitude is seen in the same sign sample, but no peak appears in the opposite sign sample.

One last check comes from looking for $\phi \to KK$ near the electrons (Fig. 37b). Obviously there is no charge correlation to look for, but we can compare the rate of ϕ mesons observed in the inclusive electron sample and a control sample, electrons from identified photon conversions (Fig. 41). A mass peak at the ϕ mass is seen in the inclusive electron sample, while it is not observed in the control sample. These tests all give confidence that the inclusive electron data sample indeed is largely from b decay.

Inclusive J/ψ Channels. B decay into inclusive J/ψ mesons, $b \to cW^* \to c\bar{c}s \to J/\psi X$, with the J/ψ detected in the $\mu^+\mu^-$ mode suffers from a very small combined branching ratio

$$BR(B \to J/\psi X) \times BR(J/\psi \to \mu^+\mu^-) = 0.011 \times 0.069 = 8 \times 10^{-4}$$

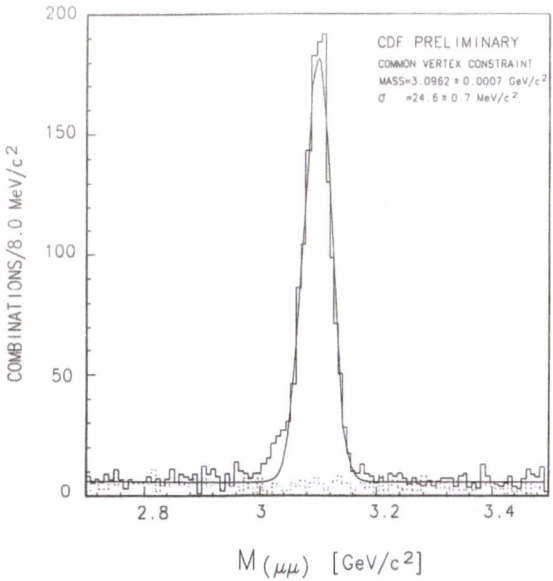

Figure 42. The CDF dimuon invariant mass distribution in the J/ψ region. The dotted histogram is the same sign dimuon data.

To compensate, there are two advantages to this mode. Dimuon detection provides a very clean J/ψ signal with little background. In addition we shall see that the majority of J/ψ mesons come from B decay. The only other significant source of J/ψ is radiative decay of QCD produced χ_c. A χ_c can be produced from the annihilation of two gluons; at least three gluons must annihilate in order to directly produce a J/ψ.

The CDF J/ψ trigger requires two muons each with $P_T > 3$ GeV/c. The dimuon mass spectrum for this data set shows a J/ψ peak with very little background (Fig. 42) [12]. The fraction of J/ψ coming from either B or χ_c decay is determined independently from inclusive J/ψ production and from exclusive final states. For the inclusive analysis, the fraction of J/ψ that comes from B decay ($\equiv F$) can be determined in a relatively unbiased way from the ratio of the inclusive J/ψ cross section to the inclusive ψ' cross section. It is assumed in this analysis that ψ' is produced entirely from B decay, since χ_c cannot decay into ψ'. Figure 43 shows the ψ' signal. There are 72 ± 17 events in the peak. This gives for the ratio of the ψ' to J/ψ production cross sections

$$\frac{\sigma(\psi')}{\sigma(J/\psi)} = (4.2 \pm 1.0) \times 10^{-2}$$

When compared with the ratio of the $B \to \psi'$ to $B \to J/\psi$ branching ratios measured by CLEO [13], $(6.8 \pm 2.5) \times 10^{-2}$, the CDF result translates into

$$F = 64\% \pm 15\% \text{ (CDF stat) } \pm 5\% \text{ (syst) } \pm 23\% \text{ (CLEO stat)}$$

for the fraction of J/ψ coming from B decay. The largest uncertainty is from the CLEO statistics on the ψ' branching ratio; the second largest is due to the CDF ψ' statistics. Both of these should greatly improve in the next year or two.

Exclusive Final States. CDF has also studied J/ψ production by reconstructing exclusive χ_c and B final states. To find the former, $\chi_c \to J/\psi + \gamma$, CDF looks for isolated

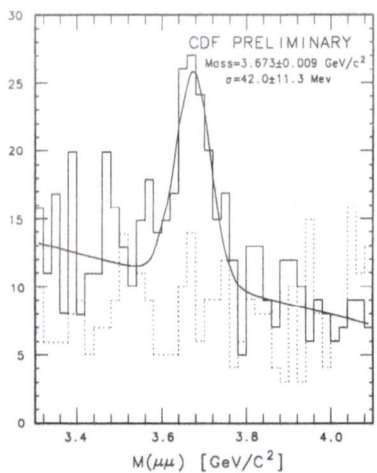

Figure 43. The CDF ψ' signal. The dotted histogram is the same sign data.

electromagnetic clusters of $E_T > 1$ GeV with a transverse shower shape consistent with that of a photon [14]. The best resolution for the $J/\psi\gamma$ resonance is obtained by plotting the difference between the $J/\psi\gamma$ and J/ψ invariant masses (Fig. 44). In this figure, the uncorrelated background is estimated by reanalyzing the event sample after reversing the direction of each J/ψ. The background not associated with J/ψ production can be estimated using the $\mu^+\mu^-$ mass sidebands above and below the J/ψ. Figure 45 shows the data and the sideband background, as well as a Monte Carlo simulation of the signal for the appropriate mixture of χ_1 and χ_2. The peak in the data clearly is due to reconstructed $\chi_c \rightarrow J/\psi + \gamma$. From the number of observed events, CDF concludes that $\sim 30\%$ of J/ψ comes from χ_c decay.

Exclusive reconstruction of B mesons is carried out for two modes [15]

$$B_d^0 \rightarrow J/\psi K^{*0} \rightarrow J/\psi K\pi$$

$$B_u^\pm \rightarrow J/\psi K^\pm$$

CDF looks in a 60^0 cone around the J/ψ direction for additional tracks. For the B_u search, all tracks with $P_T > 2.5$ GeV/c are considered K candidates. For the B_d, all opposite sign track pairs from among the three highest P_T tracks are tried; a pair is used if the $K\pi$ invariant mass is within $50~MeV/c^2$ of the K^{*0} mass. Figure 46 shows the individual mass spectra; the combined spectrum is in Figure 47. Until there is more data on final state polarization in $B \rightarrow J/\psi K^*$ decay (it affects the detection efficiency calculation), only $B \rightarrow J/\psi K^\pm$ is used to determine the B production cross section. CDF finds that $\sim 70\%$ of J/ψ comes from B, consistent with the value for F obtained in the inclusive J/ψ study. The B cross section is shown in Figure 48. As with the data points from the inclusive electron sample, the data is somewhat higher than the next to leading order theoretical prediction.

$B^\circ \rightarrow \mu^+\mu^-$

$B_{d,s}^0 \rightarrow \mu^+\mu^-$ is a flavor changing neutral decay allowed by the Standard Model via higher order electroweak diagrams (Fig. 49). The theoretical expectations are

$$BR(B_d^0 \rightarrow \mu^+\mu^-) \approx 10^{-11}$$
$$BR(B_s^0 \rightarrow \mu^+\mu^-) \approx \text{few} \times 10^{-9}$$

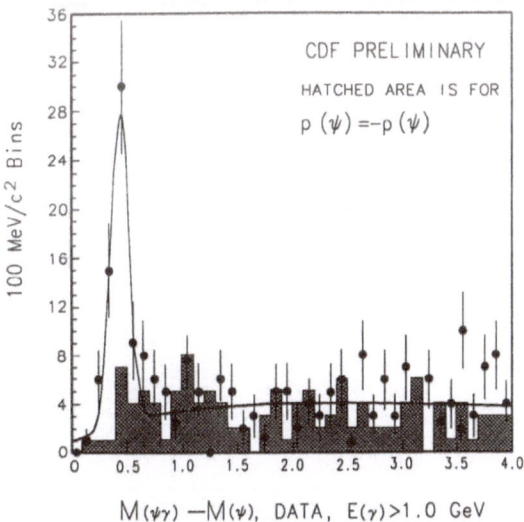

Figure 44. The CDF $J/\psi\gamma$, J/ψ invariant mass difference. The uncorrelated background is estimated by reanalyzing the events with the J/ψ momentum direction reversed.

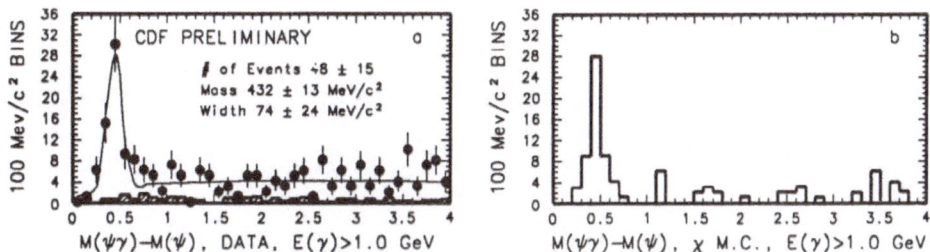

Figure 45. (a) The CDF $J/\psi\gamma$, J/ψ invariant mass difference. The solid curve is a fit to a resonance plus smooth background, and the solid histogram is the data from the J/ψ sidebands. (b) A Monte Carlo simulation of a signal from χ_1 and χ_2 production.

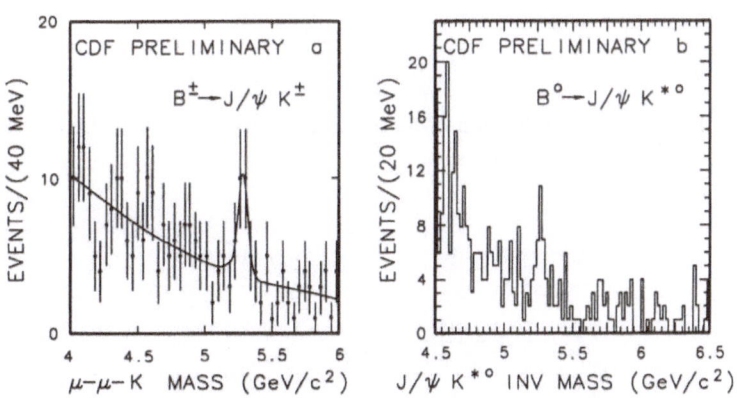

Figure 46. (a) $B^{\pm} \rightarrow J/\psi K^{\pm}$ (b) $B^0 \rightarrow J/\psi K^{*0}$ from CDF.

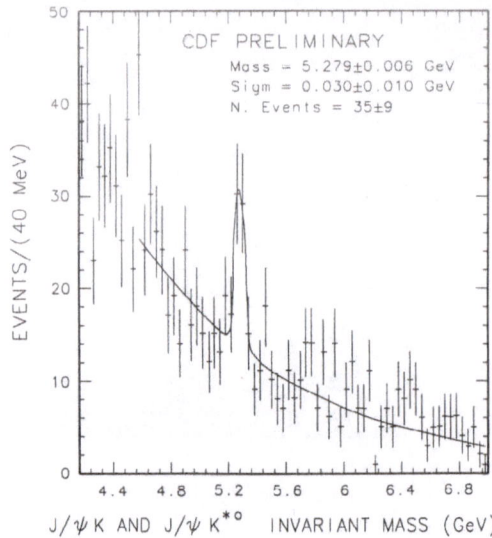

Figure 47. The combined CDF $J/\psi K$ and $J/\psi K^*$ spectrum.

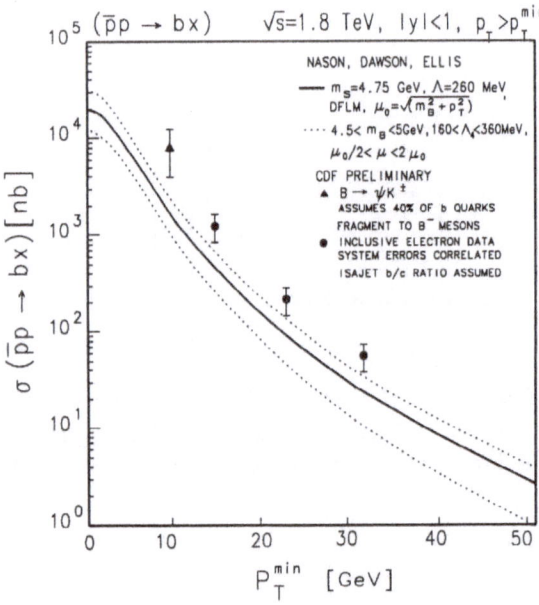

Figure 48. The CDF B production cross section results from both the inclusive electron and exclusive $B \rightarrow J/\psi K^{\pm}$ analyses.

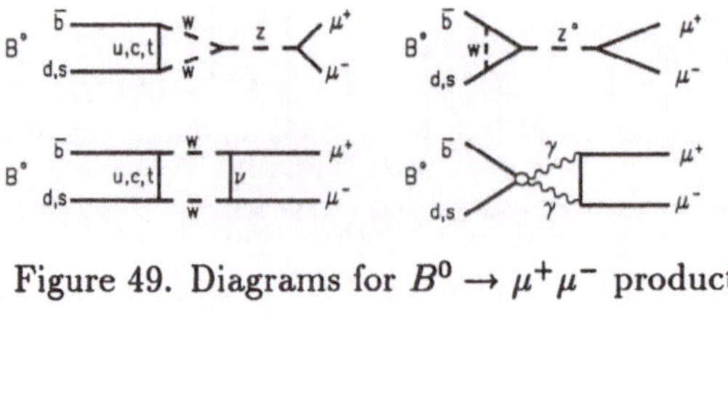

Figure 49. Diagrams for $B^0 \rightarrow \mu^+\mu^-$ production.

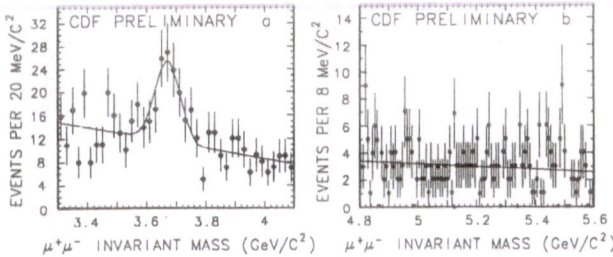

Figure 50. The CDF dimuon invariant mass spectrum (a) near the ψ', (b) in the B mass region.

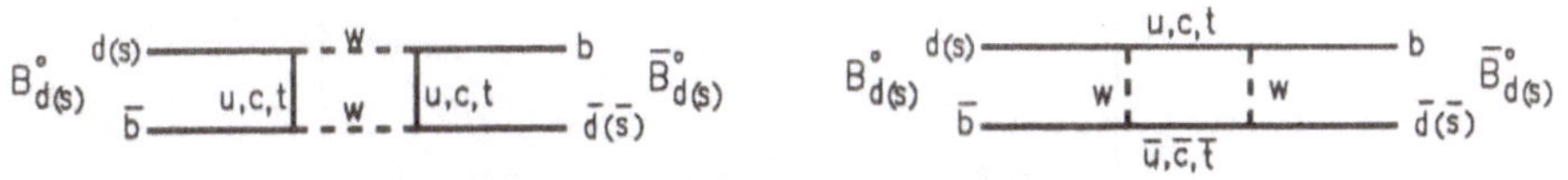

$$B^0_{d(s)} \quad \underline{\quad d(s) \quad} \quad W \quad \overline{B}^0_{d(s)}$$

Figure 51: $B^0, \overline{B}^0$ mixing diagrams.

The best published limit comes from CLEO and ARGUS [16]

$$BR(B^0_d \to \mu^+\mu^-) < 0.5 \times 10^{-4} \ @ \ 90\% \ CL$$

UA1 has a preliminary result [17]

$$BR(B^0_{d,s} \to \mu^+\mu^-) < 1.0 \times 10^{-5}$$

The CDF dimuon spectrum is shown in Figure 50 along with the ψ' peak. Given the 72 observed ψ' events and the combined $B \to \psi'X \to \mu\mu X$ branching ratio of 2.8×10^{-5}, the lack of a peak in the B region translates into a branching ratio limit of [18]

$$BR(B^0_d \to \mu^+\mu^-) < 3.2 \times 10^{-6} \ @ \ 90\% \ CL$$

The limit would be much better if it weren't for the size of the background in the B region. With the silicon vertex detector in the next CDF run, the background should be greatly reduced since candidate tracks can be required to point to a secondary vertex.

$B^\circ, \overline{B}^\circ$ Mixing

As is the case in $K^0, \overline{K}^0$ mixing, the transformation of a b quark into a $\overline{b}$ is a second order weak process (Fig. 51). The diagrams with t quark exchange dominate, and thus the difference between $B_d, \overline{B}_d$ mixing and $B_s, \overline{B}_s$ mixing comes from the CKM factors, V_{td}^2 and V_{ts}^2. Since V_{ts} is considerably larger than V_{td}, $B_s, \overline{B}_s$ mixing should be significantly larger than $B_d, \overline{B}_d$ mixing.

Mixing is characterized by

$$\chi \equiv \frac{\text{Prob}(B^0 \to \overline{B}^0)}{\text{Prob}(B^0 \to B^0) + \text{Prob}(B^0 \to \overline{B}^0)}$$

where the physical range is $0 \leq \chi_{d,s} \leq \frac{1}{2}$ (Fig. 52). The first evidence for $B\overline{B}$ mixing came from UA1 [19]. However in high energy $\overline{p}p$ colliders, both B^0_d and B^0_s are produced. What is actually measured is the combined mixing, characterized by

$$\overline{\chi} \equiv \frac{\text{Prob}(b \to \overline{B}^0 \to B^0 \to l^+)}{\text{Prob}(b \to l^\pm)} = f_d\chi_d + f_s\chi_s$$

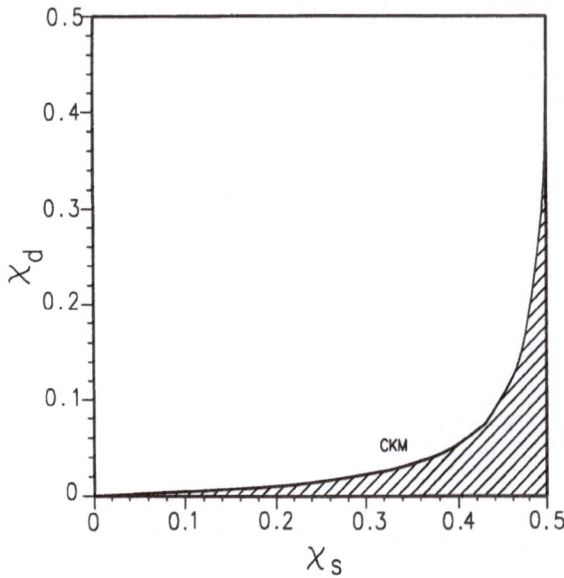

Figure 52. The shaded region is allowed by the CKM matrix when the non-mixing experimental results and 3-generation unitarity are used.

where f_d (f_s) is the fraction of B_d^0 (B_s^0) produced relative to all b mesons and baryons times $BR(B_{d,s}^0 \to lX)/BR(b \to lX)$. The Standard Model limits can be tested when the hadron collider results are combined with the measurement of χ_d from e^+e^- data on the $\Upsilon(4S)$, which gives $\chi_d = 0.16 \pm 0.04$ [20] (the $\Upsilon(4S)$ cannot decay into $B_s^0\overline{B}_s^0$). With higher statistics and smaller systematics than shown below, future $\bar{p}p$ results could extract V_{td}/V_{ts}.

Mixing is studied with a dilepton data sample, because if both B mesons in an event decay semileptonically, the lepton signs identify the parent mesons as $B\overline{B}$, BB, or $\overline{B}\,\overline{B}$. It should be noted, however, that there are other sources of leptons, most notably charm. The quantity directly measured in the $\bar{p}p$ experiments is

$$R = \frac{N(l^+l^+) + N(l^-l^-)}{N(l^+l^-)}$$

where UA1 uses their $\mu\mu$ sample [21] while CDF uses its $e\mu$ sample because of the lack of Drell Yan, J/ψ, and Υ background [22]. UA1 finds $R = 0.42 \pm 0.07 \pm 0.03$ and CDF finds $R = 0.55 \pm 0.05 \pm 0.04$, compared with the predictions of 0.26 ± 0.03 (UA1 energy) and 0.23 ± 0.06 (CDF energy) if there were no mixing. Clearly mixing is required by the data. The value of $\overline{\chi}$ can be extracted from R using

$$R = \frac{2\overline{\chi}(1 - \overline{\chi})N_f + [(1 - \overline{\chi})^2 + \overline{\chi}^2]\,N_s}{[(1 - \overline{\chi})^2 + \overline{\chi}^2]\,N_f + 2\overline{\chi}(1 - \overline{\chi})N_s + N_c}$$

N_f is the number of events in which both leptons come directly from B decay. N_s contains events where one lepton comes directly from B decay and the other comes from the sequential $b \to c \to l$ decay. N_c counts events in which the two leptons come from $c\bar{c}$ production. The equation can be understood if you note that in the numerator the coefficient in front of N_f is the probability that one and only one b mixes, while the coefficient in front of N_s is the probability that neither b mixes or both mix. Also note that $D\overline{D}$ mixing is negligible and has not been included. At present, N_s/N_f and

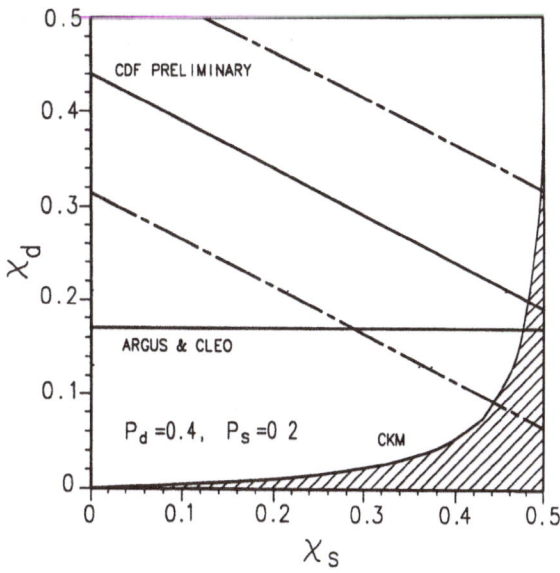

Figure 53. B mixing results from e^+e^- and CDF. The dotted and dot-dashed lines are the one sigma limits.

N_c/N_f are determined from Monte Carlo calculations. The results are

$$\text{UA1}: \quad \overline{\chi} \;=\; 0.158 \pm 0.059$$
$$\text{CDF}: \quad \overline{\chi} \;=\; 0.176 \pm 0.031(stat + syst) \pm 0.032(Monte\ Carlo)$$

where the CDF value comes from its $e\mu$ and ee samples. Figure 53 shows the CDF value rather than a combined CDF and UA1 result because the uncertainties in the two experiments are highly correlated due to the common Monte Carlo assumptions. The figure has been drawn with the assumption that b quarks form B_d, B_u, B_s, and b baryons 37.5%, 37.5%, 15%, and 10% of the time respectively. The $\overline{p}p$ and e^+e^- results overlap the allowed CKM region, but the uncertainties are big. In future Fermilab Collider runs, the large increase in the number of detected B events will allow for direct measurement of f_d and f_s from exclusive final states and a much more precise measurement of χ_s.

Search for the Λ_b

UA1 has reported [23] the observation of the decay $\Lambda_b \to \Lambda\psi$ with a Λ_b mass of $5640 \pm 50 \pm 30\ MeV/c^2$ and $F(\Lambda_b)BR(\Lambda_b \to \Lambda\psi) = (1.8 \pm 0.6 \pm 0.9) \times 10^{-3}$. Here $F(\Lambda_b)$ is the fraction of produced b quarks that fragment into a Λ_b.

CDF has also searched for this mode [24]. Among events containing a $J/\psi \to \mu^+\mu^-$ decay (Fig. 54a), a search is made for $\Lambda \to p\pi$. Figure 54b shows the invariant mass distribution for track pairs originating from a point at least 2 cm from the primary interaction vertex. The higher momentum track is assigned the proton mass, as expected for the low Q value decay of a Λ. Figure 54c shows the $\Lambda\psi$ invariant mass when these particles are in the same hemisphere of the detector. This last requirement greatly reduces contamination from a Λ and ψ each from a different b quark. There are no more than 2 events in any $\pm 40\ MeV/c^2$ ($\pm 2\sigma$) region between 5300 MeV/c^2 and 6000 MeV/c^2. This is to be contrasted with an expectation of 30 ± 23 events based on the UA1 result.

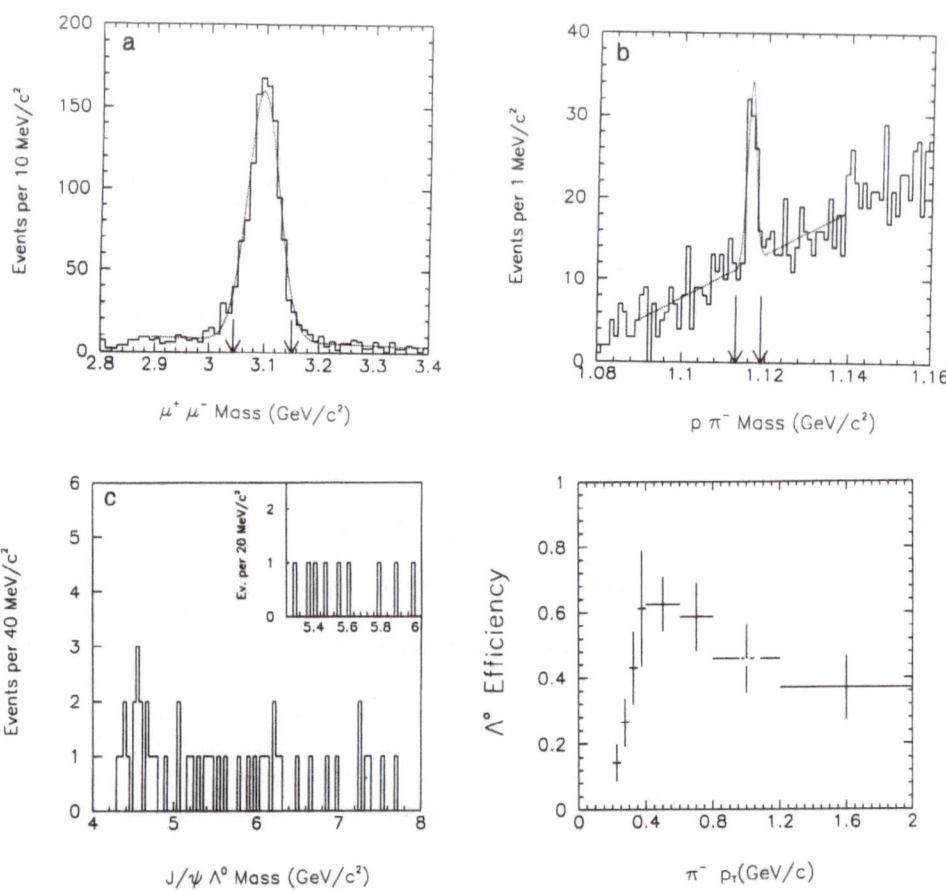

Figure 54. (a) The $\mu^+\mu^-$ invariant mass in the vicinity of the J/ψ. (b) The $p\pi^-$ (and charge conjugate) invariant mass distribution in the vicinity of the Λ^0. (c) The $J/\psi\Lambda$ invariant mass distribution in 40 MeV/c^2 bins. The insert has 20 MeV/c^2 bins. (d) The Λ^0 reconstruction efficiency as a function of the P_T of the π.

With the conservative assumption that the 2 events are signal (no background subtraction) and using the b production cross section measured by CDF using reconstructed $B^{\pm} \to J/\psi K^{\pm}$ events, CDF finds that $F(\Lambda_b)BR(\Lambda_b \to \Lambda\psi) < 0.5 \times 10^{-3}$ at the 90% CL. The only significant experimental difference between CDF and UA1 that might account for this difference in results is the Λ^0 reconstruction efficiency. The UA1 tracking chamber has single track efficiency that is good down to 0.05 GeV/c, while the CDF detector is only efficient for tracks with momentum above 0.25 GeV/c. For CDF this results in a low Λ^0 detection efficiency for very low Λ^0 transverse momentum (Fig. 54d). However for this to explain the difference in the observations, there would have to be a very nonstandard decay mechanism that produces <u>very</u> soft pions in Λ_b decay.

Since the results of the two experiments are quite different, let us make an estimate of what we would expect. To do this, compare $\Lambda_b \to \psi\Lambda$ decay with $\overline{B}_u \to \psi K^-$ decay. These processes are very similar. In both cases, a b quark decays into a c quark and a virtual W^-, which in turn decays to a $\overline{c}s$. The c and $\overline{c}$ form the ψ, and the s combines with the spectator quarks (ud for Λ_b and $\overline{u}$ for the $\overline{B}_u$) to form the Λ^0 or K^-. Thus we expect $\Gamma(\Lambda_b \to \psi\Lambda) \leq \Gamma(B_u \to \psi K)$, since three quarks have to bind to form the Λ while only two have to bind to form the K. We also expect that the lifetimes of the Λ_b and B_u are approximately equal. Using the measured $BR(B_u \to \psi K)$, we then get as an estimate $BR(\Lambda_b \to \psi\Lambda) < 1 \times 10^{-3}$. For $F(\Lambda_b)$, we note that the usual prediction in b quark fragmentation is for B_u, B_d, B_s, and $(B_c + b$ baryon) to occur 37.5%, 37.5%, 15%, and 10% of the time respectively. Thus for the single b baryon, Λ_b, we expect $F(\Lambda_b) < 0.1$. This makes the estimate of $F(\Lambda_b)BR(\Lambda_b \to \psi\Lambda) < 1 \times 10^{-4}$.

Prospects for B Physics

The cross section for $b\overline{b}$ production in the central four units of rapidity is approximately 1×10^{-28} cm^2. This means that 10^{11} $b\overline{b}$ events will be produced for a 1 fb^{-1} integrated luminosity. With an instantaneous luminosity of 5×10^{31} $cm^{-2} - sec^{-1}$, the $b\overline{b}$ production rate would be 5 KHz! Even if the acceptance range is limited to $|y| < 1$ and $P_T^b > 10$ GeV/c, the event rate would still be 200 Hz, and 4×10^9 $b\overline{b}$ events would be collected.

There are many experimental challenges that have to be met if hadron collider experiments are to make a major impact on b physics. Since the b production cross section is only $\approx 0.2\%$ of the inelastic $\overline{p}p$ cross section and the rate for writing events to magnetic tape is limited by the bandwidth of the data acquisition system, the purity and efficiency of the b trigger is critical. This means having low P_T thresholds for e, μ, and J/ψ while maintaining a high signal to noise ratio. Of enormous utility would be fast (~ 10 μsec) secondary vertex finding. Another problem is data storage for very large event samples. This requires efficient online separation of signal from background and data compaction to minimize the size of each event. Identification of b jets is also critical; it requires an efficient secondary vertex detector. But perhaps most important for the observation of CP violation is flavor tagging of the second b in an event. Techniques under study include K identification, efficient detection of moderate P_T leptons, and very efficient track finding at the secondary vertex so that the charge of the decaying B can be measured. During the last CDF run, the flavor tagging efficiency was approximately $\frac{1}{2}\%$ due to the B semileptonic branching ratio, the muon identification requirements, and the limited range of rapidity and P_T covered. If CP violation is to be studied in $\overline{p}p$ collisions, the tagging efficiency must be increased by approximately a factor of ten.

The b physics opportunities are extensive. The B_s, B_c, Λ_b, and other b hadrons

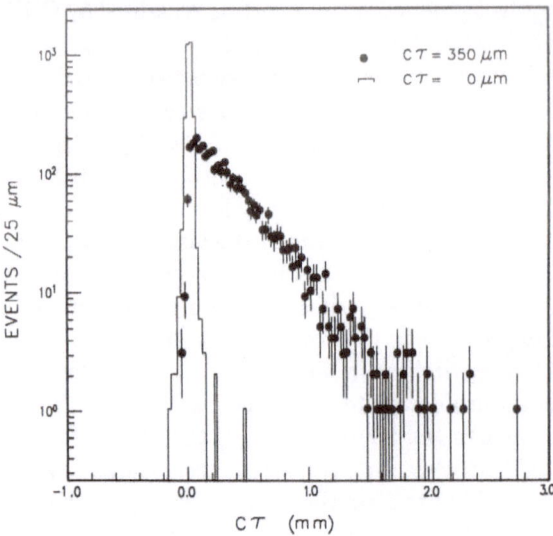

Figure 55. Monte Carlo simulation of the reconstructed proper lifetime distribution for $B^{\pm} \rightarrow J/\psi K^{\pm}$ (data points) and for a similar resonance that decays at the primary vertex.

should be observed and their masses measured (a 13 MeV/c^2 mass resolution is expected for the next run). Precision measurements will be made of the individual lifetimes for B_u, B_d, and B_s; a 3% uncertainty is expected in the next run for B_u and B_d (Fig. 55). A sensitive search for rare B decay modes can also be carried out. The predicted 10^{-9} branching ratio for $B \rightarrow \mu\mu$ could be observed. In addition, $B \rightarrow \mu\mu K$, which occurs through an electromagnetic penguin diagram, should be seen with good statistics. It is sensitive to M_{top} as well as other new massive particles and provides a measure of the CKM matrix element V_{ts}.

Direct observation of the interference effects of B_s mixing is a major b physics goal. Figure 56 shows what could be observed with dilepton events for $X_s \equiv \frac{\Delta M}{\Gamma} = 5$. Figure 56c clearly shows the oscillations due to B_s, $\overline{B}_s$ mixing. The major challenge here is to find the best estimator of the B momentum for determining the proper time of the B decay. Figure 56f shows the degradation of Figure 56c if the lepton momentum is used as the B momentum.

The ultimate goal for $\overline{\text{p}}\text{p}$ b physics is the observation of CP violation in B decay. Let us review how this can be done. The unitarity of the CKM matrix for the three known fermion generations requires

$$V_{ud}V_{ub}^* + V_{cd}V_{cb}^* + V_{td}V_{tb}^* = 0$$

Since $V_{ud} \approx 1$, $V_{tb} \approx 1$, and with the usual phase convention V_{cb} is positive real and V_{cd} is negative real, the equation becomes

$$V_{ub}^* + V_{td} = |V_{cd}V_{cb}|$$

which is a triangle in the complex plane (Fig. 57). CP violation can result if the angles are non-zero. Information on the lengths of the sides of the triangle comes from semileptonic B decay (V_{ub}, V_{cb}), opposite sign dimuon production in ν interactions (V_{cd}), and $B - \overline{B}$ mixing (V_{td}). The angles can be determined by measuring CP violating

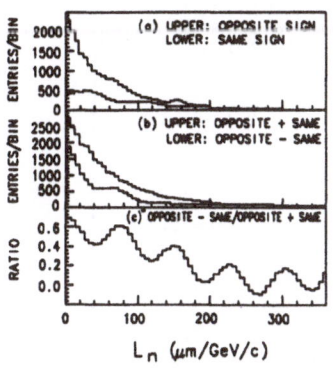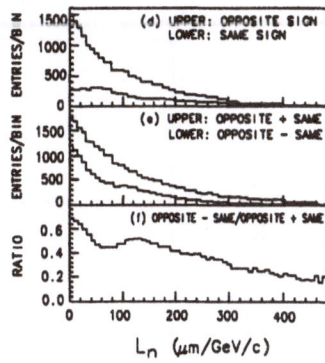

Figure 56. The distribution in $b\bar{b} \to 2$ lepton events of $L \equiv c\tau$, where τ is the proper decay time of each B. (a) The distributions for events with opposite sign or same sign leptons. (b) The sum and the difference of the curves in (a). (c) The ratio of the difference and sum curves in (b). (d), (e), and (f) are the same as (a), (b), and (c), except that the B momentum used to obtain the proper decay time is approximated by the lepton momentum.

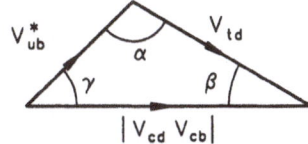

Figure 57. The CKM unitarity triangle.

asymmetries in the decay of B and $\overline{B}$ into CP eigenstates, specifically

$$\alpha : \qquad B_d \to \pi^+\pi^-$$
$$\beta : \qquad B_d \to \psi K_s$$
$$\gamma : \qquad B_s \to \rho K_s$$

To determine whether CP violation as observed in the K system is consistent with a CKM origin, one could measure the three sides and one angle. Hadron colliders could contribute to the measurement of V_{td} ($B_s - \overline{B}_s$ mixing, rare B decay, top mass) and β (CP asymmetry in $B_d \to \psi K_s$). Current data on the CKM matrix and CP violation in K decay suggest that $0.1 < sin2\beta < 1$, with 0.34 the most likely value [25]. A future measurement of this quantity with the ψK_s final state will have its accuracy limited by luminosity (number of events) and the efficiency for tagging the parent as B or $\overline{B}$. The latter is characterized by

$$d_{tag} = \epsilon_{tag}(1 - 2w)^2(1 - 2\overline{\chi})^2$$

where ϵ_{tag} is the efficiency for tagging the other B meson, w is the probability that the tag gives the wrong answer, and the last factor in the equation is due to dilution from B mixing. Figure 58 shows how this translates into uncertainty in the $sin2\beta$ determination. An uncertainty in $sin2\beta$ of 0.33 (0.11) can be expected if the b tagging efficiency can be improved by a factor of 2 (10) over what is expected in the next CDF data run.

Studying CP violation in B decay at hadron colliders will be very challenging, but it appears quite possible.

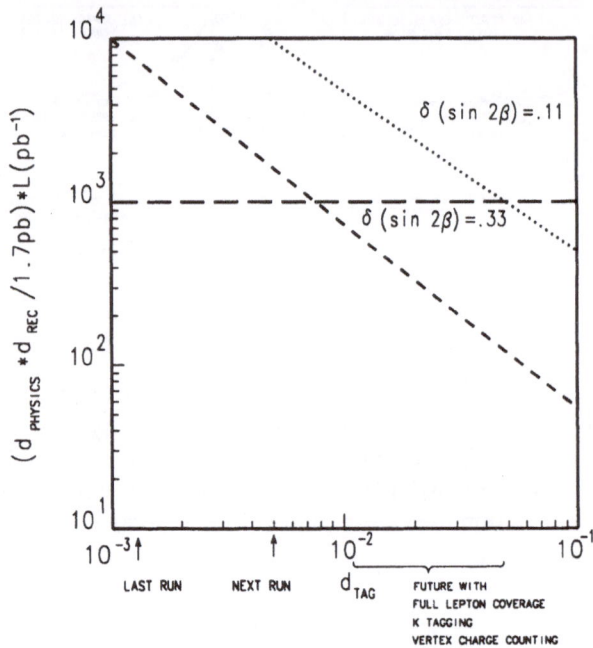

Figure 58. The luminosity required to observe a CP asymmetry as a function of d_{tag}. The two dotted lines correspond to measurement uncertainties in $sin2\beta$ of 0.11 and 0.33. The horizontal line is the integrated luminosity expected in a two year run with the upgraded Fermilab Collider.

References

[1] F. Abe et al, PRL 64 (1990), 147
    F. Abe et al, PRL 64 (1990), 142
    F. Abe et al, PR D43 (1991), 664
    F. Abe et al, PRL 68 (1992), 447
    F. Abe et al, PR D45 (1992), 3921

[2] W. Bartel et al, PL B146 (1984), 437

[3] Particle Data Group, PL B239 (1990), 1

[4] G. Altarelli et al, Nucl.Phys. B308 (1988), 724
    R.K. Ellis, Fermilab-Pub-91/30-T
    These use the next-to-leading-order calculation in
    P. Nason, S. Dawson, and R.K. Ellis, Nucl.Phys. B303 (1988), 607

[5] H. Baer et al, University of Wisconsin preprint MAD/PH/540 (1990)

[6] For example, see:
    R.K. Ellis, *Proceedings of the 7th Topical Workshop on Proton–Antiproton Collider Physics*, Fermilab (1988), 639
    P. Nason, *Proceedings of the XXIV International Conference on High Energy Physics*, Munich (1989), 962

[7] G. Altarelli et al, Nucl.Phys. B308 (1988), 724, based on the order α_s^3 calculation of P. Nason, S. Dawson, and R.K. Ellis, Nucl.Phys. B303 (1988), 607

[8] An excellent summary of the b studies carried out by the UA1 collaboration is given in N. Ellis and A. Kernan, Phys.Rep. 195 (1990), 23

[9] C. Albajar et al, PL B256 (1991), 121

[10] A. Sansoni, *Proceedings of the 25th Rencontres de Moriond*, (1991)

[11] *b Physics at CDF*, Fermilab (1991), unpublished

[12] R. Hughes et al, CDF Internal Note 1448 (1991), unpublished

[13] M.S. Alam et al, PR D34 (1986), 3279
H. Albrecht et al, PL B199 (1987), 451

[14] A. Yagil, CDF Internal Note 1526 (1991), unpublished

[15] D.A. Crane, *Proceedings of the XIth International Conference on Physics in Collision*, Colmar, France (1991)

[16] Particle Data Group, PL B239 (1990), 1

[17] C. Albajar et al, CERN-PPE/91-54, submitted to Phys.Lett.

[18] L. Pondrom, *Proceedings of the 25th International Conference on High Energy Physics*, Singapore (1990)

[19] C. Albajar et al, PL B186 (1987), 247

[20] H. Albrecht et al, PL B192 (1987), 245
M. Artuso et al, PRL 62 (1989), 2233

[21] C. Albajar et al, CERN-PPE/91-55 (1991), submitted to Phys.Lett.

[22] F. Abe et al, PRL 67 (1991), 3351

[23] C. Albajar et al, PL B273 (1991), 540

[24] F. Abe et al, submitted to PRL

[25] C.S. Kim, J.L. Rosner, and C.P. Yuan, PR D42 (1990), 96

NEUTRINO PHYSICS

Barry C. Barish

California Institute of Technology
High Energy Physics
Pasadena, CA 91125

In these lectures, I present recent developments in Neutrino Physics with an emphasis on non-accelerator experiments.

INTRODUCTION

In recent years, experimental particle physics has evolved in three general directions: the high energy frontier; precision experiments; and experiments not involving particle accelerators. The high energy frontier has traditionally been the path toward shorter distance scales and hence, has led directly to breakthroughs in our fundamental understanding of the elementary constitutions of matter and their interactions. With the great success of the "standard model", the best method to unfolding the physics beyond this picture is less clear. The high energy frontier remains the central direction with the SSC and LHC aimed at an energy scale $0(\sim 1 \text{ TeV})$, where we expect new physics. An alternate approach that can be pursued while we await these machines is to do precision experiments, where deviations from the predictions of the standard model can be the signature of new physics. These are being done, for example at, LEP, where very precise tests of the standard model are possible and extensive results of this type have been presented at this conference. Another way new physics may appear at low energies is in the detection of rare decay modes, that is those suppressed or not expected in the standard model. A series of new experiments on rare K-decays have been conducted in recent years, and ambitions plans for the future on rare decays of B-mesons, taus, muons, etc. are planned. Finally, another approach toward reaching new physics has been labeled "non-accelerator physics". This is a catch-all term for a set of experiments with fundamental particle physics goals, but requiring techniques not using particle accelerators. Such experiments range from studies of solar neutrinos, the search for WIMPS, the decay of the proton, to accurate studies of nuclear beta-decay.

Ambitious large scale laboratories and experiments have been developed for such non-accelerator experiments. The Gran Sasso Laboratory in Italy is especially impressive in both the size and sophistication of the project and its promise for this area of experimentation. Large scale experiments (IMB, Kamiokande and MACRO) and large collaborations typical of accelerator laboratories are now commonplace. However, other small scale experiments (e.g. 2β decay, neutrino mass, etc.) are also an important part of the field. Challenging

Quantitative Particle Physics, Edited by
M. Lévy *et al.*, Plenum Press, New York, 1993

technological developments (e.g. dark matter detectors, imagining $\check{C}$ counters, etc.) are crucial to progress on many important problems and are being vigorously pursued.

Non-accelerator experiments are often on the interface between particle physics and other fields, which is helping to develop strong alliances and communications between physicists in these aligned fields. In particular, on the boundaries between particle and nuclear physics (e.g. 2β decay, 17 KeV neutrino), etc. and between particle and astrophysics "particle astrophysics" (e.g. dark matter searches, high energy point sources, etc.) many subjects of joint interest are being studied.

In this presentation, I confine myself to selected critical discussions of these three areas. I do not try to give a complete review of results, but rather attempt to put into perspective both the present experimental status and future prospects.

The first general area of interest involves the problem of neutrino mass. The question of whether the neutrino has mass is not a new one, however, developments in recent years have inspired greater interest in the problem. Models of Grand Unification have stimulated the search for proton decay, magnetic monopoles and neutrino mass. All three are possible consequences of Grand Unification. The experimental search for neutrino mass involves a great variety of experiments - accelerator and non-accelerator, because of the wide range of possible mass and mixing parameters. Results include evidence relating to the possible existence of a, "17 KeV Neutrino" and secondly, measurements of the observed flux of solar neutrinos that might imply neutrino oscillations.

The second class of experiments involve perhaps the outstanding question in astrophysics: the nature of the dark matter. The favored solutions to the dark matter problem involve the existence of new particles (e.g. non-baryonic dark matter). Some possible candidates could be discovered (or eliminated) on accelerators, and in fact, LEP experimental limits on various possibilities have already ruled out some candidates and restricted the masses or couplings of others. However, some candidates or mass ranges can only be investigated using non-accelerator experiments. New limits on the existence of strange quark matter (nucleorites) and on the existence of weakly interacting massive particles (WIMP 's) from non-accelerator experiments have been presented. The search for the dark matter continues and new techniques and more sensitive experiments will enable a set of promising experiments to address this crucial question in the next few years.

Another promising type of non-accelerator experiment involves doing astronomical observations with particles (γ, μ, ν). The acceleration mechanisms for the highest energy cosmic radiation are not well understood. Candidates for point source radiation of particles include binary systems, like Cygnns X-3, where observation of γ-rays in the PeV range have been reported. The existence of such sources would be particularly exciting as electromagnetic acceleration mechanisms are no longer possible at such energies. It is there-fore believed that particle physics, acceleration and collisions of hadtons would be responsible, if indeed, such sources exist. Unfortunately, the episodic nature of these sources (even at low energies), and the lack of well instrumented detectors, have not allowed experimental clarification of these reports. Experiments for both γ's and μ's have reported results and represent the early results from a new generation of detectors which promise more sensitivity in probing, and hopefully, resolving this issue. Lastly, the possible use of neutrinos to comple-ment these studies is also getting underway.

Issues Related to Neutrino Mass

How does neutrino mass fit into the electroweak theory of Glashow-Weinberg-Salam? In the minimal theory, there is just a Higgs scalar doublet (ϕ^+, ϕ^0), and in this case, $m_\nu \equiv 0$, since there is no Dirac or Majorana mass term.

Therefore, only extended models can introduce a neutrino mass. Such an extended model could introduce right handed neutrinos (v_R), and in this case, there is a Dirac mass term in analogy to the charged lepton and quark mass terms. However, if this were the case, the question arises why m_v is so small, since one would expect $m_v \sim$ MeV's. Another possible extension is to introduce a Higg's triplet. In this case, there is a Majorana mass term. However, this implies other massless bosons that would appear in neutrinoless 2β decay that have, so far, not been observed.

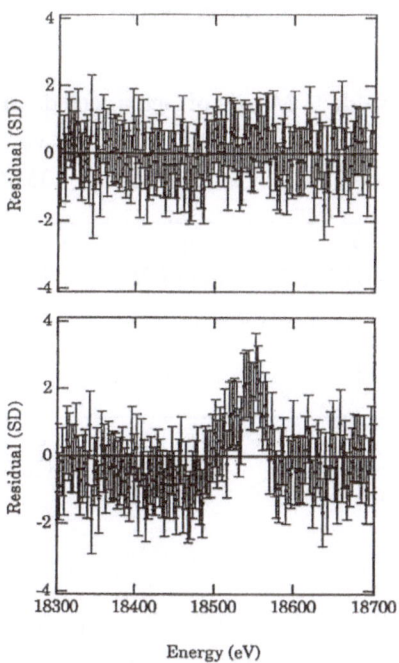

Figure 1. Analysts of the end-point spectrum for molecular tritium from Wilkerson *et al.* where the upper figure is assuming $m_{v_\mu} = 0$ eV and the lower figure is for $m_{v_\mu} = 30$ eV. The result gives $m_{v_\mu} > 9.4$ eV 95% CL.

In the simplest GUT model - SU(5), again there is no mass term and $m_v \equiv 0$. However, we now know from the present limits on proton decay that any consistent GUT theory involves a larger group. In this case, the neutrino can have mass, and for example, the See-Saw mechanism provides a Majorana mass, $m_v \sim 0(10^{-5} - 10^{-10}$eV), a rather small mass.

What do we know about the masses of the known neutrinos? Table I shows the present knowledge of masses of the quarks and leptons. The m_{v_μ} was determined from the end point of τ decay in the Argus experiment from 11 events. The m_{v_μ}. was determined from the end point of $\tau \rightarrow 5\pi v$ by the Argus experiment from 11 events. The m_{v_μ} was determined from the decay In this case, from the known masses,

$$m_\pi = 139.56871 \pm 0.00053 \; MeV,$$
$$\text{and}$$
$$m_\mu = 105.65916 \pm 0.00030 \; MeV,$$
$$\text{one predicts}$$
$$p_\mu = 29.79006 \pm 0.00080 \; MeV.$$

This is consistent with the experimental observations of $p_\mu = 29.79139 \pm 0.00083 \; MeV$.

Finally, the best limit $m_{\nu_\mu}. < 9.4$ eV 95% at CL comes from beta decay end point from molecular tritium by Wilkerson *et at.* shown in Figure 1.

Table I. Masses of leptons and quarks

ν_e	ν_μ	ν_τ
< 9.3eV	< 270 KeV	< 35 MeV
e	μ	τ
.511 MeV	105 MeV	1.78 GeV
u	c	t
~ 5 MeV	~ 1.35 GeV	~ 120 GeV (?)
d	s	b
~ 10 MeV	~ 200 MeV	5 GeV

17 KeV Neutrino

The first reported evidence for a 17 KeV was presented by Simpson [2] in 1985 using ^3H implanted in a Si(Li) detector. This result, a proposed ~ 3% admixture of a new neutrino included to fit the observed spectrum, was criticized itself, as requiting corrections, and a series of negative results did not confirm the original observation. These included 5 experiments using ^{35}S and 1 experiment using ^{65}Ni [3]. Simpson, [4] in turn, presented a critique of these experiments pointing out how each could have missed the signal or misinterpreted their results. This confused situation was followed by a new positive result by Hime and Simpson using ^{35}S with an external Si(Li) detector, also for a tritiated Ge detector and a downward correction to the original result (by changing the screening potential used in the theory). New positive results have been presented:

(1) Sur *et al.* [5] have analyzed the β spectrum from ^{14}C using a germanium detector containing a crystal with ^{14}C. They observe an effect in the spectrum interpretable as an (1.4 $\pm$ 0.45 $\pm$ 0.14)% admixture of a neutrino having a mass of 17 $\pm$ 2 KeV.

(2) Hime and Jelly (Oxford) [6] reported a new measurement on ^{35}S using a Si (Li) detector. The endpoint for ^{35}S is ~ 167 KeV and they observe a deviation at ~ 150 MeV. They interpret this as an 8 standard deviation effect requiring an admixture of a neutrino having a mass of 17.2 $\pm$ 0.5 KeV with a mixing probability - $\sin^2\Theta = 0.0085 \pm 0.0006 \pm 0.0005$.

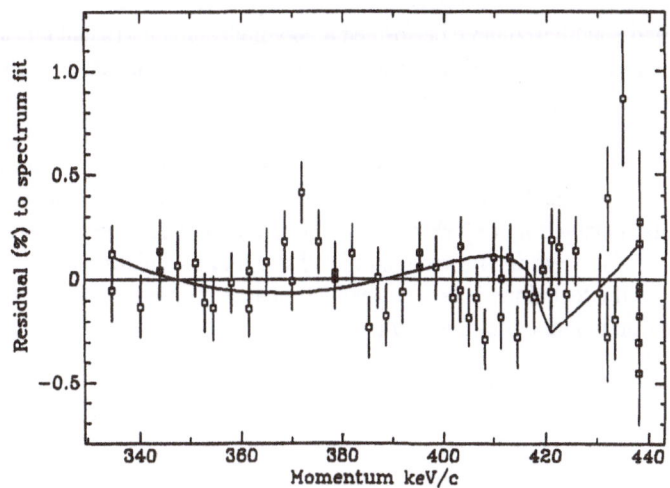

Figure 2. Recent ^{35}S magnetic spectrometer data from Seeker, *et al.* This data showing no detectable deviation from theory, but large point to point systematic errors has been used to rule out a 0.85 admixture of 17 KeV neutrino with 99.6.

Since that result was presented, a new negative result from Becket *et al.* [7] has been presented using a ^{35}S and a $\pi\sqrt{2}$ double focus magnet spectrometer. This data, shown in Figure 2, has a point to point systematic error yielding *a X^2/DOF* = 303/81 with no massive neutrino and X^2/DOF = 458/81 for 0.85% admixture of 17 KeV neutrino. In order to interpret and remove this systematic error, they correct the data for an "external error" and obtain X^2/DOF = 83.7/81 with no massive neutrino and X^2 = 124/81 for 0.85% admixture of 17 KeV neutrino. The 17 KEV neutrino has a likelihood of only 0.14%. They conclude that they rule out with 99.6% CL an 0.85% admixture of 17 KeV neutrino. However, due to the point to point scatter of the raw data, one must consider this negative result with due caution.

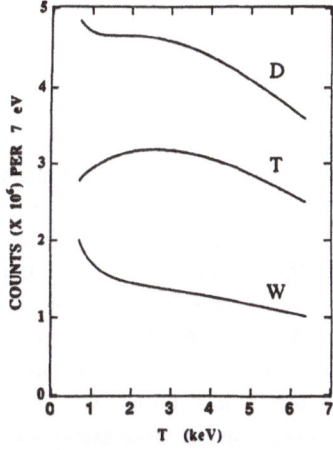

Figure 3. Bahran and Kalbfleisch high statistics tritium beta decay summed data samples, with environmental background (radioactivity and cosmic rays, not shown) subtracted. The three data samples displayed are the raw tritium data (D) and the subtracted pure tritium gas spectrum (T) obtained by subtracting the scaled wall spectrum (W).

Bahran and Kalbfleisch have performed a very high statistics (small line as 6 10[8] events) experiment on tritium β-decay. The apparatus was constructed as much as possible of hydrogen free materials (stainless steel; copper; teflon, etc.) to reduce absorption from the walls. The results are shown in Figure 3 for the observed raw spectrum (D) the scaled wall spectrum (W)and the subtracted tritium (T). The method required a large accurate subtraction which they argue can be made. They present Monte Carlo predictions of the wall that agree well with the data. Finally, Figure 4 shows the subtracted spectrum and a comparison with the expected spectrum for no added neutrino and for that with an admixture of 17 KeV ν with 99% CL, however the technique is very dependent on an accurate subtraction and a small over or under subtraction would effect the result.

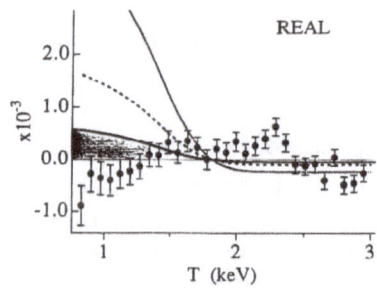

Figure 4. The deviation dK/K from the linear Kurie plot for the data of Bahran and Kalbfleisch are shown. The lower solid line indicates the 99% CL for this data and the upper curves the earlier positive results of Simpson for a 17 KeV neutrino.

Solar Neutrinos

The predictions of neutrino emission from the sun are based on the basic fusion process, in the center of the sun, starting from proton-proton interactions and evolving through a series of reactions to make $He^{4.}$ Extensive calculations of the fluxes of neutrino emission from the different reactions in the chain are shown in Figure 5 and lead to definite predictions of the detection rate in experiments at the earth's surface. The fundamental solar neutrino puzzle, which has been with us for some time is that this predicted rate is significantly higher than the observed rate from two experiments using totally different techniques [9].

The original experiment to detect solar neutrinos by Davis uses the reaction, $\sim\nu_e + {}^{37}Cl \rightarrow e^- + {}^{37}Ar$ which has a threshold = 0.814 MeV for detection of solar neutrinos. A large volume of ${}^{37}Cl$ has been placed in the Homestake mine for more than 20 years. The observed neutrino interactions above this threshold come mainly from the ${}^{8}B$ reaction. There is some theoretical uncertainty in the predictions of this rate (e.g. Bahcall and Ulrich predict 7.9 :± 2.6 SNU and Turck and Chiese predict 5.2 ± SNU). Nevertheless, both appear inconsistent with the 20 year average form Davis of 2.3 ± 0.3 SNU. There is also some possible indication of time dependence in this experiment, but it has not been confirmed, and I will not discuss it, here.

Confirmation of this observation of solar neutrinos and apparent low rate has made using the reaction, $v_e + e^- \rightarrow v_e + e^-$ in the Kamiokande 11 detector using Cerenkov techniques [10]. In this large HaO detector the signal has been observed and, using the directionality of the Cerenkov light, shown to come from the direction of the sun. For Kamiokande, the threshold $E_{th} > 7.5$ MeV gives somewhat different acceptance from the Clorine experiment, but again is mainly sensitive to ^8B neutrinos. The result is that

$$R = \frac{Experimental\ Results}{Solar\ Standard\ Model\ Predictions} = 0.46 \pm 0.05 \pm 0.06$$

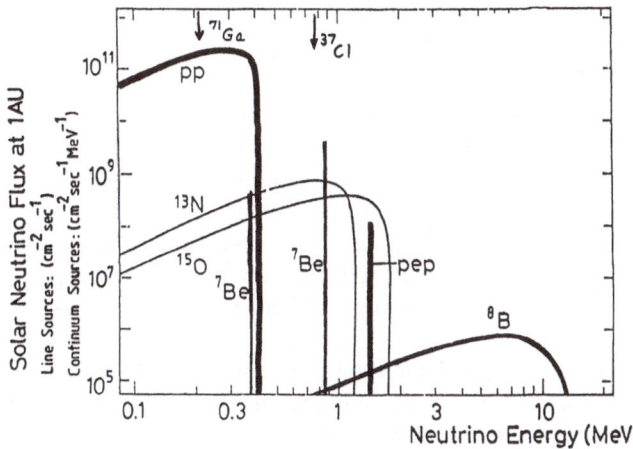

Figure 5. Solar neutrino emission spectra, with the threshold energies for ^{71}Ga and ^{37}GI indicated.

There is no indication of time dependence in the Kamiokande data.

These two results have stimulated various theoretical interpretations, including explanations involving new particle physics, like MSW oscillations (neutrino oscillations with matter effects). For some time, it has been clear that to resolve the source of the discrepancy and especially whether the effect is really due to particle physics is to measure the neutrinos from the p-p reaction. This reaction is not subject to the theoretical uncertainties due to lack of knowledge of the nuclear physics that are inherent in the ^8B channel.

The scheme for isolating the p-p reaction is to perform the equivalent of the ^{37}C1 experiment using $^{71}Ga, (v_e + ^{71}Ga \rightarrow e^- + ^{71}Ge)$, which has a threshold $E_{th} = 0.233$ MeV. Table II shows a comparison of the rates expected in ^{37}Cl and ^{71}Ga for the primary reactions and one can see the sensitivity to the p-p reaction in ^{71}Ga.

Table II. Rate in SNU's expected theoretically for ^{37}Cl and ^{71}Ga using the standard solar model.

Reaction	^{37}Cl	^{71}Ga
pp	0.0	70.8
pep	0.2	3.0
^{7}Be	1.1	34.3
^{8}B	6.1	14.0
Others	0.5	10.0
Total	7.9 ± 2.6	123^{+26}_{-17}

Two new experiments have been constructed to perform this experiment, SAGE at Baksan Laboratory and GALLEX at the Gran Sasso Laboratory. Both experiments have reported at this conference.

The SAGE experiment is a. Soviet-American collaboration using 30 tons of metallic ^{71}Ga and will eventually involve 60 tons. The ^{71}Ga must be very pure, and in particular must be free of cosmogenic ^{68}Ge, which is produced before bringing it underground. The SAGE experiment after some time, reduced these backgrounds and developed the extraction and counting techniques to begin looking for solar neutrinos. The experiment began good data taking in January 1990 and have reported on 5 months of good data. Since that time, they have been off performing various studies and have only recently resumed data taking. The procedure is to extract once per month and then use counting techniques to count Auger electrons from the ^{71}Ge K-peak (eventually they hope to also use the L-peak) over a 60 day period.

Figure 6 shows the method for identifying counts. The plot shows the rise time (ADP/Energy) vs Energy with the boxes indicating the L-peak, K-peak for Fe, and K-peak. The energy identifies the Auger electron and rise time indicates a local deposition of energy, not characteristic of backgrounds.

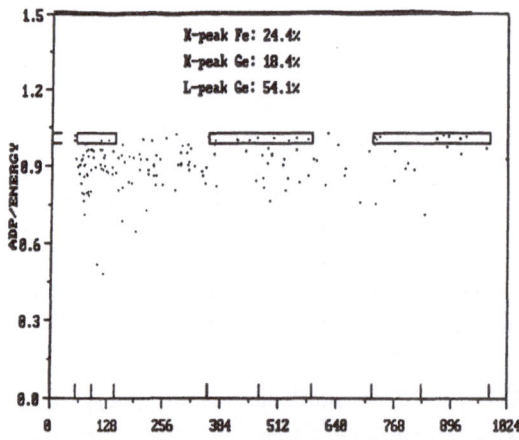

Figure 6. SAGE rise time (ADP/Energy) vs Energy with boxes indicating the L-peak, K-peak for Fe, and K-peak.

Figure 7 shows the results of a typical months extraction. A total of 8 counts were observed over 60 days with none in the first 11.5 days expected from the lifetime of Ga[71]. A detailed statistical analysis has been made for all five months, where the data after 11.5 days is used to determine the background level during the characteristic [71]Ge lifetime, and evidence for an excess signal in the first 11.5 days is sought. The results for each of the 5 months are shown in Table III along with the estimated systematic errors.

This represents the results using only the purely statistical error. The estimated systematic errors that should be included ate for counter efficiency, extraction efficiency, etc. (5 SNU at 68% CL, 14 SNU at 90% CL); time dependence of background (30 SNU at 68%, 35 SNU at 90% CL), and ADP/energy cut (10 SNU at 68% CL, 23 SNU at 90% CL).

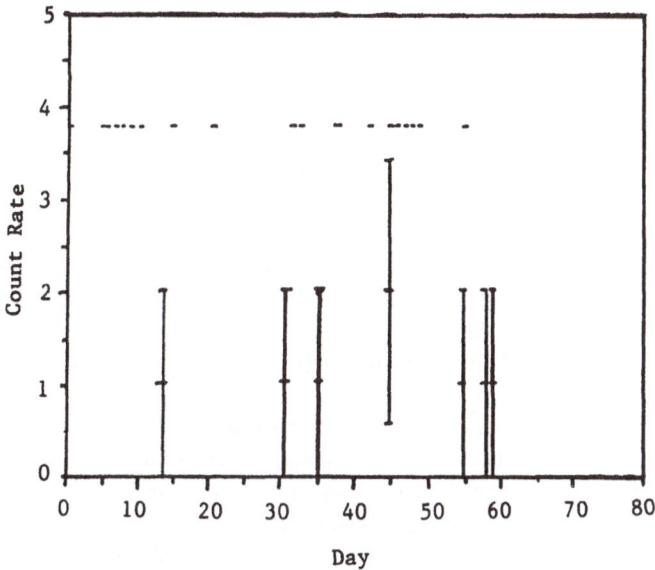

Figure 7. Counting for the extraction from January 1990 sample. 8 counts were observed over 60 days with none in the first 11.5 days expected from the lifetime of Ga[71].

Table III. SAGE results from 1990. Data taken looking for K_5peak from Ga[71] from the reaction $\nu_e + G^{71} \rightarrow e + Ge^{71}$

Extraction Date	Best Fit (SNU)	68° CL SNU	90% CL SNU
Jan 27	0	60	118
Feb 28	39	83	142
Match 29	90	175	276
April 20	0	94	174
July 24	0	148	275
Combined	20	35	60

The extraction efficiency is *assumed* to be 100% in determining the rate. This has been tested by putting a trace of ^{68}Ge into the solution and measuring the efficiency for extracting it. Although, this gives some confidence about extraction efficiency, it is best to use neutrino initiated reactions. This crucial test using neutrinos from ^{51}Cr is planned, but early tests were inconclusive and a more systematic test using an intense pure source will be required.

Assuming the extraction efficiency is 100%, the best answer is

$$\sigma \bullet \phi = 20^{+15}_{-20} \text{ (statistical)} + 32 \text{ (systematic) SNU.}$$

Restated as a limit, the (90% CL) x ϕ < 79 SNU.

Another way to Summarize the present results in terms of the simple observations is that for the sum of all five extractions done in 1990:

Total Expected Events (SSM)	=	17 atoms
Observed K-peak events	=	9 atoms
(two half lives)		
After background subtraction	=	2.6 atoms
(from longer times)		

Therefore, we can see that the background level is well below the expected signal level and using the rates at times past the lifetime of Ge^{71} (see Fig. 7) can be empirically subtracted. It appears the actual signal is significantly smaller than expectations for the SSM, even from just the p-p reaction. However, more data to obtain enough statistical significance and a direct determination of the extraction efficiency will be necessary to draw a firm conclusion from SAGE.

The second experiment, GALLEX, is being performed at the Gran Sasso laboratory in Italy. It uses 30 tons of liquid Gallium. This experiment has been installed, all extraction and counting techniques made functional, and testing was begun in June 1990. They quickly found they had a cosmogenic age background at about x20 the signal level. This background was expected to quickly dissipate as the ^{68}Ge was released, but unfortunately it did not quickly dissipate, perhaps indicated some chemical attachments keeping the ^{68}Ge. To stimulate its release the liquid was warmed from 15°C to 45°C and this rapidly brought about the release of the ^{68}Ge. They have recently reported a first result with a positive signal of about 80 SNU.

Strange Quark Matter

Witten [11] suggested in 1984 that it may be possible for certain baryon numbers to have stable forms of matter with an excess of strange quarks. If there is some stable form of quark matter then ~ 90% of the baryon number in the Universe could have been "hidden away" in such a form before the primordial nucleosynthesis. This has the attraction that it could explain the dark matter by QCD effects alone. Farhi and Jaffe [12], using a bag model, showed that for a wide range of acceptable QCD parameters, strange quark matter is stable with "a few" $\leq A \leq 10^{57}$ (black hole). However, calculations by Madsen *et al.* [13] have shown that it is not likely that such a form of matter could have been produced in stable form in the high temperature of the early Universe.

Nevertheless, it is important to look for such particles empirically. Gashow and de Rujula [14] have investigated a large number of phenomena that can be used to search for such particles. If this matter is in the range of masses from 10^{11}g < m < grams it could be detected as a rare particle among the cosmic rays incident on the earth.

At this conference, there has been a presentation from the MACRO detector at Gran Sasso for such a search. This search uses the large size of this underground detector and the abilities to measure both the velocities of slow non-relativistic particles, expected for such large masses, and the ionization properties of any incident particle. Much of the techniques developed to search for GUT monopoles are applicable for triggering on such particles and analyzing the properties of any candidate events.

The search reported from MACRO is negative and the flux limit vs β is shown in Figure 8, where the limits are shown for particle incident downward on the detector with a lower mass capable of penetrating 3000 mwe. For very heavy masses, e.g. m < 0.1 g these particles can penetrate the entire earth and upward particles are then used to set an improved limit.

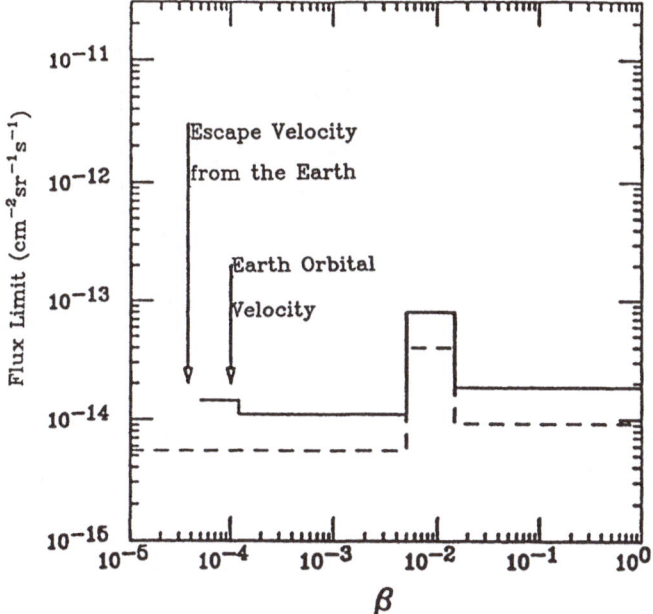

Figure 8. Nuclearite flux limit as a function $\beta = v/c$ from the combined several MACRO searches. The solid line is for nuclearites heavier than 10^{-11}g but smaller than 0.1g (able to penetrate to the MACRO depth but not able to penetrate the whole Earth). The dashed line is for nuclearites that can penetrate the whole Earth (greater than 0.1g). The velocity coverage of MACRO detector includes the Earth orbital velocity around the Sun and extends to near the escape velocity from the Earth.

This limit, along with others using track-etch and other techniques for a wide range of masses are shown in Figure 9.

High Energy Particle Astrophysics

Particle emission from astronomical objects is of fundamental interest in astrophysics and such emission, especially of very high energy particles could provide important particle physics. Two possible sources of such particles are neutron stars and binary systems, consisting of accreting compact objects and a main sequence star.

There are three known particles (γ, ν, η) that, if emitted, would maintain the directionality because they are neutral. Of the three, the lifetime of the neutron is too short to survive the great distances, even at the highest energy.

Observations of γ-rays have been actively pursued to higher energies (e.g. TeV/PeV regime) in recent years. Detectors with possible sensitivity to sources emitting high energy neutrinos are also being developed. Finally, high energy γ's or some other emitted neutral particles coming from a point source could interact at the top of the atmosphere and produce high energy muons that approximately maintain their directionality.

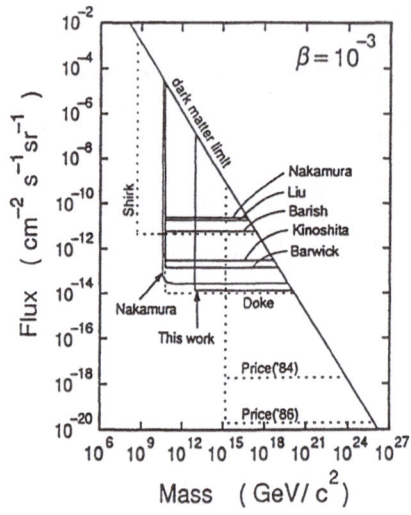

Figure 9. Compiled nuclearite flux limits as a function of the mass.

Gamma Ray Astronomy

In the field of high energy gamma ray astronomy there have been two new developments: first, the solid observation of a TeV source (Crab Nebula) for the first time, which now presents the possibility of having a "standard candle" for experiments probing to yet higher energies. The second, is the development of powerful new extensive air shower arrays (e.g. CASA, Cygnus, EASTOP, etc.). These new highly instrumented arrays offer the promise of resolving the interesting questions concerning the existence or nonexistence of γ-ray sources in the PeV range. For sources at lower energy, including the Crab Nebula, it is believed that the γ-rays are a result of electromagnetic radiation. For energies of PeV this is no longer possible and the production of such radiation, if indeed suck sources exist, is almost certainly the result of hadronlc collisions, producing pions and therefore gamma rays that result from the decay of neutral pions. This would imply that there exists some powerful accelerating mechanism for protons in these star systems. If so, there are also charge pions produced in the collision and the decay, will yield high energy neutrinos. Thus, the complementary experiments searching for high energy neutrinos coming from these sources could also be a viable probe to search for such astronomical objects.

The search for high energy gamma ray sources has been of particular interest since reports of a possible signal by the Kiel [17] and Haveford Park [18] groups coming from Cygnns Confirmation of this signal has been plagued by conflicting reports from subsequent experiments, compounded by the added fundamental problem of trying to compare results at different times from such sources because of their episodic nature even at lower energies. Observations from other sources (e.g. Hercules X-1) have also been reported, including some bursts of signals, again, however contradictory evidence leaves the situation confused.

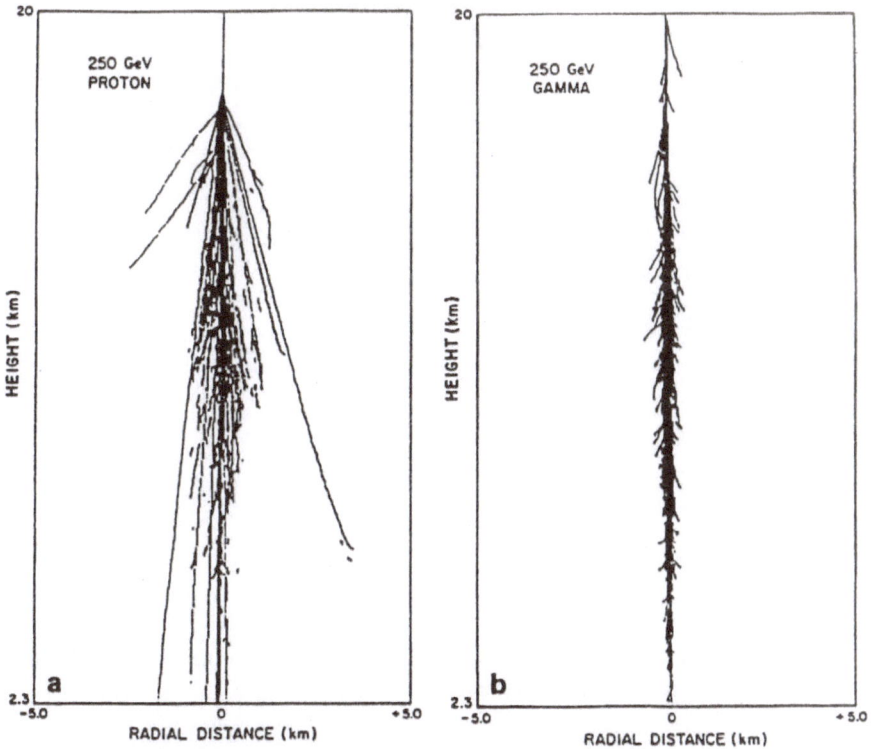

Figure 10. The comparison of the development of a high energy proton initialed shower (a) and photon initiated shower (b) in the atmosphere.

For studies beyond the TeV region, rather than $\check{C}$ techniques, extensive air shower arrays are used. The extensive air shower arrays used for these studies have thresholds well above 1 TeV, depending on the characteristics of the array, the altitude, triggering, etc. To pursue these studies in a systematic manner with high sensitivity, it is important to "tie-on" to the observation at lower energy. This is difficult due to the threshold, usually ~10 TeV, due to altitude and granularity of the arrays.

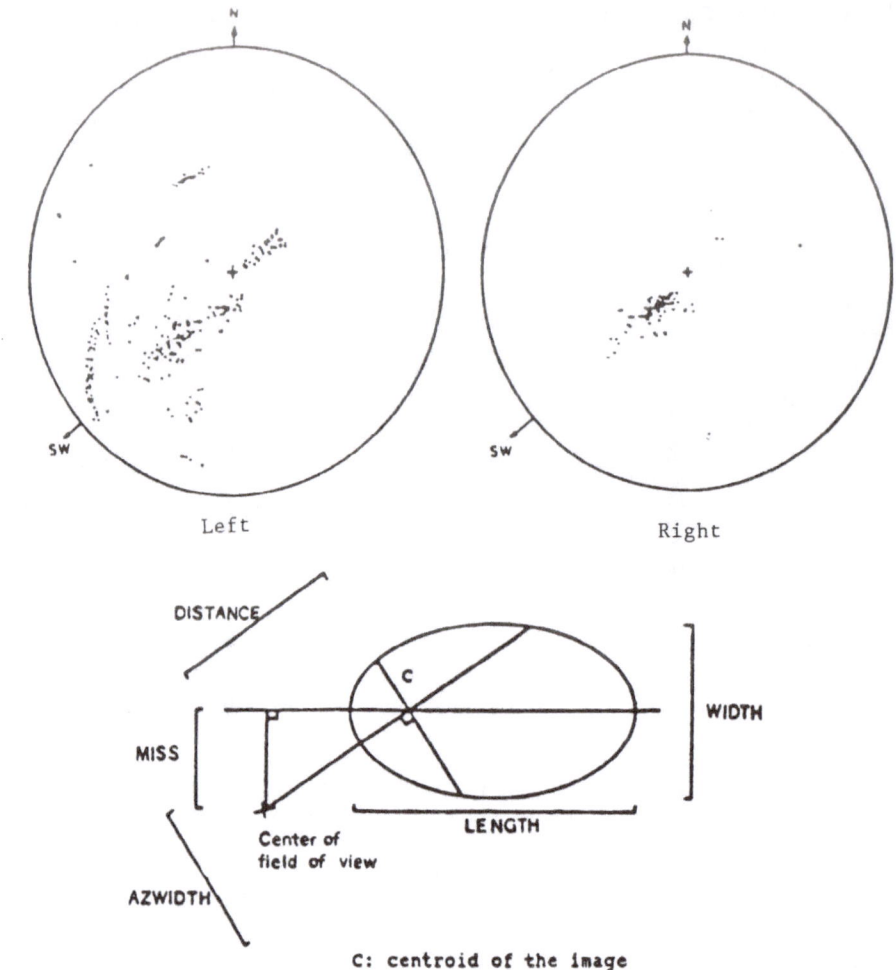

Figure 11. Comparison of the images in a Cerenkov imaging detector for proton (left) and (right) induced showers. The lower illustrates the azwidth parameter used to distinguish.

The recent observation of a strong and steady signal near 1 TeV from the Crab Nebula using the Whipple Observatory are extremely important in this sense. The Whipple Observatory is a large imaging Cerenkov telescope having excellent angular resolution and the ability to separate gamma induced events from proton induced events. The principles are illustrated in Figures 10 and 11. Primary cosmic rays are dominantly protons near 1 TeV. These protons interact and create a hadronic shower us it propagates through the atmosphere. This shower is broad due to the $< P_t > \sim 300$ MeV characteristic of hadronic collisions. In contrast a photon induced shower, which is purely electromagnetic in nature, is characteristically much narrower. The resulting images of these two different showers in the detected image is shown in Figure 11. A focused narrow image results from the γ-induced event, while a dispersed non-uniform image results from the proton. These features are characterized by determining the width, length, azwidth, etc. In particular, the azwidth parameter can be used to separate γ-induced showers from hadronic showers as shown in the Monte Carlo study in Figure 12.

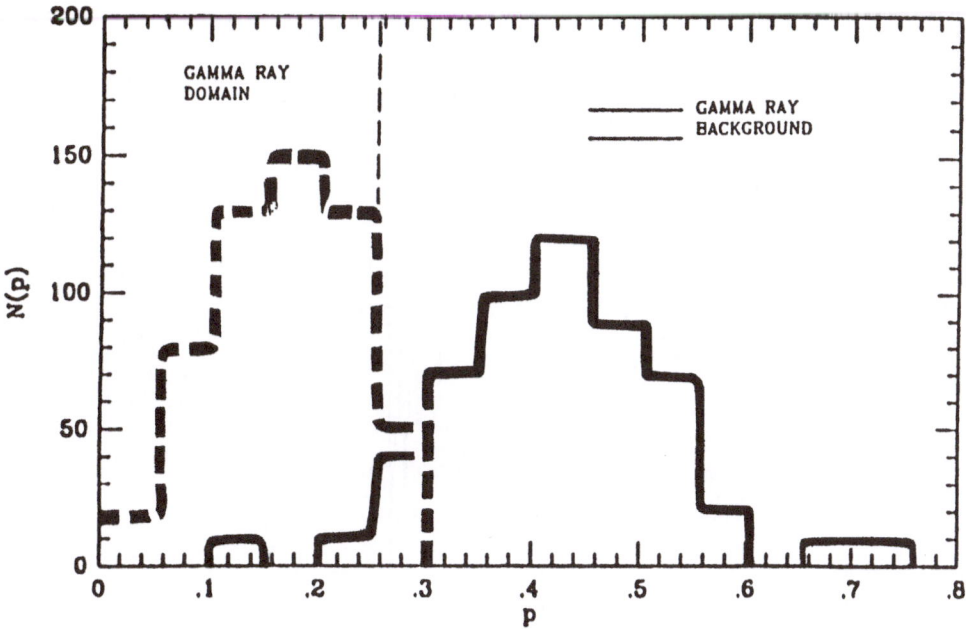

Figure 12. Separation of proton (solid) and γ (dashed) induced events using the azwidth parameter.

In the case of actual observations, the hadronic signal dominates and a subtraction is necessary. The data, using the Whipple Observatory, for the Crab Nebula is shown in Figure 13. Observations are made both in the direction of the Crab Nebula and in control background regions. Upon subtraction, a signal is observed in the expected region for γ-induced events. This signal has been steadily observed and the total significance is more than 20 σ.

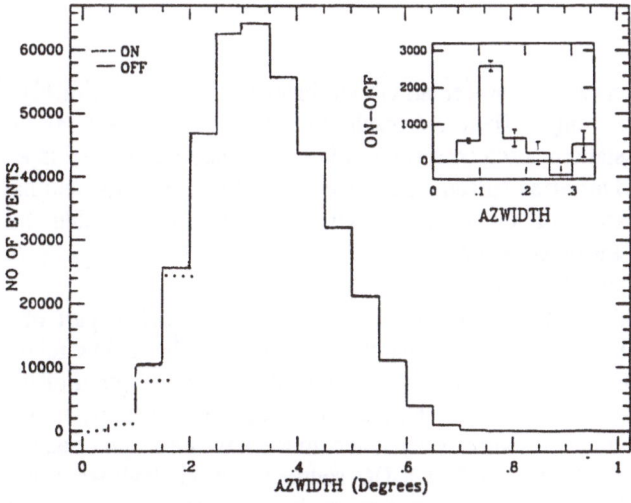

Figure 13. Observation of Crab Nebula in Whipple Observatory data. Inset shows excess due to the signal.

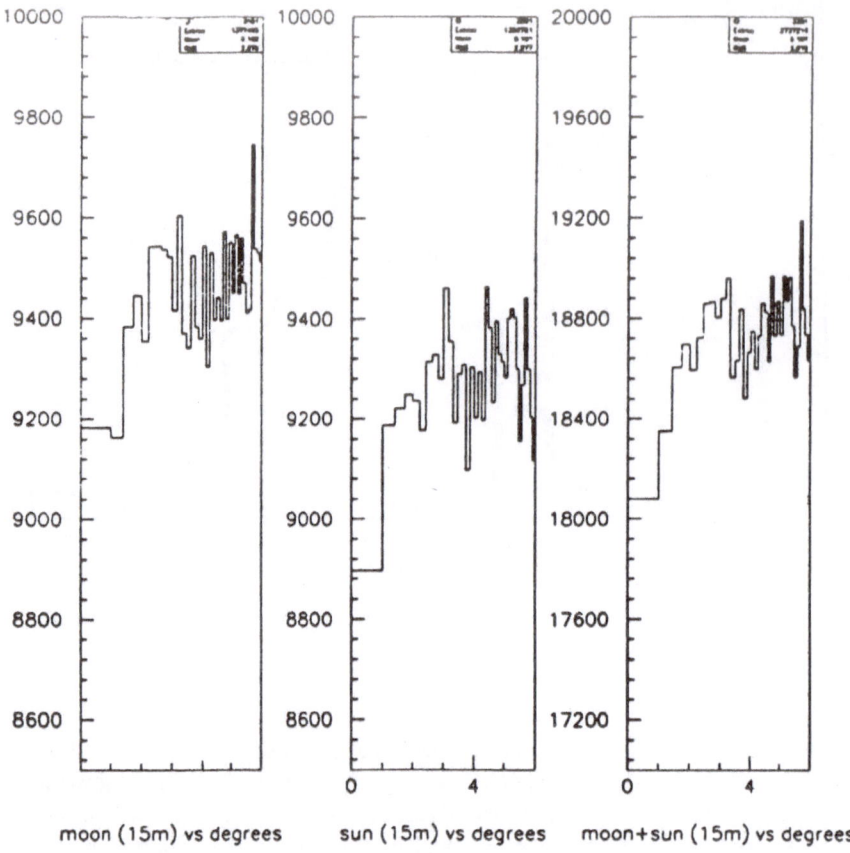

moon (15m) vs degrees   sun (15m) vs degrees   moon+sun (15m) vs degrees

Figure 14. The shadow of the Sun and Moon in CASA demonstrating pointing ability and angular resolution.

New extensive air shower arrays are beginning searches with high sensitivity for γ sources in the PeV range. There are results from the CASA/MIA array in Utah. This array consists of 1089 stations, each consisting of four scintillators separated in a square array with 15m separation. This array surrounds the Utah Fly's Eye I detector and has muon detectors buffed in patches with 64 counters/each at 16 locations within the detector. The characteristics of the detector are *Area* = 25 $10^4 m^2$, $\Delta 0 = 1°$, $E_{thresh} \sim 100\ TeV$, and has a total muon coverage of 2500m^2.

A total of 300 10^6 events have already been analyzed in part of the array already operational. The pointing ability and angular resolution are checked by following the direction of the sun and moon and observing the 'shadow' from the absorption of high energy cosmic rays coming from that direction. Figure 14 shows data demonstrating this shadow, an impressive check on the resolutions and ability to track the direction of an object in the sky.

A search for Cygnus X-3 as a DO source using this high statistics data is shown in Figure 15a, which shows no evidence of a signal. Potentially, better sensitivity can be

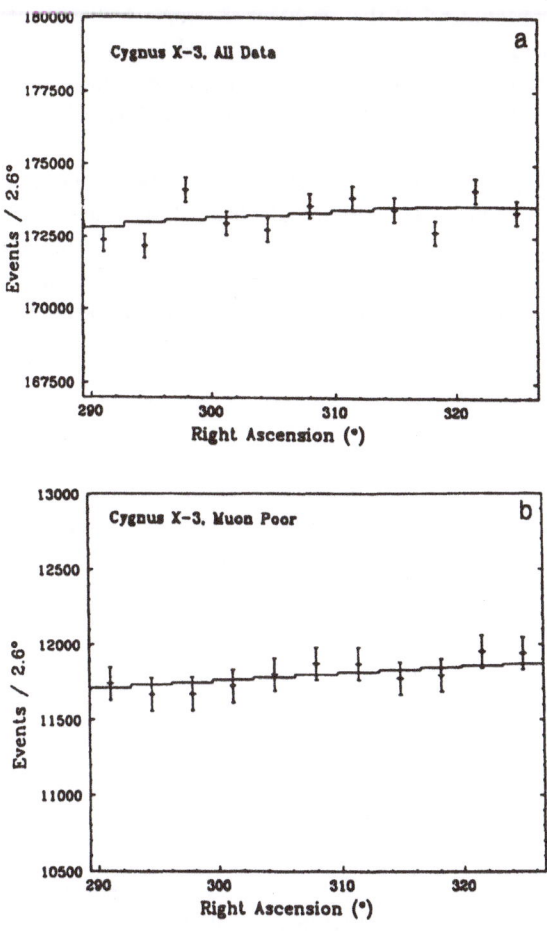

Figure 15. CASA search for DC signal from Cygnus X-3 is shown (a) and for muon poor events only (b). No signal is seen.

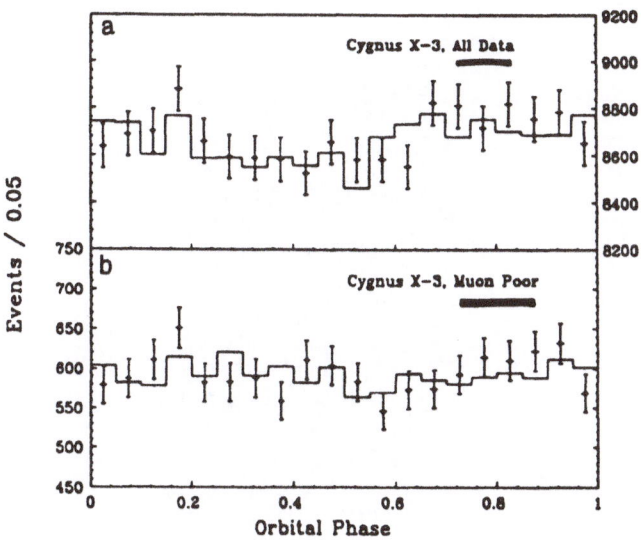

Figure 16. Periodic search for Cyguus X-3 from CASA. No signal is seen.

obtained from the 'muon content' of the observed showers. At these energies, the expected yield of muons from hadronic induced events is expected to be ~ 30 times that of γ-induced events. An analysis requiring the data to be 'muon-poor' rejects hadronic events, while retaining γ events. Such an analysis is shown in Figure 15b. Again, there is no indication of a signal.

There is one further handle to possibly increase sensitivity. Cygnus X-3 is a binary source with a period of ~ 4.8 hours. Earlier reported observations showed a signal only over a portion of this 4.8 hour period, possibly reflecting the orbiting of one body around the other and the resulting pointing of the emission. The CASA/MIA data has also been analyzed looking for this periodic structure as shown in Figure 16. Again, the results are negative.

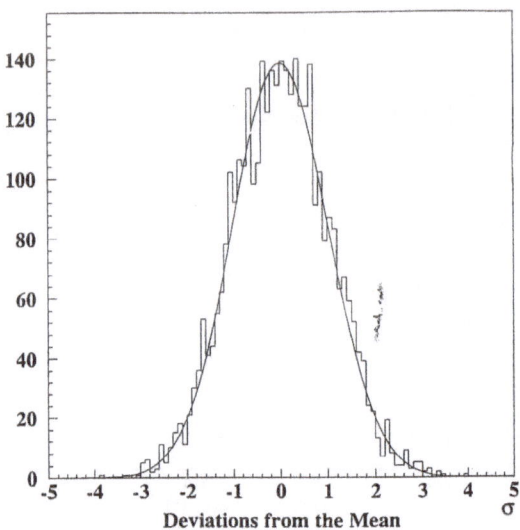

Figure 17. All sky search of muons in MACRO. It is consistent with a random distribution with no anomalous sources.

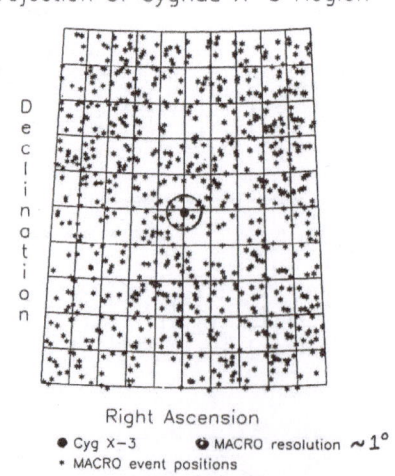

Figure 18. Events in MACRO in the vicinity of Cygnus X-3. No clustering is observed.

Although the limits use well below previously reported observations, it is not possible to draw definitive conclusions, as mentioned above, due to the fact that Cygnus is an episodic source of radiation, and is currently quiet, even at lower energies.

Muon Astronomy

A complementary technique to investigate ultra high energy neutral radiation from astrophysical sources is to detect high energy muons deep underground. The muons, created by a cascade at the top of the atmosphere will approximately point back to the sources, due to the limited transverse momentum of both the produced pions and the decay muons. The pointing accuracy depends on the resolution of the detector, multiple coulomb scattering, energy, etc. For a tracking detector, the overall intrinsic accuracy is ~ 1°. Since γ-ray induced showers are expected to be muon-poor, relative to proton induced showers, such a technique is probably only sensitive to a new type neutral radiation, that is, unless γ's have anomalous interactions at very high energies.

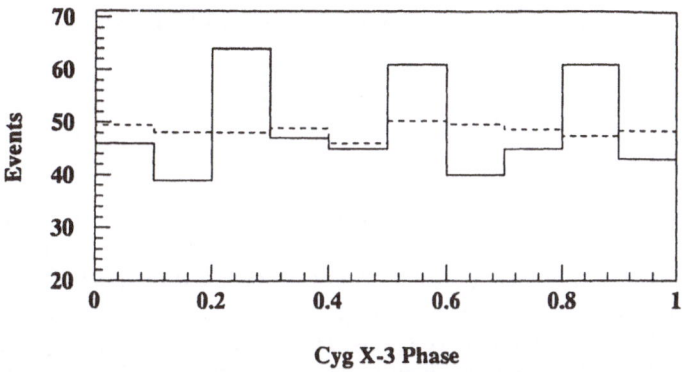

Figure 19. MACRO data analyzed for periodic signal from Cyguus X-3. No signal is observed.

At this conference, we have had a report on muon astronomy results from the MACRO detector at Gran Sasso. An all sky search (e.g. Northern Hemisphere) has been conducted for more than 1 year and sees no anomalous DC source are observed. The data, shown in Figure 17 shows no clustering of events beyond expectations of a random distribution. The sky has been divided into equal solid angle bins $\Delta\Omega$ ($\Delta d = 3.0°$, $\Delta \sin\gamma = 0.04$) and 3000 bins were searched.

The specific region of Cygnus X-3 has been studied in detail, and Figure 18 shows the distribution of events inside of a 10° x 10° region around Cygnus. No indication of clustering near Cygnus has been observed. The data has been analyzed for a periodic signal and the results, using a 1.5° half-angle cone around Cyguus, are shown in Figure 19 along with the expected back-ground. The arrival time of each event is corrected to account for the Earth's motion with respect to the solar system barycenter. No signal is observed $> 2\sigma$ above background. Finally, the MACRO result, along with that of other underground detectors is shown in Figure 20. The MACRO limit appears in conflict with the Soudan I and NUSEX observations, however, again, I must warn that the episodic nature of Cygnus X-3 make such a conclusion difficult.

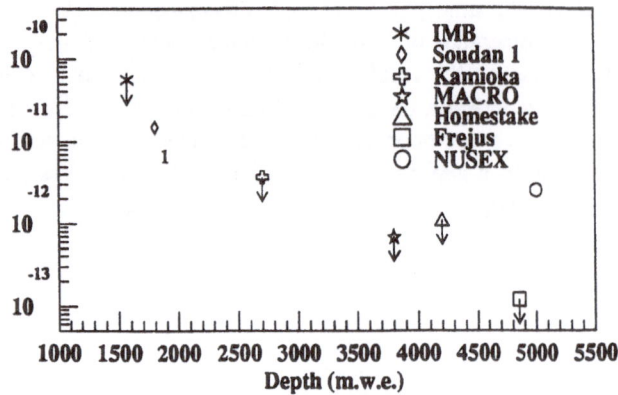

Figure 20. Summary of muon data for Cygnus X-3, Soudan 1 and NUSEX report a signal and all others, including MACRO, are limits.

Although Cygnus X-3 has been in a generally quiet period, recent radio bursts have been reported. In particular, for a burst period January 20 and 23, 1991 some indication of a correlated muon signal in Soudan (~ 1800 mwe) has been reported [19]. The corresponding Soudan flux is $\phi \sim 8 \; 1^{-10}$ cm^{-2}s^{-1}. However, both the CASA/MIA array on the Earth's surface and MACRO at 3000 mwe have reported at this conference that they see no evidence for such a burst. For CASA an upper limit $< 1.3 \; 10^{-12}$cm^{-2}s^{-1} has been reported for events in the extensive air shower array, while MACRO has reported an upper limit $3 \; 10^{-11}$ cm^{-2}s^{-1} for muons at depth 3200 mwe. These results are barely comparable with Soudan, if the spectrum falls at $E^{-2.8}$ and MACRO is inconsistent if the spectrum were flatter.

Neutrino Astronomy

Projects are under construction aimed at the detection of high energy neutrinos from astronomical print sources. Already operational, the MACRO detector is ~ 1000m^2 and will begin searching for neutrino point sources within the next year with the full detector. Although this detector has both good background rejection and pointing ability, most flux estimates for candidate sources indicate it is either marginal or too small to observe such sources.

There are two new projects to reach much larger areas being constructed. Both detectors use the principal of using a Cerenkov array deep under water. They will study upward muons resulting from the interactions of neutrinos coming from the other side of the earth and interacting below the detector. One array will be in Lake Balkal in Siberia and the other, DUMAND, off the Island of Hawaii. The plans for the DUMAND array were reported at this conference. It will consist of 9 strings of 24 phototubes each, covering an area of 100m diameter and 200m high. The first three strings are planned for 1992 and the completion of the array is expected in 1993. It is at a depth of ~ 4500m and is read out through a string bottom controller and a cable to shore. The design resolution $\sigma_0 \sim 1°$ and for a 15 GeV μ, the number of detectors recording a signal is $< \eta > \sim 13$.

This new generation of neutrino detectors along with the enlarged extensive air shower arrays may help determine existence of high energy sources.

Detection Of Supernova Neutrinos In Macro

The detection of neutrinos of stellar origin by underground experiments has opened a new field of observational astronomy - low energy neutrino astrophysics. Neutrinos represent the first non-electromagnetic probe used for astronomical investigation and have the feature that they probe the physics from deep within stellar interiors. Solar neutrinos provide information on the fundamental burning or fusion processes in the sun, whereas supernova neutrinos give information regarding the evolutionary stages of very massive stars.

The $\bar{\nu}_e$ burst from SN1987A detected by the KAMIOKANDE-II and IMB experiments[1] dramatically verified the predicted emission of neutrinos from a collapsing star. Despite the large fiducial volumes of these water Cerenkov experiments the number of neutrino events was small due to the large distance (50 kpc) to the Large Magellanic Cloud. A future collapse, nearer the center of our galaxy, (~ 8 kpc) will have much larger event rates which will provide detailed information bearing directly on both astrophysical models of stellar collapse and elementary particle physics.

In this lecture, I discuss MACRO[2], a large liquid scintillator and streamer tube experiment, which is sufficiently complete now to operate as a new Galactic supernova observatory. Liquid scintillator experiments, though smaller volume, have complementary capabilities to water Cerenkov detectors, including superior energy resolution, potentially lower energy thresholds, and the ability to detect an additional signature from the produced neutron capture.

The sensitive mass of MACRO will be over 600 tons when the complete detector is operational in 1993. A final design mass of 1000 tons could be reached in a future extension, still under study. The data I present here were collected by the first operational 45 tons of target, during approximately 2.5 years of operation, and represent a proof in principle and illustration of the capabilities of the full detector, as well as a negative search of some significance.

Supernova Theory And Neutrino Burst Production

The evolution of a massive star, including its final gravitational collapse, has been extensively simulated with theoretical models.[23] The detection of the SN1987A dramatically confirmed the emission of $\bar{\nu}_e$ at a level near expectations. The results from KAMIOKANDE II for SN1987A had eight out of the eleven observed events occurring in a 2 s interval.

To characterize the expected neutrino flux $\Phi(E_\nu)$ we adopt a simplified constant temperature thermal neutrino spectrum of the Fermi-Dirac type for the cooling stage:

$$\Phi(E_\nu) = \frac{A_0 E_\nu^2}{1 + \exp[E_\nu / T]} \tag{1}$$

The values of the parameters have been taken from the statistical analysis of SN1987A data by Bludman and Schinder. They give temperatures of 3.3 MeV for ν_e and $\overline{\nu}_e$ and 6.6 MeV for $\nu_{\mu,\tau}$.

The constants A_0 for the different neutrino types are related by $A_0(\nu_{\mu,\tau}) = A_0(\nu_e)/16$ with $A_0(\nu_e) = 5.21 \times 10^{55}$ MeV^{-3}. The total energy radiated through neutrinos of all types is $E = 6 \times 5.862\, A_0\, T^4 \approx 3.4 \times 10^{53}$ ergs.

Neutrino Detection In Macro

MACRO is located in Hall B of the Gran Sasso Laboratory (120 km east of Rome) at a minimum depth of 3200 m water equivalent. At this depth, the flux of cosmic ray muons reaching the experiment is reduced by a factor greater than 10^6 relative to the surface. The detector is composed of six supermodules and when completed will have dimensions 72 m x 12 m x 9 m. Its large area makes MACRO well suited for: the study of high energy downward-going muons produced by primary cosmic rays, interacting at the top of the atmosphere; upward-going muons originating in high energy neutrino interactions in the rock below; as well as for its primary goal, the search for rare penetrating heavy particles where it has sensitivity to the magnetic monopole flux well beyond the Parker bound. Finally, it has a large liquid scintillator mass and thus is an important new low energy neutrino observatory for the study of supernovæ.

MACRO detects neutrinos through their interaction with liquid scintillator, divided into many individual counters. A schematic drawing of the completed lower half of MACRO is shown in Fig. 21.

Here I shall examine the capabilities of the detector for stellar collapse neutrino detection. The detector is divided into six supermodules, where each supermodule contains three layers of 16 horizontal scintillation counters, each with an active volume of 73.2 x 19 x 1120 cm^3; each vertical wall is covered with 14 counters, each with an active volume of 21.7 x 46.2 x 1107 cm^3. The liquid scintillator consists of 96.4% mineral oil, $\approx$ 3.6% pseudocumene and small concentrations of scintillation fluors; it has density of 0.86 g/cm^3 and a refractive index of 1.48. The number densities of ^{12}C and ^{1}H nuclei are 3.7 x 10^{22} cm^{-3} and 7.3 x 10^{22} cm^{-3} respectively. The inside of the counters are lined with teflon (n = 1.33) for total internal reflection and there is an air gap above the liquid. Light is collected at each end of the counters by two photomultiplyer tubes (PMT) and focusing mirrors, (one PMT at each end for vertical counters) which are assembled in end compartments filled with mineral oil and separated from the scintillator by clear PVC windows.

To calculate the response of the detector to neutrinos from a stellar collapse we have used neutrino cross sections on electrons and free protons from Bahcall[24], charged current (CC) and neutral current (NC) neutrino cross sections on ^{12}C and in particular those for the superallowed (isovector, axial vector) transition to the (1^+, T = 1) 15.1 MeV level have been calculated by several authors[26]; and the recent calculation of this transition[27] based on data from inelastic scattering of neutrons on ^{12}C[28,29]. Recent measurements by the KARMEN group[30] confirm these calculations. Table I gives the number of events expected from a supernova at the Galactic center (8.5 kpc) in the complete MACRO detector containing 600 tons of liquid scintillator and with electron energy threshold of 7 MeV.

The dominant detectable reaction is $\bar{\nu}_e + p \rightarrow n + e^+$, where the positron energy provides the primary trigger for the detector. This reaction is followed, after neutron moderation by neutron capture $n + p \rightarrow \gamma + d$ with $E_\gamma = 2.2$ MeV. The average neutron moderation time is 10 μsec, and the average capture time is 180 μsec. The detection of the 2.2 MeV γ-rays as an additional signature of this neutrino reaction is possible in MACRO due to the relatively large number of observed photoelectrons (~20/MeV).

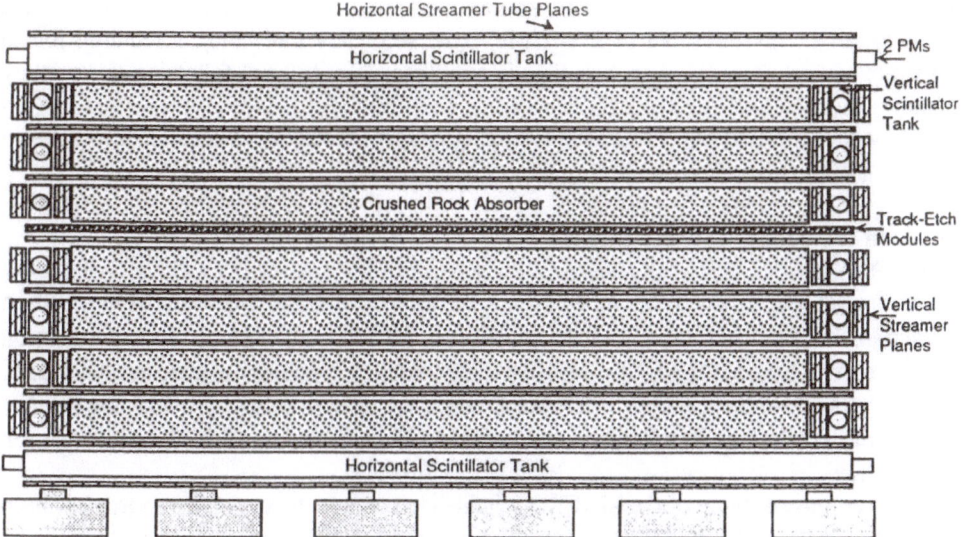

Figure 21. A schematic drawing of the lower half of one supermodule of the MACRO experiment.

The Gravitational Collapse Triggers and Data Acquisition Systems

The problem of triggering on low energy events induced by a supernova neutrino burst is simplified by the pulsed character of the supernova. The only background are clusters to statistical fluctuations in the random backgrounds, dominantly due to cosmic ray muons, and from natural radioactive decay from the environment surrounding the detectors.

Cosmic ray muons are observed as events with an average energy loss of about 40 MeV with a rate of 2 mHz scintillation counter. Most are rejected by their large energy release, (~ 35MeV) in a given scintillation counter, and position matching with tracks reconstructed in the streamer tube system. The rate of cosmic rays not eliminated by these signatures was less than 0.1 mH$_Z$ scintillation detector.

The background from natural radioactivity originates primarily from the decay of radioactive isotopes present in the Gran Sasso rock, in the concrete used in the laboratory, and in the materials used within the experiment[25]. The gamma rays emitted by the radioactive isotopes of the ^{238}U and ^{232}Th decay chains or by ^{40}K are detected in the scintillation counters after being partially degraded by preceding interactions. Neutrons are a further source of γ radiation, with energies extending up to nuclear binding energies. Neutrons can be captured both within the scintillator and in other materials interspersed within the detector. In a single scintillation counter, the background rates are 5 kHz for energy deposition greater than 1 MeV and 1 Hz for energy deposition greater than 4 MeV.

The large background due to radioactivity is rejected using special trigger electronics, the rejection of radioactivity over a given energy threshold requires a fast energy reconstruction using the pulse height and timing information from both sides of the counter. The trigger circuits for stellar gravitational collapse detection use a primary energy threshold in the range of 5-10 MeV. To detect the 2.2 MeV photon from n-capture on protons with reasonable efficiency, an energy threshold of E $\approx$ 1 MeV (the secondary threshold) is required for an interval of several times the n-capture time in scintillator ($\tau \approx 180$ μs) following this primary trigger.

For the neutral current (NC) and elastic scattering (ES) reactions, ν_x is equivalent to the sum of all ν and $\bar{\nu}$ neutrino flavors.

Detection of Neutron Capture Gamma Rays

A special effort has been in MACRO to detect both the positron from neutrino interaction, $\bar{\nu}_e + p \rightarrow e^+ + n$, and the subsequent 2.2 MeV gamma ray from the neutron capture reaction $n + p \rightarrow d + \gamma$. To test this ability, Am/Be, a neutron emitter has been used. First, the γ-ray from the 4.4 MeV ^{12}C* level, which we refer to as γ_4, provides the primary trigger, which causes the thresholds to be lowered to allow detection of the 2.2 MeV neutron capture γ-ray.

The response of three of the MACRO scintillation counters to an Am/Be source has been studied in detail. The full energy spectrum observed agrees well with a Monte Carlo simulation (see Fig. 22). The numbers of detected γ-rays from both the ^{12}C* cascade (4.4 MeV) and neutron capture (2.2 MeV) agree within 10% with the values predicted.[33] Furthermore, the distribution of the time between the primary trigger and the secondary one (1 < E < 3 MeV) is exponential with a characteristic time of 180 μs, as expected for neutron capture. With these data we estimate that the efficiency for detecting the neutron capture

Table IV. Number of Events from a Stellar Collapse at the Galactic Center (8.5 kpc) in MACRO (600 tons of Liquid Scintillator) E^e_{th}= 7 MeV

Current Type	Reaction Type	Neutronization Burst	Cooling Stage
CC	$\bar{\nu}_e + p \rightarrow n + e^+$	n/a	182
NC	$\nu_x + {}^{12}C \rightarrow \nu_x$ $+ {}^{12}C^*(15MeV)$	0	9
ES	$\nu_x + e \rightarrow \nu_x + e$	0	2

following a primary $\bar{\nu}_e$ event in the same counter (mimicked by a primary γ_4-ray from the Am/Be source) is about 25% for a secondary energy threshold of 1 MeV.

The final handle used to detect the late neutron signal is the spatial correlation between the primary $(\bar{\nu}_e + p \rightarrow n + e^+)$ signal and the secondary n-absorption signal. The ~ 1 MeV resolution in longitudinal position along the 12 m long scintillation counter for the n-capture γ-rays from the (uncollimated) Am/Be source measurements is $\sigma_Z < 1$ m. When position correlation, time correlation, and energy cuts are applied, signal/background is expected to be ≈ 2. This capability of detecting neutron capture γ-rays with a secondary threshold of 1 MeV is now being implemented.

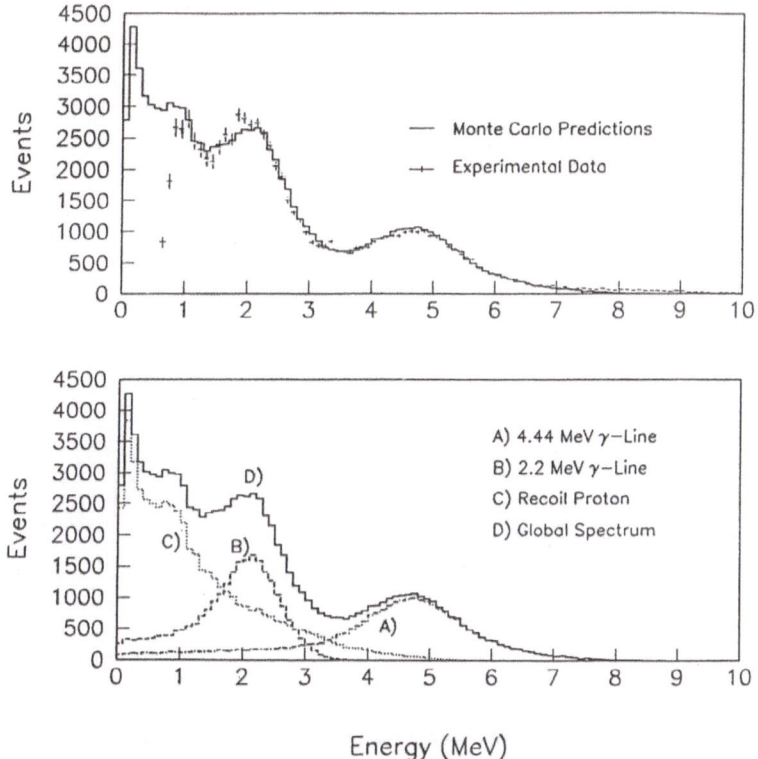

Figure 22. The camparison between experimental data and Monte Carlo calculations for the spectrum induced by an Am/Be source in a MACRO counter. The lower figures show Monte Carlo predictions for the separate components.

Determination of the Absolute Energy Scale

Cosmic ray muons that pass through the detector provide a calibration point for high energizer (~ 34 MeV energy loss for a vertical muon). The trajectory of muons crossing the scintillation counters can be reconstructed from the information obtained by the scintillation counter system alone or, with higher accuracy, by the streamer tube tracking system. The muon event rate is 2 m^{-2}h^{-1}, thus the energy scale of each counter can be determined to approximately 5% accuracy in a few days of running. Muon events are also used to measure the response vs. position in each counter.

In addition, we obtain a calibration at low energies from the Gran Sasso rock (and concrete) which contains ^{208}Tl and emits a 2.614 MeV γ-ray. When measured with a large NaI detector, the differential energy spectrum from background γ-rays in the Gran Sasso experimental halls appears as a rapidly decreasing continuum on which the ^{208}Tl γ-ray line is clearly visible (see Fig. 23). Large liquid scintillation counters have poorer energy resolution

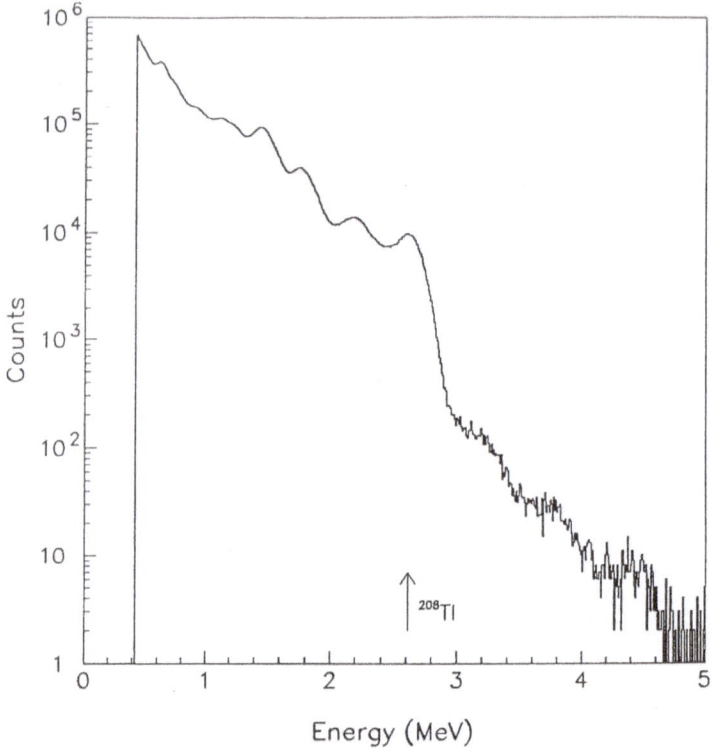

Figure 23. The background radioactivity energy spectrum at the experimental site measured with a NaI detector.

because of a lower number of photoelectrons per event and because of position dependent light attenuation effects. The position of an event is reconstructed using the difference in light transit time to the two ends of the counter, and the position compensated spectrum shows a distinct slope change at the ^{208}Tl line (see Fig. 24).

Using this line we are able to perform an energy calibration for each of the supermodules within approximately ±10% without altering the data acquisition of the rest of the apparatus. This line thus provides a fast and valuable reference for setting primary and secondary thresholds for the stellar gravitational collapse data acquisition.

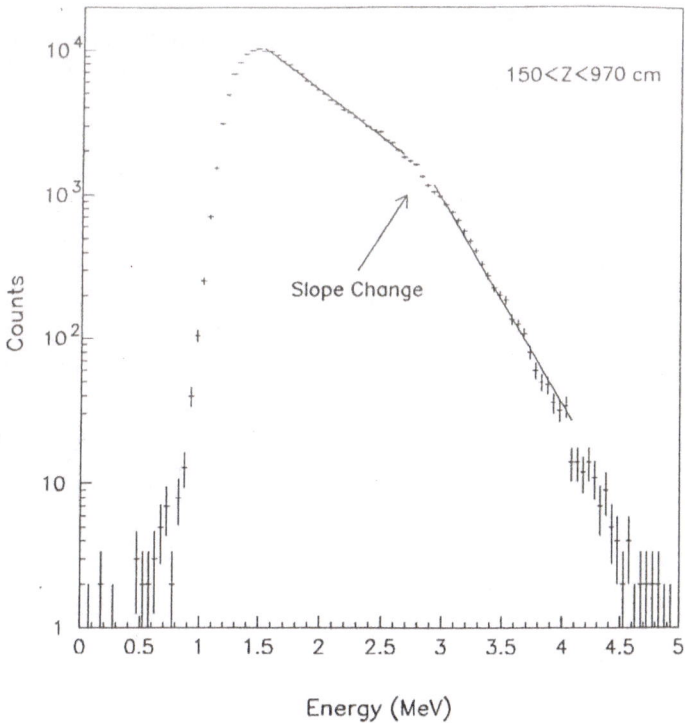

Figure 24. The background radioactivity energy spectrum measured with a MACRO scintillation counter after compensation for position dependent effects.

Analysis Methods Used in the Search for a Stellar Gravitational Collapse $\bar{\nu}_e$ Burst

The Gran Sasso Laboratory time standard is a commercial (ESAT RAD100) rubidium atomic clock. The 1 kHz master clock signal is distributed to the various underground experiments by fiber optic cables.

It is expected that in a core collapse supernova, $\bar{\nu}_e$ burst detection should occur several hours before the detection of light emission. For that reason, we are implementing a "fast alarm system" to signal possible SN candidates by a fast on-line analysis, and then to rapidly analyze the detailed information.

A detailed search has been performed for a stellar collapse during the 2.5 year period beginning in October, 1989 both to prove the techniques and watch for a burst.

A search for event clusters has been performed by applying software energy cuts (E = 7, 8, 9, 10 and 12 MeV) and burst durations from 62.5 ms to 32 s. Since sensitivity to $\bar{\nu}_e$ bursts depends on the integral background radioactivity, we present data here for an energy cut at 10 MeV. This energy cut results in a good signal to noise ratio for 45 tons of liquid scintillator for a Galactic supernova. As larger masses of liquid scintillator become active and the signal to noise ratio is improved, the software energy threshold can be reduced. For the full period, Fig. 25 shows the number of clusters vs. the burst duration for several cluster multiplicities.

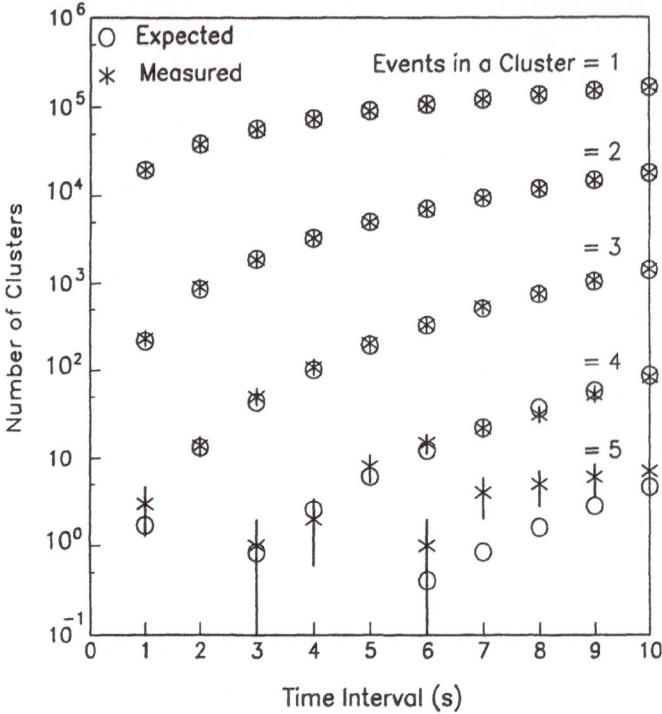

Figure 25. Number of background event clusters vs. time interval for multiplicities 1, 2, 3, 4 and 5. Data (stars) are compared with expectations (circles)

No evidence of a supernova is observed as the agreement with Poisson expectations using our measured random event rate is good.

The sensitivity of this partial detector to stellar collapse can be seen more clearly in Fig. 26. This plot shows the allowable background radioactivity rate defined by specified probabilities (10^{-1}, 10^{-3} and 10^{-5}) of having a false supernova detection in ten years for a specified supernova burst size (number of primary events) and durations (2 sec). The average background rate is ≈ 15 mHz for an energy cut of 10 MeV. For reference, using this energy cut, the expected number of detected primary events is 11 for a supernova at the Galactic center.

A supernova search involves a tradeoff between signal and noise. Increasing the time interval beyond a few seconds will capture most of the signal at the expense of increased background. Reducing the time interval will minimize the background, but the signal will be reduced by an efficiency factor. For SN1987A as seen by the KAMIOKANDE-II detector, this efficiency would be 73% (8 out of 11 events) for a two-second interval. For a similar event occurring at the Galactic center, we would expect to see 8 events in 2 s with 45 tons of liquid scintillator using a primary trigger threshold of 10 MeV. Once a supernova candidate is found, the time interval can be expanded for further analysis.

Once a candidate event cluster is selected, it is analyzed further to determine of the time sequence and energies if the events in the cluster are consistent with the signature of a $\bar{\nu}_e$ burst.

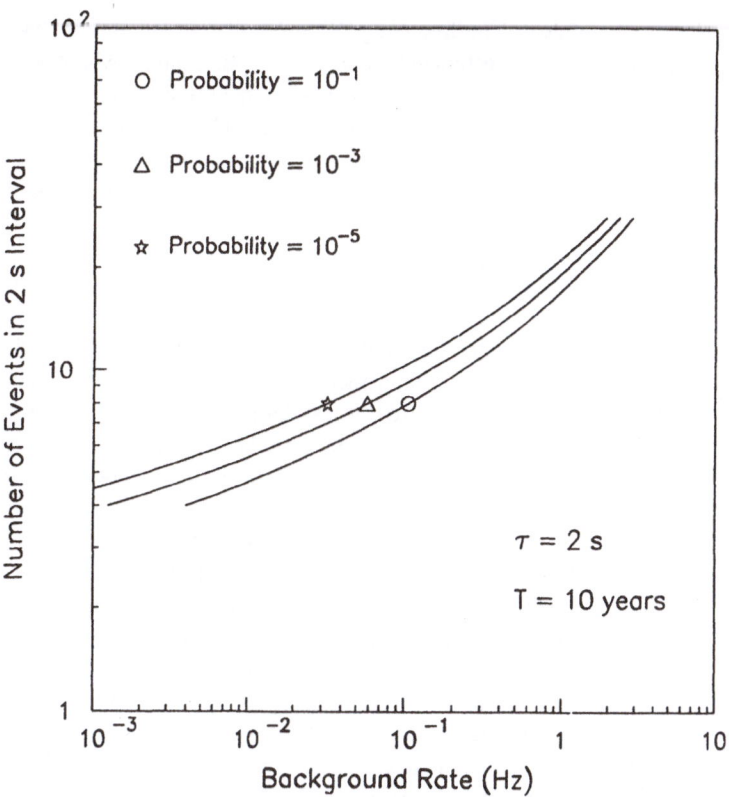

Figure 26. Probabilities for observing a given number of events in a 2 second interval for a running time of 10 years vs. Background Rate.

Discussion And Conclusions

The first 45 tons MACRO detector has achieved operation as a stellar collapse neutrino observatory. The trigger and data acquisition systems developed and implemented allow us to take advantage of some of the unique signatures that a gravitational collapse event would produce in liquid scintillator. In particular, we have studied our ability to observe the delayed neutron capture γ-rays following the initial burst of positrons that would signal such an event. Our operational sensitivity now extends to most of the stars in our Galaxy and will cover the entire Galaxy when MACRO becomes fully operational.

During the two year period of this search, we did not see any indication of a neutrino burst giving more than four events in a two second period. With an efficiency of 73% and a threshold energy of 10 MeV, MACRO would have observed five events within a two second interval for an event similar to SN1987A occurring at a distance of 11 kpc. Although such an event could not be accepted as a supernova on the basis of probability arguments alone, an off-line analysis would enable us to distinguish a genuine collapse.

To put these results in perspective, it is useful to determine the fraction of stars in our Galaxy sampled by a search sensitive to a stellar collapse within a given distance from our solar system. Based on the results of Bahcall and Piran,[26] approximately 62% of the stars are within 11 kpc. In terms of a potential gravitational collapse sample, our 25 months search of 62% of the Galaxy is equivalent to a 15.5 month search of the full Galaxy.

In the coming years MACRO will increase its active scintillator mass. A program to minimize dead time is well underway, and we will continue to monitor event clusters for potential supernova candidates. The frequency of core collapse supernovæ (Type Ib and II) for our Galaxy has been recently estimated to be 7.3 h^2 per century where h is the Hubble parameter in units of 100 km s^{-1}Mpc^{-1}. Assuming a ten year lifetime, the probability that MACRO will observe such a supernova is between 17% and 52% for h between 0.5 and 1. We expect that a Galactic supernova will produce enough neutrino events in the full detector to allow us to examine the time development of the neutrino sphere in the few seconds following core collapse.

Macro Search For A Wimp Annihilation Signal From The Center Of The Earth

Weakly Interacting Massive Particles (WIMPs) are one of the leading candidates for the local dark matter. Detecting them or ruling them out as dark matter candidates has great importance to both particle physics and astrophysics. Direct detection is very difficult, due to WIMP's very weak interaction with ordinary matter and the very small recoil energy obtained when such interactions do occur. Many ambitious projects are underway to develop detectors capable of a sensitive search, but it will be several years, at least, before such a detector is available. Therefore, it is of considerable interest to search for them indirectly, by looking for high energy neutrinos coming from the center of astrophysical bodies, such as the sun and the earth.[27-30] In this case, WIMPs are gravitationally attracted to the sun or the earth and after collisions are captured forming a high density near the center, where annihilation with anti-WIMPs becomes probable. One of the products of this annihilation are high energy neutrinos, and these can be searched for in MACRO by observing upward muons from high energy neutrino charged current interactions in the rocks below the detector.

In order to convert the experimental data on the yield or limit into a result on the amount of dark matter that can be attributed to various WIMPs, it is necessary to have a detailed calculation that relates the density of WIMPs in our galaxy to the flux of upward muons produced. Several authors have made estimates and calculations for the WIMP capture rate, the neutrino flux, and the neutrino spectrum from the sun or earth due to WIMP annihilations.[30-31].

In this presentation, I concentrate on the phenomenology and an early limit from MACRO on the yield from annihilations in the earth. This both illustrates MACRO sensitivity and presents some phenomenology we have done for detection from the earth. Andrew Gould[6] has derived the most accurate formula for WIMP capture and calculated the capture rate of Dirac neutrinos by the earth. He found that, for the most interesting range of Dirac neutrino mass (20GeV-80GeV), the capture rate by the earth is greatly enhanced and using the earth offers a higher sensitivity than using the sun. This is true for any WIMP that couples to the vector current. For Majorana WIMPs that couple to only axial-vector current, for instance, photinos, higgsinos, Majorana neutrinos, only the nuclei that have non-zero spins contribute to the capture cross section. This makes most of the elements in the Earth useless. Gould did not calculate the capture rate for Majorana WIMPs, since he has used a crude earth model in which many elements that could contribute considerably to the capture of Majorana WIMPs have been ignored.

I present here a more detailed earth model[32] as the input to Gould's formula to calculate the capture rate for both vector coupling and axial-vector coupling WIMPs. We also used an updated formula derived by J. Engel and P. Vogel[33] for the WIMP-nuclei elastic scattering cross section. We then used a Monte Carlo program that is similar to Ritz and Seckel's[30] to calculate the neutrino spectra and the upward muon flux produced by WIMP annihilations in the center of the Earth. Finally, I discuss the results of these calculations to interpret the results from the first supermodule of MACRO detector.

Capture Rate By The Earth

The complete formula of Gould for the capture rate is quite complicated[31]. The picture, however, is very simple. A WIMP with galactic velocity travels toward an astrophysical body, such as the sun or the earth, being focused by its gravitational field and finally hits the astrophysical body. Occasionally, it scatters on a nucleus, and if its velocity after scattering is less than the escape velocity at that point, the WIMP is considered to be captured. In the derivation of Gould's formula, he assumed that the WIMPs in our galaxy have Maxwellian velocity distribution and the movement of the astrophysical body relative to the rest frame of such a distribution has been taken into account.

In order to use Gould's formula to calculate the WIMP capture rate by the earth, one has to find the escape velocity everywhere in the earth, the WIMP-nuclei scattering cross section for every type of nuclei and their distributions in the earth.

From the earth density table of Ref. 32, we calculated several physical quantities of the earth as a function of the distance from the center of the earth.

Using the earth model in the same reference and the known natural abundances of isotopes, we obtained the data on all the nuclei that may contribute to the capture rate. It should be pointed out that the natural isotope abundances on the earth surface do not necessarily reflect the isotope abundance deep inside the earth, since heavier isotopes tend to sink deeper towards the center of the earth. This effect is not important for Dirac neutrinos and sneutrinos since the difference in their couplings to different isotopes of the same element is very small. However, for Majorana type of WIMPs, such as Majorana neutrino, photinos and higgsinos, this effect may be important, because only the isotopes with non-zero spin contribute to the elastic scattering cross section. I wrote that all the non-zero spin nuclei listed are either themselves the most abundant isotope or are heavier than the most abundant isotope. Because of this feature, using the isotope abundance on the earth surface in our calculation puts the error on the conservative side. In other words, there could be more non-zero spin nuclei inside the earth than we have assumed and this would make the capture rate of Majorana WIMPs even higher and the limits on their density more stringent.

For the WIMP-nuclei elastic scattering cross section, we used the tabulated formulas in Table 2 of Ref. 34. For the axial part of the cross section, however, we did not use the shell model. Instead, we used the following formula derived by J. Engel and P. Vogel[33] that relates the axial coupling of a nucleus to its magnetic moment:

$$g_A = 2\lambda_N \sum_q T_q^3 \Delta q \qquad (2)$$

where g_A is the axial part of the coupling constant, T_q^3 is the third component of weak isospin, Δq is the fraction of nucleon spin that is carried by quark q and λ_N is related to the spin J and magnetic moment μ of the nucleus by:

$$\lambda_N = 2\sqrt{\frac{J+1}{3J}} \frac{\mu - g_{odd}^l J}{g_{odd}^s - g_{odd}^l} \qquad (3)$$

where g_{odd}'s should be replaced by either g_n or g_p depending on whether the unpaired nucleon is a neutron or a proton. g_n's and g_p's are given by:

$$g_p^l = 1; \quad g_n^l = 0; \quad g_p^s = 5.586; \quad g_n^s = -3.826 \tag{4}$$

The g_A and λ_N calculated using equation (2) and (3) are then applied to the following formulas to calculate the WIMP-nuclei scattering cross section.

For Dirac neutrinos:

$$\sigma_{el} = \frac{G_F^2}{8\pi} K_N (g_V^2 + 3g_A^2) \tag{5}$$

For Majorana neutrinos:

$$\sigma_{el} = \frac{3G_F^2}{2\pi} K_N g_A^2 \tag{6}$$

For higgsinos:

$$\sigma_{el} = \frac{3G_F^2}{2\pi} K_N g_A^2 \cos^2(2\beta) \tag{7}$$

For photinos:

$$\sigma_{el} = \frac{3\lambda_N^2}{\pi} K_N \left[\sum_q \left(\frac{eQ_q}{m_{sq}}\right)^2 \Delta q \right]^2 \tag{8}$$

For sneutrinos:

$$\sigma_{el} = \frac{G_F^2}{2\pi} K_N g_V^2 \tag{9}$$

where G_F is Fermi's coupling constant,

$$K_N = \left(\frac{M_{WIMP} M_{nuc}}{M_{WIMP} M_{nuc}}\right)^2, \quad g_V = N - (1 - 4\sin^2(\theta_w)) \tag{10}$$

Z, N and Z are the number of neutrons and the charge of the nucleus respectively, β is related to the ratio of the two components of the vacuums expectation value by $\tan(\beta) = v_2 / v_1$, eQ_q is the charge of quark q and m_{sq} is the mass of the corresponding squark.

Since the formula of J. Engel and P. Vogel is derived with minimum assumptions about the wave functions of nuclei, using this formula and the measured magnetic moment of a nucleus to derive the axial-coupling should be more accurate than using the shell model. For the quark spin contents of a nucleon (Δq), we used the EMC[21] results. Other assumptions used in the calculations are:

Mass of squarks: all are 84GeV.
Vacuum expectation values: $\cos^2\beta = 0.5$.

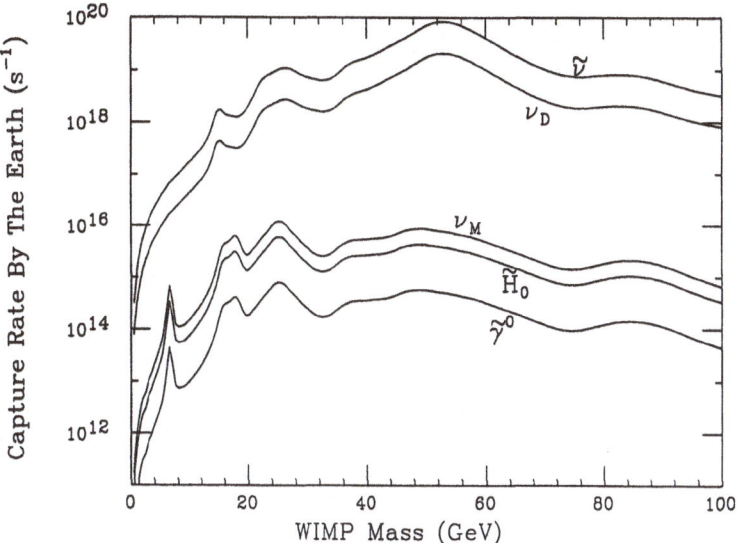

Figure 27. WIMP capture rates by the earth, calculated using Gould's formula.

The WIMP-nuclei scattering cross sections were calculated and used as input to Gould's formula. The earth capture rates for several WIMPs have been computed as a function of the WIMP mass. The results are shown in Fig. 27. Some other assumptions used in the computation are:

Local WIMP density: 0.3 GeV/cm^3

Average WIMP velocity: $\bar{v} = 300 km/s$

Earth velocity in galactic coordinates: $v_{earth} = \bar{v}$

The curve for Dirac neutrinos agrees with Gould's within the uncertainties of the calculation. The "peaks" in our Dirac neutrino curve are as high as Gould's but the "valleys" are shallower. This is because we have put many more earth elements that fill the gaps between the major peaks.

The captured WIMPs will accumulate in the earth until their density is high enough and annihilation occurs. The relation between the annihilation rate and the capture rate is given by:

$$A = \frac{C}{2}[\tanh(t / \tau)]^2 \tag{11}$$

where A is the annihilation rate, C is the capture rate, t is the age of the earth and τ is the time constant for equilibrium to occur. If one ignores small temperature and density variations near the center of the earth, τ can be calculated by:

$$\tau = (\frac{3kT}{mG\rho})^{3/4}(2\langle\sigma v\rangle C)^{-1/2} \tag{12}$$

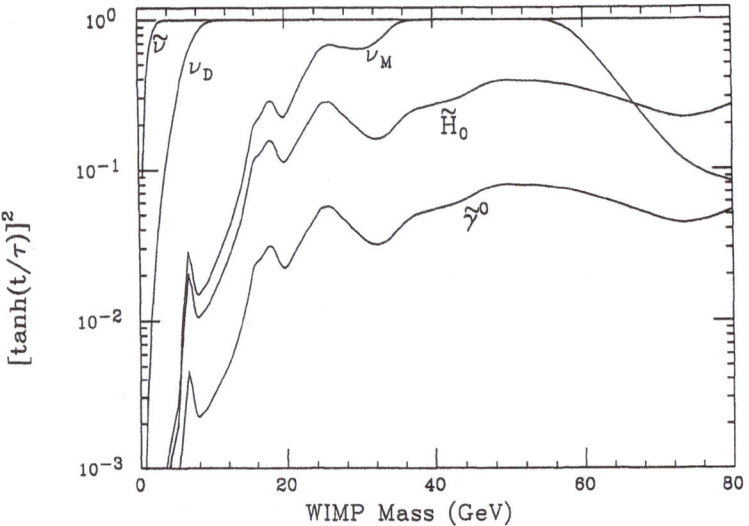

Figure 28. The factor $[\tanh(t/\tau)]^2$ for various WIMPs as a function of the WIMP mass. When this factor is close to 1, the capture and annihilation of WIMPs have reached equilibrium and WIMP signal will be at "full" signal.

where k is the Boltzmann constant, T is the temperature at the center of the earth, m is the mass of the WIMP, G is the gravitational constant, ρ is the density at the center of the earth and $\langle\sigma v\rangle$ is the thermally averaged annihilation cross section.

The factor $[\tanh(t/\tau)]^2$ we calculated for various WIMPs are shown in Fig. 28 as a function of the WIMP mass. It is clear, in most of the mass range, Dirac neutrinos and sneutrinos will have "full signal", while the signal for Majorana type WIMPs in most mass region is reduced because the capture and annihilation has not reached equilibrium yet.

Any annihilation channel which has a product that can decay and produce ν_μ and $\bar{\nu}_\mu$ can contribute to the muon neutrino flux. To produce upward muons, however, the muon neutrinos must have high enough energy. This requires that the annihilation products be short lived and decay before they lose their energy or have interactions with the earth medium.

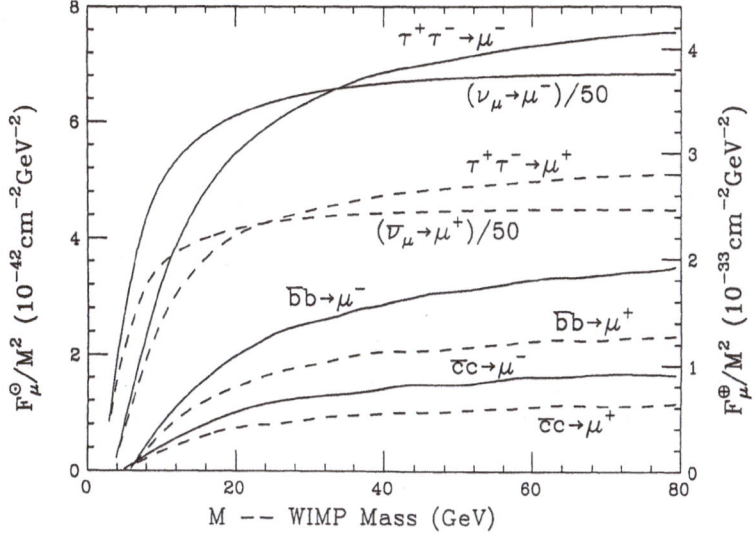

Figure 29. The upward muon flux produced by ONE annihilation in the sun or in the earth for different annihilation channels. The left vertical scale shows the quantity F_μ^Θ / M^2, where F_μ^Θ is the flux produced by ONE annihilation in the sun and M is the mass of the WIMP. The right vertical scale shows the similar quantity for the earth. The curves for the direct muon neutrinos are scaled down by a factor of 50 in order to be shown in the same graph.

Only the annihilation channels that produce $\tau^\pm, b\bar{b}$ or $c\bar{c}$ jets and the channel that directly produces a pair of muon neutrinos contribute to the upward muon flux (for WIMPs with mass below the W mass).

We used a LUND Monte Carlo program to calculate the neutrino and upward muon yield for these channels.

We then used the following approximate formula of T.K. Gaisser and T. Stanev[22] to simulate the upward muon production in the rocks below the detector:

$$\frac{d\sigma_\nu}{dE_\mu} = \left[0.72 + 0.26 \left(\frac{E_\mu}{E_\nu} \right)^2 \right] x 10^{-38} cm^2 GeV^{-1} \qquad (13)$$

$$\frac{d\sigma_{\bar{\nu}}}{dE_\mu} = \left[0.09 + 0.69 \left(\frac{E_\mu}{E_\nu} \right)^2 \right] x 10^{-38} cm^2 GeV^{-1} \qquad (14)$$

For the energy loss of the muons in the rock, we used the formula:

$$\frac{dE_\mu}{dX} = -\alpha - \beta E \qquad (15)$$

where $\alpha = 2 MeV \; cm^2/g$ and $\beta = 3.9 \; x \qquad 10^{-6} cm^2/g$.

The results of the calculations for all the contributing annihilation channels are compiled in Fig. 29. The graph shows the upward muon flux produced at the detector by ONE annihilation in the sun or in the earth, as a function of the WIMP mass. The curves for the sun and the earth are the same but have different vertical scales. The vertical scale is suppressed by a factor of the square of the WIMP mass in order for the graph to be drawn in linear scale showing more details.

Using the data in Figures 27 - 29 and the WIMP annihilation branching ratios into various channels, which we derived from Ref. 37 and Ref. 38, we obtained the flux of upward muons expected from the center of the earth assuming WIMPs are the major component of the dark matter in our galaxy. The results are shown in Fig. 30 as a function of the WIMP mass.

Interpretation Of Macro Results

We have used the data from the first supermodule (1/12 of the MACRO detector) running for 1.5 years to make an initial search for WIMPs from the earth. The streamer tubes record muon tracks with a tracking accuracy of 0.5 degree. The scintillator counters measure the time of flight of muons with about 1ns time resolution and therefore easily distinguish upward muon from downward ones.

During this run, none of the detected upward muons points back to the center of the earth.

In order to obtain the effective Exposure of this run, the near vertical downward muons are used to calibrate the efficiency of the detector. Using this data, we obtained the effective exposure $A \cdot T = 9.2 \times 10^8 m^2 s$.

Based on this number, we obtained an upper limit on the upward muon flux from the center of the earth. The limit is shown in Fig. 30 compared with expected signal of various WIMPs assuming they are the major components of the dark matter in our galaxy. Dirac neutrinos over a wide mass range are ruled out as major component of dark matter, unless the number of Dirac neutrinos and the number of anti-Dirac neutrinos are highly asymmetric. Muon sneutrinos are also ruled out unless the Self annihilation channel is highly suppressed. Full MACRO running for a few years will reach a limit two orders of magnitude lower than the present one, as indicated in Fig. 30.

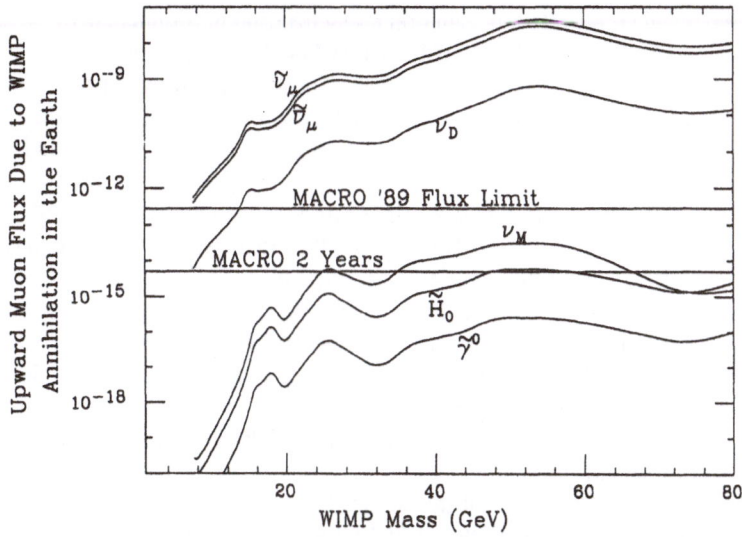

Figure 30. The flux of upward muons expected form the center of the earth assuming the WIMPs are the major component of the dark matter in our galaxy. The present MACRO flux limit and future MACRO capability are also shown on the graph.

REFERENCES

1. H. Albrecht *et al.*, An improved upper limit on the tau-neutrino mass from the decay tau $\rightarrow \pi^- \pi^+$ tau neutrino, Phys. Lett. ~82B (1988) 149.

2. J.J. Simpson, Evidence of heavy neutrino emission in beta decay, Phys. Rev. Lett. 54 (1985)1891.

3. V.M. Datar *et al.*, Search for a heavy neutrino in the beta decay of S^{-35} Nature 318 (1985) 547; T. Ohs *et al.*, Search for heavy neutrinos in the beta decay of S^{-35} as evidence against the 17-KeV heavy neutrinos. Phys. Lett. 160B (185) 322; 3. Maxkey and F. Boeh; Experimental search for a heavy neutrino in the beta spectrum of S^{-35}, Phys. Rev. C32 (1985) 2215; T. Altsitsoglou *et al.*, Search for heavy neutrino in beta decay, Phys. Rev. Lett. 55 (1985) 799; A. Apahkov *et al.*, Pis'maZh. Eksp. Teor. Fiz. 42 (1985)233 [JETP Lett. 42 (1985) 289].

4. A. Hime and J.J. Simpson, Evidence of the 17 KeV neutrino in the beta spectrum of H_{-3}, Phys. Rev. D39 (1989) 1837; J.J. Simpson and A. Hime, Phys. Rev. D39 (1989) 1825.

5. B. Sur *et al.*, Evidence for the emission of a 17 KeV neutrino in the beta decay of C^{-14}, Phys. Rev. Lett. 66 (1991) 2444.

6. A, Hime and N.A., New evidence for the 17 KeV neutrino, Jelley, Phys. Lett. B257 441.

7. H.W. Becket, D.A. Imel, H. Henrikson, Novikov and F. Boehm, Search for evidence of the 17 keV neutrino in the Spectrum of ^{5}S (California Institute of Technology, (1991), preprint, unpublished.

8. M. Bahran and G. Kalbfleisch, Limit on heavy neutrino in tritium beta decay, Univ. of Oklahoma, preprint OKHEP-91-005 July 91, unpublished.

9. J.N. Bahcall and ILK. Ulrich, Solar models, neutrino experiments and helioseismology, Rev. Mod.1 Phys. 60 (1988) 297; J.N. Bahcall, Neutrino Astrophysics, Cambridge Univ. Press (1989); See also S. Turck-Chiez et al., Astrophys. J. (1988) 415.

10. K. Hirata et al., Observation of β-8 solar neutrinos in the KAMIOKANDE-II detector, Phys. 63 (1989) 16; K Hirata et al., Results from one thousand days of real time directional solar neutrino data, Phys. Rev. Lett. 65 (1990) 1297

11. E. Witten, Cosmic separation of phases, Phys. Rev. D30 (1984) 272.

12. E. Fahri. and R. L. Jaffc, Strange matter, Phys. Rev. D30 (1984) 2379.

13. J. Madsen, H. Heiselberg and K. Riisager, Does strange matter evaporate in the early universe?, Phys. Rev. D34 (1986) 2974.

14. A. De RuJula and S.L. Glashow, Does dark matter affect the solar neutrino flux?, Nature A434 (1985) 605.

15. S. Ritz, D. Seckel, Detailed neutrino spectra from cold dark matter annihilations in the sun, Nucl. Phys. B304 (1988) 877; William H. Press and David N. Spergel, Resonant enhancements in wimp capture by the earth, Astrophys.J., 296 (1985) 679; Andrew Gould, Astrophys. J. 321 (1987) 571.

16. L. Rosskowski, Light neutralino as dark matter supersymmetric dark matter below the W mass, CERN-TH6041/91 and CERN-TH6042/91 and this conference.

17. M. Samorsld and M. Stamm, Detection of 2×10^{15} ev to 2×10^{16} ev gamma rays from Cygnus X-3, Ap. J. 268 (1983) L17; and Proc. 18th International Cosmic Ray Conference (Bongalare) 11, (1983) 244.

18. J. Lloyd-Evans et al., Observation of gamma rays $> 10^{15}$ ev from Cygnus X-3, Nature 305 (1983) 784.

19. M.A. Tomson et al., The observation of underground muons from the direction of Cygnus X-3 during the January 1991 radio flare, OUNP-91-17 July 1991, submitted to Physics Letters B.

20. R.M. Bionta et al., Observation of a neutrino burst in coincidence with supernova SN1987A in the large magellanic cloud, Phys. Rev. Lett. 58, (1987) 1494.

21. M. Calicchio et al. (The MACRO Collaboration), Status report of the MACRO experiment at Gran Sasso, Nucl. Instr. and Meth. A 264, (1988) 18.

22. M.S. Turner et al., Magnetic Monopoles and the survival of galactic magnetic fields, Phys. Rev. D26, (1982) 1296.

23. W.D. Arnett et al., Supernova SN 1987A, Annu. Rev. Astron. Astrophys. 27, (1989) 629.

24. J.N. Bahcall, Neutrino Astrophysics, Cambridge Univ. Press, 1989.

25. P. Belli et al., Deep underground neutrino flux measurement with large BF-3 Counters, Nuovo Cimento 101A, (1989) 959.

26. J.N. Bahcall and T. Piran, Stellar collapses in the galaxy, Ap. J. Lett. 267, (1983) L77.

27. M. Srednicki et al., High energy neutrinos from the sun and cold dard matter, Nucl. Phys. B279, (1987) 894.

28. J. Silk et al., The photino, the sun and high-energy neutrinos, Phys. Rev. Lett. 55, (1985) 259.

29. K.-W. Ng, et al., Dark matter induced neutrinos from the sun: theory versus experiment, Phys. Lett. 188B, (1987) 138.

30. S. Ritz and D. Seckel, Detailed neutrino spectro from cold dark matter annihilations in the sun, Nucl. Phys. B304, (1988) 877.

31. A. Gould, Resonant enhancedments in WIMP capture by the earth, Astrophys. J. 321, (1987) 571.
32. D. Anderson, "Theory of the Earth", Blackwell Scientific Publications (1989).
33. J. Engel and P. Vogel, Spin dependent cross-sections of weakly interacting massive particles on nuclei, Phys. Rev. D40, (1989) 3132.
34. J.R. Primack *et al.*, Detection of cosmic dark matter, Ann. Rev. Nucl. Part 38, (1988) 751.
35. J. Ashman *et al.*, A measurement of the spin asymetry and determination of the structure function G(1) in deep inelastic muon-proton scattering, Phys. Lett. B206, (1988) 346.
36. T.K. Gaisser and T. Stanev, Neutrino induced muon flux deep underground and search for neutrino oscillations, Phys. Rev. D30, (1984) 985.
37. G.L. Kane and I. Kani, Cross-sections for dark matter particles and implications for allowed masses, interactions and detections, Nucl. Phys. B277, (1986) 525.
38. T.K. Gaisser *et al.*, Limits on cold dark matter candidates from deep underground detectors, Phys. Rev. D34, (1986) 2206.

INFLATION AFTER COBE:
LECTURES ON INFLATIONARY COSMOLOGY *

Michael S. Turner

Departments of Physics and Astronomy & Astrophysics, Enrico Fermi Institute, The University of Chicago, Chicago, IL 60637-1433
NASA/Fermilab Astrophysics Center,
Fermi National Accelerator Laboratory, Batavia, IL 60510-0500

Abstract

In these lectures I review the standard hot big-bang cosmology, emphasizing its successes, its shortcomings, and its major challenge—a detailed understanding of the formation of structure in the Universe. I then discuss the motivations for—and the fundamentals of—inflationary cosmology, particularly emphasizing the quantum origin of metric (density and gravity-wave) perturbations. Inflation addresses the shortcomings of the standard cosmology and provides the "initial data" for structure formation. I conclude by addressing the implications of inflation for structure formation, evaluating the various cold dark matter models in the light of the recent detection of temperature anisotropies in the cosmic background radiation by COBE. In the near term, the study of structure formation offers a powerful probe of inflation, as well as specific inflationary models.

1 Hot Big Bang: Successes and Challenges

1.1 Successes

The hot big-bang model, more properly the Friedmann-Robertson-Walker (FRW) cosmology or standard cosmology, is spectacularly successful: In short, it provides a reliable and tested accounting of the history of the Universe from about 0.01 sec after the bang until today, some 15 billion years later. The primary pieces of evidence that support the model are: (1) The expansion of the Universe; (2) The cosmic background radiation; and (3) The primordial abundances of the light elements D, ^{3}He, ^{4}He, and ^{7}Li [1].

1.1.1 The expansion

Although the precise value of the Hubble constant is not known to better than a factor

---

*Supported in part by the DOE (at Chicago and Fermilab) and by the NASA through grant NAGW-2381 (at Fermilab).

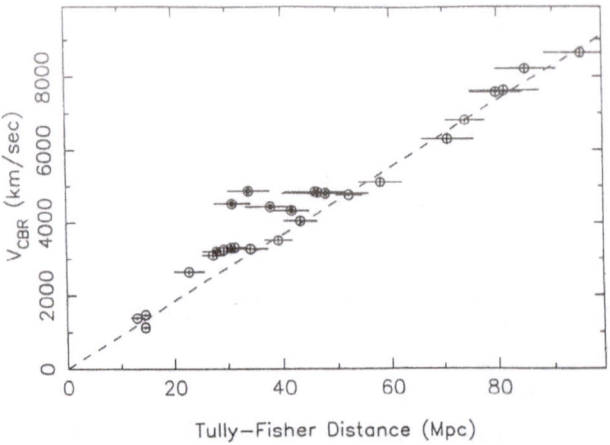

Figure 1. Hubble diagram (from [3]). The deviation from a linear relationship around 40 Mpc is due to peculiar velocities.

of two, $H_0 = 100h\,\mathrm{km\,sec^{-1}\,Mpc^{-1}}$ with $h = 0.4 - 1$, there is little doubt that the expansion obeys the "Hubble law" out to red shifts approaching unity [2, 3]; see Fig. 1. As is well appreciated, the fundamental difficulty in determining the Hubble constant is the calibration of the cosmic-distance scale as "standard candles" are required [4, 5].

The Hubble law allows one to infer the distance to an object from its red shift z: $d = zH_0^{-1} \simeq 3000z\,h^{-1}\,\mathrm{Mpc}$ (for $z \ll 1$, the galaxy's recessional velocity $v \simeq zc$), and hence "maps of the Universe" constructed from galaxy positions and red shifts are referred to as red-shift surveys. Ordinary galaxies and clusters of galaxies are seen out to red shifts of order unity; more unusual and rarer objects, such as radio galaxies and quasars, are seen out to red shifts of almost five (the current record holder is a quasar with red shift 4.9). Thus, we can probe the Universe with visible light to within a few billion years of the big bang.

1.1.2 The cosmic background radiation

The spectrum of the cosmic background radiation (CBR) is consistent that of a black body at temperature 2.73 K over more than three decades in wavelength ($\lambda \sim 0.03\,\mathrm{cm} - 100\,\mathrm{cm}$); see Fig. 2. The most accurate measurement of the temperature and spectrum is that by the FIRAS instrument on the COBE satellite which determined its temperature to be 2.726 ± 0.01 K [6]. It is difficult to come up with a process other than an early hot and dense phase in the history of the Universe that would lead to such a precise black body [7]. According to the standard cosmology, the surface of last scattering for the CBR is the Universe at a red shift of about 1100 and an age of about $180,000\,(\Omega_0 h^2)^{-1/2}$ yrs. It is possible that the Universe became ionized again after this epoch, or due to energy injection never recombined; in this case the last-scattering surface is even "closer," $z_{\mathrm{LSS}} \simeq 10[\Omega_B h/\sqrt{\Omega_0}]^{-2/3}$.

The temperature of the CBR is very uniform across the sky, to better than a part in 10^4 on angular scales from tens of arcseconds to 90 degrees; see Fig. 3. Three forms of temperature anisotropy—two spatial and one temporal—have now been detected: (1) A dipole anisotropy of about a part in 10^3, generally believed to be due to the motion of galaxy relative to the cosmic rest frame, at a speed of about $620\,\mathrm{km\,sec^{-1}}$ [9]; (2) A yearly modulation in the temperature in a given direction on the sky of about a part in 10^4, due to our orbital motion around the sun at $30\,\mathrm{km\,sec^{-1}}$, see Fig. 4 [10]; and (3) The

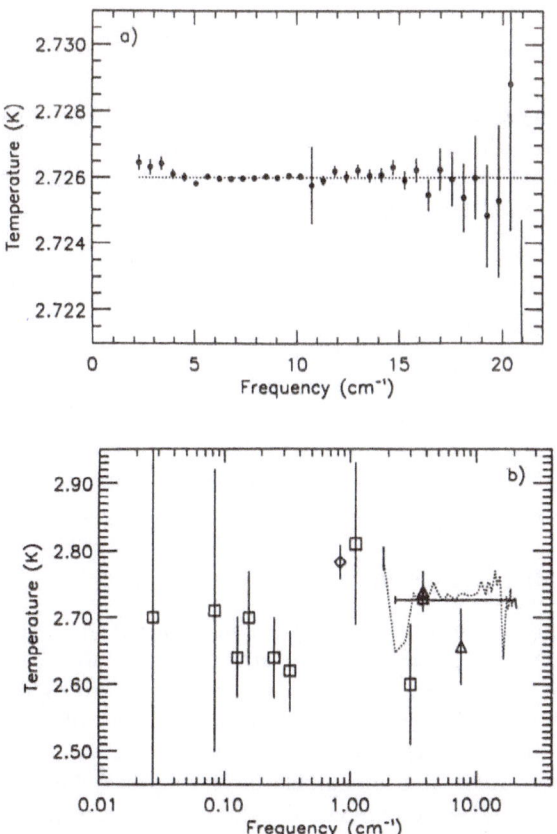

Figure 2 (a) COBE FIRAS measurements of the CBR temperature, (b) Summary of other CBR temperature measurements (from [6]), the dotted curve indicates the data from the other high-precision measurement, by the UBC rocket borne COBRA instrument [8]

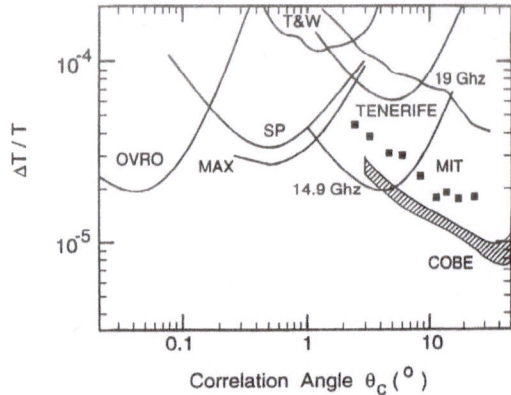

Figure 3. Summary of recent high-sensitivity CBR anisotropy measurements; with the exception of COBE all results are upper limits (from [12, 13]). The solid boxes (MIT balloon experiment) have recently been reanalyzed and shown to be a detection which is consistent with the COBE DMR result [14].

temperature anisotropies detected by the Differential Microwave Radiometer (DMR) on the Cosmic Background Explorer (COBE) satellite, $\langle(\Delta T/T)^2\rangle^{1/2}_{10°} = 1.1 \pm 0.2 \times 10^{-5}$ and $(\Delta T/T)_Q = 6 \pm 2 \times 10^{-6}$, where the first measurement refers to the *rms* temperature fluctuation averaged over the entire sky as measured by a beam of width 10°, and the second is the magnitude of the quadrupole temperature anisotropy [11]. The 10° and quadrupole anisotropies provide strong evidence for primeval density inhomogeneities of the same magnitude. which amplified by gravity, grew into the structures that we see today: galaxies, clusters of galaxies, superclusters, voids, walls, etc.

1.1.3   Primordial nucleosynthesis

Last, but certainly not least, there are the abundance of the light elements. According to the standard cosmology, when the age of the Universe was measured in seconds, the temperatures were of order MeV, and the conditions were right for nuclear reactions which ultimately led to the synthesis of significant amounts of D, ^{3}He, ^{4}He, and ^{7}Li. The yields of primordial nucleosynthesis depend upon the baryon density, quantified as the baryon-to-photon ratio η, and the number of very light ($\lesssim$ MeV) particle species, often quantified as the equivalent number of light neutrino species, N_ν. The predictions for the primordial abundances of all four light elements agree with their measured abundances provided that $3 \times 10^{-10} \lesssim \eta \lesssim 5 \times 10^{-10}$ and $N_\nu \lesssim 3.4$; see Fig. 5 [15].

Accepting the success of the standard model of nucleosynthesis, our precise knowledge of the present temperature of the Universe allows us to convert η to a mass density, and by dividing by the critical density, $\rho_{\rm crit} \simeq 1.88h^2 \times 10^{-29}\,{\rm g\,cm}^{-3}$, to the fraction of critical density contributed by ordinary matter:

$$0.011 \lesssim \Omega_B h^2 \lesssim 0.019; \qquad \Rightarrow \quad 0.011 \lesssim \Omega_B \lesssim 0.12; \tag{1}$$

this is the most accurate determination of the baryon density. Note, the uncertainty in the value of the Hubble constant leads to most of the uncertainty in Ω_B.

The nucleosynthesis bound to N_ν, and more generally to the number of light degrees of freedom in thermal equilibrium at the epoch of nucleosynthesis, is consistent with precision measurements of the properties of the Z^0 boson, which give $N_\nu = 3.0 \pm 0.05$; further, the cosmological bound predates these accelerator measurements! The nucleosynthesis bound provides a stringent limit to the existence of new, light particles (even

344

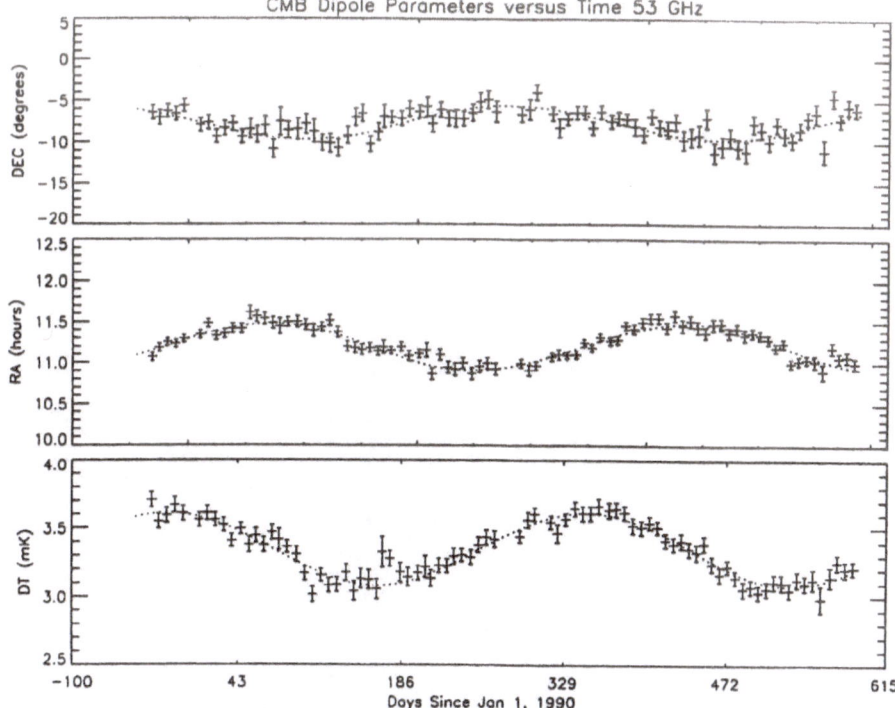

Figure 4. Yearly modulation of the CBR temperature—the earth really orbits the sun(!) (from [10]).

beyond neutrinos), and even provides a bound to the mass the tau neutrino, excluding a tau-neutrino mass between 0.5 MeV and 25 MeV [16]. Primordial nucleosynthesis provides a beautiful illustration of the powers of the Heavenly Laboratory, though it is outside the focus of these lectures.

The remarkable success of primordial nucleosynthesis gives us confidence that the standard cosmology provides an accurate accounting of the Universe at least as early as 0.01 sec after the bang, when the temperature was about 10 MeV.

1.1.4   Et cetera—and the age crisis?

There are additional lines of reasoning and evidence that support the standard cosmology [7]. I mention two: the age of the Universe and structure formation. I will discuss the basics of structure formation a bit later; for now it suffices to say that the standard cosmology provides a basic framework for understanding the formation of structure—amplification of small primeval density inhomogeneities through gravitational instability—which has recently been confirmed by COBE [11]. Here I focus on the age of the Universe.

The expansion age of the Universe—time back to zero size—depends upon the present expansion rate, energy content, and equation of state: $t_{exp} = f(\rho, p)H_0^{-1} \simeq 9.8h^{-1}f(\rho, p)$ Gyr. For a matter-dominated Universe, f is between 1 and 2/3 (for Ω_0 between 0 and 1), so that the expansion age is somewhere between 7 Gyr and 20 Gyr. There are other independent measures of the age of the Universe, e.g., based upon long-lived radioisotopes, the oldest stars, and the cooling of white dwarfs. These "ages," ranging from 13 to 18 Gyr, span the same interval(!). This wasn't always the case; as late as the early 1950's it was believe that the Hubble constant was 500 km sec^{-1} Mpc^{-1},

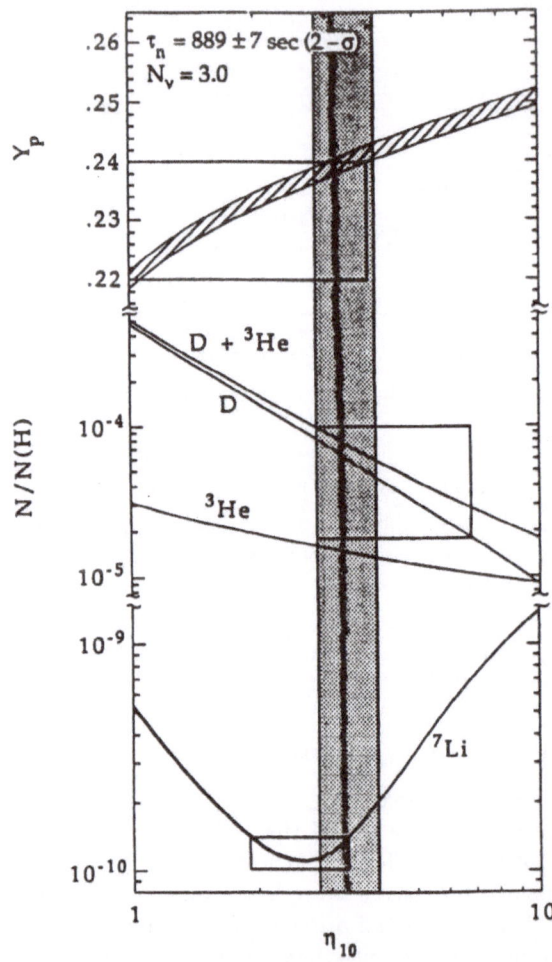

Figure 5 Predicted light element abundances and inferred abundances (from [15]) The measured primordial abundances are indicated and the concordance region is shaded

implying an expansion age of at most 2 Gyr—less than the age of the earth. This discrepancy was an important motivation for the steady-state cosmology.

While there is *general* agreement between the expansion age and other determinations of the age of the Universe, some cosmologists are worried that cosmology is on the verge of another age crisis [5]. Let me explain, while Sandage and a few others continue to obtain values for the Hubble constant around $50\,\mathrm{km\,s^{-1}\,Mpc^{-1}}$ [2], a variety of different techniques seem to be converging on a value around $80 \pm 10\,\mathrm{km\,s^{-1}\,Mpc^{-1}}$ [5]. If $H_0 = 80\,\mathrm{km\,s^{-1}\,Mpc^{-1}}$, then $t_{\exp} = 12 f(\rho, p)\,\mathrm{Gyr}$, and for $\Omega_0 = 1$, $t_{\exp} = 8\,\mathrm{Gyr}$, which is clearly inconsistent with other measures of the age. If $H_0 = 80\,\mathrm{km\,s^{-1}\,Mpc^{-1}}$, one is almost forced to consider the radical alternative of a cosmological constant. For example, even with $\Omega_0 = 0.2$, $f \simeq 0.85$, corresponding to $t_{\exp} \simeq 10\,\mathrm{Gyr}$; on the other hand, for a flat Universe with $\Omega_\Lambda = 0.8$, $f \simeq 1.1$ and the expansion age $t_{\exp} \simeq 13.5\,\mathrm{Gyr}$. As I shall discuss later, structure formation provides another motivation for a cosmological constant. My own gut-level feeling is that when the dust settles, we will find that $H_0 = 50\,\mathrm{km\,s^{-1}\,Mpc^{-1}}$; then again, maybe not.

1.2 Basics of the Big Bang Model

The standard cosmology is based upon the maximally spatially symmetric Robertson-Walker line element

$$ds^2 = dt^2 - R(t)^2 \left[\frac{dr^2}{1 - kr^2} + r^2(d\theta^2 + \sin^2\theta\, d\phi^2) \right];$$

$$(2)$$

where $R(t)$ is the cosmic-scale factor, $R_{\mathrm{curv}} \equiv R(t)|k|^{-1/2}$ is the curvature radius, and $k/|k| = -1, 0, 1$ is the curvature signature. All three models are without boundary: the positively curved model is finite and "curves" back on itself; the negatively curved and flat models are infinite in extent (though finite versions of both can be constructed by imposing a periodic structure: identifying all points in space with a fundamental cube). The Robertson-Walker metric embodies the observed isotropy and homogeneity of the Universe. It is interesting to note that this form of the line element was originally introduced for sake of mathematical simplicity; we now know that it is well justified at early times or today on large scales ($\gg 10\,\mathrm{Mpc}$), at least within our Hubble volume.

The coordinates, r, θ, and ϕ, are referred to as comoving coordinates: A particle at rest in these coordinates remains at rest, i.e., constant r, θ, and ϕ. A freely moving particle eventually comes to rest these coordinates, as its momentum is red shifted by the expansion, $p \propto R^{-1}$. Motion with respect to the comoving coordinates (or cosmic rest frame) is referred to as peculiar velocity; unless "supported" by the inhomogeneous distribution of matter peculiar velocities decay away as R^{-1}. Thus the measurement of peculiar velocities, which is not easy as it requires independent measures of both the distance and velocity of an object, can be used to probe the distribution of mass in the Universe.

Physical separations (i.e., measured by meter sticks) between freely moving particles scale as $R(t)$; or said another way the physical separation between two points is simply $R(t)$ times the coordinate separation. The momenta of freely propagating particles decrease, or "red shift," as $R(t)^{-1}$, and thus the wavelength of a photon stretches as $R(t)$, which is the origin of the cosmological red shift. The red shift suffered by a photon emitted from a distant galaxy $1 + z = R_0/R(t)$; that is, a galaxy whose light is red shifted by $1 + z$, emitted that light when the Universe was a factor of $(1 + z)^{-1}$ smaller. Thus, when the light from the most distant quasar yet seen ($z = 4.9$) was emitted the Universe was a factor of almost six smaller; when CBR photons last scattered the Universe was about 1100 times smaller.

1.2.1 Friedmann equation and the First Law

The evolution of the cosmic-scale factor is governed by the Friedmann equation

$$H^2 \equiv \left(\frac{\dot{R}}{R}\right)^2 = \frac{8\pi G\rho_{\text{tot}}}{3} - \frac{k}{R^2};\tag{3}$$

where ρ_{tot} is the total energy density of the Universe, matter, radiation, vacuum energy, and so on. A cosmological constant is often written as an additional term ($= \Lambda/3$) on the rhs; I will choose to treat it as a constant energy density ("vacuum-energy density"), where $\rho_{\text{vac}} = \Lambda/8\pi G$. (My convention in this regard is not universal.) The evolution of the energy density of the Universe is governed by

$$d(\rho R^3) = -p\,dR^3;\tag{4}$$

which is the First Law of Thermodynamics for a fluid in the expanding Universe. (In the case that the stress energy of the Universe is comprised of several, noninteracting components, this relation applies to each separately; e.g., to the matter and radiation separately today.) For $p = \rho/3$, ultra-relativistic matter, $\rho \propto R^{-4}$; for $p = 0$, very nonrelativistic matter, $\rho \propto R^{-3}$; and for $p = -\rho$, vacuum energy, $\rho = \text{const}$. If the rhs of the Friedmann equation is dominated by a fluid with equation of state $p = \gamma\rho$, it follows that $\rho \propto R^{-3(1+\gamma)}$ and $R \propto t^{2/3(1+\gamma)}$.

We can use the Friedmann equation to relate the curvature of the Universe to the energy density and expansion rate:

$$\frac{k/R^2}{H^2} = \Omega - 1; \qquad \Omega = \frac{\rho_{\text{tot}}}{\rho_{\text{crit}}};\tag{5}$$

and the critical density today $\rho_{\text{crit}} = 3H^2/8\pi G = 1.88h^2\,\text{g cm}^{-3} \simeq 1.05 \times 10^4\,\text{eV cm}^{-3}$. There is a one to one correspondence between Ω and the spatial curvature of the Universe: positively curved, $\Omega_0 > 1$; negatively curved, $\Omega_0 < 1$; and flat ($\Omega_0 = 1$). Further, the "fate of the Universe" is determined by the curvature: model universes with $k \leq 0$ expand forever, while those with $k > 0$ necessarily recollapse. The curvature radius of the Universe is related to the Hubble radius and Ω by

$$R_{\text{curv}} = \frac{H^{-1}}{|\Omega - 1|^{1/2}}.\tag{6}$$

In physical terms, the curvature radius sets the scale for the size of spatial separations where the effects of curved space become "pronounced." And in the case of the positively curved model it is just the radius of the 3-sphere.

The energy content of the Universe consists of matter and radiation (today, photons and neutrinos). Since the photon temperature is accurately known, $T_0 = 2.73 \pm 0.01$ K, the fraction of critical density contributed by radiation is also accurately known: $\Omega_{\text{rad}}h^2 = 4.18 \times 10^{-5}$. The matter content is another matter.

1.2.2 A short diversion concerning the present mass density

The matter density today, i.e., the value of Ω_0, is not nearly so well known [17]. Stars contribute less than 1% of critical density; based upon nucleosynthesis, we can infer that baryons contribute between 1% and 10% of critical. The dynamics of various systems allow astronomers to infer their gravitational mass. With their telescopes they measure the amount of light, and form a mass-to-light ratio. Multiplying this by the measured luminosity density of the Universe gives a determination of the mass density. (The critical mass-to-light ratio is $1200h\,M_\odot/\mathcal{L}_\odot$.)

The motions of stars and gas clouds in spiral galaxies indicate that most of the mass of spiral galaxies exists in the form of dark (i.e., no detectable radiation), extended halos, whose full extent is still not known. Many cite the flat rotation curves of spiral galaxies, which indicate that the halo density decreases as r^{-2}, as the best evidence that most of the matter in the Universe is dark. Taking the mass-to-light ratio inferred for spiral galaxies to be typical of the Universe as a whole and remembering that the full extent of the dark matter halos is not known, one infers $\Omega_{halo} \gtrsim 0.03 - 0.1$.

The masses of clusters of galaxies can be estimated using the virial theorem, and these mass estimates too indicate the presence of large amounts of dark matter. Taking cluster mass-to-light ratios to be typical of the Universe as a whole, in spite of the fact that only about 1 in 10 galaxies resides in a cluster, one infers $\Omega_{cluster} \sim 0.1 - 0.3$.

Most galaxies are found in associations of a few galaxies known as small groups. Estimating the masses of these systems using dynamics is tricky because of the problem of "interlopers," galaxies that happen to be in the same part of the sky, but are not associated with the group [18]. This Fall, however, ROSAT detected the weak x-ray emission from the hot gas in the small group NGC 2300 [19]; from their measurements they were able to infer the shape of the gravitational potential—and hence total mass of the group—as well as the mass of the x-ray emitting gas and the visible mass in galaxies. They found that the total mass of the group was about 20 times that in ordinary matter(!). If one takes this to be a universal ratio of the total amount of matter to that in baryons and $\Omega_B \sim 0.05$, one concludes that $\Omega_0 \sim 1$.

Not one of these methods is wholly satisfactory: Rotation curves of spiral galaxies are still "flat" at the last measured points, indicating that the mass is still increasing; likewise, cluster virial mass estimates are insensitive to material that lies beyond the region occupied by the visible galaxies—and moreover, only about one galaxy in ten resides in a cluster. What one would like is a measurement of the mass of a very big sample of the Universe, say a cube of $100h^{-1}$ Mpc on a side, which contains tens of thousands of galaxies.

Over the past five years or so progress has been made toward such a measurement. It involves the peculiar motion of our own galaxy, at a speed of about 620 km sec^{-1} in the general direction of Hydra-Centaurus. This motion is due to the lumpy distribution of matter in our vicinity. By using gravitational-perturbation theory (actually, not much more than Newtonian physics) and the distribution of galaxies in our vicinity (as determined by the IRAS catalogue of infrared selected galaxies), one can infer the average mass density in a very large volume and thereby Ω_0.

The basic physics behind the method is simple: the net gravitational pull on our galaxy depends both upon how inhomogeneous the distribution of galaxies is and how much mass is associated with each galaxy; by measuring the distribution of galaxies and our peculiar velocity one can infer the "mass per galaxy" and Ω_0.

The value that has been inferred is big(!): $\Omega_{IRAS} \sim 1 \pm 0.2$ [20]. Moreover, the measured peculiar velocities of other galaxies in this volume, more than thousand, have been used in a similar manner and indicate a similarly large value for Ω_0 [21]. While this technique is very powerful, it does have its drawbacks: One has to make simple assumptions about how accurately mass is traced by light (the observed galaxies); one has to worry whether or not a significant portion of our galaxy's velocity is due to galaxies outside the IRAS sample—if so, this would lead to an overestimate of Ω_0; and so on. This technique is not only very promising—but provides the "correct" answer (in my opinion!).

The so-called classical kinematic tests—Hubble diagram, angle-red shift relation, galaxy count-red shift relation—can, in principle, provide a determination of Ω_0 [22]. However, all these methods require standard candles, rulers, or galaxies, and for this reason have proved inconclusive. However, that hasn't stopped efforts to use these tests, particularly the galaxy number-count test [23], and one or more of these classical tests may one day provide a definitive measurement.

To summarize this aside on the mass density of the Universe:

1. Most of the matter is dark.

2. Baryons provide between about 1% and 10% of the mass density.

3. Ω_0 could conceivably be as small as 0.1—in which case all the dark matter could be baryons (e.g., neutron stars, "jupiters," and so on).

4. If asked for the value of Ω_0, a typical astronomer would respond with a number in the interval 0.2 ± 0.1.

5. The evidence continues to mount for a gap between Ω_B and Ω_0—in which case nonbaryonic dark matter is required.

The current prejudice—and certainly that of this author—is a flat Universe ($\Omega_0 = 1$) with nonbaryonic dark matter, $\Omega_X \sim 1 \gg \Omega_B$. However, I shall continue to display the Ω_0 dependence of important quantities.

1.2.3 The early, radiation-dominated Universe

In any case, at present, matter outweighs radiation by a wide margin. However, since the energy density in matter decreases as R^{-3}, and that in radiation as R^{-4} (the extra factor due to the red shifting of the energy of relativistic particles), at early times the Universe was radiation dominated—indeed the calculations of primordial nucleosynthesis provide excellent evidence for this. Denoting the epoch of matter-radiation equality by subscript 'EQ,' and using $T_0 = 2.73\,\mathrm{K}$, it follows that

$$R_{\mathrm{EQ}} = 4.18 \times 10^{-5}\,(\Omega_0 h^2)^{-1}; \qquad T_{\mathrm{EQ}} = 5.62(\Omega_0 h^2)\,\mathrm{eV}; \tag{7}$$

$$t_{\mathrm{EQ}} = 4.17 \times 10^{10}(\Omega_0 h^2)^{-2}\,\mathrm{sec}. \tag{8}$$

At early times the expansion rate and age of the Universe were determined by the temperature of the Universe and the number of relativistic degrees of freedom:

$$\rho_{\mathrm{rad}} = g_*(T)\frac{\pi^2 T^4}{30}; \qquad H \simeq 1.67 g_*^{1/2} T^2/m_{\mathrm{Pl}}; \tag{9}$$

$$\Rightarrow R \propto t^{1/2}; \qquad t \simeq 2.42 \times 10^{-6} g_*^{-1/2}(T/\,\mathrm{GeV})^{-2}\,\mathrm{sec}; \tag{10}$$

where $g_*(T)$ counts the number of ultra-relativistic degrees of freedom ($\approx$ the sum of the internal degrees of freedom of particle species much less massive than the temperature) and $m_{\mathrm{Pl}} \equiv G^{-1/2} = 1.22 \times 10^{19}\,\mathrm{GeV}$ is the Planck mass. For example, at the epoch of nucleosynthesis, $g_* = 10.75$ assuming three, light ($\ll$ MeV) neutrino species; taking into account all the species in the standard model, $g_* = 106.75$ at temperatures much greater than 300 GeV; see Fig. 6.

A quantity of importance related to g_* is the entropy density in relativistic particles,

$$s = \frac{\rho + p}{T} = \frac{2\pi^2}{45} g_* T^3,$$

and the entropy per comoving volume,

$$S \ \propto \ R^3 s \ \propto \ g_* R^3 T^3.$$

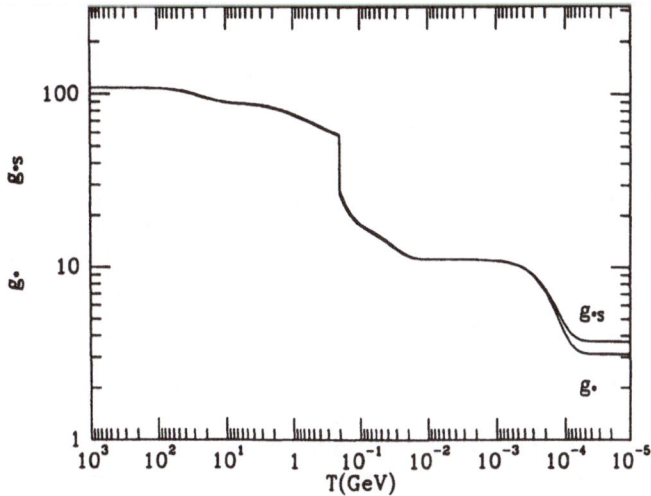

Figure 6. The total effective number of relativistic degrees of freedom $g_*(T)$ in the standard model of particle physics as a function of temperature.

By a wide margin most of the entropy in the Universe exists in the radiation bath. The entropy density is proportional to the number density of relativistic particles. At present, the relativistic particle species are the photons and neutrinos, and the entropy density is a factor of 7.04 times the photon-number density: $n_\gamma = 413\,\text{cm}^{-3}$ and $s = 2905\,\text{cm}^{-3}$.

In thermal equilibrium—which provides a good description of most of the history of the Universe—the entropy per comoving volume S remains constant. This fact is very useful. First, it implies that the temperature and scale factor are related by

$$T \propto g_*^{-1/3} R^{-1}, \tag{11}$$

which for $g_* = \text{const}$ leads to the familiar $T \propto R^{-1}$.

Second, it provides a way of quantifying the net baryon number (or any other particle number) per comoving volume:

$$N_B \equiv R^3 n_B = \frac{n_B}{s} \simeq (4-7) \times 10^{-11}. \tag{12}$$

The baryon number of the Universe tells us two things: (1) the entropy per particle in the Universe is extremely high, about 10^{10} or so compared to about 10^{-2} in the sun and a few in the core of a newly formed neutron star. (2) The asymmetry between matter and antimatter is very small, about 10^{-10}, since at early times quarks and antiquarks were roughly as abundant as photons. One of the great successes of particle cosmology is baryogenesis, the idea that B, C, and CP violating interactions occurring out-of-equilibrium early on allow the Universe to develop a net baryon number of this magnitude [24].

Finally, the constancy of the entropy per comoving volume allows us to characterize the size of comoving volume corresponding to our present Hubble volume in a very physical way: by the entropy it contains,

$$S_U = \frac{4\pi}{3} H_0^{-3} s \simeq 10^{90}. \tag{13}$$

1.2.4 The earliest history

The standard cosmology is tested back to times as early as about 0.01 sec; it is only natural to ask how far back one can sensibly extrapolate. Since the fundamental particles of Nature are point-like quarks and leptons whose interactions are perturbatively weak at energies much greater than 1 GeV, one can imagine extrapolating as far back as the epoch where general relativity becomes suspect, i.e., where quantum gravitational effects are likely to be important: the Planck epoch, $t \sim 10^{-43}$ sec and $T \sim 10^{19}$ GeV. Of course, at present, our firm understanding of the elementary particles and their interactions only extends to energies of the order of 100 GeV, which corresponds to a time of the order of 10^{-11} sec or so. We can be relatively certain that at a temperature of $100 \, \text{MeV} - 200 \, \text{MeV}$ ($t \sim 10^{-5}$ sec) there was a transition (likely a second-order phase transition) from quark/gluon plasma to very hot hadronic matter, and that some kind of phase transition associated with the symmetry breakdown of the electroweak theory took place at a temperature of the order of 300 GeV ($t \sim 10^{-11}$ sec).

It is interesting to look at the progress that has taken place since Weinberg's classic text on cosmology was published in 1972 [25]; at that time many believed that the Universe had a limiting temperature of the order of several hundred MeV, due to the exponentially rising number of particle states, and that one could not speculate about earlier times. Today, based upon our present knowledge of physics and powerful mathematical tools (e.g., gauge theories, grand unified theories, and superstring theory) we are able to make quantitative speculations back to the Planck epoch—and even earlier. Of course, these speculations could be totally wrong, based upon a false sense of confidence (arrogance?). As I shall discuss, inflation is one of these well defined—and well motivated—speculations about the history of the Universe well after the Planck epoch, but well before primordial nucleosynthesis.

1.2.5 The matter and curvature dominated epochs

After the equivalence epoch, the matter density exceeds that of radiation. During the matter-dominated epoch the scale factor grows as $t^{2/3}$ and the age of the Universe is related to red shift by

$$t = 2.06 \times 10^{17} (\Omega_0 h^2)^{-1/2} (1 + z)^{-3/2} \text{ sec.} \tag{14}$$

If $\Omega_0 < 1$, the matter-dominated epoch is followed by a "curvature-dominated" epoch where the rhs of the Friedmann equation is dominated by the $|k|/R^2$ term. When the Universe is curvature dominated it is said to expand freely, no longer decelerating since the gravitational effect of matter has become negligible: $\ddot{R} \approx 0$ and $R \propto t$. The epoch of curvature dominance begins when the matter and curvature terms are equal:

$$R_{\text{CD}} = \frac{\Omega_0}{1 - \Omega_0} \longrightarrow \Omega_0; \qquad z_{\text{CD}} = \Omega_0^{-1} - 2 \longrightarrow \Omega_0^{-1}; \tag{15}$$

where the limits shown are for $\Omega_0 \to 0$. By way of comparison, in a flat Universe with a cosmological constant, the Universe becomes "vacuum dominated" when $R = R_{\text{vac}}$:

$$R_{\text{vac}} = \left(\frac{\Omega_0}{1 - \Omega_0}\right)^{1/3} \longrightarrow \Omega_0^{1/3}; \qquad z_{\text{vac}} = \left(\frac{1 - \Omega_0}{\Omega_0}\right)^{1/3} - 1 \longrightarrow \Omega_0^{-1/3}. \tag{16}$$

For a given value of Ω_0, the transition occurs much more recently, which has important implications for structure formation since small density perturbations only grow during the matter-dominated era.

1.2.6 One last thing: horizons

In spite of the fact that the Universe was vanishingly small at early times, the rapid expansion precluded causal contact from being established throughout. Photons travel on null paths characterized by $dr = dt/R(t)$; the physical distance that a photon could have traveled since the bang until time t, the distance to the horizon, is

$$
\begin{aligned}
d_H(t) & = R(t) \int_0^t \frac{dt'}{R(t')} \\
& = t/(1-n) = nH^{-1}/(1-n) \qquad \text{for } R(t) \propto t^n, \ n < 1.
\end{aligned} \tag{17}
$$

Note, in the standard cosmology the distance to the horizon is finite, and up to numerical factors, equal to the age of the Universe or the Hubble radius, H^{-1}. For this reason, I will use horizon and Hubble radius interchangeably.[1]

An important quantity is the entropy within a horizon volume: $S_{\text{HOR}} \sim H^{-3}T^3$; during the radiation-dominated epoch $H \sim T^2/m_{\text{Pl}}$, so that

$$
S_{\text{HOR}} \sim \left(\frac{m_{\text{Pl}}}{T}\right)^3 ; \tag{18}
$$

from this we conclude that at early times the comoving volume that encompasses all that we can see today (characterized by an entropy of 10^{90}) was comprised of a very large number of causally disconnected regions.

1.3 The challenge: development of structure

This brings us to what I believe is the major challenge of the standard cosmology at present: a detailed understanding of the formation of structure in the Universe. We have every indication that the Universe at early times, say $t \ll 300,000\,\text{yrs}$, was very homogeneous; however, today inhomogeneity (or structure) is ubiquitous: stars ($\delta\rho/\rho \sim 10^{30}$), galaxies ($\delta\rho/\rho \sim 10^5$), clusters of galaxies ($\delta\rho/\rho \sim 10 - 10^3$), superclusters, or "clusters of clusters" ($\delta\rho/\rho \sim 1$), voids ($\delta\rho/\rho \sim -1$), great walls, and so on.

For some 25 years the standard cosmology has provided a general framework for understanding this: Once the Universe becomes matter dominated (around 1000 yrs after the bang) primeval density inhomogeneities ($\delta\rho/\rho \sim 10^{-5}$) are amplified by gravity and grow into the structure we see today [26]. The fact that a fluid of self-gravitating particles is unstable to the growth of small inhomogeneities was first pointed out by Jeans and is known as the Jeans instability. The existence of these inhomogeneities was confirmed in spectacular fashion by the COBE DMR discovery of CBR anisotropy this past spring: The temperature anisotropies detected almost certainly owe their existence to primeval density inhomogeneities, as causality precludes microphysical processes from producing anisotropies on angular scales larger than about 1°, the angular size of the horizon at last scattering.

At last, the basic picture has been put on firm ground (whew!). Now the challenge is to fill in the details—origin of the density perturbations, precise evolution of the structure, and so on. As I shall emphasize, such an understanding may well be within reach, and offers a window on the early Universe.

---

[1] In inflationary models the horizon and Hubble radius are not roughly equal as the horizon distance grows exponentially relative to the Hubble radius; in fact, at the end of inflation they differ by e^N, where N is the number of e-folds of inflation. However, I will slip and use "horizon" and "Hubble radius" interchangeably, though I will always mean Hubble radius.

1.3.1 The general picture: gravitational instability

Let us begin by expanding the perturbation to the matter density in plane waves

$$\frac{\delta \rho_M(\mathbf{x}, t)}{\rho_M} = \frac{1}{(2\pi)^3} \int d^3 k \, \delta_k(t) e^{-i\mathbf{k}\cdot\mathbf{x}}, \tag{19}$$

where $\lambda = 2\pi/k$ is the comoving wavelength of the perturbation and $\lambda_{\text{phys}} = R\lambda$ is the physical wavelength. The comoving wavelengths of perturbations corresponding to bright galaxies, clusters, and the present horizon scale are respectively: about 1 Mpc, 10 Mpc, and $3000h^{-1}$ Mpc, where 1 Mpc $\simeq 3.09 \times 10^{24}$ cm $\simeq 1.56 \times 10^{38}$ GeV^{-1}.

The growth of small matter inhomogeneities of wavelength smaller than the Hubble scale ($\lambda_{\text{phys}} \lesssim H^{-1}$) is governed by a Newtonian equation:

$$\ddot{\delta}_k + 2H\dot{\delta}_k + v_s^2 k^2 \delta_k / R^2 = 4\pi G \rho_M \delta_k, \tag{20}$$

where $v_s^2 = dp/d\rho_M$ is the square of the sound speed. Competition between the pressure term and the gravity term on the rhs determine whether or not pressure can counteract gravity: Perturbations with wavenumber larger than the Jeans wavenumber, $k_J^2 = 4\pi G R^2 \rho_M / v_s^2$, are Jeans stable and just oscillate; perturbations with smaller wavenumber are Jeans unstable and can grow. For cold dark matter $v_s \simeq 0$ and all scales are Jeans unstable; even for baryonic matter, after decoupling k_J corresponds to a baryon mass of only about $10^5 M_\odot$. All the scales of interest here are Jeans unstable and we will ignore the pressure term.

Let us discuss solutions to this equation under different circumstances. First, consider the Jeans problem, evolution of perturbations in a static fluid, i.e., $H = 0$. In this case Jeans unstable perturbations grow exponentially, $\delta_k \propto \exp(t/\tau)$ where $\tau = 1/\sqrt{4G\pi\rho_M}$. Next, consider the growth of Jeans unstable perturbations in a matter-dominated Universe, i.e., $H^2 = 8\pi G \rho_M / 3$ and $R \propto t^{2/3}$. Because the expansion tends to "pull particles away from one another," the growth is only power law, $\delta_k \propto t^{2/3}$; i.e., at the same rate as the scale factor. Finally, consider a radiation or curvature dominated Universe, i.e., $8\pi G \rho_{\text{rad}}/3$ or $|k|/R^2$ much greater than $8\pi G \rho_M/3$. In this case, the expansion is so rapid that matter perturbations grow very slowly, as $\ln R$ in radiation-dominated epoch, or not at all $\delta_k = $ const in the curvature-dominated epoch.

The growth of nonlinear perturbations is another matter; once a perturbation reaches an overdensity of order unity or larger it "separates" from the expansion— i.e., becomes its own self-gravitating system and ceases to expand any further. In the process of virial relaxation, its size decreases by a factor of two—density increases by a factor of 8; thereafter, its density contrast grows as R^3 since the average matter density is decreasing as R^{-3}, though smaller scales could become Jeans unstable and collapse further to form smaller objects of higher density, stars, etc.

From this we learn that structure formation begins when the Universe becomes matter dominated and ends when it becomes curvature dominated (at least the growth of linear perturbations). The total growth available for linear perturbations is $R_{\text{CD}}/R_{\text{EQ}} \simeq 2.4 \times 10^4 \, \Omega_0^2 h^2$; since nonlinear structures have evolved by the present epoch, we can infer that primeval perturbations of the order $(\delta\rho_M/\rho_M)_{\text{EQ}} \sim 4 \times 10^{-5} (\Omega_0 h)^{-2}$ are required. Note that in a low-density Universe larger initial perturbations are necessary as there is less time for growth ("the low Ω_0 squeeze"). Further, in a baryon-dominated Universe things are even more difficult as perturbations in the baryons cannot begin to grow until after decoupling since matter is tightly coupled to the radiation. (In a flat, low-Ω_0 model with a cosmological constant the growth of linear fluctuations continues until almost today since $z_\Lambda \sim \Omega_0^{-1/3}$, and so the total growth factor is about $2.4 \times 10^4 (\Omega_0 h^2)$. We will return to this model later.)

1.3.2 CBR temperature fluctuations

The existence of density inhomogeneities has another important consequence: fluctuations in the temperature of the CBR of a similar amplitude [27]. The temperature difference measured between two points separated by a large angle ($\gtrsim 1°$) arises due to a very simple physical effect:[2] The difference in the gravitational potential between the two points on the last-scattering surface, which in turn is related to the density perturbation, determines the temperature anisotropy on the angular scale subtended by that length scale,

$$\left(\frac{\delta T}{T}\right)_\theta = -\left(\frac{\delta\phi}{3}\right)_\lambda \approx \frac{1}{2}\left(\frac{\delta\rho}{\rho}\right)_{\mathrm{HOR},\lambda} ; \tag{21}$$

where the scale $\lambda \sim 100h^{-1}\,\mathrm{Mpc}(\theta/\deg)$ subtends an angle θ on the last-scattering surface. This is known as the Sachs-Wolfe effect [28].

The quantity $(\delta\rho/\rho)_{\mathrm{HOR},\lambda}$ is the amplitude with which a density perturbation crosses inside the horizon, i.e., when $R\lambda \sim H^{-1}$. Since the fluctuation in the gravitational potential $\delta\phi \sim (R\lambda/H^{-1})^2(\delta\rho/\rho)$, the horizon-crossing amplitude is equal to the gravitational potential (or curvature) fluctuation. The horizon-crossing amplitude $(\delta\rho/\rho)_{\mathrm{HOR}}$ has several nice features: (i) during the matter-dominated era the potential fluctuation on a given scale remains constant, and thus the potential fluctuations at decoupling on scales that crossed inside the horizon after matter-radiation equality, corresponding to angular scales $\lesssim 0.1°$, are just given by their horizon-crossing amplitude; (ii) because of its relationship to $\delta\phi$ it provides a dimensionless, geometrical measure of the size of the density perturbation on a given scale, and its effect on the CBR; (iii) by specifying perturbation amplitudes at horizon crossing one can effectively avoid discussing the evolution of density perturbations on scales larger than the horizon, where a Newtonian analysis does not suffice and where gauge subtleties (associated with general relativity) come into play; and finally (iv) the density perturbations generated in inflationary models are characterized by $(\delta\rho/\rho)_{\mathrm{HOR}} \simeq \mathrm{const}$.

On angular scales smaller than about 1° two other physical effects lead to CBR temperature fluctuations: the motion of the last-scattering surface (Doppler) and the intrinsic fluctuations in the local photon temperature. These fluctuations are much more difficult to compute, and depend on microphysics—the ionization history of the Universe and the damping of perturbations in the photon-baryon fluid due to photon streaming. Not only are the Sachs-Wolfe fluctuations simpler to compute, but they accurately mirror the primeval fluctuations since at the epoch of decoupling microphysics is restricted to angular scales less than about a degree.

In sum, on large angular scales the Sachs-Wolfe effect dominates; on the scale of about 1° the total CBR fluctuation is about twice that due to the Sachs-Wolfe effect; on smaller scales the Doppler and intrinsic fluctuations dominate. CBR temperature fluctuations on scales smaller than about 0.1° are severely reduced by the smearing effect of the finite thickness of last-scattering surface.

Details aside, in the context of the gravitational instability scenario density perturbations of sufficient amplitude to explain the observed structure lead to temperature fluctuations in the CBR of characteristic size,

$$\frac{\delta T}{T} \approx 10^{-5}\,(\Omega_0 h)^{-2}. \tag{22}$$

---

[2]Large angles mean those larger than the angle subtended by the horizon-scale at decoupling, $\theta \sim H_{\mathrm{DEC}}^{-1}/H_0^{-1} \sim z_{\mathrm{DEC}}^{-1/2} \sim 1°$.

To be sure I have brushed over important details, but this equation conveys a great deal. First, the overall amplitude is set by the inverse of the growth factor, which is just the ratio of the radiation energy density to matter density at present. Next, it explains why theoretical cosmologists were so relieved when the COBE DMR detected temperature fluctuations of this amplitude, and conversely why one heard offhanded remarks before the COBE DMR detection that the standard cosmology was in trouble because the CBR temperature was too uniform to allow for the observed structure to develop. Finally, it illustrates one of the reasons why cosmologists who study structure formation have embraced the flat-Universe model with such enthusiasm: If we accept the Universe that meets the eye, $\Omega_0 \sim 0.1$ and baryons only, then the simplest models of structure formation predict temperature fluctuations of the order of 10^{-3}, far too large to be consistent with observation. Later, I will mention Peebles' what-you-see-is-what-you-get model [29], also known as PIB for primeval baryon isocurvature fluctuation, which is still viable because the spectrum of perturbations decreases rapidly with scale so that the perturbations that give rise to CBR fluctuations are small (which is no mean feat). Historically, it was fortunate that one started with a low-Ω_0, baryon-dominated Universe: the theoretical predictions for the CBR fluctuations were sufficiently favorable that experimentalist were stirred to try to measure them—and then, slowly, theorists lowered their predictions. Had the theoretical expectations begun at 10^{-5}, experimentalists might have been too discouraged to even try!

1.3.3 An initial data problem

With the COBE DMR detection in hand we can praise the success of the gravitational instability scenario; however, the details now remain to be filled in. The structure formation problem is now one of initial data, namely

1. The quantity and composition of matter in the Universe, Ω_0, Ω_B, and Ω_{other}.

2. The spectrum of initial density perturbations: for the purist, $(\delta\rho/\rho)_{\text{EQ}}$, or for the simulator, the Fourier amplitudes at the epoch of matter-radiation equality.

In a statistical sense, these initial data provide the "blueprint" for the formation of structure.

The initial data are the challenge and the opportunity. Although the gravitational instability picture has been around since the discovery of the CBR itself, the lack of specificity in initial data has impeded progress. With the advent of the serious study of the earliest history of the Universe a new door was opened. We now have several well motivated early-Universe blueprints: Inflation-produced density perturbations and nonbaryonic dark matter; cosmic-string produced perturbations and nonbaryonic dark matter [30]; texture produced density perturbations and nonbaryonic dark matter [31]; a baryon-dominated Universe with isocurvature fluctuations[3] [29]. Structure formation also provides the opportunity to probe the earliest history of the Universe, by testing these interesting, if not bold, blueprints. I will be focusing on the blueprints motivated by inflation.

---

[3]Isocurvature baryon-number fluctuations correspond at early times to fluctuations in the local baryon number but not the energy density. At late times, when the Universe is matter dominated, they become fluctuations in the mass density of a comparable amplitude.

2 Inflation: An Overview

2.1 Shortcomings of the Standard Cosmology

By now the shortcomings of the standard cosmology are well appreciated: the horizon or large-scale smoothness problem; the small-scale inhomogeneity problem (origin of density perturbations); the flatness or oldness problem; and the monopole problem. I will only briefly review them here. They do not indicate any logical inconsistencies of the standard cosmology; rather, that very special initial data seem to be required for evolution to a universe that is qualitatively similar to ours today. Nor is inflation the first attempt to address these shortcomings: Over the past two decades cosmologists have pondered this question and proposed other solutions [33]. Inflation is a solution based upon well-defined, albeit speculative, early-Universe microphysics describing the post-Planck epoch.

The uniformity of the CBR temperature, to better than a part in 10^4, implies that the Universe on the largest scales (say $\gtrsim 100h^{-1}\,\mathrm{Mpc}$) is very smooth as density inhomogeneities induce temperature fluctuations of a similar magnitude. The existence of particle horizons in the standard cosmology precludes explaining the smoothness as a result of microphysical events: The horizon at decoupling, the last time one could imagine temperature fluctuations being smoothed by particle interactions, corresponds to an angular scale on the sky of about $1°$, which precludes temperature variations on larger scales from being erased. In terms of entropy, the presently observed Universe, corresponds to a comoving volume containing an entropy of order 10^{90}; during the early radiation dominated epoch the horizon volume contained an entropy of order $(m_{\mathrm{Pl}}/T)^3$, implying that at early times our current Hubble volume consisted of countless, causally distinct regions.

To account for the small-scale lumpiness of the Universe today, density perturbations with horizon-crossing amplitudes of 10^{-5} on scales of $1\,\mathrm{Mpc}$ to $10^4\,\mathrm{Mpc}$ or so are required. As can be seen in Fig. 7, in the standard cosmology the physical size of a perturbation, which grows as the scale factor, begins larger than the horizon and relatively late in the history of the Universe crosses inside the horizon,

$$
\begin{aligned}
t_{\mathrm{HOR}} &\simeq 3 \times 10^8 \, (\lambda/\,\mathrm{Mpc})^2 \, \sec & \lambda &\lesssim 13h^{-2}\,\mathrm{Mpc}; \\
&\simeq 3 \times 10^7 \, (\lambda/\,\mathrm{Mpc})^3 \, \sec & \lambda &\gtrsim 13h^{-2}\,\mathrm{Mpc}.
\end{aligned}
\tag{23}
$$

This precludes a causal microphysical explanation for the origin of the required density perturbations.[4]

The fact that Ω_0 is of order unity means that the curvature radius is comparable to the Hubble radius. Had that been the case at the initial epoch, the Universe would be a very different place today: Since the curvature term in the Friedmann equation decreases only as R^{-2}, while the matter and radiation densities decrease as R^{-3} and R^{-4} respectively, a curvature radius comparable to the Hubble radius early on would have led to a Universe that quickly became curvature dominated. For positive curvature, recollapse would follow quickly, and for negative curvature, a coasting phase that would lead to a Universe that cools too quickly (for $t_{\mathrm{initial}} \sim 10^{-43}\,\sec$, the temperature reaches

---

[4]Of course, it is possible to produce the perturbations at very late times, when the relevant scale has already crossed inside the horizon [32]; the motivation for the nonstandard microphysics required to do so is lacking at present. It is also possible for microphysics to produce isocurvature perturbations by producing a pressure wave that eventually propagates to large scales; this is the type of perturbation that is generated by cosmic strings or textures.

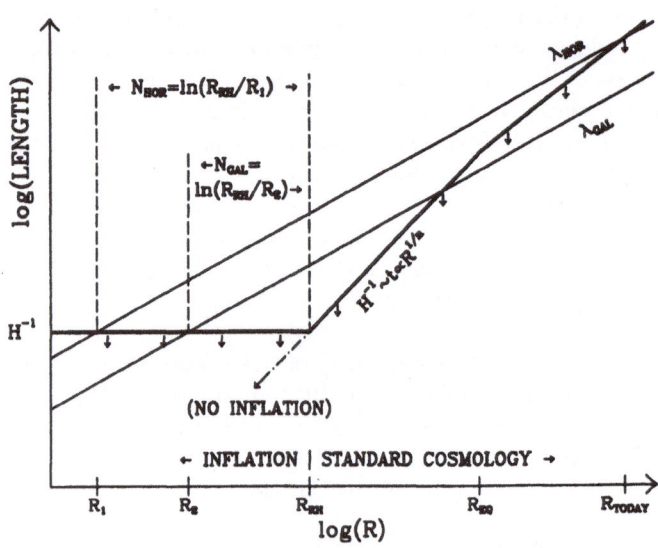

Figure 7. The physical wavelength of a density perturbation and the horizon size H^{-1} as a function of scale factor; λ_{GAL} indicates a galactic sized perturbation ($\lambda \sim 1\,\mathrm{Mpc}$) and λ_{HOR} corresponds to the present Hubble radius (horizon). Microphysics operates on scales $\lesssim H^{-1}$; without inflation scales cross the Hubble radius but once.

3 K at an age of 10^{-11} sec). Put another way, Ω is an unstable fixed point:

$$\Omega(t) = \frac{1}{1 - x(t)}$$

$$x(t) = \frac{k/R^2}{8\pi G\rho/3}; \tag{24}$$

the deviation of $\Omega(t)$ from unity increases as $x(t) \propto R^n$, $n = 2$ (radiation-dominated epoch), $n = 1$ (matter-dominated epoch). In order that Ω still be close to unity today, it must have extremely close to unity early on; for $t_{\text{initial}} \sim 10^{-43}$ sec, $|\Omega(t_{\text{initial}}) - 1| \lesssim 10^{-60}$ is necessary. Thus, for most of its history the Universe must have been extremely flat, i.e., $R_{\text{curv}} \gg H^{-1}$; if Ω_0 is not equal to unity, then the Universe just today is beginning to exhibit its curvature. Why now?

Last, I mention the monopole problem: The simplest grand unified theories and the standard cosmological lead to a disastrous prediction, the extreme overproduction of magnetic monopoles [34]. This overproduction traces to the smallness of the horizon at very early times: magnetic monopoles are produced as defects of the GUT phase transition at an abundance of about 1 per horizon volume which corresponds to a present monopole to photon ratio of order $(T_{\text{GUT}}/m_{\text{Pl}})^3$.

The first three problems do not involve logical inconsistencies: The initial data for a perturbed FRW model that is extremely flat exist. Rather, it is the fact that such initial data are "very special" which is disturbing. Collins and Hawking quantified it: The set of initial data that evolve to a state qualitatively similar to our Universe is of measure zero [35]. Maybe the Creator had a lucky day! Or better yet, perhaps the present state of the Universe traces to events that took place early on. Inflation provides an interesting example of the latter.

2.2 Generic Aspects of Inflation

Inflationary cosmology has become a very mature subject in the decade since Guth wrote his influential paper [36] that launched the inflationary cosmology boom. While there are a multitude of different kinds of inflation (see below), two features are common to all models of inflation [37]

- Superluminal expansion

- Massive entropy production

Superluminal expansion refers to accelerated growth of the scale factor ($\ddot{R} > 0$ which implies $R \propto t^n$ with $n > 1$), and its necessity is easy to understand. In order that the physical size of a comoving scale, $d_{\text{phys}} \propto R(t)$, begin sub-Hubble size and and become super-Hubble size, $R(t)$ must increase faster than t since $H^{-1} \propto R(t)^{1/n}$. Thus, "superluminal" expansion is a necessary kinematic requirement if one is to both solve the horizon and create density perturbations (see Fig. 7).

The reason for the second requirement is equally simple: In the absence of entropy production the entropy per comoving volume $S \propto (RT)^3$ remains constant; rapid expansion can create a "very large" smooth patch, but the entropy within that patch remains constant. As discussed above, at early times the entropy within a horizon-sized patch is very small, too small to account for the entropy within our present Hubble volume. Only massive entropy production can change this [37].

To illustrate, consider Guth's original model of inflation based upon a first-order phase transition [36]. The basic idea is that the Higgs field responsible for the spontaneous breakdown of the GUT symmetry gets "hung up" in a local, high-energy,

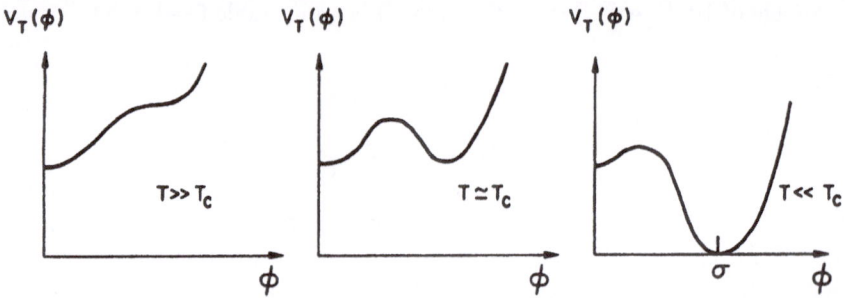

Figure 8. The free-energy density as a function of temperature for a first-order phase transition.

minimum of its potential (more precisely, free-energy density). At high temperatures the state of minimum free energy is characterized by $\phi = 0$, indicating that the full GUT symmetry is manifest; as the temperature drops below the critical temperature, the state of minimum free energy is characterized by $\phi \neq 0$, the state that exhibits broken symmetry. In a first-order transition $\phi = 0$ can remain a local minimum of the free energy, separated from the global minimum by a potential barrier; see Fig. 8. During the time that ϕ is hung up the large vacuum-energy density, $\rho = V(\phi = 0) \equiv \mathcal{M}^4$, drives very rapid expansion ($\mathcal{M}$ is the energy scale that characterizes of the symmetry breaking).

For definiteness, take $\mathcal{M} = 10^{14}$ GeV, a typical scale for inflation; the Hubble time associated with the false-vacuum energy $H^{-1} \sim 10^{-34}$ sec. The size of a region that one might expect to be smooth is of order $ct \sim 10^{-23}$ cm; the entropy within such a patch is of order 10^{14}. While the Higgs field is trapped in the false vacuum, the temperature of the Universe continues to decrease as R^{-1}; very soon the thermal energy density becomes insignificant compared to the constant false-vacuum energy density. At this point, the Universe enters a de Sitter phase of exponential expansion since $\rho \sim \mathcal{M}^4 = $ const; this is the superluminal expansion. As the Universe expands, it cools exponentially with the entropy per comoving volume remaining constant; the smooth horizon-sized patch continues to contain an entropy of only 10^{14} as it grows exponentially in size.

During inflation the scale factor undergoes many e-folds; the precise number is determined by how long the Higgs field is hung up: $N = H\Delta t$. Again, for definiteness, suppose that the Universe gets hung up for a mere 10^{-32} sec; then, during inflation the patch grows in linear size by $e^{100} \sim 10^{43}$ and its temperature drops by the same factor. Thus far, inflation has done little. When the Higgs field does make its way to the true vacuum, the enormous false-vacuum energy is released and ultimately thermalized, reheating the patch to a temperature of order $\mathcal{M} \sim 10^{14}$ GeV, thereby increasing the entropy of the patch by a factor of $e^{3N} \sim 10^{129}$. This is the massive entropy production. After "reheating" the patch contains an entropy of order 10^{143}, and can easily contain the comoving volume that corresponds to our present Hubble radius, which is characterized by an entropy of "only" 10^{90}.

It is clear that the smoothness problem has been solved. The kinematic requirement for producing density perturbations on astrophysically interesting scales has been satisfied; the mechanism that produces density perturbations, quantum fluctuations in the ϕ field, will be discussed later. What about the flatness problem? Suppose for definiteness that the curvature radius at the beginning of inflation is of order the Hubble radius (which corresponds to Ω just beginning to deviate from unity); at the end of inflation the curvature radius has grown by a factor of e^N, while the energy density has remained constant. This means that $\Omega_{\text{end}} = 1/[1 - (k/R^2)/(8\pi G\rho/3)] \simeq 1 \pm e^{-2N}$ has been reset to a value exponentially close to unity. Using our fiducial numbers, at the end of inflation the curvature radius is order 10^{20} cm; from then until today it grows by a factor of $\mathcal{M}/3\,\text{K} \sim 10^{27}$, reaching a present size of order 10^{47} cm. This is enormous compared to the present Hubble radius and implies that Ω is still very close to unity today. The flatness problem has clearly been solved and a flat Universe predicted.

Consider the fate of monopoles—or any other "pollutant" in the pre-inflationary Universe. The number of monopoles within the patch ($= N_M$) remains constant; however, the number per comoving volume, $n_M/s = N_M/S$, decreases by a factor of $e^{3N} \sim 10^{129}$ due to the massive entropy production. Undesirables are diluted away! Of course, this also implies that the baryon number of the Universe, $n_B/s \sim 10^{-10}$, must be produced after inflation.

Finally, a simple exercise; what is the minimum amount of inflation needed to solve the smoothness problem? Start with a Hubble-sized patch at the beginning of inflation; it contains an entropy of $S_{\text{initial}} \sim H^{-3}T^3 \sim (m_{\text{Pl}}/\mathcal{M})^3$. Assuming perfect conversion of vacuum energy to radiation, after inflation the entropy contained within the patch is $e^{3N}S_{\text{initial}} \sim e^{3N}m_{\text{Pl}}^3/\mathcal{M}^3$. To solve the smoothness problems this must be greater than 10^{90}, which implies

$$N \gtrsim N_{\text{min}} = 56 + \ln(\mathcal{M}/10^{14}\,\text{GeV}). \tag{25}$$

Equivalently, one can express the size of the patch today relative to the present Hubble radius,

$$d_{\text{patch}} = \exp(N - N_{\text{min}})\,H_0^{-1}. \tag{26}$$

What about the flatness problem? It is simple to show the present value of Ω is related to that at the beginning of inflation and the size of the patch today:

$$|\Omega_0 - 1| = \left(\frac{H_0^{-1}}{d_{\text{patch}}}\right)^2 |\Omega_{\text{preinflation}} - 1|. \tag{27}$$

Remarkably enough, the amount of inflation required to solve the flatness and smoothness problems is the same. Put another way, if one comfortably solves the smoothness problem, Ω_0 is necessarily very, very close to unity. This means that a flat Universe is an unequivocal prediction of inflation.

2.3   Current Status of Inflationary Models

2.3.1   Types of inflation

In this very brief overview I divide models of inflation into three broad classes: old, slow rollover, and first-order (or extended). By old inflation I mean Guth's original model, which I forgot to mention was a nonstarter! Let me explain; once trapped in the false vacuum, the Higgs field must quantum-mechanically tunnel to the true vacuum; in order to ensure a sufficient amount of inflation, this transition must not occur until 60 or so Hubble times after inflation has begun. As we shall see this is essentially impossible to arrange.

The decay of the false vacuum is well understood [38]: It proceeds via the nucleation of bubbles of true vacuum that expand outward at the speed of light. For a given potential the bubble nucleation rate (per unit volume) Γ is straightforward to calculate [38]. Roughly speaking, bubbles convert all of space into the true vacuum when Γ/H^4, the number of bubbles nucleated in a Hubble volume in a Hubble time exceeds order unity; since each bubble nucleated during a Hubble time liberates about a Hubble volume, $\Gamma/H^4 \sim 1$ ensures that all of space is converted to true-vacuum in a Hubble time (before the expansion "creates" more false vacuum). The false-vacuum energy is converted into "heat" by the collision of vacuum bubbles [39].

The recipe for successful old inflation is for Γ/H^4 to remain less than unity for 60 or so Hubble times and then increase to greater than unity. Unfortunately, shortly after inflation begins Γ, like the expansion rate becomes constant, as the temperature of the Universe rapidly approaches zero and become irrelevant. This is the fundamental problem with old inflation; Γ/H^4 is constant. The Universe can either get hung up in the false vacuum and inflate, or make the transition to the true vacuum, not both!

Slow-rollover inflation solved this problem, but at a price. The fix, suggested independently by Linde [40], and Albrecht and Steinhardt [41], is for inflation to occur as the scalar field slowly rolls the potential. They proposed using very flat potentials with small or nonexistent barriers between the false and true vacuum states; the vacuum-driven expansion takes place as the scalar field slowly (timescale $\gtrsim 60H^{-1}$), but inevitably rolls toward the true-vacuum state. When the scalar field responsible for inflation (often called the inflaton) reaches the true minimum of its potential it oscillates about it, the large vacuum energy having been converted into coherent inflaton oscillations. These oscillations ultimately decay into light-particle states reheating the Universe. From the quantum view, these coherent field oscillations correspond to zero-momentum inflaton particles; the decay of the scalar-field oscillations corresponds to the decay of massive inflaton particles [42].

Slow rollover led to the first viable models of inflation. There was, however, a price: In all models of slow-rollover the inflaton field must be very weakly coupled (dimensionless self coupling of order 10^{-14} or so); as we shall see this is dictated by achieving density perturbations of size 10^{-5} or so. Because of this fact, the inflaton cannot be directly responsible for GUT symmetry breaking as loop corrections from the inflaton-gauge interaction would spoil the flatness of the potential. The decoupled nature of the scalar field responsible for inflation gave birth to its name.

In the broadest sense, slow-rollover inflation refers to any model of inflation where a scalar field is displaced from the minimum of its potential and slowly rolls toward it, with the Universe inflating as it does. The slowly rolling scalar field responsible for inflation can begin near the origin at a high-energy local minimum of the potential (often referred to as "new inflation"), as it might with potentials used for spontaneous symmetry breaking, or it can begin far from the origin, away from a local minimum of the potential (often referred to as "chaotic inflation"), as it would with a potential such as $V(\phi) = \lambda\phi^4$ or $V(\phi) = m^2\phi^2/2$.

The latest and perhaps most interesting development in inflationary models is first-order (or extended) inflation [43]. In many ways it combines the best features of old inflation—intimate connection to particle physics phenomenology—and slow-rollover inflation—it works! As the name suggests, these models are associated with a first-order phase transition; how then do these models solve the Guth dilemma—the constancy of Γ/H^4? The first model of this type was due to La and Steinhardt [44]; their new twist was to use the Brans-Dicke theory of gravity rather than general relativity. In Brans-Dicke the gravitational constant $G_{\text{eff}} = \Phi^{-2}$, and evolves as the Brans-Dicke field Φ evolves. Because of this, for constant energy density *the scale factor only increases as a power of time*, $R(t) \propto t^{\omega+1/2}$ *and H decreases with time;* here ω is the coefficient of the kinetic-energy term for Φ. Thus, the efficiency of bubble nucleation $\Gamma/H^4 \propto t^4$ increases during inflation; at early times it can be much less than unity (so that the Universe remains trapped in the false vacuum) and then exceeds unity triggering the end of inflation via the nucleation and percolation of bubbles of true vacuum.

Models based on variations of this idea have been proposed. For example, if the

Higgs field couples to other fields which are evolving during inflation, then Γ will vary during inflation, leading to the variation of Γ/H^4 [45]. In first-order inflation models the Higgs field plays a relatively passive role, remaining trapped in the false vacuum during inflation; further, it need not be weakly coupled, nor is the shape of its potential particularly relevant.

By means of a conformal transformation extended inflation can be recast as slow-rollover inflation with an exponential potential with $\ln \Phi$ field playing the role of the inflaton [46]. In first-order inflation models there is another problem one has to worry about: If Γ/H^4 does not change rapidly enough, then too many bubbles will be nucleated long before the end of inflation; these bubbles eventually grow to astrophysical size and can have disastrous consequences (large anisotropies in the CBR, interference with primordial nucleosynthesis, and so on) [47]. To avoid "the big-bubble problem" in extended inflation ω must be less than about 20; that there be an upper limit to ω is not surprising since in the limit $\omega \to \infty$, Brans-Dicke goes to general relativity.

2.3.2 Viable models

There is no standard model of inflation; nor is there a model of inflation without some flaw. There are a number of "proof of existence" models, models that successful implement inflation, but are only beautiful in the eyes of their creators. Of course, this situation should be viewed in light of our general ignorance about physics at energy scales $\gg 10^3$ GeV (most inflation models involve an energy scale of order 10^{14} GeV). Moreover, the same criticism—lack of a standard model—applies to baryogenesis, and applied to primordial nucleosynthesis until the early 1970's!

Slow-rollover. There are numerous viable models; I will mention but a representative few. There is an almost decade old model based upon an ordinary GUT due to Shafi and Vilenkin [48], and Pi [49]. This model has the virtue that the inflaton field does more than cause inflation; it also breaks Peccei-Quinn symmetry and induces GUT symmetry breaking (by producing a negative mass-squared for the GUT Higgs field). After inflation the Universe reheats to a temperature of order 10^7 GeV, and a scenario for baryogenesis is included. In short, it is a complete model.

In passing, let me mention a similar model just proposed by Knox and myself [50]. The new twist is that the scale of inflation can be as small as the electroweak scale(!), and the inflaton field can be used to induce electroweak-symmetry breaking and other low-energy phenomena (e.g., righthanded neutrino masses). In principle, this model can be tested in laboratory experiments. Of course, this model is only viable provided one believes that the baryon asymmetry of the Universe can be produced at the weak scale or below.

There are many supersymmetric implementations of slow-rollover inflation [51]; a particularly elegant one is that of Holman, Ramond, and Ross [52]. The superpotential for their inflaton is very simple, $W(\phi) = (\Delta^2/M)(\phi - M)^2$; here $M = m_{\mathrm{Pl}}/\sqrt{8\pi}$ and Δ is the GUT scale. In this model, the self coupling of the inflaton in its scalar potential is given by the fourth power of the ratio of the GUT to Planck scales, $(\Delta/M)^4$, and the canonical small number arises because of the discrepancy between the GUT and Planck scales. The reheat temperature in this model is order 10^6 GeV, and the details of baryogenesis are spelled out.

There is a model called (by the authors) "natural inflation" [53]. The primary purpose of this model is to address the small self-coupling of the inflaton. To wit, the inflaton is a pseudo Nambu-Goldstone boson akin to the axion; a Nambu-Goldstone boson has an absolutely flat potential, i.e. is massless, and becomes a pseudo Nambu-Goldstone boson due to explicit symmetry breaking. The potential, $V(\phi) = \Lambda^4[1 + \cos(\phi/f)]$, has two energy scales: $f \sim m_{\mathrm{Pl}}$, the scale of the spontaneous symmetry breaking and $\Lambda \sim 10^{-5} f$, the scale of explicit symmetry breaking (GUT scale?). (In the axion analogy, $\Lambda = \Lambda_{\text{QCD}} \simeq 200\,\text{MeV}$ and f is the PQ symmetry-breaking scale.) Some superstring adherents have taken interest in this model as superstring theories often have pseudo Nambu-Goldstone bosons with Planck-scale symmetry breaking.

There is a broad class of slow-rollover models referred to as chaotic inflation; they illustrate the simplicity of inflation and were pioneered by Linde [54]. The simplest such models are based upon potentials that have nothing to do with spontaneous symmetry breaking, $V(\phi) = \lambda \phi^4$ or $V(\phi) = m^2 \phi^2/2$, with the inflaton initially displaced far from the origin, $\phi_{\text{initial}} \gtrsim 5 m_{\text{Pl}}$. (Chaotic inflation can also be implemented with more complicated potentials whose minima are not at the origin [55].) As with all slow-rollover models, there is a small, dimensionless number: $\lambda \sim 10^{-14}$ or $m^2 \simeq 10^{-12} m_{\text{Pl}}^2$. While the simplest models of chaotic inflation are not tied to specific particle-physics theories, some have been [56].

There are models where the inflaton field is not actually a scalar field; e.g., where it is related to the size of the compactified dimensions in models with extra dimensions [57], or is related to the scalar curvature $\mathcal{R}$ in higher derivative theories of gravity) [58].

The common undesirable feature of all slow-rollover models is a small, dimensionless number of order 10^{-14}, typically the self coupling of the inflaton; as we shall discuss, this small number is necessary to guarantee density perturbations of the appropriate size. To ensure the stability of the flatness of the potential against quantum (radiative) corrections the inflaton must be weakly coupled to the "rest of the world," and in this since, *all* the models mentioned are natural. However, weak coupling works at cross purposes with reheating and baryogenesis. Slow-rollover models liberate only a tiny fraction of the false-vacuum energy to radiation and have a relatively low reheat temperatures, which is problematic for baryogenesis as it must proceed after inflation. The second problem lies in the name "inflaton;" because the field responsible for inflation is so weakly coupled, without heroic efforts it is difficult to make it an integral part of a more encompassing particle physics theory.

First-order. These models have the potential (no pun intended) to incorporate the best aspects of both slow-rollover and old inflation. Inflation is again intimately connected to a cosmological phase transition at a scale of order the GUT scale and no special flatness is required of the Higgs potential. Moreover, reheating proceeds via vacuum-bubble collisions which guarantees good reheating and a unique signature of first-order inflation, a background of gravitational waves proceeded by bubble collisions, $\Omega_{\text{GW}} \sim 10^{-8}$ at a frequency determined by the scale of inflation, $f_{\text{GW}} \sim 10^6\,\text{Hz}\,(\mathcal{M}/10^{12}\,\text{GeV})$ [59].

The simplest first-order inflation model is extended inflation. First the good news: Brans-Dicke gravity exhibits conformal (scale) invariance (the Planck scale is replaced by a field). Conformal invariance is "the Hallmark" of superstring theory, which has stimulated new interest in Brans-Dicke like theories. Now the bad news; in order to avoid "the big-bubble problem," the Brans-Dicke parameter ω must be less than about 20, while solar-system tests set a *lower limit* of about 500 [60]. In its simplest form, extended inflation is not viable. Several variants have been put forth [43]; the simplest fix is to give the Brans-Dicke field a mass [46]. (A mass for the Brans-Dicke field anchors at the right value and makes the immune to solar-system tests.) Any mass less than about $10^9\,\text{GeV}$ and greater than a tiny fraction of an eV will do. Moreover, this simple fix involves something that string theorists must do anyway: break conformal invariance (the world is not conformally invariant, it has a multitude of energy scales).

In sum, inflation provides a very attractive early Universe paradigm. Models of inflation are based upon well defined, albeit very speculative, physics at energy scales well below the Planck scale. At present there is no standard model, or even a particularly compelling model; there are, however, a variety of models that work. Given our general ignorance about physics at energy scales $\gg 10^3\,\mathrm{GeV}$, perhaps that should be enough for the time being. In any case, while elegance, simplicity, and mathematical beauty often provide guidance to the theorist, in the end, experiment and observation are the final arbiters. As I will discuss toward the end, observations involving structure formation are starting to do just that.

2.4   Initial Conditions: No-hair theorems

Inflation is cosmologically attractive because it promises to account for our present nearly FRW space-time starting from very general initial conditions. Somewhat para- doxically, inflation is usually analyzed in the context of the isotropic and homogeneous FRW cosmology. I will now explain the *apparent* paradox and discuss to what extent inflation lessens the dependence of the present state of the Universe upon its initial state.

To begin consider the anisotropic but homogeneous (Bianchi) models; the mean expansion rate of the Universe can be written as

$$H^2 \equiv (\dot{\overline{R}}/\overline{R})^2 = \frac{8\pi G\rho}{3} + F(\dot{\overline{R}}, \overline{R}); \tag{28}$$

where $\overline{R}$ is the mean scale factor and ρ is the usual energy density and the function F accounts for the additional terms that arise due to anisotropy. In general, the function F decreases at least as rapidly as $1/\overline{R}^2$, that is, as rapidly as the spatial curvature term in the FRW cosmology or faster. The false-vacuum energy density appears in the energy density term and is of course constant. Provided that F is positive, the Universe will eventually become vacuum-energy dominated; once it does, the $F(\dot{\overline{R}}, \overline{R})$ term will quickly decrease and become insignificant and the space time becomes isotropic.[5] This justifies the usual FRW analysis of inflation.

Not all anisotropic space-times will inflate; if F is sufficiently large and negative it will prevent inflation; the simplest noninflating model is a very positively curved FRW model that recollapses before it can inflate. The strongly positively curved models preclude a true cosmological no-hair theorem; however, it has been shown that all spatially homogeneous, but anisotropic models eventually inflate, except for the very positively curved models [62]. And further, it has been shown that "smooth regions" of inhomogeneous models of sufficient size and that are negatively curved will inflate [63, 64]. While not all spacetimes will inflate, the class of spacetimes that do is not special, but very generic [64]. Thus, inflation does indeed lessen the dependence of the present state of the Universe on its initial state.

Does inflation render a generic space-time isotropic and homogeneous forever? The answer is clearly no; the most one can expect in an inhomogeneous space-time is that negatively curved regions inflate. Further, once inflation is over, inhomogeneity and anisotropy will "grow back." Consider spatial curvature; if the Universe was not flat before inflation it will not be flat after inflation. However, inflation exponentially postpones the epoch when spatial curvature becomes important because the value of Ω after inflation becomes exponential close to unity. Likewise, in the exponential distant future our Hubble volume will become larger than the generic smooth patch created

---

[5]There is one worry; namely that the inflaton field will evolve to the minimum of its potential before the vacuum-dominated phase begins. In general, this does not occur as anisotropy increases the expansion rate, and thus the friction term in the equation of motion for the inflaton; see [61].

by inflation and we will in principle we able to see the inhomogeneity beyond our inflationary patch [65].

Finally, there are the initial data for the scalar field responsible for inflation itself. In first-order inflation, as in old inflation, this is a dynamical issue: the initial value of the scalar field is determined by thermal considerations. However, in slow-rollover inflation the story is very different; the initial value of the inflaton field (and its spatial and temporal derivatives) are not so determined, and at the classical level must be considered to be initial data. While this has become a subject unto itself, some very general statements can be made. First, the inflaton field must be smooth on a scale comparable to the Hubble radius, otherwise the energy density associated with spatial gradients will dominate over the vacuum energy preventing inflation. Second, the value of the scalar field must be small enough in models of "new inflation" or large enough in models of "chaotic inflation" so that it takes the field more than 60 Hubble times to roll to the bottom of the potential. Finally, the initial velocity of the inflaton (i.e., $\dot{\phi}$) must be small enough so that it does not rapidly speed to the bottom of the potential. For a given inflationary model, all of these considerations can be studied and quantitative statements made about the necessary initial data for the inflaton field [66]; further, attempts have been made using the wavefunction of the Universe to quantify the quantum expectation for the initial state of the inflaton field [67].

In the final analysis it cannot be said that all initial spacetimes undergo inflation and become isotropic and homogeneous for all time; further, the initial data for the inflaton itself must now be considered. The strongest statement that one can make is to say that inflation greatly lessens the dependence of the present state of the Universe upon its initial state. In my mind, that's no mean feat and inflation should be considered a great success.

3 Inflation: The Fundamentals

In this Section I discuss how to analyze an inflationary model, given the scalar potential. In two sections hence I will work through a number of examples. The focus will be on the metric perturbations—density fluctuations [68] and gravity waves [69]—that arise due to quantum fluctuations, and the CBR temperature anisotropies that result from them.[6] Perturbations on all astrophysically interesting scales, say $1\,\mathrm{Mpc}$ to $10^4\,\mathrm{Mpc}$, are produced during an interval of about 8 e-folds around 50 e-folds before the end of inflation, when these scales crossed outside the horizon during inflation. I will show how the density perturbations and gravity waves can be related to three features of the inflationary potential: its value V_{50}, its steepness $x_{50} \equiv (m_{\mathrm{Pl}} V'/V)_{50}$, and the change in its steepness x'_{50}, evaluated in the region of the potential where the scalar field was about 50 e-folds before the end of inflation. In principle, cosmological observations, most importantly CBR anisotropy, can be used to determine the characteristics of the density perturbations and gravitational waves and thereby V_{50}, x_{50}, and x'_{50}.

All viable models of inflation are of the slow-rollover variety, or can be recast as such [70]. In slow-rollover inflation a scalar field that is initially displaced from the minimum of its potential rolls slowly to that minimum, and as it does the cosmic-scale factor grows very rapidly. Once the scalar field reaches the minimum of the potential it oscillates about it, so that the large potential energy has been converted into coherent scalar-field oscillations, corresponding to a condensate of nonrelativistic scalar particles. The eventual decay of these particles into lighter particle states and

---

[6]Isocurvature perturbations can arise due to quantum fluctuations in other massless fields, e.g., the axion field, if it exists [71].

their subsequent thermalization lead to the reheating of the Universe to a temperature $T_{\mathrm{RH}} \simeq \sqrt{\Gamma m_{\mathrm{Pl}}}$, where Γ is the decay width of the scalar particle [42, 70]. Here, I will focus on the classical evolution of the inflaton field during the slow-roll phase and the small quantum fluctuations in the inflaton field which give rise to density perturbations and those in the metric which give rise to gravity waves.

To begin, let us assume that the scalar field driving inflation is minimally coupled so that its stress-energy tensor takes the canonical form,

$$T_{\mu\nu} = \partial_\mu \phi \partial_\nu \phi - \mathcal{L} g_{\mu\nu}; \tag{29}$$

where the Lagrangian density of the scalar field $\mathcal{L} = \frac{1}{2} \partial_\mu \phi \partial^\mu \phi - V(\phi)$. If we make the usual assumption that the scalar field ϕ is spatially homogeneous, or at least so over a Hubble radius, the stress-energy tensor takes the perfect-fluid form with energy density, $\rho = \frac{1}{2}\dot{\phi}^2 + V(\phi)$, and isotropic pressure, $p = \frac{1}{2}\dot{\phi}^2 - V(\phi)$. The classical equations of motion for ϕ can be obtained from the first law of thermodynamics, $d(R^3 \rho) = -p\,dR^3$, or by taking the four-divergence of $T^{\mu\nu}$:

$$\ddot{\phi} + 3H\dot{\phi} + V'(\phi) = 0; \tag{30}$$

the $\Gamma \dot{\phi}$ term responsible for reheating has been omitted since we shall only be interested in the slow-rollover phase. In addition, there is the Friedmann equation, which governs the expansion of the Universe,

$$H^2 = \frac{8\pi}{3 m_{\mathrm{Pl}}^2} \left(V(\phi) + \frac{1}{2}\dot{\phi}^2 \right) \simeq \frac{8\pi V(\phi)}{3 m_{\mathrm{Pl}}^2}; \tag{31}$$

where we assume that the contribution of all other forms of energy density, e.g., radiation and kinetic energy of the scalar field, and the curvature term (k/R^2) are negligible. The justification for discussing inflation in the context of a flat FRW model with a homogeneous scalar field driving inflation were discussed earlier (and at greater length in Ref. [72]); including the ϕ kinetic term increases the righthand side of Eq. (31) by a factor of $(1 + x^2/48\pi)$, a small correction for viable models.

In the next Section I will be more precise about the amplitude of density perturbations and gravitational waves; for now, let me briefly discuss how these perturbations arise and give their characteristic amplitudes. The metric perturbations produced in inflationary models are very nearly "scale invariant," a particularly simple spectrum which was first discussed by Harrison and Zel'dovich [73], and arise due to quantum fluctuations. In deSitter space all massless scalar fields experience quantum fluctuations of amplitude $H/2\pi$. The graviton is massless and can be described by two massless scalar fields, $h_{+,\times} = \sqrt{16\pi G}\phi_{+,\times}$ (+ and $\times$ are the two polarization states). The inflaton by virtue of its flat potential is for all practical purposes massless.

Fluctuations in the inflaton field lead to density fluctuations because of its scalar potential, $\delta\rho \sim HV'$; as a given mode crosses outside the horizon, the density perturbation on that scale becomes a classical metric perturbation. While outside the horizon, the description of the evolution of a density perturbation is beset with subtleties associated with the gauge freedom in general relativity; there is, however, a simple gauge-invariant quantity, $\zeta \simeq \delta\rho/(\rho + p)$, which remains constant outside the horizon. By equating the value of ζ at postinflation horizon crossing with its value as the scale crosses outside the horizon it follows that $(\delta\rho/\rho)_{\mathrm{HOR}} \sim HV'/\dot{\phi}^2$ (note: $\rho + p = \dot{\phi}^2$); see Fig. 7.

The evolution of a gravity-wave perturbation is even simpler; it obeys the massless Klein-Gordon equation

$$\ddot{h}_k^i + 3H\dot{h}_k^i + k^2 h_k^i / R^2 = 0; \tag{32}$$

where k is the wavenumber of the mode and $i = +, \times$. For superhorizon sized modes, $k \lesssim RH$, the solution is simple: $h_k^i = \mathrm{const}$. Like their density perturbation counterparts, gravity-wave perturbations become classical metric perturbations as they cross outside the horizon; they are characterized by an amplitude $h_k^1 \simeq \sqrt{16\pi G}(H/2\pi) \sim H/m_{Pl}$. At postinflation horizon crossing their amplitude is unchanged.

Finally, let me write the horizon-crossing amplitudes of the scalar and tensor metric perturbations in terms of the inflationary potential,

$$(\delta\rho/\rho)_{HOR,\lambda} = c_S \left(\frac{V^{3/2}}{m_{Pl}^3 V'} \right)_1 ; \tag{33}$$

$$h_{HOR,\lambda} = c_T \left(\frac{V^{1/2}}{m_{Pl}^2} \right)_1 ; \tag{34}$$

where $(\delta\rho/\rho)_{HOR,\lambda}$ is the amplitude of the density perturbation on the scale λ when it crosses the Hubble radius during the post-inflation epoch, $h_{HOR,\lambda}$ is the dimensionless amplitude of the gravitational wave perturbation on the scale λ when it crosses the Hubble radius, and c_S, c_T are numerical constants of order unity. Subscript 1 indicates that the quantity involving the scalar potential is to be evaluated when the scale in question crossed outside the horizon during the inflationary era. The metric perturbations produced by inflation are characterized by almost scale-invariant horizon-crossing amplitudes; the slight deviations from scale invariance result from the variation of V and V' during inflation which enter through the dependence upon t_1. [In Eq. (33) I got ahead of myself and used the slow-roll approximation (see below) to rewrite the expression, $(\delta\rho/\rho)_{HOR,\lambda} \simeq HV'/\dot\phi$, in terms of the potential only.]

Eqs. (30-33) are the fundamental equations that govern inflation and the production of metric perturbations. It proves very useful to recast these equations using the scalar field as the independent variable; we then express the scalar and tensor perturbations in terms of the value of the potential, its steepness, and the rate of change of its steepness when the interesting scales crossed outside the Hubble radius during inflation, about 50 e-folds in scale factor before the end of inflation, defined by

$$V_{50} \equiv V(\phi_{50}); \qquad x_{50} \equiv \frac{m_{Pl}V'(\phi_{50})}{V(\phi_{50})}; \qquad x'_{50} = \frac{m_{Pl}V''(\phi_{50})}{V(\phi_{50})} - \frac{m_{Pl}[V'(\phi_{50})]^2}{V^2(\phi_{50})}.$$

To evaluate these three quantities 50 e-folds before the end of inflation we must find the value of the scalar field at this time. During the inflationary phase the $\ddot\phi$ term is negligible (the motion of ϕ is friction dominated), and Eq. (30) becomes

$$\dot\phi \simeq \frac{-V'(\phi)}{3H}; \tag{35}$$

this is known as the slow-roll approximation [74]. While the slow-roll approximation is almost universally applicable, there are models where the slow-roll approximation cannot be used; e.g., a potential where during the crucial 8 e-folds the scalar field rolls uphill, "powered" by the velocity it had when it hit the incline.

The conditions that must be satisfied in order that $\ddot\phi$ be negligible are:

$$|V''| < 9H^2 \simeq 24\pi V/m_{Pl}^2; \tag{36}$$

$$|x| \equiv |V'm_{Pl}/V| < \sqrt{48\pi}. \tag{37}$$

The end of the slow roll occurs when either or both of these inequalities are saturated, at a value of ϕ denoted by ϕ_{end}. Since $H \equiv \dot R/R$, or $Hdt = d\ln R$, it follows that

$$d\ln R = \frac{8\pi}{m_{Pl}^2} \frac{V(\phi)d\phi}{-V'(\phi)} = -\frac{8\pi d\phi}{m_{Pl} x}. \tag{38}$$

Now express the cosmic-scale factor in terms of is value at the end of inflation, R_{end}, and the number of e-foldings before the end of inflation, $N(\phi)$,

$$R = \exp[-N(\phi)]\, R_{\text{end}}.$$

The quantity $N(\phi)$ is a time-like variable whose value at the end of inflation is zero and whose evolution is governed by

$$\frac{dN}{d\phi} = \frac{8\pi}{m_{\text{Pl}}\, x}. \tag{39}$$

Using Eq. (39) we can compute the value of the scalar field 50 e-folds before the end of inflation ($\equiv \phi_{50}$); the values of V_{50}, x_{50}, and x'_{50} follow directly.

As ϕ rolls down its potential during inflation its energy density decreases, and so the growth in the scale factor is not exponential. By using the fact that the stress-energy of the scalar field takes the perfect-fluid form, we can solve for evolution of the cosmic-scale factor. Recall, for the equation of state $p = \gamma\rho$, the scale factor grows as $R \propto t^q$, where $q = 2/3(1 + \gamma)$. Here,

$$\gamma = \frac{\frac{1}{2}\dot\phi^2 - V}{\frac{1}{2}\dot\phi^2 + V} = \frac{x^2 - 48\pi}{x^2 + 48\pi}; \tag{40}$$

$$q = \frac{1}{3} + \frac{16\pi}{x^2}. \tag{41}$$

Since the steepness of the potential can change during inflation, γ is not in general constant; the power-law index q is more precisely the logarithmic rate of the change of the logarithm of the scale factor, $q = d\ln R/d\ln t$.

When the steepness parameter is small, corresponding to a very flat potential, γ is close to -1 and the scale factor grows as a very large power of time. To solve the horizon problem the expansion must be "superluminal" ($\ddot R > 0$), corresponding to $q > 1$, which requires that $x^2 < 24\pi$. Since $\frac{1}{2}\dot\phi^2/V = x^2/48\pi$, this implies that $\frac{1}{2}\dot\phi^2/V(\phi) < \frac{1}{2}$, justifying neglect of the scalar-field kinetic energy in computing the expansion rate for all but the steepest potentials. (In fact there are much stronger constraints; the COBE DMR data imply that $n \gtrsim 0.5$, which restricts $x_{50}^2 \lesssim 4\pi$, $\frac{1}{2}\dot\phi^2/V \lesssim \frac{1}{12}$, and $q \gtrsim 4$.)

Next, let us relate the size of a given scale to when that scale crosses outside the Hubble radius during inflation, specified by $N_1(\lambda)$, the number of e-folds before the end of inflation. The physical size of a perturbation is related to its comoving size, $\lambda_{\text{phys}} = R\lambda$; with the usual convention, $R_{\text{today}} = 1$, the comoving size is the physical size today. When the scale λ crosses outside the Hubble radius $R_1\lambda = H_1^{-1}$. We then assume that: (1) at the end of inflation the energy density is $\mathcal{M}^4 \simeq V(\phi_{\text{end}})$; (2) inflation is followed by a period where the energy density of the Universe is dominated by coherent scalar-field oscillations which decrease as R^{-3}; and (3) when value of the scale factor is R_{RH} the Universe reheats to a temperature $T_{\text{RH}} \simeq \sqrt{m_{\text{Pl}}\Gamma}$ and expands adiabatically thereafter. The "matching equation" that relates λ and $N_1(\lambda)$ is:

$$\lambda = \frac{R_{\text{today}}}{R_1} H_1^{-1} = \frac{R_{\text{today}}}{R_{\text{RH}}} \frac{R_{\text{RH}}}{R_{\text{end}}} \frac{R_{\text{end}}}{R_1} H_1^{-1}. \tag{42}$$

Adiabatic expansion since reheating implies $R_{\text{today}}/R_{\text{RH}} \simeq T_{\text{RH}}/2.73\,\text{K}$; and the decay of the coherent scalar-field oscillations implies $(R_{\text{RH}}/R_{\text{end}})^3 = (\mathcal{M}/T_{\text{RH}})^4$. If we define $\bar q = \ln(R_{\text{end}}/R_1)/\ln(t_{\text{end}}/t_1)$, the mean power-law index, it follows that $(R_{\text{end}}/R_1)H_1^{-1} = \exp[N_1(\bar q - 1)/\bar q]H_{\text{end}}^{-1}$, and Eq. (42) becomes

$$N_1(\lambda) = \frac{\bar q}{\bar q - 1}\left[48 + \ln \lambda_{\text{Mpc}} + \frac{2}{3}\ln(\mathcal{M}/10^{14}\,\text{GeV}) + \frac{1}{3}\ln(T_{\text{RH}}/10^{14}\,\text{GeV})\right]; \tag{43}$$

In the case of perfect reheating, which probably only applies to first-order inflation, $T_{\mathrm{RH}} \simeq \mathcal{M}$.

The scales of astrophysical interest today range roughly from that of galaxy size, $\lambda \sim$ Mpc, to the present Hubble scale, $H_0^{-1} \sim 10^4$ Mpc; up to the logarithmic corrections these scales crossed outside the horizon between about $N_1(\lambda) \sim 48$ and $N_1(\lambda) \simeq 56$ e-folds before the end of inflation. *That is, the interval of inflation that determines its all observable consequences covers only about 8 e-folds.*

Except in the case of strict power-law inflation q varies during inflation; this means that the $(R_{\mathrm{end}}/R_1)H_1^{-1}$ factor in Eq. (42) cannot be written in closed form. Taking account of this, the matching equation becomes a differential equation,

$$\frac{d\ln \lambda_{\mathrm{Mpc}}}{dN_1} = \frac{q(N_1) - 1}{q(N_1)}; \tag{44}$$

subject to the "boundary condition:"

$$\ln \lambda_{\mathrm{Mpc}} = -48 - \frac{4}{3}\ln(\mathcal{M}/10^{14}\,\mathrm{GeV}) + \frac{1}{3}\ln(T_{\mathrm{RH}}/10^{14}\,\mathrm{GeV})$$

for $N_1 = 0$, the matching relation for the mode that crossed outside the Hubble radius at the end of inflation. Equation (44) allows one to obtain the precise expression for when a given scale crossed outside the Hubble radius during inflation. To actually solve this equation, one would need to supplement it with the expressions $dN/d\phi = 8\pi/m_{\mathrm{Pl}}x$ and $q = 16\pi/x^2$. For our purposes we need only know: (1) The scales of astrophysical interest correspond to $N_1 \sim$ "50 ± 4," where for definiteness we will throughout take this to be an equality sign. (2) The expansion of Eq. (44) about $N_1 = 50$,

$$\Delta N_1(\lambda) = \left(\frac{q_{50} - 1}{q_{50}}\right) \Delta \ln \lambda_{\mathrm{Mpc}}; \tag{45}$$

which, with the aid of Eq. (39), implies that

$$\Delta\phi = \left(\frac{q_{50} - 1}{q_{50}}\right) \frac{x_{50}}{8\pi} \Delta\lambda_{\mathrm{Mpc}}. \tag{46}$$

We are now ready to express the perturbations in terms of V_{50}, x_{50}, and x'_{50}. First, we must solve for the value of ϕ, 50 e-folds before the end of inflation. To do so we use Eq. (39),

$$N(\phi_{50}) = 50 = \frac{8\pi}{m_{\mathrm{Pl}}^2} \int_{\phi_{\mathrm{end}}}^{\phi_{50}} \frac{V\,d\phi}{V'}. \tag{47}$$

Next, with the help of Eq. (46) we expand the potential V and its steepness x about ϕ_{50}:

$$V \simeq V_{50} + V'_{50}(\phi - \phi_{50}) = V_{50}\left[1 + \frac{x_{50}^2}{8\pi}\left(\frac{q_{50}}{q_{50} - 1}\right)\Delta\ln\lambda_{\mathrm{Mpc}}\right]; \tag{48}$$

$$x \simeq x_{50} + x'_{50}(\phi - \phi_{50}) = x_{50}\left[1 + \frac{m_{\mathrm{Pl}}x'_{50}}{8\pi}\left(\frac{q_{50}}{q_{50} - 1}\right)\Delta\ln\lambda_{\mathrm{Mpc}}\right]; \tag{49}$$

of course these expansions only make sense for potentials that are smooth. We note that additional terms in either expansion are $\mathcal{O}(\alpha_i^2)$ and beyond the accuracy we are seeking.

Now recall the equations for the amplitude of the scalar and tensor perturbations,

$$(\delta\rho/\rho)_{\mathrm{HOR},\lambda} = c_S \left(\frac{V^{1/2}}{m_{\mathrm{Pl}}^2 x}\right)_1; \tag{50}$$

$$h_{\mathrm{HOR},\lambda} = c_T \left(\frac{V^{1/2}}{m_{\mathrm{Pl}}^2}\right)_1; \tag{51}$$

where subscript 1 means that the quantities are to be evaluated where the scale λ crossed outside the Hubble radius, $N_1(\lambda)$ e-folds before the end of inflation. The origin of any deviation from scale invariance is clear: For tensor perturbations it arises due to the variation of the potential; and for scalar perturbations it arises due to the variation of both the potential and its steepness.

Using Eqs. (45-50) it is now simple to calculate the power-law exponents α_S and α_T that quantify the deviations from scale invariance,

$$\alpha_T = \frac{x_{50}^2}{16\pi} \frac{q_{50}}{q_{50} - 1} \simeq \frac{x_{50}^2}{16\pi}; \tag{52}$$

$$\alpha_S = \alpha_T - \frac{m_{Pl} x_{50}'}{8\pi} \frac{q_{50}}{q_{50} - 1} \simeq \frac{x_{50}^2}{16\pi} - \frac{m_{Pl} x_{50}'}{8\pi}; \tag{53}$$

where

$$q_{50} = \frac{1}{3} + \frac{16\pi}{x_{50}^2} \simeq \frac{16\pi}{x_{50}^2}; \tag{54}$$

$$h_{HOR,\lambda} = c_T \left(\frac{V_{50}^{1/2}}{m_{Pl}^2}\right) \lambda_{Mpc}^{\alpha_T}; \tag{55}$$

$$(\delta\rho/\rho)_{HOR,\lambda} = c_S \left(\frac{V_{50}^{1/2}}{x_{50} m_{Pl}^2}\right) \lambda_{Mpc}^{\alpha_S}. \tag{56}$$

The spectral indices α_i are defined as, $\alpha_S = [d\ln(\delta\rho/\rho)_{HOR,\lambda}/d\ln\lambda_{Mpc}]_{50}$ and $\alpha_T = [d\ln h_{HOR,\lambda}/d\ln\lambda_{Mpc}]_{50}$, and in general vary slowly with scale. Note too that the deviations from scale invariance, quantified by α_S and α_T, are of the order of x_{50}^2, $m_{Pl} x_{50}'$. In the expressions above we retained only lowest-order terms in $\mathcal{O}(\alpha_i)$. The next-order contributions to the spectral indices are $\mathcal{O}(\alpha_i^2)$; those to the amplitudes are $\mathcal{O}(\alpha_i)$ and are given two sections hence. The justification for truncating the expansion at lowest order is that the deviations from scale invariance are expected to be small—and are required by astrophysically data to be small.

As I discuss in more detail two sections hence, the more intuitive power-law indices α_S, α_T are related to the indices that are usually used to describe the power spectra of scalar and tensor perturbations, $P_S(k) = |\delta_k|^2 = Ak^n$ and $P_T(k) = |h_k|^2 = A_T k^{n_T}$,

$$n = 1 - 2\alpha_S = 1 - \frac{x_{50}^2}{8\pi} + \frac{m_{Pl} x_{50}'}{4\pi}; \tag{57}$$

$$n_T = -2\alpha_T = -\frac{x_{50}^2}{8\pi}. \tag{58}$$

$$\tag{59}$$

CBR temperature fluctuations on large-angular scales ($\theta \gtrsim 1°$) due to metric perturbations arise through the Sachs-Wolfe effect; very roughly, the temperature fluctuation on a given angular scale θ is related to the metric fluctuation on the length scale that subtends that angle at last scattering, $\lambda \sim 100h^{-1} Mpc(\theta/deg)$,

$$\left(\frac{\delta T}{T}\right)_\theta \sim \left(\frac{\delta\rho}{\rho}\right)_{HOR,\lambda}; \tag{60}$$

$$\left(\frac{\delta T}{T}\right)_\theta \sim h_{HOR,\lambda}; \tag{61}$$

where the scalar and tensor contributions to the CBR temperature anisotropy on a given scale add in quadrature. Let me be more specific about the amplitude of the quadrupole

CBR anisotropy. For small α_S, α_T the contributions of each to the quadrupole CBR temperature anisotropy:

$$\left(\frac{\Delta T}{T_0}\right)^2_{Q-S} \approx \frac{32\pi}{45} \frac{V_{50}}{m_{\mathrm{Pl}}{}^4 x^2_{50}}; \tag{62}$$

$$\left(\frac{\Delta T}{T_0}\right)^2_{Q-T} \approx 0.61 \frac{V_{50}}{m_{\mathrm{Pl}}{}^4}; \tag{63}$$

$$\frac{T}{S} \equiv \frac{(\Delta T/T_0)^2_{Q-T}}{(\Delta T/T_0)^2_{Q-S}} \approx 0.28 x^2_{50}; \tag{64}$$

where expressions have been evaluated to lowest order in x^2_{50} and $m_{\mathrm{Pl}} x'_{50}$. These quantities represent the ensemble averages of the scalar and tensor contributions to the quadrupole temperature anisotropy, which in terms of the spherical-harmonic expansion of the CBR temperature anisotropy on the sky are given by $5\langle|a_{2m}|^2\rangle/4\pi$. Further, the scalar and tensor contributions to the *measured* quadrupole anisotropy add in quadrature, and are subject to "cosmic variance." (Cosmic variance refers to the dispersion in the values measured by different observers in the Universe.)

Before going on, some general remarks [80]. The steepness parameter x^2_{50} must be less than about 24π to ensure superluminal expansion. For "steep" potentials, the expansion rate is "slow," i.e., q_{50} closer to unity, the gravity-wave contribution to the quadrupole CBR temperature anisotropy becomes comparable to, or greater than, that of density perturbations, and both scalar and tensor perturbations exhibit significant deviations from scale invariance. For "flat" potentials, i.e., small x_{50}, the expansion rate is "fast," i.e., $q_{50} \gg 1$, the gravity-wave contribution to the quadrupole CBR temperature anisotropy is much smaller than that of density perturbations, and the tensor perturbations are scale invariant. Unless the steepness of the potential changes rapidly, i.e., large x'_{50}, the scalar perturbations are also scale invariant.

3.1   Metric perturbations and CBR anisotropy

I was purposefully vague when discussing the amplitudes of the scalar and tensor modes, except when specifying their contributions to the quadrupole CBR temperature anisotropy; in fact, the spectral indices α_S and α_T, together with the scalar and tensor contributions to the CBR quadrupole serve to provide all the information necessary. Here I will fill in more details about the metric perturbations.

The scalar and tensor metric perturbations are expanded in harmonic functions, in the flat Universe predicted by inflation, plane waves,

$$h_{\mu\nu}(\mathbf{x},t) = \frac{1}{(2\pi)^3} \int d^3k\, h^{\mathrm{i}}_{\mathbf{k}}(t)\, \varepsilon^{\mathrm{i}}_{\mu\nu}\, e^{-\imath \mathbf{k}\cdot\mathbf{x}}; \tag{65}$$

$$\frac{\delta\rho(\mathbf{x},t)}{\rho} = \frac{1}{(2\pi)^3} \int d^3k\, \delta_{\mathbf{k}}(t)\, e^{-\imath \mathbf{k}\,\mathbf{x}}; \tag{66}$$

where $h_{\mu\nu} = R^{-2}g_{\mu\nu} - \eta_{\mu\nu}$, $\varepsilon^{\mathrm{i}}_{\mu\nu}$ is the polarization tensor for the gravity-wave modes, and $i = +, \times$ are the two polarization states. Everything of interest can be computed in terms of $h^{\mathrm{i}}_{\mathbf{k}}$ and $\delta_{\mathbf{k}}$. For example, the *rms* mass fluctuation in a sphere of radius r is obtained in terms of the window function for a sphere and the power spectrum $P_S(k) \equiv \langle|\delta_{\mathbf{k}}|^2\rangle$ (see below),

$$\langle(\delta M/M)^2\rangle_r = \frac{9}{2\pi^2 r^2} \int_0^\infty [j_1(kr)]^2\, P_S(k)dk; \tag{67}$$

where $j_1(x)$ is the spherical Bessel function of first order. If $P_S(k)$ is a power law, it follows roughly that $(\delta M/M)^2 \sim k^3|\delta_k|^2$, evaluated on the scale $k = r^{-1}$. This is what I meant by $(\delta\rho/\rho)_{\mathrm{HOR},\lambda}$: the *rms* mass fluctuation on the scale λ when it crossed inside the horizon. Likewise, by $h_{\mathrm{HOR},\lambda}$ I meant the *rms* strain on the scale λ as it crossed inside the Hubble radius, $(h_{\mathrm{HOR},\lambda})^2 \sim k^3|h_k^i|^2$.

In the previous discussions I have chosen to specify the metric perturbations for the different Fourier modes when they crossed inside the horizon, rather than at a common time. I did so because scale invariance is made manifest, as the scale independence of the metric perturbations at post-inflation horizon crossing. Recall, in the case of scalar perturbations $(\delta\rho/\rho)_{\mathrm{HOR}}$ is up to a numerical factor the fluctuation in the Newtonian potential, and, by specifying the scalar perturbations at horizon crossing, we avoid the discussion of scalar perturbations on superhorizon scales, which is beset by the subtleties associated with the gauge noninvariance of δ_k.

It is, however, necessary to specify the perturbations at a common time to carry out most calculations; e.g., an N-body simulation of structure formation or the calculation of CBR anisotropy. To do so, one has to take account of the evolution of the perturbations after they enter the horizon. After entering the horizon tensor perturbations behave like gravitons, with h_k decreasing as R^{-1} and the energy density associated with a given mode, $\rho_k \sim m_{\mathrm{Pl}}^2 k^5|h_k|^2/R^2$, decreasing as R^{-4}. The evolution of scalar perturbations is slightly more complicated; modes that enter the horizon while the Universe is still radiation dominated remain essentially constant until the Universe becomes matter dominated (growing only logarithmically); modes that enter the horizon after the Universe becomes matter dominated grow as the scale factor. (The gauge noninvariance of δ_k is not an important issue for subhorizon size modes; here a Newtonian analysis suffices, and there is only one growing mode, corresponding to a density perturbation.)

The method for characterizing the scalar perturbations is by now standard: The spectrum of perturbations is specified at the present epoch (assuming linear growth for all scales); the spectrum at earlier epochs can be obtained by multiplying δ_k by $R(t)/R_{\mathrm{today}}$. The inflationary metric perturbations are gaussian; thus δ_k is a gaussian, random variable. Its statistical expectation value is

$$\langle \delta_k\, \delta_q \rangle = P_S(k)(2\pi)^3 \delta^{(3)}(\mathbf{k} - \mathbf{q}); \tag{68}$$

where the power spectrum today is written as

$$P_S(k) \equiv A k^n T(k)^2; \tag{69}$$

$n = 1 - 2\alpha_S$ ($= 1$ for scale-invariant perturbations), and $T(k)$ is the "transfer function" which encodes the information about the post-horizon crossing evolution of each mode and depends upon the matter content of the Universe, e.g., baryons plus cold dark matter, baryons plus hot dark matter, baryons plus hot and cold dark matter, and so on. The transfer function is defined so that $T(k) \to 1$ for $k \to 0$ (long-wavelength perturbations); an analytic approximation to the cold dark matter transfer function is given by [76]

$$T(k) = \frac{\ln(1 + 2.34q)/2.34q}{[1 + (3.89q) + (16.1q)^2 + (5.46q)^3 + (6.71q)^4]^{1/4}}; \tag{70}$$

where $q = k/(\Omega_0 h^2\,\mathrm{Mpc}^{-1})$. Inflationary power spectra for different dark matter possibilities are shown in Fig. 9.

The overall normalization factor

$$A = \frac{1024\pi^3}{75H_0^{3+n}}\,\frac{V_{50}}{m_{\mathrm{Pl}}^4 x_{50}^2}\,\frac{[1 + \frac{7}{6}n_T - \frac{1}{3}(n-1)]\left\{\Gamma[\frac{3}{2} - \frac{1}{2}(n-1)]\right\}^2}{2^{n-1}[\Gamma(\frac{3}{2})]^2}\,k_{50}^{1-n}; \tag{71}$$

where the $\mathcal{O}(\alpha_i)$ correction to A has been included [77]. The quantity $n_T = -2\alpha_T = -x_{50}^2/8\pi$, $n-1 = -2\alpha_S = n_T + x_{50}'/4\pi$, k_{50} is the comoving wavenumber of the scale that crossed outside the horizon 50 e-folds before the end of inflation. All the formulas below simplify if this scale corresponds to the present horizon scale, specifically, $k_{50} = H_0/2$. [Eq. (71) can be simplified by expanding $\Gamma(\frac{3}{2} + x) = \Gamma(3/2)[1 + x(2 - 2\ln 2 - \gamma)]$, valid for $|x| \ll 1$; $\gamma \simeq 0.577$ is Euler's constant.]

From this expression it is simple to compute the Sachs-Wolfe contribution of scalar perturbations to the CBR temperature anisotropy; on angular scales much greater than

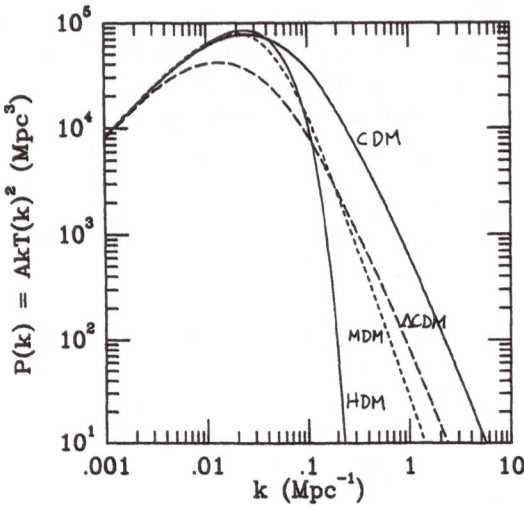

Figure 9. Power spectra for cold dark matter (CDM), hot dark matter (HDM), mixed dark matter (MDM = 30% hot + 70% cold), and cold dark matter with a cosmological constant (ΛCDM = 20% CDM + 80% Λ). All spectra are normalized to the COBE DMR quadrupole temperature anisotropy; $h = 0.5$ for all models except ΛCDM ($h = 0.8$).

about $1°$ (corresponding to multipoles $l \ll 100$) it is the dominant contribution. If we expand the CBR temperature on the sky in spherical harmonics,

$$\frac{\delta T(\theta, \phi)}{T_0} = \sum_{l \geq 2, m=-l}^{l=\infty, m=l} a_{lm} Y_{lm}(\theta, \phi); \tag{72}$$

where $T_0 = 2.73\,\mathrm{K}$ is the CBR temperature today, then the ensemble expectation for the multipole coefficients is given by

$$\langle |a_{lm}|^2 \rangle = \frac{H_0^4}{2\pi} \int_0^\infty k^{-2} P_S(k) |j_l(kr_0)|^2 \, dk; \tag{73}$$

$$\simeq \frac{A H_0^{3+n} r_0^{1-n}}{16} \frac{\Gamma(l + \frac{1}{2}n - \frac{1}{2})\Gamma(3 - n)}{\Gamma(l - \frac{1}{2}n + \frac{5}{2})[\Gamma(2 - \frac{1}{2}n)]^2}; \tag{74}$$

where $r_0 \approx 2H_0^{-1}$ is the comoving distance to the last scattering surface, and this expression is for the Sachs-Wolfe contribution from scalar perturbations only. For n not too different from one, the ensemble expectation for the quadrupole CBR temperature anisotropy is

$$\left(\frac{\Delta T}{T_0}\right)^2_{Q-S} \equiv \frac{5|a_{2m}|^2}{4\pi} \approx \frac{32\pi}{45} \frac{V_{50}}{m_{\text{Pl}}^4 x_{50}^2} (k_{50}r_0)^{1-n}. \tag{75}$$

(By choosing $k_{50} = r_0^{-1} = \frac{1}{2}H_0$, the last factor becomes unity.)

The ensemble expectation values for the multipole amplitudes are often referred to as the angular power spectrum. Further, the *rms* temperature fluctuation on a given angular scale is related to the multipole amplitudes

$$\left(\frac{\Delta T}{T}\right)^2_\theta \sim l^2 \langle |a_{lm}|^2 \rangle \qquad \text{for} \quad l \simeq 200°/\theta. \tag{76}$$

The procedure for specifying the tensor modes is similar, cf. Refs. [78, 79]. For the modes that enter the horizon after the Universe becomes matter dominated, $k \lesssim 0.1h^2\,\text{Mpc}$, which are the only modes that contribute significantly to CBR anisotropy on angular scales greater than a degree,

$$h_{\mathbf{k}}^i(\tau) = a^i(\mathbf{k}) \left(\frac{3j_1(k\tau)}{k\tau}\right); \tag{77}$$

where $\tau = r_0(t/t_0)^{1/3}$ is conformal time. [For the modes that enter the horizon during the radiation-dominated era, $k \gtrsim 0.1h^2\,\text{Mpc}^{-1}$, the factor $3j_1(k\tau)/k\tau$ is replaced by $j_0(k\tau)$ for the remainder of the radiation era. In either case, the factor involving the spherical Bessel function quantifies the fact that tensor perturbations remain constant while outside the horizon, and after horizon crossing decrease as R^{-1}.]

The tensor perturbations too are characterized by a gaussian, random variable, here written as $a^i(\mathbf{k})$; the statistical expectation

$$\langle h_{\mathbf{k}}^i h_{\mathbf{q}}^j \rangle = P_T(k)(2\pi)^6 \delta^{(3)}(\mathbf{k} - \mathbf{q})\delta_{ij}; \tag{78}$$

where the power spectrum

$$P_T(k) = A_T k^{n_T-3} \left[\frac{3j_1(k\tau)}{k\tau}\right]^2; \tag{79}$$

$$A_T = \frac{8}{3\pi} \frac{V_{50}}{m_{\text{Pl}}^4} \frac{(1 + \frac{5}{6}n_T)[\Gamma(\frac{3}{2} - \frac{1}{2}n_T)]^2}{2^{n_T}[\Gamma(\frac{3}{2})]^2} k_{50}^{-n_T}; \tag{80}$$

where the $\mathcal{O}(\alpha_i)$ correction to A_T has been included. Note that $n_T = -2\alpha_T$ is zero for scale-invariant perturbations.

Finally, the contribution of tensor perturbations to the multipole amplitudes, which arise solely due to the Sachs-Wolfe effect [28, 78, 79], is given by

$$\langle |a_{lm}|^2 \rangle \simeq 36\pi^2 \frac{\Gamma(l+3)}{\Gamma(l-1)} \int_0^\infty k^{n_T+1} A_T |F_l(k)|^2 \, dk; \tag{81}$$

where

$$F_l(k) = -\int_{r_D}^{r_0} dr \, \frac{j_2(kr)}{kr} \left[\frac{j_l(kr_0 - kr)}{(kr_0 - kr)^2}\right]; \tag{82}$$

and $r_D = r_0/(1 + z_D)^{1/2} \approx r_0/35$ is the comoving distance to the horizon at decoupling (= conformal time at decoupling). Equation (81) is approximate in that very short wavelength modes, $kr_0 \gg 100$, that crossed inside the horizon before matter-radiation equality have not been properly taken into account; to take them into account, the integrand must be multiplied by a transfer function,

$$T(k) \simeq 1.0 + 1.44(k/k_{\mathrm{EQ}}) + 2.54(k/k_{\mathrm{EQ}})^2; \tag{83}$$

where $k_{\mathrm{EQ}} \equiv H_0/(2\sqrt{2}-2)R_{\mathrm{EQ}}^{1/2}$ is the scale that entered the horizon at matter radiation equality [80]. In addition, for $l \gtrsim 1000$, the finite thickness of the last-scattering surface must be taken into account.

The tensor contribution to the quadrupole CBR temperature anisotropy for n_T not too different from zero is

$$\left(\frac{\Delta T}{T_0}\right)^2_{Q-T} \equiv \frac{5|a_{2m}|^2}{4\pi} \simeq 0.61 \frac{V_{50}}{m_{\mathrm{Pl}}^4} (k_{50}r_0)^{-n_T}; \tag{84}$$

where the integrals in the previous expressions have been evaluated numerically.

Both the scalar and tensor contributions to a given multipole are dominated by wavenumbers $kr_0 \sim l$. For scale-invariant perturbations and small l, both the scalar and tensor contributions to $(l + \frac{1}{2})^2 \langle |a_{lm}|^2 \rangle$ are approximately constant. The contribution of scalar perturbations to $(l + \frac{1}{2})^2 \langle |a_{lm}|^2 \rangle$ begins to decrease for $l \sim 150$ because the scalar contribution to these multipoles is dominated by modes that entered the horizon before matter domination (and hence are suppressed by the transfer function). The contribution of tensor modes to $(l + \frac{1}{2})^2 \langle |a_{lm}|^2 \rangle$ begins to decrease for $l \sim 30$ because the tensor contribution to these multipoles is dominated by modes that entered the horizon before decoupling (and hence decayed as R^{-1} until decoupling). Figure 10 shows the contribution of scalar and tensor perturbations to the CBR anisotropy multipole amplitudes (and includes both the tensor and scalar transfer functions); the expected variance in the CBR multipoles is given by the sum of the scalar and tensor contributions.

3.2   Worked examples

In this Section I apply the formalism developed in the two previous sections to four specific models. So that I can, where appropriate, solve numerically for model parameters, I will: (1) Assume that the astrophysically interesting scales crossed outside the horizon 50 e-folds before the end of inflation; and (2) Use the COBE DMR quadrupole measurement, $\langle (\Delta T)^2_Q \rangle^{1/2} \approx 16 \pm 2\mu K$ [11], to normalize the scalar perturbations; using Eq. (62) this implies

$$V_{50} \approx 1.6 \times 10^{-11} \, m_{\mathrm{Pl}}^4 \, x_{50}^2. \tag{85}$$

Of course it is entirely possible that a significant portion of the quadrupole anisotropy is due to tensor-mode perturbations, in which case this normalization must be reduced by a factor of $(1 + T/S)^{-1}$. And, it is straightforward to change "50" to the number appropriate to a specific model, or to normalize the perturbations another way.

Before going on let us use the COBE DMR quadrupole anisotropy to bound the tensor contribution to the quadrupole anisotropy and thereby the energy density that drives inflation:

$$V_{50} \lesssim 6 \times 10^{-11} m_{\mathrm{Pl}}^4. \tag{86}$$

Thus, the tensor contribution to the CBR quadrupole implies that the vacuum energy that drives inflation must be much less than the Planck energy density, strongly suggesting that inflation is not a quantum-gravitational phenomenon.

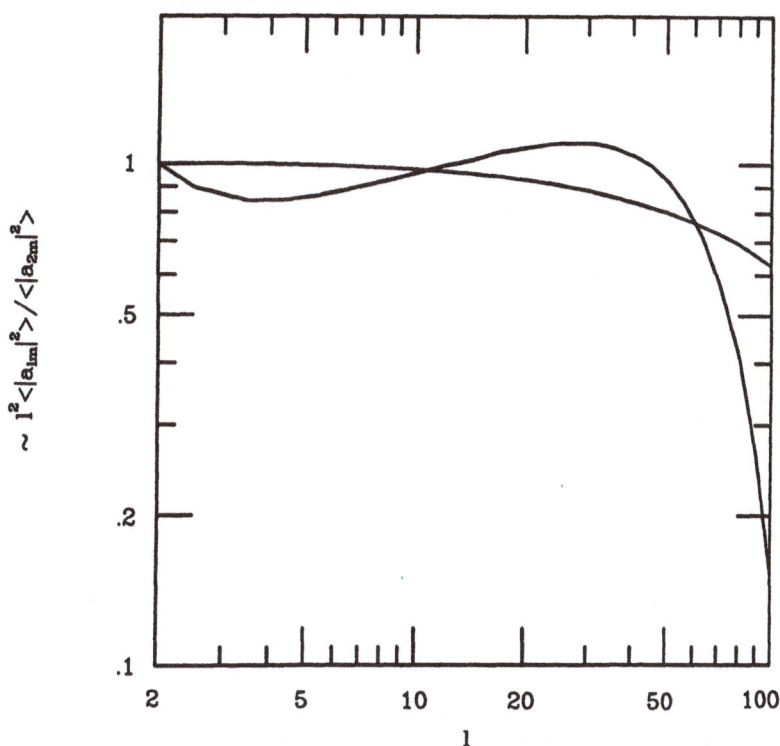

Figure 10. Scalar and tensor contributions to the CBR multipole moments: $l(l + 1)\langle|a_{lm}|^2\rangle/6\langle|a_{2m}|^2\rangle$ for the scalar and $l(l + \frac{1}{2})\langle|a_{lm}|^2\rangle/5\langle|a_{2m}|^2\rangle$ for the tensor. The tensor contribution begins to fall off for $l \sim 30$; here $n - 1 = n_T = 0$, $z_{\mathrm{DEC}} = 1000$, and $h = 0.5$ (from [81]).

3.2.1 Exponential potentials

There are a class of models that can be described in terms of an exponential potential,

$$V(\phi) = V_0 \exp(-\beta\phi/m_{\text{Pl}}).$$ (87)

This type of potential was first invoked in the context of power-law inflation [82], and has recently received renewed interest in the context of extended inflation [44]. In the simplest model of extended, or first-order, inflation, that based upon the Brans-Dicke-Jordan theory of gravity [44], β is related to the Brans-Dicke parameter: $\beta^2 = 64\pi/(2\omega + 3)$.

For such a potential the slow-roll conditions are satisfied provided that $\beta^2 \lesssim 24\pi$; thus inflation does not end until the potential changes shape, or in the case of extended inflation, until the phase transition takes place. In either case we can relate ϕ_{50} to ϕ_{end},

$$N(\phi_{50}) = 50 = \frac{8\pi}{m_{\text{Pl}}^2} \int_{\phi_{50}}^{\phi_{\text{end}}} \frac{V d\phi}{-V'}; \quad \Rightarrow \quad \phi_{50} = \phi_{\text{end}} - 50\beta/8\pi.$$ (88)

Since ϕ_{end} is in effect arbitrary, the overall normalization of the potential is irrelevant. The two other parameters, x_{50} and x'_{50}, are easy to compute:

$$x_{50} = -\beta; \qquad x'_{50} = 0.$$ (89)

Using the COBE DMR normalization, we can relate V_{50} and β:

$$V_{50} = 1.6 \times 10^{-11} \, m_{\text{Pl}}^4 \beta^2.$$ (90)

Further, we can compute q, α_S, α_T, and T/S:

$$q = 16\pi/\beta^2; \qquad T/S = 0.28\beta^2; \qquad \alpha_T = \alpha_S = 1/(q-1) \simeq \beta^2/16\pi.$$ (91)

Note, for the exponential potential, q, $\alpha_T = \alpha_S$ are independent of epoch. In the case of extended inflation, $\alpha_S = \alpha_T = 4/(2\omega + 3)$; since ω must be less than about 20 [47], this implies significant tilt: $\alpha_S = \alpha_T \gtrsim 0.1$.

3.2.2 Chaotic inflation

The simplest chaotic inflation models are based upon potentials of the form:

$$V(\phi) = a\phi^b;$$ (92)

$b = 4$ corresponds to Linde's original model of chaotic inflation and a is dimensionless [54], and $b = 2$ is a model based upon a massive scalar field and $m^2 = 2a$ [83]. In these models ϕ is initially displaced from $\phi = 0$, and inflation occurs as ϕ slowly rolls to the origin. The value of ϕ_{end} is easily found: $\phi_{\text{end}}^2 = b(b-1)m_{\text{Pl}}^2/24\pi$, and

$$N(\phi_{50}) = 50 = \frac{8\pi}{m_{\text{Pl}}^2} \int_{\phi_{\text{end}}}^{\phi_{50}} \frac{V d\phi}{V'};$$ (93)

$$\Rightarrow \quad \phi_{50}^2/m_{\text{Pl}}^2 = 50b/4\pi + b^2/48\pi \simeq 50b/4\pi;$$ (94)

the value of ϕ_{50} is a few times the Planck mass.

For purposes of illustration consider $b = 4$; $\phi_{\text{end}} = m_{\text{Pl}}/\sqrt{2\pi} \simeq 0.4m_{\text{Pl}}$, $\phi_{50} \simeq 4m_{\text{Pl}}$, $\phi_{46} \simeq 3.84m_{\text{Pl}}$, and $\phi_{54} \simeq 4.16m_{\text{Pl}}$. In order to have sufficient inflation the initial value of ϕ must exceed about $4.2m_{\text{Pl}}$; inflation ends when $\phi \approx 0.4m_{\text{Pl}}$; and the scales of astrophysical interest cross outside the horizon over an interval $\Delta\phi \simeq 0.3m_{\text{Pl}}$.

The values of the potential, its steepness, and the change in steepness are easily found,

$$V_{50} = a \, m_{\rm Pl}^b \left(\frac{50b}{4\pi}\right)^{b/2}; \qquad x_{50} = \sqrt{\frac{4\pi b}{50}}; \qquad m_{\rm Pl} x_{50}' = \frac{-4\pi}{50}; \qquad (95)$$

$$q_{50} = 200/b; \qquad T/S = 0.07b; \qquad \alpha_T \simeq b/200; \qquad \alpha_S = \alpha_T + 0.01. \qquad (96)$$

Unless b is very large, scalar perturbations dominate tensor perturbations [84], α_T, α_S are very small, and q is very large. Further, when α_T, α_S become significant, they are equal. Using the COBE DMR normalization we find:

$$a = 1.6 \times 10^{-11} b^{1-b/2} (4\pi/50)^{b/2+1} \, m_{\rm Pl}^{4-b}. \qquad (97)$$

For the two special cases of interest: $b = 4$, $a = 6.4 \times 10^{-14}$; and $b = 2$, $m^2 \equiv 2a = 2.0 \times 10^{-12} m_{\rm Pl}^2$.

3.2.3 New inflation

These models entail a very flat potential where the scalar field rolls from $\phi \approx 0$ to the minimum of the potential at $\phi = \sigma$. The original models of slow-rollover inflation [85] were based upon potentials of the Coleman-Weinberg form

$$V(\phi) = B\sigma^4/2 + B\phi^4 \left[\ln(\phi^2/\sigma^2) - \frac{1}{2}\right]; \qquad (98)$$

where B is a very small dimensionless coupling constant. Other very flat potentials also work (e.g., $V = V_0 - \alpha\phi^4 + \beta\phi^6$ [74]). As before we first solve for ϕ_{50}:

$$N(\phi_{50}) = 50 = \frac{8\pi}{m_{\rm Pl}^2} \int_{\phi_{\rm end}}^{\phi_{50}} \frac{V d\phi}{V'}; \qquad \Rightarrow \qquad \phi_{50}^2 = \frac{\pi\sigma^4}{100 |\ln(\phi_{50}^2/\sigma^2)| m_{\rm Pl}^2}; \qquad (99)$$

where the precise value of $\phi_{\rm end}$ is not relevant, only the fact that it is much larger than ϕ_{50}. Provided that $\sigma \lesssim m_{\rm Pl}$, both ϕ_{50} and $\phi_{\rm end}$ are much less than σ; we then find

$$V_{50} \simeq B\sigma^4/2; \qquad x_{50} \simeq -\frac{(\pi/25)^{3/2}}{\sqrt{|\ln(\phi_{50}^2/\sigma^2)|}} \left(\frac{\sigma}{m_{\rm Pl}}\right)^2 \ll 1; \qquad (100)$$

$$m_{\rm Pl} x_{50}' \simeq -24\pi/100; \qquad q_{50} \simeq \frac{2.5 \times 10^5 |\ln(\phi_{50}^2/\sigma^2)|}{\pi^2} \left(\frac{m_{\rm Pl}}{\sigma}\right)^4 \gg 1; \qquad (101)$$

$$\alpha_S \simeq \frac{1}{q_{50}} \ll 1; \qquad \alpha_T = \alpha_S + 0.03; \qquad \frac{T}{S} \simeq \frac{6 \times 10^{-4}}{|\ln(\phi_{50}^2/\sigma^2)|} \left(\frac{\sigma}{m_{\rm Pl}}\right)^4. \qquad (102)$$

Provided that $\sigma \lesssim m_{\rm Pl}$, x_{50} is very small; this means that q is very large, gravity-waves and density perturbations are very nearly scale invariant, and T/S is small. Finally, using the COBE DMR normalization, we can determine the dimensionless coupling constant B:

$$B \simeq 6 \times 10^{-14}/|\ln(\phi_{50}^2/\sigma^2)| \approx 3 \times 10^{-15}. \qquad (103)$$

3.2.4 Natural inflation

This model is based upon a potential of the form [53]

$$V(\phi) = \Lambda^4 [1 + \cos(\phi/f)]. \qquad (104)$$

The flatness of the potential (and requisite small couplings) arise because the ϕ particle is a pseudo-Nambu-Goldstone boson (f is the scale of spontaneous symmetry breaking and Λ is the scale of explicit symmetry breaking; in the limit that $\Lambda \to 0$ the ϕ particle is a massless Nambu-Goldstone boson). It is a simple matter to show that ϕ_{end} is of the order of πf.

This potential is difficult to analyze in general; however, there are two limiting regimes: (i) $f \gg m_{\text{Pl}}$; and (ii) $f \lesssim m_{\text{Pl}}$ [74]. In the first regime, the 50 or so relevant e-folds take place close to the minimum of the potential, $\sigma = \pi f$, and inflation can be analyzed by expanding the potential about $\phi = \sigma$,

$$V(\psi) \simeq m^2 \psi^2 / 2; \tag{105}$$

$$m^2 = \Lambda^4 / f^2; \qquad \psi = \phi - \sigma. \tag{106}$$

In this regime natural inflation is equivalent to chaotic inflation with $m^2 = \Lambda^4 / f^2 \simeq 2 \times 10^{-12} m_{\text{Pl}}^2$.

In the second regime, $f \lesssim m_{\text{Pl}}$, inflation takes place when $\phi \lesssim \pi f$, so that we can make the following approximations: $V \simeq 2\Lambda^4$ and $V' = -\Lambda^4 \phi / f^2$. Taking $\phi_{\text{end}} \sim \pi f$, we can solve for $N(\phi)$:

$$N(\phi) = \frac{8\pi}{m_{\text{Pl}}^2} \int_\phi^{\pi f} \frac{V \, d\phi}{-V'} \simeq \frac{16\pi m_{\text{Pl}}^2}{f^2} \ln(\pi f / \phi); \tag{107}$$

from which it is clear that achieving 50 e-folds of inflation places a lower bound to f, very roughly $f \gtrsim m_{\text{Pl}}/3$ [74, 53].

Now we can solve for ϕ_{50}, V_{50}, x_{50}, and x'_{50}:

$$\phi_{50} / \pi f \simeq \exp(-50 m_{\text{Pl}}^2 / 16\pi f^2) \lesssim \mathcal{O}(0.1); \qquad V_{50} \simeq 2\Lambda^4; \tag{108}$$

$$x_{50} \simeq \frac{1}{2} \frac{m_{\text{Pl}}}{f} \frac{\phi_{50}}{f} \lesssim \mathcal{O}(0.1); \qquad x'_{50} \simeq -\frac{1}{2} \left(\frac{m_{\text{Pl}}}{f} \right)^2. \tag{109}$$

Using the COBE DMR normalization, we can relate Λ to f / m_{Pl}:

$$\Lambda / m_{\text{Pl}} = 6.7 \times 10^{-4} \sqrt{\frac{m_{\text{Pl}}}{f}} \exp(-25 m_{\text{Pl}}^2 / 16\pi f^2). \tag{110}$$

Further, we can solve for T/S, α_T, and α_S:

$$\frac{T}{S} \simeq 0.07 \left(\frac{m_{\text{Pl}}}{f} \right)^2 \left(\frac{\phi_{50}}{f} \right)^2 \lesssim \mathcal{O}(0.1); \tag{111}$$

$$\alpha_T = \frac{1}{16\pi} \frac{q_{50}}{q_{50} - 1} \left(\frac{1}{4} \frac{m_{\text{Pl}}^2}{f^2} \frac{\phi_{50}^2}{f^2} \right) \approx \frac{1}{64\pi} \left(\frac{m_{\text{Pl}}}{f} \right)^2 \left(\frac{\phi_{50}}{f} \right)^2 \ll 0.1; \tag{112}$$

$$\alpha_S = \frac{1}{16\pi} \frac{q_{50}}{q_{50} - 1} \left(\frac{1}{4} \frac{m_{\text{Pl}}^2}{f^2} \frac{\phi_{50}^2}{f^2} + \frac{m_{\text{Pl}}^2}{f^2} \right) \approx \frac{1}{16\pi} \left(\frac{m_{\text{Pl}}}{f} \right)^2; \tag{113}$$

$$q_{50} = 64\pi \left(\frac{f}{m_{\text{Pl}}} \right)^2 \left(\frac{f}{\phi_{50}} \right)^2 \gg 1. \tag{114}$$

Regime (ii) provides the exception to the rule that $\alpha_S \approx \alpha_T$ and large α_S implies large T/S. For example, taking $f = m_{\text{Pl}}/2$, we find:

$$\phi_{50} / f \sim 0.06; \qquad x_{50} \sim 0.06; \qquad x'_{50} = -2; \qquad q_{50} \sim 10^4; \tag{115}$$

$$\alpha_T \sim 10^{-4}; \qquad \alpha_S \sim 0.08; \qquad T/S \sim 10^{-3}. \tag{116}$$

The gravitational-wave perturbations are very nearly scale invariant, while the density perturbations deviate significantly from scale invariance. I note that regime (ii), i.e., $f \lesssim m_{Pl}$, occupies only a tiny fraction of parameter space because f must be greater than about $m_{Pl}/3$ to achieve sufficient inflation; further, regime (ii) is "fine tuned" and "unnatural" in the sense that the required value of Λ is exponentially sensitive to the value of f/m_{Pl}.

Finally, I note that the results for regime (ii) apply to any inflationary model whose Taylor expansion in the inflationary region is similar; e.g., $V(\phi) = -m^2\phi^2 + \lambda\phi^4$, which was originally analyzed in Ref. [74].

3.2.5   Lessons

To summarize the general features of our results. In all examples the deviations from scale invariance enhance perturbations on large scales. The only potentials that have significant deviations from scale invariance are either very steep or have rapidly changing steepness. In the former case, both the scalar and tensor perturbations are tilted by a similar amount; in the latter case, only the scalar perturbations are tilted.

For "steep" potentials, the expansion rate is "slow," i.e., q_{50} close to unity, the gravity-wave contribution to the CBR quadrupole anisotropy becomes comparable to, or greater than, that of density perturbations, and both scalar and tensor perturbations are tilted significantly. For flat potentials, i.e., small x_{50}, the expansion rate is "fast," i.e., $q_{50} \gg 1$, the gravity-wave contribution to the CBR quadrupole is much smaller than that of density perturbations, and unless the steepness of the potential changes significantly, large x'_{50}, both spectra very nearly scale invariant; if the steepness of the potential changes rapidly, the spectrum of scalar perturbations can be tilted significantly. The models that permit significant deviations from scale invariance involve exponential or low-order polynomial potentials; the former by virtue of their steepness, the latter by virtue of the rapid variation of their steepness. Exponential potentials are of interest because they arise in extended inflation models; potentials with rapidly steepness include $V(\phi) = -m^2\phi^2 + \lambda\phi^4$ or $\Lambda^4[1 + \cos(\phi/f)]$.

Finally, to illustrate how observational data could used to determine the properties of the inflationary potential and test the consistency of the inflationary hypothesis, suppose observations determined the following:

$$(\Delta T)_Q \simeq 16\mu K; \qquad T/S = 0.24; \qquad n = 0.9; \qquad (117)$$

that is, the COBE DMR quadrupole anisotropy, a four to one ratio of scalar to tensor contribution to the CBR quadrupole, and spectral index of 0.9 for the scalar perturbations. From T/S, we determine the steepness of the potential: $x_{50} \simeq 0.94$. From the steepness and the quadrupole anisotropy the value of the potential: $V_{50}^{1/4} \simeq 2.4 \times 10^{16}\,\mathrm{GeV}$. From the spectral index the change in steepness: $x'_{50} \simeq -0.81/m_{Pl}$. These data can also be expressed in terms of the value of the potential and its first two derivatives:

$$V_{50} = 1.4 \times 10^{-11} m_{Pl}{}^4; \qquad V'_{50} = 1.5 \times 10^{-11} m_{Pl}{}^3; \qquad V''_{50} = 1.0 \times 10^{-12} m_{Pl}{}^2. \quad (118)$$

Further, they the lead to the prediction: $n_T = -0.035$, which, when "measured," can be used as a consistency check for inflation.

4   Structure Formation After COBE

Filling in the details of structure formation is one of the pressing challenges of the standard cosmology. In order to do so one must have the "initial data" for the structure formation problem: the spectrum of density perturbations and the quantity and composition of matter in the Universe. With initial data in hand one can hope to carry out detailed numerical simulations which can be compared to the observations.[7] While neither the observational data nor the simulations are perfect, the situation in both regards is improving rapidly. In particular, the discovery of CBR anisotropy by the COBE DMR has provided the first direct evidence for the existence of density perturbations and thereby opened the door for their study.

Over the past decade or so many cosmologists have come to believe that required initial data trace to events that took place during the earliest history of the Universe ($t \ll 10^{-2}\,\text{sec}$). Thus, the study of structure formation has the potential to test theories of the early Universe and the underlying particle physics. Inflation leads to two limiting scenarios: hot dark matter and cold dark matter, both with scale-invariant density perturbations.

In the hot dark matter scenario the streaming of neutrinos from regions of higher density to lower density erases perturbations on small scales ($\lesssim 13h^{-2}\,\text{Mpc}$); therefore structure forms from the the top down: superclusters must form first and fragment into smaller objects. Therein lies the fundamental problem: Since we know that superclusters are just forming today, galaxies form too late to be consistent with the abundance of galaxies observed at red shifts of unity or so [86].

Cold dark matter looks much more promising; cold dark matter refers to dark matter particles that move very slowly, either by virtue of their large mass (e.g., 10 GeV to 2 TeV neutralino) or the fact that they were born cold ($10^{-5}\,\text{eV}$ axion). This means that perturbations on small scales are not erased and that structure forms from the bottom up. Cold dark matter has been subject to intense scrutiny over the past decade and has thus far survived, albeit with a number of scratches and bruises [87]. CDM models will be the focus of this Section.

That is not to say that cold dark matter models are the only promising possibilities. There are scenarios where the density perturbations arise due to topological (and non-topological) defects such as strings [30], global monopoles, and textures [31] with hot or cold dark matter. Scenarios have been discussed where the density perturbations arise in a rather recent phase transition (since decoupling!), due to new physics in the neutrino sector [32].

Finally, perhaps the most interesting alternative is Peebles' PIB model or what-you-see-is-what-you-get model [29]. In PIB $\Omega_0 = \Omega_B \sim 0.2$, $h \sim 0.8$, and the density perturbations are isocurvature perturbations (variations in the local baryon-to-photon ratio and not the energy density). PIB is not motivated by what early Universe theorists would like, rather by "what we see" (though it violates the primordial nucleosynthesis bound by large factor since $\Omega_B h^2 \simeq 0.13 \gg 0.02$). Remarkably, the scenario is still viable, though measurements of CBR anisotropy on scales of 1° to 90° are really putting it to the test: normalizing to the COBE 10° measurement, its predictions for the quadrupole are a factor of two small, while its predictions of scales of about five degrees exceed current upper limits [88].

4.1   The Universe observed

By now we know a lot—and a little—about the structure that exists in the Universe today. A resurgence of interest in structure formation, brought about in part by the very intriguing early Universe suggestions for initial data, has resulted in an explosion of

---

[7]Since the fluctuations predicted by inflation and other theories are only specified in a statistical sense this comparison can only be done statistically; in the case of inflation, the fluctuations are gaussian and so all predictions can be specified in terms of the power spectrum, $\langle |\delta_k|^2 \rangle$.

observations that bear on the issue over the past decade. They include red-shift surveys (large-angle, pencil-beam and sparsely sampled surveys), the spectrum and spatial variation of the CBR temperature, peculiar-velocity measurements, QSO absorption line systems, studies of clusters and superclusters, determinations of the distribution and quantity of dark matter, studies of galactic evolution, catalogues of millions of galaxies on the sky, and on and on.

To place things in perspective, we know much about the distribution of light (bright galaxies)—as opposed to mass (which is what theorists like to discuss); the largest red-shift survey, the CfA_2 slices of the Universe, contains only about 20,000 galaxies with median red shift of about 0.02 [89]; and the total number of red shifts measured for all purposes is only about 50,000. We have no definitive evidence as to the epoch of galaxy formation, or how the neutral hydrogen left between galaxies became ionized (if it weren't, it we would not be able to see emission from distant QSOs shortward of Lyman alpha, 1215Å in the rest frame of the QSO). We probably only know the mean density of galaxies to within 20%; we have no fair sample of clusters; and so on.

Let me briefly try to summarize some of the data that can be used to test models of structure formation. Within the spirit of my broad brush description, I will group the observations into three classes: Small-scale, observations that probe the Universe on scales less than order $30h^{-1}$ Mpc or so; Intermediate-scale, observations that probe the Universe on scales of $30h^{-1}$ Mpc $- 300h^{-1}$ Mpc or so; and Large-scale, observations that probe the Universe on the very largest scales accessible.[8]

- **Small-scale Structure** ($\lambda \lesssim 30h^{-1}$ Mpc): Our knowledge of these scales is the most extensive and well developed, though largely restricted to the distribution of bright galaxies like our own. These are also the scales on which astrophysical effects—star formation, blast waves, and so on—are potentially most important and poorly understood. The Universe on these scales is organized into galaxies and clusters, whose properties have been studied and quantified; for galaxies, number density and morphology—i.e., spiral, elliptical, etc.—rotation curves, and so on; and for clusters, number density, velocity dispersions, richness class, and so on. Both galaxies and clusters cluster, with measured two-point correlation functions, $\xi_{gg}(r) \simeq (r/5h^{-1}\,\mathrm{Mpc})^{-1.8}$ and $\xi_{cc}(r) \simeq (r/25h^{-1}\,\mathrm{Mpc})^{-1.8}$, though the cluster correlation function is less well known and depends upon cluster richness [90]. At some level we know the distribution of dark matter: spiral galaxies have large halos with unknown spatial extent and the bulk of the mass in clusters is dark [17]. We also know the pairwise galaxy velocity dispersion (line-of-sight velocity dispersion), $\langle (v_1 - v_2)^2 \rangle^{1/2}|_{10\,\mathrm{Mpc}} \simeq 300 - 400\,\mathrm{km\,s^{-1}}$ [91]. (As discussed earlier, the peculiar motions of galaxies depend upon the amplitude of density perturbations and the amount of matter in the Universe, and thus are indicative of such.) On scales less than about $8h^{-1}$ Mpc the Universe is nonlinear: specifically, the *rms* fluctuation in the number density of bright galaxies measured in a sphere of radius of $8h^{-1}$ Mpc is unity.

- **Intermediate-scale Structure** ($30h^{-1}$ Mpc $- 300h^{-1}$ Mpc): These are the scales on which our knowledge is the most fragmentary and often more qualitative than quantitative.[9] Observations include the the voids and "Great Wall" seen in the CfA_2 red shift survey; the reoccurring walls seen in the pencil-beam survey of

---

[8]I warn the reader that my nomenclature is not universal; many refer to what I call intermediate scales as large scales.

[9]I often call these the *NY Times* scales, as new observations and their extravagant interpretation are reported there almost weekly!

Broadhurst et al. [92]; the angular-correlation function of galaxies $w(\theta)$, which is related to $\xi_{gg}(r)$, measured by Efstathiou et al. [93] in the APM catalogue of 2 million galaxies on the sky (effective depth of $400h^{-1}$ Mpc); the peculiar velocities of galaxies measured by the Seven Samurai and others [94], about $400\,\mathrm{km\,s^{-1}}$ on the scale of $50h^{-1}$ Mpc; Great Attractors, and on and on. From red-shift surveys like the CfA$_2$ slices of the Universe, the IRAS 1.2 Jy survey of infrared-selected galaxies [95], and the APM-Stromlo 1 in 20 red-shift survey [96], the fluctuations in the galaxy number density have been measured on scales out to a few hundred Mpc; see Fig. 11a. By the year 2000 the Sloan Digital Sky Survey [97] will produce a "Map of the Universe," from the red shifts of a million galaxies (mean red shift of about 0.15 and survey depth of $500h^{-1}$ Mpc). With the exception of the peculiar-velocity measurements all these observations probe the distribution of light not mass. CBR anisotropy measurements on angular scales of a few degrees down to a few arcminutes also have the potential to probe the distribution of matter on these scales, as the CBR anisotropy on a given angular scale is related to the fluctuations in the mass density on a ranges of length scales around the characteristic length scale that subtends that angular size on the last scattering surface: $\lambda \sim 100h^{-1}\,\mathrm{Mpc}(\theta/\mathrm{deg})$. Very sensitive experiments are being done on these angular scales; with the important exception of the COBE DMR detection, there are now only upper limits, at the level of a few times 10^{-5}; see Fig. 3. I believe that more detections are just around the corner!

- **Large-scale Structure** ($\gtrsim 300h^{-1}$ Mpc): These scales are probed primarily by CBR anisotropy, though the Sloan Digital Sky Survey should provide some information about the distribution of galaxies on these scales. On angular scales much greater than about 1° the anisotropy arises due to the fluctuations in the gravitational potential on the last-scattering surface (Sachs-Wolfe effect), while on small-angular scales the situation is more complicated as the velocity of the matter, temperature fluctuations intrinsic to the radiation, and the ionization history of the Universe become important. On large-angular scales it is very simple to relate the CBR anisotropy to the "virgin spectrum" of density fluctuations. It is these scales that were probed by the COBE DMR detection, proving the first direct information about the existence of the density inhomogeneity that seeded structure formation.

4.1.1   Normalization: the great leap forward!

Lacking a definite prediction for the overall normalization for inflationary density perturbations, those who study formation of structure have historically used data on small-scales to normalize the spectrum of density perturbations, typically on the scale of $8h^{-1}$ Mpc. In so doing it is useful to define

$$\sigma_8 \equiv \langle (\delta M/M)^2 \rangle^{1/2}_{8h^{-1}\,\mathrm{Mpc}};\qquad(119)$$

which is the *rms* mass fluctuation in spheres of radius $8h^{-1}$ Mpc. The simplest (and most naive) procedure is to assume that light faithfully traces mass, i.e., $\delta\rho/\rho = \delta n_{\mathrm{GAL}}/n_{\mathrm{GAL}}$, and set $\sigma_8 = 1$ since the *rms* fluctuation in galaxy number in spheres of radius $8h^{-1}$ Mpc is unity; I will refer to this minimal cold dark matter (MCDM). I should remark that there is no a priori reason to expect light to trace mass, except on the very largest scales where only gravity is important.

Because this normalization leads to a galaxy pairwise velocity dispersion that is about a factor of two too large, the concept of "biasing" was introduced; namely, that light is a biased tracer of mass [98]. If light doesn't trace mass, the simplest *ansatz* is a linear factor between the two:

$$\delta n_{GAL}/n_{GAL} = b(\delta\rho/\rho). \qquad (120)$$

Of course there is every reason to expect that the real relationship is more complicated, $b = b(\lambda)$. In biased CDM models (BCDM), $\sigma_8 = b^{-1}$. In principle, the bias factor $b(\lambda)$ can be measured on scales where there is information about both the distribution of galaxies and of mass, cf. Fig. 11.

[Another way of understanding the motivation for bias is the so-called Ω problem: Why do the dynamical measurements indicate $\Omega_0 = 0.2$, if Ω_0 is really unity? The biasing explanation is that most of the mass in the Universe is in low surface-brightness galaxies that are too faint to see and that are less strongly clustered than the bright galaxies. Bright galaxies are more strongly clustered and account for only 20% of the mass density.]

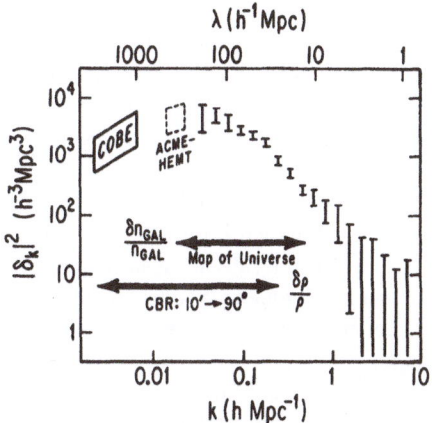

Figure 11. Summary of observational knowledge of the power spectrum $|\delta_k|^2$ based upon the IRAS 1.2 Jy red shift survey and CBR anisotropy measurements (from [95]). ACME-HEMT indicates the South Pole experiment that has detected anisotropy that may or may not be intrinsic to the CBR.

Until the COBE DMR detection, a bias factor to 1.5 to 2 was in vogue to resolve the discrepancy in the galaxy pairwise velocity dispersion; $b \sim 1.5 - 2$ was known as the *standard* CDM model. Since the peculiar velocities of galaxies arise due to the lumpy distribution of matter; larger b implies a smoother mass distribution and thus smaller peculiar velocities. (Likewise, reducing the matter content, or Ω_0, can help.) Unfortunately, the predictions of BCDM on intermediate-scales could not account for the level of inhomogeneity seen—voids, galaxy-galaxy angular-correlation function, peculiar velocities, and so on—since the mass distribution was smoother. Thus, CDM was faulted for predicting too little power on "large scales" (in my nomenclature, intermediate scales).

The COBE DMR detection of CBR anisotropy changed the situation overnight by providing a new, more direct normalization of the density perturbations! Assuming the correctness of the result, we now have a measurement of the inhomogeneity in the mass distribution on large-scales—and at last a "physics normalization." Remarkably, the COBE normalization (with scale-invariant perturbations) corresponds to the simplest CDM model: $\sigma_8 = 1.2 \pm 0.2$ [99], i.e., no biasing.[10]

The COBE DMR normalization has changed the way we view inflation and structure formation: Intermediate (and large) scales seem to be OK; the problem is with small scales. Addressing this problem is the focus of the brief discussion of CDM models that follows.

4.2   CDM models

The initial data for structure formation include: (i) spectrum of primeval density perturbations—amplitude on a given scale (normalization) and spectral index n; (ii) composition of the Universe—Ω_i, $i =$ baryons, cold dark matter, hot dark matter, vacuum energy and so on; and (iii) Hubble constant which sets the time/length scale for the Universe. With these in hand one can compute the spectrum of density perturbations at the equivalence epoch and let gravity run its course. Of course, astrophysics—cooling of baryons, star formation, etc.—is important too, but more difficult to model. Progress here too is being made with large N-body codes that include both gravity and hydrodynamics for the baryons [100]. The list of wanted cosmological parameters for a numerical simulation is: σ_8 (in the simple biasing prescription $\sigma_8 = b^{-1}$), n, Ω_B, Ω_{other}, and h.

What predictions does inflation make for these parameters? The firmest is a flat Universe, in my notation $\Omega_0 = 1.0$, which implies nonbaryonic dark matter dominates. As mentioned earlier, hot dark matter (30 eV or so neutrinos) was ruled out early on; and so the cold dark matter scenario *appeared* to be the unique inflationary blueprint for structure formation [101]. Let me explain; to get the age of the Universe right we must have $h \sim 0.5$. This fact together with the primordial nucleosynthesis determination of $\Omega_B h^2$ implies $\Omega_B \simeq 0.04 - 0.10$. In *most* inflationary models the density perturbations are very nearly scale invariant, implying $n = 1$. Finally, the variance in galaxy counts on $8h^{-1}$ Mpc suggests $\sigma_8 = 1$. This is the minimal cold dark matter model (MCDM); it is certainly the simplest CDM scenario, though it is no longer the unique CDM model.

Partly due to problems with MCDM, partly due to the improvement in the observations that test models of structure formation, and partly due to the passage of time we now realize that there are other possibilities, some just as well motivated, some less well motivated. I will characterize the different models by their values for the key cosmological parameters for structure formation: σ_8, n, Ω_{other}, h, and Ω_B.

4.2.1   MCDM

This is the simplest and the original CDM model; it is characterized by $b = 1$, $n = 1$, $\Omega_{\text{other}} = \Omega_{\text{cold}} \simeq 0.9$, $\Omega_B \sim 0.1$, and $h = 0.5$. It is consistent with the COBE DMR data, which for $n = 1$ imply $\sigma_8 = 1.2 \pm 0.2$, and intermediate-scale structure. However, it has too much power on small scales, quantified by a galaxy-pairwise velocity dispersion of about $1000 \, \text{km} \, \text{s}^{-1}$ compared to the observed $400 \, \text{km} \, \text{s}^{-1}$. A comparison of MCDM power spectrum with the observations is shown in Fig. 12.

Since MCDM is the simplest and most well motivated model perhaps inflationists should sit tight and wait for the data (or their interpretations) to change. After all, the

---

[10] For HDM the COBE DMR normalization implies $\sigma_8 = 0.7$. This drives another nail in the coffin, as it implies that only about 1% of the material in the Universe is in nonlinear structures.

disagreement is on small scales where the Universe is highly nonlinear and astrophysics can play an important role.

From Fig. 12 we see that concordance with the data can be achieved by three different "symmetry operations": translation of the predictions of the model in the vertical direction, corresponding to biasing; rotation, corresponding to deviation from scale invariance; and horizontal translation, corresponding to a change in the transfer function.

4.2.2 BCDM

This is the CDM model with biasing, imposed to solve the problem of too much power on small scales. The parameters of this model are: $b \sim 1.5 - 2$, $n = 1$, $\Omega_{\text{other}} = \Omega_{\text{cold}} \sim 0.9$, $\Omega_B \sim 0.1$, and $h = 0.5$. This model is disfavor for two reasons: (1) The COBE DMR results imply $b = 0.8 \pm 0.2$; and (2) (apparent) insufficient power on intermediate scales to account for peculiar velocities, the galaxy-galaxy angular correlation function, etc.

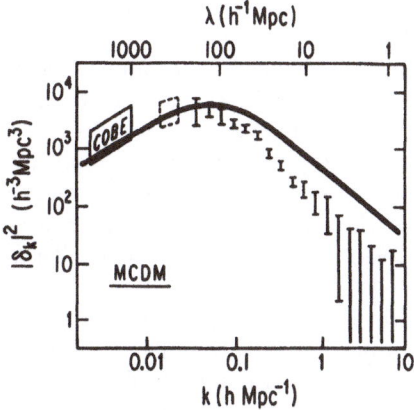

Figure 12. MCDM vs. observation.

However, one should keep in mind that the COBE DMR results are new and may still change, and that our knowledge of intermediate scales is the least secure. Perhaps the truth is somewhere in between MCDM and BCDM; both are well motivated. Shifting the MCDM power spectrum in Fig. 12 downward by a factor of $2 - 4$ corresponds to $\sigma_8 \sim 0.5 - 0.7$.

4.2.3 Tilt

Tilted CDM (TCDM) models are characterized by: $\sigma_8 \sim 0.5$, $n \sim 0.8$, $\Omega_{\text{other}} = \Omega_{\text{cold}} \sim 0.9$, $\Omega_B \sim 0.1$, and $h = 0.5$ [102]. From the beginning it was realized that the inflationary perturbations were not precisely scale-invariant, typically with more power on large scales ($n < 1$) [74], and so tilted models too are well motivated. Relative to scale-invariant perturbations ($n = 1$) the density perturbation in a tilted model is

$$\left(\frac{\delta\rho}{\rho}\right) \propto \left(\frac{\delta\rho}{\rho}\right)_{n=1} \lambda^{(1-n)/2}. \qquad (121)$$

The COBE DMR result provides a normalization on very large scales, $\lambda \sim 10^4 \, \mathrm{Mpc}$; relative to MCDM, the density perturbations on intermediate scales, $\lambda \sim 300 \, \mathrm{Mpc}$, are only a factor of about 1.4 smaller, while on small scales, $\lambda \sim 10 \, \mathrm{Mpc}$, they are about a factor of 2 smaller; see Fig. 13a.

If tilt is the truth, two kinds of inflationary potentials are singled out; exponential and low-order polynomial potentials [103]. Further, for exponential potentials, the contribution of gravity waves to the CBR anisotropy on large-angular scales is significant, which lowers the overall normalization of density perturbations further, by a factor of $(1 + T/S)^{1/2}$ [12].

4.2.4 Best-fit models

These models address the problem of too much small-scale power by changing the transfer function. The models considered thus far are: cold dark matter with a cosmological constant (ΛCDM), $n = 1$, $\Omega_B \sim 0.05$, $\Omega_{\mathrm{cold}} \sim 0.15$, $\Omega_\Lambda \sim 0.8$, and $h \sim 0.8$ [104]; and mixed dark matter (MDM)—"the neutrino cocktail—"$n = 1$, $\Omega_B \sim 0.1$, $\Omega_{\mathrm{cold}} \sim 0.6$, $h = 0.5$, and $\Omega_{\mathrm{hot}} \sim 0.3$, corresponding to a 7 eV or so mass neutrino [105].

How is the transfer function changed? It is simplest to see in MDM; since part of the dark matter is in the form of neutrinos which freestream out of density fluctuations on small scales, perturbations on small scales are depressed (see Fig. 9), just what the doctor ordered. In ΛCDM the story is a little more complicated; the bend in the transfer function is set by the scale that crosses inside the horizon at matter-radiation equality, $k_{\mathrm{EQ}} \sim 0.5(\Omega_{\mathrm{matter}} h^2) \, \mathrm{Mpc}$, where $\Omega_{\mathrm{matter}} = \Omega_B + \Omega_{\mathrm{cold}} \simeq 0.2$. Relative to MCDM, k_{EQ} is a factor of two smaller, shifting the spectrum to smaller k and decreasing power on small scales (see Fig. 9).

[The ΛCDM model, which I once called "the best-fit Universe," has a number of other nice features. It automatically solves the Ω problem since 80% of the energy density is in vacuum energy which is uniformly distributed and thus does not "show up" in dynamical measurements of the mass density. It allows one to accommodate the higher values of the Hubble constant which are favored by many measurements. Likewise, MDM also address the Ω problem: there is not enough phase space in galaxies for neutrinos to account for halo masses; further, neutrinos probably move to fast to be captured even in clusters. This would explain why dynamical measurements of Ω_0 based on galactic rotation curves or cluster virial masses do not lead to values of Ω_0 close to unity.]

Because both ΛCDM and MDM have less power on small scales than MCDM, when normalized to the COBE DMR result, they are a much better fit to the small scale data, and on large scales they are very similar to MCDM. They fit the present data very well; see Fig. 13b. One should, however, recall the words of Francis Crick (of DNA fame); loosely quoted,

> A theory that agrees with all of the data at any given time is necessarily wrong, since at any given time some of the data are incorrect.

The weak point of the "best-fit models" is motivation; however, let me try to make the best case for each. ΛCDM: A cosmological constant that contributes an energy density of about $10^{-46} \, \mathrm{GeV}^4$ would be very surprising. Since there is no physical mechanism known that explains why the present vacuum energy isn't of order $m_{\mathrm{Pl}}{}^4$

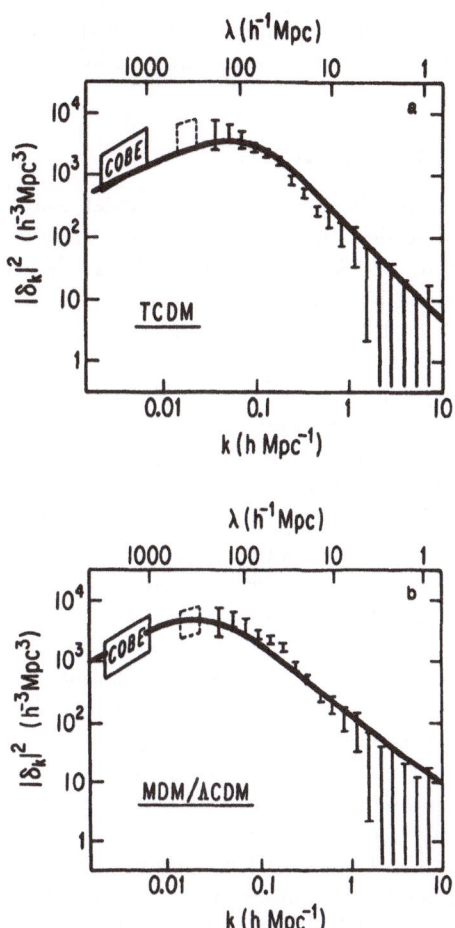

Figure 13. (a) TCDM vs. observation; (b) MDM/ΛCDM vs. observation.

(perhaps with the help of supersymmetry only of order G_F^{-2}), one cannot rigorously say that $\rho_{vac} \simeq 10^{-46}\,\mathrm{GeV}^4$ is be fine-tuned in the technical sense. MDM: Cold dark matter has so many good features that it must be *part* of the truth; neutrinos exist—in three varieties—and the see-saw mechanism suggests nondegenerate masses in "the eV range" (meaning $10^{-6}\,\mathrm{eV}$ to tens of eV), it could well be that one of the three neutrinos has a mass of order $10\,\mathrm{eV}$.

Certainly neither case for motivation is strong. Why is the cosmological constant just today becoming dynamically significant (recall, $\rho_{vac}/\rho_{matter} \propto R^3$)? And if history is any guide, cosmologists beginning with Einstein have too often invoked a cosmological constant to solve their problems. For MDM, one has to posit two kinds of nonbaryonic dark matter that each contribute comparably to the energy density of the Universe. If nonbaryonic dark matter exists it is already puzzling that baryons and dark matter each contribute similar amounts to the mass density [106].

4.3   The scorecard and future

Whereas cosmologists used to talk about "the CDM model" and its "uniqueness" (words that were also once used to describe the superstring), there is now a menu of CDM models. How do they stand, and which measurements can discriminate between them? In discussing their models, it is often said that theorists have many hands; on the one hand, on the other hand, on the other other hand and so on. Let me try my hand at it.

Occam's razor points to the simplest model, MCDM. Moreover, it was vindicated by COBE and only differs from the observational data by a factor of two or so on small scales where complicated astrophysics can be very important. Perhaps theorists should sit tight and wait. On the other hand (here we go), biasing at some level is likely to be a fact of life, arguing for BCDM; BCDM resolves the small-scale problems of CDM, but COBE indicates that $b \sim 1$. Maybe the truth is somewhere in between MCDM and BCDM; the COBE normalization could come down a bit, making $b \sim 1.3$ or so viable.

On the other other hand, deviation from scale-invariance was in the cards from the beginning, and so TCDM is well motivated too. Moreover, the tilt required points to a smaller class of inflationary models, exponential potentials and low-order polynomial potentials, which can be discriminated between by the size of their tensor perturbations.

On my final hands are the best-fit models, ΛCDM and MDM. They are not as well motivated, but agree better with the data at hand. Of the two, my first final hand has to go to MDM, and my second final hand to ΛCDM.

There will be a variety of observations that can used to discriminate between the different CDM models; I will focus on CBR measurements on the $0.5° - 2°$ scale, as there are several experiments with the sensitivity to probe CDM models which will be announcing results soon [107]. These experiments add roughly another order of magnitude to the range of scales probed by CBR anisotropy (recall, COBE probes $10°$ to $90°$). For reference, the MCDM prediction for these angular scales is $\delta T/T \sim few \times 10^{-5}$; the current upper limits are just above this level! Let me describe possible outcomes.

Scenario 1: The upper limits become detections. MCDM, ΛCDM, and MDM are in; BCDM and TCDM is out. (Because the tilted models have less power on small scales than MCDM, their predictions for small-angle anisotropy are smaller.)

Scenario 2: Detections are announced below the 10^{-5} level. BCDM and TCDM are in; the rest are out. If the detections are much below the 10^{-5} level, exponential potentials are strongly favored as in these tilted models much of the COBE signal is due to tensor perturbations whose contribution to CBR anisotropy falls dramatically around a few degrees [81, 12, 108].

Finally, let me mention a very different test, the value of the Hubble constant. Suppose all parties agree on the currently popular value $h = 0.8$; all CDM models except ΛCDM fall by the way side, based on the age of the Universe. Conversely, suppose that evidence for $h = 0.5$ becomes overwhelming; ΛCDM is out.

5   Concluding Remarks

Inflation is an extremely attractive cosmological paradigm; in spite—no, because—of its beauty it must be put to the ultimate test: confrontation with observation. In testing inflation one must focus on its robust predictions: In order of robustness, spatially flat Universe ($\Omega_0 = 1$); very nearly scale-invariant spectrum of density perturbations; and nearly scale-invariant spectrum of gravitational waves. In addition, for first-order inflation there should be a "spike" in the stochastic background of gravitational waves of very significant energy density, $\Omega_{GW} \sim 10^{-9}$ or so, at a frequency $f \sim 10^4 \, \mathrm{Hz}(\mathcal{M}/10^{10} \, \mathrm{GeV})$ [59]. It is also possible that fluctuations in other fields lead to primeval magnetic fields [109] or isocurvature perturbations (e.g., in axions or baryons) [71].

There are a variety of means of testing these predictions. For example, there are kinematic and dynamic techniques for measuring Ω_0. The density perturbations lead to temperature fluctuations in the CBR. The tensor (gravity-wave) perturbations also lead to CBR anisotropies, or may be detected directly by the next generation of gravity wave detectors, Laser Interferometer Gravitational-wave Observatories (LIGOs) [110].

From the "primary predictions," a series of secondary predictions follow. For example, since primordial nucleosynthesis restricts $\Omega_B \lesssim 0.1$, nonbaryonic dark matter is a necessity, and a host of experiments are under way to search for nonbaryonic dark matter [111]. In a flat, matter-dominated Universe $H_0 t_0 = \frac{2}{3}$ or $t_0 \simeq 6.5 h^{-1} \, \mathrm{Gyr}$, which implies that h must be greater than 0.65 to ensure that the Universe is older than 10 Gyr (the absolute minimum age that is consistent with other measures of the age of the Universe). The spectrum of density perturbations, together with the matter content, provide the initial data for the structure formation problem, leading to another test.

In the near term I believe that structure formation will provide the most powerful test of inflation and probe of inflationary models. On balance, the inflation-inspired CDM models are doing quite well so far compared to the alternatives: Texture and cosmic-string models required a high level of biasing ($b \sim 4$) to be compatible with COBE DMR results [112]; and PIB not only strongly violates the primordial nucleosynthesis constraint to Ω_B but also seems to be inconsistent with CBR anisotropy bounds on the scale of five degrees [88]. Measurements of CBR anisotropy on the degree scale will further discriminate between CDM and the alternatives; relative to CDM, PIB predicts a very large anisotropy, and the texture/cosmic string models predict a very small anisotropy whose properties are very non gaussian (most of the CBR is very quiet with a few hot and cold spots).

Great efforts have been made and are being made to further test the CDM scenarios; they involve many different techniques, CBR anisotropy, red-shift surveys, peculiar velocity measurements, and so on. These observations not only have the power to falsify CDM, but could also reveal much about the inflationary potential: the value of the potential, its steepness, and the change in steepness, which in turn can used to learn about the underlying model. For example, suppose that density perturbations do deviate significantly from scale invariance, then two classes of models are ruled out—chaotic and new inflation—and two types of models are ruled in—exponential potentials (as found in extended inflation) or low-order polynomial potentials (as found in natural inflation). The ratio of tensor to scalar perturbations can further narrow the field: large tensor contribution to the CBR quadrupole points to exponential potentials and small-tensor contribution points to low-order polynomial potentials.

The moment of truth for inflation may be near!

References

[1] For a textbook treatment of the standard cosmology see e.g., S. Weinberg, *Gravitation and Cosmology* (Wiley, NY, 1972); E.W. Kolb and M.S. Turner, *The Early Universe* (Addison-Wesley, Redwood City, CA, 1990).

[2] A. Sandage, *Physica Scripta* **T43**, 22 (1992).

[3] J. Mould et al., *Astrophys. J.*, in press (1993).

[4] See e.g., M. Rowan-Robinson, *The Cosmological Distance Ladder* (Freeman, San Francisco, 1985).

[5] M. Fukugita, C.J. Hogan, and P.J.E. Peebles, *Nature*, in press (1993).

[6] J. Mather et al., *Astrophys. J.*, in press (1993).

[7] P.J.E. Peebles, D.N. Schramm, E. Turner, and R. Kron, *Nature* **352**, 769 (1991).

[8] H. Gush, M. Halpern, and E.H. Wishnow, *Phys. Rev. Lett.* **65**, 537 (1990).

[9] G.F. Smoot et al., *Astrophys. J.* **396**, L1 (1992); D.J. Fixsen et al., *ibid*, in press (1993).

[10] G.F. Smoot, in *First Course in Current Topics in Astrofundamental Physics*, eds. N. Sanchez and A. Zichichi (World Scientific, Singapore, 1992), p. 192.

[11] G.F. Smoot et al., *Astrophys. J.* **396**, L1 (1992); E.L. Wright, *ibid* **396**, L3 (1992).

[12] R. Davis et al., *Phys. Rev. Lett.* **69**, 1856 (1992).

[13] E.L. Wright, *ibid* **396**, L3 (1992).

[14] S.S. Meyer, E.S. Cheng, and L.A. Page, *Astrophys. J.* **371**, L1 (1991); K. Ganga, S.S. Meyer, E.S. Cheng, and L.A. Page, *Astrophys. J.*, in press (1993).

[15] T.P. Walker et al., *Astrophys. J.* **376**, 51 (1991).

[16] E.W. Kolb et al., *Phys. Rev. Lett.* **67**, 533 (1991).

[17] For recent reviews of dark matter see e.g., M.S. Turner, *Physica Scripta* **T36**, 167 (1991); P.J.E. Peebles, *Nature* **321**, 27 (1986); V. Trimble, *Ann. Rev. Astron. Astrophys.* **25**, 425 (1987); J. Kormendy and G. Knapp, *Dark Matter in the Universe* (Reidel, Dordrecht, 1989); K. Ashman, *Proc. Astron. Soc. Pac.* **104**, 1109 (1992); S. Faber and J. Gallagher, *Ann. Rev. Astron. Astrophys.* **17**, 135 (1979).

[18] S. Faber and J. Gallagher, *Ann. Rev. Astron. Astrophys.* **17**, 135 (1979).

[19] J.S. Mulchaey, D.S. Davis, R.F. Mushotzky, and D. Burstein, *Astrophys. J.*, in press (1993).

[20] M. Rowan-Robinson et al., *Mon. Not. R. astr. Soc.* **247**, 1 (1990); N. Kaiser et al., *ibid* **252**, 1 (1991); M. Strauss et al., *Astrophys. J.* **385**, 444 (1992).

[21] E. Bertschinger and A. Dekel, *Astrophys. J.* **336**, L5 (1989); A. Dekel et al., *Astrophys. J.*, in press (1993); M. Strauss et al., *ibid* **397**, 395 (1992).

[22] A. Sandage, *Astrophys. J.* **133**, 355 (1961); *Physica Scripta* **T43**, 7 (1992); Refs. [1].

[23] E. Loh and E. Spillar, *Astrophys. J.* **307**, L1 (1986); M. Fukugita et al., *ibid* **361**, L1 (1990).

[24] See e.g., E.W. Kolb and M.S. Turner, *Ann. Rev. Nucl. Part. Sci.* **33**, 645 (1983); A. Dolgov, *Phys. Repts.*, in press (1993); A. Cohen, D. Kaplan, and A. Nelson, *Ann. Rev. Nucl. Part. Sci.*, in press (1993).

[25] S. Weinberg, *Gravitation and Cosmology* (Wiley, NY, 1972).

[26] For a more complete pedagogical discussion of structure formation see e.g., Refs. [1]; P.J.E. Peebles, *The Large-scale Structure of the Universe* (Princeton Univ. Press, Princeton, 1980); G. Efstathiou, in *The Physics of the Early Universe*, eds. J.A. Peacock, A.F. Heavens, and A.T. Davies (Adam-Higler, Bristol, 1990).

[27] For a pedagogical discussion of CBR anisotropy see e.g., G. Efstathiou, in *The Physics of the Early Universe*, eds. J.A. Peacock, A.F. Heavens, and A.T. Davies (Adam-Higler, Bristol, 1990). Also see, J.R. Bond and G. Efstathiou, *Mon. Not. R. astr. Soc.* **226**, 655 (1987); J.R. Bond et al., *Phys. Rev. Lett.* **66**, 2179 (1991).

[28] R.K. Sachs and A.M. Wolfe, *Astrophys. J.* **147**, 73 (1967).

[29] P.J.E. Peebles, *Nature* **327**, 210 (1987); *Astrophys. J.* **315**, L73 (1987); R. Cen, J.P. Ostriker, and P.J.E. Peebles, *ibid*, in press (1993).

[30] See e.g., A. Vilenkin, *Phys. Repts.* **121**, 263 (1985); A. Albrecht and A. Stebbins, *Phys., Rev. Lett.* **69**, 2615 (1992); D. Bennett, A. Stebbins, and F. Bouchet, *Astrophys. J.* **399**, L5 (1992).

[31] See e.g., N. Turok, *Phys. Rev. Lett.* **63**, 2652 (1989); A. Gooding, D. Spergel, and N. Turok, *Astrophys. J.* **372**, L5 (1991).

[32] J. Fry, C.T. Hill, and D.N. Schramm, *Comments on Nucl. Part. Phys.* **19**, 25 (1989); A. Gupta et al., *Phys. Rev. D* **45**, 441 (1992).

[33] See e.g., C.W. Misner, *Astrophys. J.* **151**, 431 (1968); R. Penrose, in *General Relativity: An Einstein Centenary Survey*, eds. S.W. Hawking and W. Israel (Cambridge Univ. Press, Cambridge, 1979); R.H. Dicke and P.J.E. Peebles, *ibid*.

[34] J. Preskill, *Ann. Rev. Nucl. Part. Sci.* **34**, 461 (1984).

[35] C.B. Collins and S.W. Hawking, *Astrophys. J.* **180**, 317 (1973).

[36] A.H. Guth, *Phys. Rev. D* **23**, 347 (1981).

[37] Y. Hu, M.S. Turner, and E.J. Weinberg, *Phys. Rev. Lett.*, in press (1993).

[38] S. Coleman, *Phys. Rev. D* **15**, 2929 (1977); S. Coleman and R. De Luccia, *ibid* **21**, 3305 (1980).

[39] R. Watkins and L. Widrow, *Nucl. Phys. B* **374**, 446 (1992); S.W. Hawking, J. Stewart, and I. Moss, *Phys. Rev. D* **26**, 2681 (1982).

[40] A.D. Linde, *Phys. Lett. B* **108**, 389 (1982).

[41] A. Albrecht and P.J. Steinhardt, *Phys. Rev. Lett.* **48**, 1220 (1982).

[42] A. Albrecht et al., *Phys. Rev. Lett.* **48**, 1437 (1982); L. Abbott and M. Wise, *Phys. Lett. B* **117**, 29 (1982); A.D. Linde and A. Dolgov, *ibid* **116**, 329 (1982).

[43] For a review of first-order inflation see e.g., E.W. Kolb, *Physica Scripta* **T36**, 199 (1991).

[44] D. La and P.J. Steinhardt, *Phys. Rev. Lett.* **62**, 376 (1989).

[45] F. Adams and K. Freese, *Phys. Rev. D* **43**, 353 (1991).

[46] E.W. Kolb, D. Salopek, and M.S. Turner, *Phys. Rev. D* **42**, 3925 (1990).

[47] E.J. Weinberg, *Phys. Rev. D* **40**, 3950 (1989); M.S. Turner, E.J. Weinberg, and L. Widrow, *ibid* **46**, 2384 (1992).

[48] Q. Shafi and A. Vilenkin, *Phys. Rev. Lett.* **52**, 691 (1984).

[49] S.-Y. Pi, *Phys. Rev. Lett.* **52**, 1725 (1984).

[50] L. Knox and M.S. Turner, *Phys. Rev. Lett.* **70**, 371 (1993).

[51] K.A. Olive, *Phys. Repts.* **190**, 307 (1990).

[52] R. Holman, P. Ramond, and G.G. Ross, *Phys. Lett. B* **137**, 343 (1984).

[53] K. Freese, J. Frieman, and A. Olinto, *Phys. Rev. Lett.* **65**, 3233 (1990).

[54] A.D. Linde, *Phys. Lett. B* **129**, 177 (1983).

[55] A.D. Linde, *Phys. Lett. B* **132**, 317 (1983); A.B. Goncharov and A.D. Linde, *Phys. Lett. B* **139**, 27 (1984).

[56] H. Murayama, H. Suzuki, T. Yanagida, and J. Yokoyama, *Phys. Rev. Lett*, in press (1993).

[57] Q. Shafi and C. Wetterich, *Phys. Lett. B* **129**, 387 (1983); *ibid* **152**, 51 (1985).

[58] A.A. Starobinski. *Phys. Lett. B* **91**, 99 (1980); M.B. Mijic, M.S. Morris, and W.-M. Suen, *Phys. Rev. D* **34**, 2934 (1986).

[59] M.S. Turner and F. Wilczek, *Phys. Rev. Lett.* **65**, 3080 (1990); A. Kosowsky, M.S. Turner, and R. Watkins *ibid* **69**, 2026 (1992).

[60] R.D. Reasenberg et al., *Astrophys. J.* **234**, L219 (1979).

[61] M.S. Turner and L. Widrow, *Phys. Rev. Lett.* **57**, 2237 (1986).

[62] R.M. Wald, *Phys. Rev. D* **28**, 2118 (1983); L. Jensen and J. Stein-Schabes, *ibid* **34**, 931 (1986).

[63] L. Jensen and J. Stein-Schabes, *Phys. Rev. D* **35**, 1146 (1987).

[64] A.A. Starobinskii, *JETP Lett.* **37**, 66 (1983).

[65] J.A. Frieman and M.S. Turner, *Phys. Rev. D* **30**, 265 (1984).

[66] D. Goldwirth and T. Piran, *Phys. Repts.* **214**, 223 (1992).

[67] See e.g., A.D. Linde, *Inflation and Quantum Cosmology* (Academic Press, San Diego, CA, 1990).

[68] A.H. Guth and S.-Y. Pi, *Phys. Rev. Lett.* **49**, 1110 (1982); A.A. Starobinskii, *Phys. Lett. B* **117**, 175 (1982); S.W. Hawking, *ibid* **115**, 295 (1982); J.M. Bardeen, P.J. Steinhardt, and M.S. Turner, *Phys. Rev. D* **28**, 679 (1983).

[69] V.A. Rubakov, M. Sazhin, and A. Veryaskin, *Phys. Lett. B* **115**, 189 (1982); R. Fabbri and M. Pollock, *ibid* **125**, 445 (1983); L. Abbott and M. Wise, *Nucl. Phys. B* **244**, 541 (1984); B. Allen, *Phys. Rev. D* **37**, 2078 (1988).

[70] At first sight, first-order inflation might seem very different from slow-rollover inflation, as reheating occurs through the nucleation of percolation of true-vacuum bubbles. However, such models can be recast as slow-rollover inflation by means of a conformal transformation, and the analysis of metric perturbations proceeds as in slow rollover inflation. See e.g., E.W. Kolb, D. Salopek, and M.S. Turner, *Phys. Rev. D* **42**, 3925 (1990).

[71] See e.g., A.D. Linde, *Phys. Lett. B* **158**, 375 (1985); D. Seckel and M.S. Turner, *Phys. Rev. D* **32**, 3178 (1985); M.S. Turner, A. Cohen, and D. Kaplan, *Phys. Lett. B* **216**, 20 (1989).

[72] E.W. Kolb and M.S. Turner, *The Early Universe* (Addison-Wesley, Redwood City, CA, 1990), Ch. 8.

[73] E.R. Harrison, *Phys. Rev. D* **1**, 2726 (1970); Ya.B. Zel'dovich, *Mon. Not. R. astr. Soc.* **160**, 1p (1972).

[74] P.J. Steinhardt and M.S. Turner, *Phys. Rev. D* **29**, 2162 (1984).

[75] The material presented in this Section is a summary of work completed during this school and will be published elsewhere, M.S. Turner, *Phys. Rev. D*, in press (1993).

[76] J.M. Bardeen et al., *Astrophys. J.* **304**, 15 (1986).

[77] D.H. Lyth and E.D. Stewart, *Phys. Lett. B* **274**, 168 (1992); E.D. Stewart and D.H. Lyth, *ibid*, in press (1993).

[78] L. Abbott and M. Wise, *Nucl. Phys. B* **244**, 541 (1984).

[79] M. White, *Phys. Rev. D* **46**, 4198 (1992).

[80] M.S. Turner, *Phys. Rev. D*, in press (1993).

[81] M.S. Turner and J.E. Lidsey, *Phys. Rev. D*, in press (1993).

[82] L. Abbott and M. Wise, *Nucl. Phys. B* **244**, 541 (1984); F. Lucchin and S. Mattarese, *Phys. Rev. D* **32**, 1316 (1985); R. Fabbri, F. Lucchin, and S. Mattarese, *Phys. Lett. B* **166**, 49 (1986).

[83] V. Belinsky, L. Grishchuk, I. Khalatanikov, and Ya.B. Zel'dovich, *Phys. Lett. B* **155**, 232 (1985); L. Jensen, unpublished (1985).

[84] A.A. Starobinskii, *Sov. Astron.* **11**, 133 (1985).

[85] A.D. Linde, *Phys. Lett. B* **108**, 389 (1982); A. Albrecht and P.J. Steinhardt, *Phys. Rev. Lett.* **48**, 1220 (1982).

[86] S.D.M. White, C. Frenk, and M. Davis, *Astrophys. J.* **274**, L1 (1983); *ibid* **287**, 1 (1983); J. Centrella and A. Melott, *Nature* **305**, 196 (1982).

[87] J.P. Ostriker, *Ann. Rev. Astron. Astrophys.* **31**, in press (1993).

[88] L. Knox and M.S. Turner, *Astrophys. J.*, in press (1993).

[89] V. De Lapparent, M. Geller, and J. Huchra, *Astrophys. J.* **302**, L1 (1986); *ibid* **332**, 44 (1988); M. Geller and J. Huchra, *Science* **246**, 897 (1989).

[90] N. Bahcall, *Ann. Rev. Astron. Astrophys.* **26**, 631 (1988).

[91] M. Davis et al., *Astrophys. J.* **292**, 371 (1985).

[92] T. Broadhurst et al., *Nature* **343**, 726 (1990).

[93] S.J. Maddox et al., *Mon. Not. R. astr. Soc.* **242**, 43p (1990).

[94] A. Dressler et al., *Astrophys. J.* **313**, L37 (1987); E. Bertschinger et al., *ibid* **364**, 370 (1990) and references therein.

[95] K. Fisher et al., *Astrophys. J.* **389**, 188 (1992); also see, M.S. Vogeley et al., *ibid* **391**, L5 (1992); W. Saunders et al., *Nature* **349**, 42 (1991).

[96] J. Loveday et al., *Astrophys. J.* **390**, 338 (1992); *ibid* **400**, L43 (1992).

[97] The Sloan Digital Sky Survey is a collaboration between The University of Chicago, Fermilab, Johns Hopkins University, Princeton University and the Institute for Advanced Study.

[98] N. Kaiser, *Astrophys. J.* **284**, L9 (1984); in *Inner Space/Outer Space*, eds. E.W. Kolb et al. (Univ. of Chicago Press, Chicago, 1986), p. 258.

[99] G. Efstathiou, J.R. Bond, and S.D.M. White, *Mon. Not. R. astr. Soc.* **258**, 1p (1992).

[100] R.Y. Cen and J.P. Ostriker, *Astrophys. J.* **393**, 22 (1992); A.E. Evrard, F.J. Summers, and M. Davis, *Astrophys. J.*, in press (1993); E. Bertschinger and J. Gelb, *Computers in Physics* **5**, 164 (1991); J. Gelb and E. Bertschinger, *Astrophys. J.*, in press (1993); N. Katz, L. Hernquist, and D. Weinberg, *Astrophys. J.* **399**, L109 (1992); G. Evrard, *Mon. Not. R. astr. Soc.* **235**, 911 (1988); C. Park and J.R. Gott *ibid* **249**, 288 (1991); C. Park, *ibid* **242**, 59p (1990); C. Frenk et al., *Astrophys. J.* **351**, 10 (1990).

[101] For a synopsis of structure formation in the CDM scenario see e.g., G.R. Blumenthal et al., *Nature* **311**, 517 (1984).

[102] See e.g., J.P. Ostriker, *Ann. Rev. Astron. Astrophys.* **31**, in press (1993); F. Adams et al., *Phys. Rev. D* **47**, 426 (1993); J. Gelb et al., *Astrophys. J.* **403**, L5 (1993); R. Cen, N. Gnedin, L. Kofman, and J.P. Ostriker, *ibid* **399**, L11 (1992).

[103] R. Davis et al., *Phys. Rev. Lett.* **69**, 1856 (1992); F. Lucchin, S. Mattarese, and S. Mollerach, *Astrophys. J.* **401**, L49 (1992); D. Salopek, *Phys. Rev. Lett.* **69**, 3602 (1992); A. Liddle and D. Lyth, *Phys. Lett. B* **291**, 391 (1992); J.E. Lidsey and P. Coles, Queen Mary College preprint (1992); A. Dolgov and J. Silk, unpublished (1992); T. Souradeep and V. Sahni, *Mod. Phys. Lett. A* **7**, 3541 (1992).

[104] M.S. Turner, G. Steigman, and L. Krauss, *Phys. Rev. Lett.* **52**, 2090 (1984); M.S. Turner, *Physica Scripta* **T36**, 167 (1991); P.J.E. Peebles, *Astrophys. J.* **284**, 439 (1984); G. Efstathiou et al., *Nature* **348**, 705 (1990); L. Kofman and A.A. Starobinskii, *Sov. Astron. Lett.* **11**, 271 (1985).

[105] Q. Shafi and F. Stecker, *Phys. Rev. Lett.* **53**, 1292 (1984); S. Achilli, F. Occhionero, and R. Scaramella, *Astrophys. J.* **299**, 577 (1985); S. Ikeuchi, C. Norman, and Y. Zahn, *Astrophys. J.* **324**, 33 (1988); A. van Dalen and R.K. Schaefer, *Astrophys. J.* **398**, 33 (1992); M. Davis, F. Summers, and D. Schlegel, *Nature* **359**, 393 (1992); J. Holtzman and J.A. Primack, *Astrophys. J.*, in press (1993); A. Klypin et al., *Astrophys. J.*, in press (1993); D. Pogosyan and A.A. Starobinsky, DAMTP/IOA/MRAO preprint (1993).

[106] M.S. Turner and B.J. Carr, *Mod. Phys. Lett. A* **2**, 1 (1987).

[107] T. Gaier et al., *Astrophys. J.* **398**, L1 (1992); D.C. Alsop et al., *ibid* **387**, 146 (1992); A.C.S. Readhead et al., *ibid* **346**, 566 (1989); P. De Bernardis et al., *ibid* **396**, L57 (1992); R.A. Watson et al., *Nature* **357**, 660 (1992); S.S. Meyer, E.S. Cheng, and L.A. Page, *Astrophys. J.* **371**, L1 (1991); J.O. Gundersen et al., *ibid*, in press (1993); P.R. Meinhold et al., *ibid*, in press (1993).

[108] R. Crittenden, J.R. Bond, R. Davis, G. Efstathiou, and P.J. Steinhardt, *Phys. Rev. Lett.*, in press (1993).

[109] M.S. Turner and L.M. Widrow, *Phys. Rev. D* **37**, 2743 (1988); B. Ratra, *Astrophys. J.* **391**, L1 (1992).

[110] A. Abramovici et al., *Science* **256**, 325 (1992); K.S. Thorne, in *300 Years of Gravitation*, eds. S.W. Hawking and W. Israel (Cambridge Univ. Press, Cambridge, 1987), p. 330.

[111] See e.g., J.R. Primack, D. Seckel, and B. Sadoulet, *Ann. Rev. Nucl. Part. Sci.* **38**, 751 (1988); D.O. Caldwell, *Mod. Phys. Lett. A* **5**, 1543 (1990); P.F. Smith and J.D. Lewin, *Phys. Repts.* **187**, 203 (1990).

[112] U.-L. Pen, D.N. Spergel, and N. Turok, Princeton Observatory Preprint 485 (1993); D. Coulson, D. Spergel, and N. Turok, work in progress (1993).

OBLIQUE ELECTROWEAK PARAMETERS

AND ADDITIONAL FERMION GENERATIONS

Gautam Bhattacharyya

Department of Pure Physics, University of Calcutta,
92 Acharya Prafulla Chandra Road, Calcutta 700 009.
India

1 INTRODUCTION

The search for physics beyond the Standard Model (SM) through oblique electroweak corrections [1,2,3] has in recent years become a wave. Although the precision tests at the CERN e^+e^- collider LEP, based on the 1990 data of $\sim$ 5,50,000 Z-events, have further squeezed possibilities of physics beyond the SM, nonetheless it is perhaps prudent to still keep an open mind and continue to examine extensions of it by confronting them with the high-statistics data that have now become available. The so called "oblique parameters" provide sensitive hunting ground for new physics. Their key property is that they can smell physics at a very high energy scale ($\sim$ TeV) addressing, for example, the issues of technicolour theories afresh or looking for a hypothetical multiplet of heavy fermions. In our discussion we do not, however, discuss technicolour theories at all, but concentrate mainly on the issues of heavy fermions which can be tackled in the perturbative approach.

2 THE OBLIQUE PARAMETERS

Oblique parameters are some finite combinations of the 2-point gauge boson self-energies. The gauge group of our consideration is $SU(2) \otimes U(1)$. Altogether there are four 2-point functions: $\Pi_{\gamma\gamma}, \Pi_{\gamma Z}, \Pi_{ZZ}$ and Π_{WW}. There are two important energy scales at which there are experimental measurements, namely $s = M_Z^2$, and $s = 0$, where we set our renormalisation conditions. So taking into consideration each Π-function at the two energy scales we have in total eight such different functions. Out of which $\Pi_{\gamma\gamma}(0)$ and $\Pi_{\gamma Z}(0)$ are zero because of our choice of renormalisation conditions satisfying the electromagnetic Ward identities. Of the other six, three are absorbed in the renormalisation of α, G_μ and M_Z.

Quantitative Particle Physics, Edited by
M. Lévy *et al.*, Plenum Press, New York, 1993

The remaining three objects (in fact three independent linear combinations) survive after renormalisation and cast impact on the observables. These are the three oblique electroweak parameters defined as follows [1,3]:

$$
\begin{aligned}
S &= \frac{16\pi}{M_Z^2}[\Pi_{33}(M_Z^2) - \Pi_{33}(0) - \Pi_{3Q}(M_Z^2)] \\
&= \frac{8\pi}{M_Z^2}[\Pi_{3Y}(0) - \Pi_{3Y}(M_Z^2)]
\end{aligned}
$$
$$\text{where} \quad Q = t_3 + Y/2, \tag{1}$$

$$T = \frac{4\pi}{s^2 c^2 M_Z^2}[\Pi_{11}(0) - \Pi_{33}(0)], \tag{2}$$

and

$$U = \frac{16\pi}{M_W^2}[\Pi_{11}(M_W^2) - \Pi_{11}(0)] - \frac{16\pi}{M_Z^2}[\Pi_{33}(M_Z^2) - \Pi_{33}(0)]. \tag{3}$$

At this point we take the opportunity of discriminating contributions from the standard model and beyond. Since contributions from different sources of physics add linearly to the Π-functions, for every Π one can define a $\tilde{\Pi}$ given by $\tilde{\Pi} \equiv \Pi - \Pi^{SM}$. We, therefore, use the variants $\tilde{S}, \tilde{T}$ and $\tilde{U}$ as three independent oblique electroweak parameters defined so as to be non vanishing only for physics beyond the SM. $\tilde{T}$ is basically the shift of the ρ parameter from its SM value due to higher order effects and $\tilde{S}$ is the renormalisation of the quantum mixing of the $SU(2)$ gauge boson with the $U(1)$ gauge boson from zero scale to the Z-mass scale. They enter the expressions of the electroweak observables through the modifications of the ρ parameter and the effective weak angle as follows:

$$\rho = \rho^{SM} + \alpha\tilde{T}, \tag{4}$$

and

$$\sin^2\bar{\theta}_W = (\sin^2\bar{\theta}_W)^{SM} + \frac{\alpha}{4(c^2 - s^2)}\left[\tilde{S} - 4c^2 s^2 \tilde{T}\right] \tag{5}$$
$$\text{where}$$
$$c^2 \equiv M_W^2/M_Z^2 \quad \text{and} \quad s^2 = 1 - c^2$$

$\tilde{U}$, on the other hand, appears in the renormalisation of the W-mass in the following way:

$$
2c^2\left[1 + \left\{1 - \frac{4\pi\alpha(M_Z^2)}{\sqrt{2}G_\mu M_Z^2}\right\}^{1/2}\right]^{-1} = 1 + \frac{3\alpha}{16\pi(c^2 - s^2)s^2}\left(\frac{m_t}{M_Z}\right)^2 \\
+ \frac{\alpha}{4(c^2 - s^2)s^2}\left[4c^2 s^2 \tilde{T} - 2s^2 \tilde{S} + (c^2 - s^2)\tilde{U}\right] \tag{6}
$$

It is noteworthy that while $\tilde{T}$ and $\tilde{U}$ get contributions from an isospin breaking effects, $\tilde{S}$ is nonvanishing even for isospin symmetric situation. A simultaneous fit[3] to the leptonic and hadronic cross sections and the leptonic forward backward charge-asymmetries measured at LEP[4] put bounds on the allowed region of the $\tilde{S}.\tilde{T}$ parameter space. The fitted values of $\tilde{S}$ and $\tilde{T}$ and the consequently derived number for

$\tilde{U}$ using the TEVATRON result[5] $M_W = 80.14 \pm 0.31$ GeV, and for $m_t = 140$ GeV, $m_H = 100$ GeV and $\alpha_S = 0.12$, are the following:

$$\tilde{S} = -0.76 \pm 0.71,$$
$$\tilde{T} = -0.70 \pm 0.49,$$
$$\tilde{U} = -0.11 \pm 1.07.$$

$$(7)$$

3 THE ADDITIONAL GENERATIONS

3.1 A chiral generation

A chiral fermion generation is one whose left- and right-handed (LH and RH) multiplets transform differently under the single gauged $SU(2)$. Thus each standard model fermion belongs to a chiral generation transforming as a doublet in the LH sector and a singlet in the RH sector. The LEP measurements [4] on the Z-decay widths do not allow any further light chiral fermion beyond the standard ones. However, one can always take in addition into consideration a replica of a standard generation, but heavy enough not to be produced at the current phase of LEP measurements. The idea behind taking a complete (lepton plus quark) generation is to keep the theory anomaly-free. A very important consequence is that the bound on $\tilde{S}$ can be translated to a maximum number of chiral mass degenerate heavy fermion generations to· be allowed. The contribution to $\tilde{S}$ from a degenerate chiral fermion multiplet is the following:

$$\tilde{S} = \frac{1}{\pi} \sum_f n_c^f x_f \left[2 - 4\sqrt{x_f - \frac{1}{4}} \cot^{-1} \left(2\sqrt{x_f - \frac{1}{4}} \right) \right], \qquad (8)$$

where $x_f = M_f/M_Z$, M_f being the common mass of the multiplet and n_c^f is the colour factor. And in the limit when M_f is extremely heavy, the contribution to $\tilde{S}$ from a complete generation is $2/3\pi$. Since a degenerate multiplet does not contribute to $\tilde{T}$, we get the fitted value of $\tilde{S}$, fixing $\tilde{T} = 0$, as $\tilde{S}=0.04\pm0.44$ which restricts the number of additional heavy chiral fermion multiplets to three.

In such a limit, in fact, the contribution can be expressed in a more meaningful way:

$$\tilde{S} = \frac{1}{6\pi} \sum_f n_c^f \text{Tr} \left(T_{3L} - T_{3R} \right)^2, \qquad (9)$$

which is positive semi-definite (T_{3L} and T_{3R} being the third components of weak isospin in the LH and RH sectors respectively).

At this point one comment is worth mentioning. If we remove the degeneracy between the members of the multiplet, the mass splitting [6] will be severely constrained from the parameter $\tilde{T}$. The contribution to $\tilde{S}$ will still be there but will be smaller in size. In this lecture we, however, focus our attention mainly on the parameter $\tilde{S}$ and hence have chosen a mass degenerate situation.

The reason behind such a non-decoupling for chiral fermions (i.e., $T_{3L} \neq T_{3R}$) is the following. The multiplets which are chiral with respect to the gauged $SU(2)$ contain fermions which get masses through spontaneous symmetry breaking and hence their masses are not gauge invariant. As a result, their contributions to $\tilde{S}$ do not decouple.

3.2 The vector-like multiplets

Vector-like multiplets are those which transform identically in the LH and RH sectors under the gauged $SU(2)$. They are called "vector" because they do not have any axial coupling with the Z-boson. It is obvious from eq. (9) that the heavy vector multiplets do not contribute to the parameter $\tilde{S}$ (in fact also to $\tilde{T}$) in the unmixed situation. But if they mix with their standard counterparts then through such mixing a non-trivial contribution to $\tilde{S}$ arises[7]. In this lecture we take into consideration the mixing effects of leptons only. In general, such leptons make non-vanishing contributions to the anomaly and the theory needs additional fermions to be well-behaved. For these reasons, we choose only anomaly-free additional multiplets, which are also unconstrained by the $\tilde{S}$, $\tilde{T}$ and $\tilde{U}$ variables. These, for example, could be singlet (i.e., sterile) neutrinos, or leptons appearing in vector multiplets. After spontaneous symmetry breaking these fermions can mix with the sequential fermions and give non-zero contributions to $\tilde{S}$, $\tilde{T}$ and $\tilde{U}$.

In general, exotic leptons of arbitrary representations may be allowed to mix with the corresponding SM counterparts by postulating the existence of appropriate Higgs representations in addition to the scalar doublet of the SM. Here we restrict ourselves to minimal extensions of the SM whilst only incorporating extra lepton generations, keeping the scalar sector just as in the SM. This limits our choice of the exotic lepton generations to singlet, doublet and triplet only. These multiplets are, in fact, present in many popular extensions of the SM. The neutral singlet and the vector doublet leptons are contained in the fundamental $\underline{27}$ of E_6, whilst the vector triplet appears in a supersymmetrized left-right model where triplet scalars are employed for left-right symmetry breaking and/or neutrino mass generation.

We now discuss the three cases of singlet neutrinos, vector doublet leptons and vector triplet leptons in turn. In each case we observe that there are two types of mass terms: first $SU(2)$-breaking mass terms (m, m') driven by the doublet Higgs and then an $SU(2)$-invariant mass term (M) and we take $M \gg m, m'$.

The standard sequential lepton multiplet is:

$$\psi_L = \begin{pmatrix} \nu \\ l^- \end{pmatrix}_L : \quad l_L^{c+}. \tag{10}$$

Introducing a sterile LH neutrino, N, is possibly one of the simplest extensions of the SM. Its mixing with the SM neutrino produces a light state which is almost massless and a heavy state whose mass can be arbitrarily large. It may be noted that the light state gets a predominant contribution from ν while the heavy state corresponds mainly to N. This is the standard see-saw mechanism where the mixing angle (ϕ) is proportional to m/M and the eigenvalues are m^2/M and M. Introducing $x = (M/M_Z)^2$ and for $x > 1$, $\tilde{S}$ can be expressed as:

$$\tilde{S} = \frac{\sin^2 \phi}{\pi} \left[5/18 + (x/2) \ln x + J \right] \tag{11}$$

where

$$J = \left[(12 - 18x) \ln x + 6(x-1)^2 (x+2) \ln\{(x-1)/x\} + 6x^2 + 3x - 10 \right] / 36 \tag{12}$$

To obtain the predictions for $\tilde{S}$ from the above we must fix m and M. Keeping in mind the upper bound of 35 MeV on the ν_τ mass we choose $m = 1.8$ GeV and find that for

$M = M_Z$, $\tilde{S} = 3.2 \times 10^{-5}$ while for $M = 1$ TeV (10 TeV), $\tilde{S} = 2.2 \times 10^{-6}$ (3.8×10^{-8}). It is always positive and decreases monotonically, asymptotically approaching zero as $\sim \ln x / x$. The contribution to $\tilde{T}$, which is an analogous variant of T, vanishes identically at the level of our approximation.

Vector doublet leptons transform as $SU(2)$ doublets both in the LH and RH sectors, e.g.:

$$\begin{pmatrix} N_1 \\ E^- \end{pmatrix}_L ; \quad \begin{pmatrix} E^{c+} \\ N_2 \end{pmatrix}_L . \tag{13}$$

The neutrinos of the vector doublet cannot acquire Majorana masses due to the absence of an appropriate Higgs. In general, there can be two $SU(2)$-invariant mass terms involving the fields in eqs.(10) and (13). However, by a suitable redefinition, it is possible to arrange that only the doublets of eq.(13) are coupled. This forbids any possibility of mixing in the neutrino sector. The situation with the charged leptons is different, owing to the presence of the singlet in eq.(10). The effective (2×2) charged lepton mass matrix in this redefined basis is:

$$\begin{array}{c} \\ e_R \\ E_R \end{array} \begin{array}{cc} e_L & E_L \\ \begin{pmatrix} m' & m \\ 0 & M \end{pmatrix} \end{array} \tag{14}$$

where $M \gg m, m'$, the two $SU(2)$ breaking masses. It is rather easy to check from the structure of the above mass matrix that the mixing angle (ϕ) in the RH sector is $\sim (m/M)$, whilst in the LH sector it is $\sim (m \, m'/M^2)$. The mass eigenvalues are approximately m' and $\sqrt{M^2 + m^2}$. In our calculations we retain terms $\sim \mathcal{O}(1/M^2)$ and find that at this level the LH mixing does not contribute. The expression for $\tilde{S}$ is found to be identical to the one given for mixing with a sterile neutrino (eqs. (11) and (12)). The contribution to $\tilde{T}$ vanishes as in the previous case. For numerical estimates, in this case we take $m = 10$ GeV which is consistent with the experimental bounds on charged lepton masses and mixings and find that for $M = M_Z$, $\tilde{S} = 1.0 \times 10^{-3}$. For $M = 1$ TeV (10 TeV) we get $\tilde{S} = 6.6 \times 10^{-5}$ (1.1×10^{-6}).

$$\begin{pmatrix} E_1^+ \\ N_1 \\ E_2^- \end{pmatrix}_L : \quad \begin{pmatrix} E_2^{c+} \\ N_2 \\ E_1^{c-} \end{pmatrix}_L \tag{15}$$

These triplets are of hypercharge zero. Thus while constructing the $SU(2) \otimes U(1)$ invariant neutral and charged mass matrices, one must include the general Majorana type mass terms from the coupling of the above multiplets to their own charge-conjugate ones. In general the standard lepton doublet can couple to the two triplets in different strengths through the doublet Higgs. To the extent one can neglect the difference between these two symmetry breaking mass terms compared to the gauge invariant pieces, it is straight-forward to show that the theory respects a discrete symmetry which interchanges the above two triplets not affecting the standard doublet. This decouples one neutral and one charged member which, barring accidental cancellation, acquire mass $\sim M$ and the mass matrices take the following 2×2 forms in the neutral and charged sectors respectively:

$$\begin{array}{c} \\ \nu \\ N \end{array} \begin{array}{cc} \nu & N \\ \begin{pmatrix} 0 & m/\sqrt{2} \\ m/\sqrt{2} & M \end{pmatrix} \end{array} \tag{16}$$

and

$$
\begin{array}{c}
\quad\quad e_L \quad\quad E_L \\
\begin{array}{c} e_R \\ E_R \end{array}
\left(\begin{array}{cc} m' & 0 \\ m & M \end{array} \right)
\end{array}
\tag{17}
$$

where $N = (N_1 + N_2)/\sqrt{2}$ and $E = (E_2 + E_1^c)/\sqrt{2}$.

The mixing angle in the neutral sector is $\phi \sim (m/\sqrt{2}M)$. In the charged sector the mixing angle in the left-handed state is $\sqrt{2}\phi$ while in the right-handed state it is suppressed by another factor of (m/M) and we neglect it. To a good approximation the eigenstates in the neutral sector constitute one very light state and one heavy Majorana state of mass $\sqrt{M^2 + m^2/2}$. The charged sector consists of a light fermion of mass m' and a heavy fermion of mass M. The expression for $\tilde{S}$ in this case is:

$$
\tilde{S} = \frac{\sin^2 \phi}{2\pi} \left[-5/9 + 3x \ln x + 6J - 4I \right]
\tag{18}
$$

where J is given in eq. (12) and the integral I can be expressed in terms of x and $y = 2\sqrt{x - 1/4}$ as

$$
I = (1/3) \ln x - 5/9 - 4x/3 + \frac{3y + y^3}{3} \arctan(1/y) \quad \text{for} \quad x > (1/4)
\tag{19}
$$

Choosing $m = 10$ GeV, in this case we find that for $M = 1$ TeV (10 TeV), $\tilde{S} = 3.0 \times 10^{-5}$ (5.4×10^{-7}). $\tilde{S}$ decreases in magnitude, vanishing asymptotically as $\sim \ln x/x$, as M increases but unlike the other two cases it can be negative for $M \sim M_Z$. As in the previous cases $\tilde{T}$ does not receive any contribution from the above mixing.

4 CONCLUSION

We have examined contributions of additional fermion generations to the oblique electroweak parameters. We have first considered heavy degenerate chiral generations as a straight-forward extension of the standard scenario concluding that one cannot take into consideration at will any number of such pieces, the number being limited to a maximum of three from present data. We have also considered the mixing of sequential leptons with three different kinds of exotic multiplets, which are by themselves anomaly-free and in the unmixed situation make vanishing contributions to the oblique electroweak parameters $\tilde{S}$ and $\tilde{T}$. In all cases $\tilde{T}$ is zero at our level of approximation. On the other hand, $\tilde{S}$ receives a positive contribution in all cases – excepting for $M \sim M_Z$ in the triplet case – the magnitude depending on the extent of mixing. The mixing angle, in turn, decreases with increasing M and $\tilde{S}$ vanishes asymptotically as $\sim \ln x/x$. We find that $\tilde{S}$ is usually small and can at most reach 1.0×10^{-3} probing which seems to be a remote possibility at the present juncture. Thus it is unlikely that the different scenarios of vector like multiplets considered here can be distinguished in the near future. These will, however, be disfavoured if upcoming results assert a statistically significant non-zero $\tilde{S}$ larger than the above prediction.

Acknowledgements

I would like to thank Prof. Probir Roy for an enjoyable and fruitful collaboration. I am grateful to my supervisor Dr. Amitava Raychaudhuri who has been my mentor for the last three years and has collaborated with me on a part of this work. I would also like to acknowledge the hospitlity of the Theory Division, CERN where part of the work was done and to thank the organisers of the Cargese School for providing an enjoyable atmosphere. I also acknowledge fruitful discussions with Dr. D. Choudhuri and Prof. R. Barbieri. Finally let me express my appreciation of the beauty of the village of Cargese and the charm of the Mediterranian.

References

[1] M. Peskin and T. Takeuchi, Phys. Rev. Lett. **65** (1990) 964: G. Altarelli and R. Barbieri. Phys. Lett. **B253** (1990) 161.

[2] B. Holdom and J. Terning, Phys. Lett. **B247** (1990) 88; M. Golden and L. Randall. Nucl. Phys. **B361** (1991) 3; M. Dugan and L. Randall. Phys. Lett. **B264** (1991) 154; A. Dobado et al., ibid. **B255** (1991) 405; S. Bertolini and A. Sirlin, ibid. **B257** (1991) 179; D. Kennedy and P. Langacker, Phys. Rev. Lett. **65** (1990) 2967; W. Marciano and J. Rosner, ibid. **65** (1990) 2963; E. Gates and J. Terning, ibid. **67** (1991) 1840; E. Ma and P. Roy, ibid. **68** (1992) 2879: G. Altarelli. R. Barbieri and S. Jadach, Nucl. Phys. **B369** (1992) 3.

[3] G. Bhattacharyya, S. Banerjee and P. Roy, Phys. Rev. **D45** (1992) R729.

[4] ALEPH collab., D. Decamp et al. Z. Phys. **C48**, 365 (1990): DELPHI collab.. P. Abreu et al., preprint DELPHI 90-62 PHYS 80. contributed to the Aspen conference, January 1991: L3 collab.. B. Adeva et al., L3 preprint 28, February 1991; OPAL collab., G. Alexandar et al., CERN-PPE/91-67, April 1991.

[5] CDF Collaboration, F. Abe et al. Phys. Rev. Lett. **65** (1990) 2243: UA2 Collaboration, J. Alliti et al. Phys. Lett. **B241** (1990) 150.

[6] M. Veltman, Nucl. Phys. **B123** (1977) 89.

[7] G. Bhattacharyya and A. Raychaudhuri, submitted to Phys. Lett. **B**.

ELECTROWEAK SYMMETRY BREAKING
FROM THE TOP

Nick Evans

Physics Department
University of Southampton
Southampton SO9 5NH
UK

INTRODUCTION

The last decade of high energy collider experiments has verified the high accuracy of the Standard Model (SM) at describing physics below $\sim 90 GeV$ and thus placed strong constraints on extensions to the SM. The mechanism by which electroweak symmetry (EWS) is broken, however, has not been experimentally investigated and remains an open question. In the minimal SM of Glashow, Weinberg and Salam the fermion, $W^{\pm}$ and Z^0 mass terms that break the $SU(2)_L \otimes U(1)_Y$ symmetry are replaced by gauge invariant interactions with a fundamental scalar Higgs field that acquires a non zero vacuum expectation value (vev) and breaks EWS. However, the fermion masses are replaced by Yukawa couplings to the Higgs field giving no understanding of the mass structure of the SM. Similarly the vev of the Higgs field that minimizes its potential is an arbitrary parameter of the theory. This lack of predictive power suggests that the Higgs model is at best a low energy effective theory of physics at the EW scale.

There has been interest recently in models in which the Higgs is a bound fermion pair. Consider a fermion doublet, $f = (h, l)$, the left hand components of which, f_L, transforms under the fundamental representation of $SU(2)_L$ and the right hand components of which, h_R and l_R, are singlets. An attractive force between the fermions and antifermions which becomes sufficiently strongly interacting at some energy scale, Λ, will cause the formation of the fermion condensates, $< \bar{f}f > \neq 0$ (the binding energy between a fermion and antifermion pair is greater than their combined rest energy so empty space will spontaneously create bound fermion pairs). The bilinear

$$\bar{f}f = \bar{h}_L h_R + \bar{l}_L l_R + h.c. \tag{1}$$

is not $SU(2)_L$ invariant and thus the condensates breaks EWS. The fermion, f, becomes massive through interactions with the condensate; there is only one energy scale in the

Quantitative Particle Physics, Edited by
M. Lévy *et al.*, Plenum Press, New York, 1993

problem, Λ, and so

$$m_f \sim \Lambda \tag{2}$$

QCD is an example of such a theory in which the $SU(3)_c$ gauge group is asymptotically free. The quarks become confined at the scale Λ_{QCD} where the coupling becomes large and quark condensates form. QCD breaks EWS. However, the characteristic scale of QCD is $\Lambda_{QCD} \sim 200MeV$ whilst EWS is broken at scales $\mathcal{O}(250GeV)$ so additional symmetry breaking is required. The scenario can also be realized by a strong force acting over a short range given by $\frac{1}{\Lambda}$.

MINIMAL TOP MODE STANDARD MODEL (TMSM)

Collider experiments have, so far, failed to find the top quark; direct searches[1] show the mass of the top, $m_t \geq 90GeV$ and indirect calculation[2] from EW radiative corrections suggest $m_t \leq 200GeV$. The top mass is, therefore, of the same order of magnitude as the EWS breaking energy scale (the Higgs vev $v = 246GeV$). Nambu[3] has suggested that this indicates that the top is playing the role of the fermion in the introductory discussion and that the Higgs might be a $\bar{t}t$ boundstate. The minimal top condensate model[3,4,5] introduces a point-like four Fermi interaction for the top quark

$$\mathcal{L} = G_t \bar{\psi}_L^c t_R^c \bar{t}_R^{c'} \psi_L^{c'} + h.c. \tag{3}$$

where $\psi_L = (t, b)_L$ and c and c' run over QCD colours. The dynamical self energy of the top is given by the solution to the appropriate Schwinger-Dyson (SD) equation[6]. To be able to numerically solve the equation the full vertex in the SD equation is truncated to the first order perturbative vertex (an uncontrolled approximation). In the large number of colours limit the Gap Equation[5] is

$$\tag{4}$$

The non-trivial solution to Eq.(4) yields

$$\frac{1}{G_t} = \frac{N_c}{8\pi^2}\left[\Lambda_0^2 - m_t^2 ln\left(\frac{\Lambda_0^2}{m_t^2}\right)\right] \tag{5}$$

where N_c is the number of QCD colours and Λ_0 is a numerical cut off.

There is a non zero solution for the top mass if $G \geq G_c = \frac{8\pi^2}{N_c\Lambda_0^2}$ (i.e. the force must be strong enough to cause condensation). Similarly the mass of the Higgs (a $\bar{t}t$ boundstate), m_H, can be calculated from a truncated Bethe-Salpeter equation[7]. For a given numerical cut off, G_t is tuned so that the Higgs vev $v = 246GeV$ and m_t and m_H are then predictions of the theory. The numerical results including QCD corrections[6] are shown in Table 1.

Table 1. Numerical results for m_t and m_H as a function of Λ_0 in minmimal TMSM

Λ_0/GeV	m_t/GeV	m_H/GeV
10^4	591	618
10^5	448	769
10^{15}	268	

These predictions violate the relation $m_t \sim \mathcal{O}(\Lambda)$ because the coupling G is fine tuned to $\mathcal{O}(m_t^2/\Lambda_0^2)$. If the four Fermi interaction represents massive gauge boson exchange from a broken gauge symmetry then there are two mass scales in the problem, the scale at which the interaction becomes strong, Λ, and the scale at which the interaction is broken, measured by the mass of the gauge boson M_{GB}. The fine tuning corresponds to adjusting the ratio M_{GB}/Λ .

Below the scale Λ_0 the effective theory of TMSM can be shown to be precisely the SM. The Goldstone Bosons "eaten" by the $W^\pm$ and Z^0 fields are bound states of top and bottom quarks.[5]

The minimal model predictions for the top mass are well above the EW radiative correction bounds even for high cut offs (which involve fine tuning G_t to ~ 26 decimal places!). However, the numerical calculations are performed with uncontrolled approximations; equivalent calculations in QCD are, though, surprisingly accurate. If the new dynamics is at scales close to the EWS breaking scale we might also expect the structure of the interaction vertex to deviate from the four Fermi point vertex approximation.

A complete model of EWS breaking must account for all the fermion masses in the SM; in the minimal TMSM it is assumed that the lighter fermion masses "feed down" from the top condensate through interactions at or above the scale Λ. It is interesting to develop models of the dynamics at the scale Λ both to shed light on the fermion mass structure and to study the stability of the top mass predictions in such models. The theory above the scale Λ must also presumably be renormalizable. The top mode moose[8] is such a model developed from Georgi's moose models of technicolour theories[9]. The model makes use of the physics of chiral symmetry breaking in QCD.

CHIRAL SYMMETRY BREAKING IN MOOSE NOTATION

The "moose"[9] is a very convenient notation for models of this kind. A circle with a number n inside represents a gauge group $SU(n)$, and directional lines represent left-handed Weyl fermions which transform under the gauge groups. A line leaving (entering) a circle with a number n inside represents a left-handed fermion transforming under the $n(\bar{n})$ representation of that group. External lines labelled by a number k correspond to k copies of the representation of the gauge group, and, therefore, have a global symmetry $SU(k) \otimes U(1)$.

A simple example of a moose is shown in Fig. 1 which represents two flavour QCD.

Figure 1. A QCD Moose.

The fermion lines represent the massless up and down quark fields $\psi = (u, d)$. The right-handed quarks, $\psi_R = \frac{1}{2}(1 + \gamma_5)\psi$ have been written as a left-handed two component field $\psi_R^c \equiv \chi^L$ where $\psi_R^c = \sigma_2 \psi_R^*$ transforms as a $\bar{3}$ under $SU(3)_c$ and also transforms under a global $SU(2)_R$ symmetry. The left-handed quarks ψ_L transform under a global $SU(2)_L$ symmetry. When the QCD group $SU(3)_c$ becomes strong at Λ_{QCD} the quarks form condensates

$$< i(\chi_L^T \sigma_2 \psi_L) + h.c. > \neq 0 \qquad (6)$$

henceforth for brevity we shall write such condensates as simply $< \chi \psi >$. The condensate breaks the chiral $SU(2)_L \otimes SU(2)_R$ to the vector symmetry $SU(2)_{L+R}$. There are three massless Goldstone Bosons, the pions, associated with the three broken axial generators.

THE TOP MODE MOOSE

The masses in the quark sector of the SM break an $[SU(3) \otimes U(1)]^3$ flavour symmetry corresponding to independent rotations on $U_R = (u, c, t)_R$, $D_R = (d, s, b)_R$ and $\psi_L = (Q_1, Q_2, Q_3)_L$ where $Q_1 = (u, d)$, $Q_2 = (c, s)$ and $Q_3 = (t, b)$ (i.e. there are three families of quarks with differing masses). It is this symmetry that gives rise to the GIM mechanism that protects the SM from flavour changing neutral currents(FCNCs). The intimate connection between fermion masses and flavour symmetry leads us to propose that broken gauged flavour interactions are the origin of the four Fermi interaction in TMSM. Since we wish to have a GIM mechanism to suppress FCNCs from the new dynamics we gauge , in the quark sector, the chiral flavour group

$$SU(3)_R^D \otimes SU(3)_L^Q \otimes SU(3)_R^U \tag{7}$$

The basic dynamical mechanism of the top mode moose is demonstrated by the toy model in Fig.2 which describes massless charged 2/3 quarks $U = (u, c, t)$ in the absence of EW forces. In this toy world the flavour symmetry of the standard model is just $SU(3)_L \otimes SU(3)_R$ corresponding to separate rotations on U_L and U_R.

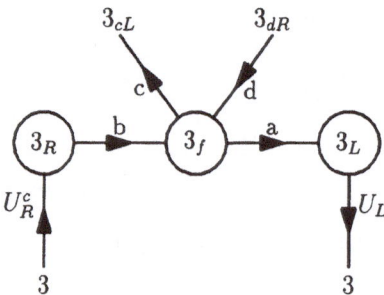

Figure2. The toy moose.

The gauged (ga) $SU(3)_L^{ga} \otimes SU(3)_R^{ga}$ chiral family group will be broken to the global (gl) family group of the standard model by the asymptotically free, unbroken group $SU(3)_f^{ga}$ (which acts like QCD). The $SU(3)_f^{ga}$ group becomes strongly interacting below some scale Λ_f and the fermions that transform under it, a,b,c and d ("preons") form condensates. The new fermions a and b also transform under the gauged flavour groups whilst c and d transform only under global flavour groups $SU(3)_{cL}^{gl}$ and $SU(3)_{dR}^{gl}$. When $SU(3)_f^{ga}$ becomes strong there is a choice of vacua that depends on the family gauge couplings. **IF** the condensates $< ad >$ and $< bc >$ form then $SU(3)_L^{ga} \otimes SU(3)_{dR}^{gl}$ breaks to $SU(3)_L$ and $SU(3)_R^{ga} \otimes SU(3)_{cL}^{gl}$ breaks to $SU(3)_R$. The diagonal subgroups $SU(3)_L$ and $SU(3)_R$ would be the global symmetries of the toy standard model. The choice of vacuum is a non-perturbative problem since there are three strongly interacting gauge groups acting on the preons. Although an assumption of the model, a Chiral Lagrangian analysis of the problem[10] in which we naively extend perturbative results into the non-perturbative regime supports this choice of vacuum. See also reference 11.

The quark masses of the SM break the global flavour symmetry. The $SU(3)_L \otimes SU(3)_R$ symmetry is broken in the model by the introduction of a fermion mass matrix $\tilde{M}^U$ which couples the c and d preons together. $\tilde{M}^U$ is generated by physics above the family symmetry breaking scale. Heavy flavour gauge boson exchange then gives rise to effective four Fermi operators in the low energy theory that generate the quark masses. The $SU(3)_L^{ga}$ and $SU(3)_R^{ga}$ gauge bosons mix through the loops of preons shown in Fig.3.

This mixing occurs because the preons mix with each other through condensates and the mass matrix $\tilde{M}^U$. The elements of $\tilde{M}^U$ are chosen so that the operator in Eq.3 is sufficiently strong to cause top condensation. The charm and up masses are then fed down from the top condensate by the other family interactions which are suppressed relative to the top-top interaction by the smaller elements of $\tilde{M}^U$. The top mass has been calculated in this model[12] by solving the appropriate Gap Equation and the results are given in Table 2, these results are similar to those in Table 1.

CONCLUSION

TMSM is a very tidy explanation of EWS breaking since it introduces no new particles that transform under the SM gauge group and explains why the top mass is so large. More elaborate dynamical models can be developed from TMSM in which the full fermion mass structure is accomodated; such models are necessarily complicated if they are to reproduce the complicated flavour symmetry breaking of the SM.

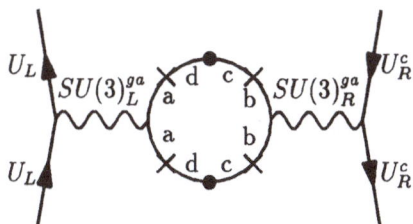

Figure 3. The mixing of gauge bosons of $SU(3)_L^{ga}$ and $SU(3)_R^{ga}$ through the fermion condensates in conjunction with the explicit mass terms. The blobs represent insertions of the mass matrix, $\tilde{M}^U$, and the crosses represent insertions of fermion condensates.

Table 2. Numerical results for m_t as a function of Λ_f in the moose model.

Λ_f/GeV	m_t/GeV
10^4	475
10^5	403
10^{15}	269

Acknowledgements

This work has been performed in collaboration with Steve King and Doug Ross. The author is grateful to SERC for financial support whilst performing this work, to the Hermann Jahn Memorial Award Committee for financing his attendence of the school and to the organizers for allowing him to present this talk.

REFERENCES

1. F.Abe et al., Phys. Rev. Lett. 68 (1992) 447.
2. M.Peskin and T.Takeuchi, SLAC-PUB-5681 (1991).
3. Y. Nambu, New Theories in Physics, Proc. XI Warsaw Symposium on Elementary Particle Physics, ed. Z. Adjuk et al., publ. World Scientific, Singapore (1989).
4. V.A.Miransky, M.Tanabashi and M.Yamawaki, Phys. Lett. 221B (1989) 177.
5. W.A.Bardeen, C.T.Hill and M.Lindner, Phys. Rev. D41 (1990) 1647.
6. S.F.King and S.H.Mannan, Phys. Lett. B241 (1990) 249.
7. H.M.Chesterman and S.F.King, Nucl. Phys. B358 (1991) 59.
8. S.F.King, Phys. Rev. D45 (1992) 990.
9. H.Georgi, Technicolour and Families, Proc. 1990 International Workshop on Strong Coupling Gauge Theories and Beyond, ed. T.Muta and K.Yamawaki, publ. World Scientific, Sigapore (1991). R.S.Chivukula and H.Georgi, Phys. Lett. 188B (1987) 59. R.S.Chivukula, H.Georgi and H.L.Randall, Nucl. Phys. B292 (1987) 93.
10. N.J.Evans, S.F.King and D.A.Ross, Southampton Preprint, SHEP 91/92 - 11.
11. M.A.Luty, Lawrence Berkeley Lab. Preprint, LBL-32299.
12. T.Elliott and S.F.King, Phys. Lett. B283 (1992) 371.

HIGGS MASS LIMITS FROM
ELECTROWEAK BARYOGENESIS

Stanley Myint

Department of Physics
Boston University
Boston, MA 02215, USA

INTRODUCTION

The possibility of creating baryon asymmetry of the Universe has attracted a lot of attention lately. In my opinion, the main reason this has occurred is that from a simple experimental fact that almost all Universe we know consists of matter, instead of anti-matter, it is possible to obtain very strong constraints upon parameters of various particle physics models. These constraints are experimentally verifiable or will be so in the near future.

To start with, we will discuss three basic laws of baryogenesis stated by Sakharov [1] in 1967. Possibility of large anomalous baryon number violation (B-violation) in the early Universe was raised in 1977-78 by Linde and by Dimopoulos and Susskind [2]. In 1984, Klinkhamer and Manton [3] discovered field configurations which they named "sphalerons" and which are relevant in B-violating processes. A major contribution by Kuzmin, Rubakov and Shaposhnikov [4] in 1985 was to suggest that large B-violation could happen during a first order electroweak phase transition (EWPT).

In the scenario that is accepted today, baryogenesis at electroweak phase transition consists of three phases:

1. Before the EWPT any previously existing B is washed out (unless there is primordial B-L asymmetry).

2. During the EWPT B can be created again.

3. After the EWPT B can survive, provided certain conditions are satisfied.

There are different mechanisms for the step 2., the creation of B-asymmetry at EWPT in various extensions of the SM [5]-[8], and I will not discuss them here. Instead, this talk will be concentrated on the condition of step 3., non-erasal of B-asymmetry [6], [9]. The models I will consider are the standard model (SM) and the minimal supersymmetric model (MSUSY).

NECESSARY REQUIREMENTS FOR BARYOGENESIS

When we say "baryon asymmetry of the Universe" (BAU), we refer to the fact that in the Universe we know there are more baryons than anti-baryons:

$$n_B \equiv n_b - n_{\bar{b}} \neq 0, \tag{1}$$

and that this asymmetry is maximal, i.e. Universe is essentially all matter with a very small percentage of anti-matter. [1]

Instead of an explanation of this fact we might conjecture that this is the way our Universe was created. A much more ambitious task would be to show how BAU arises from an initially B-symmetric universe as a result of particle physics laws.

Sakharov made the first attempt in this direction in 1967 [1] by postulating the three requirements which must be satisfied for BAU to evolve from initially B-symmetric conditions:

1. B-violation.

2. C and CP violation.

3. Non equilibrium conditions.

It is easy to see the need for the first requirement: if initially Universe was B-symmetric, and all the processes were B-conserving, then today it would still be B-symmetric. Therefore, B-violating processes are necessary to produce net B-asymmetry.

Secondly, the baryon number is odd under both C and CP operations. If we start from a B-symmetric initial state, even if we have B-violating processes, C and CP conservation would ensure that an equal number of baryons and anti-baryons would be created and there would be no asymmetry. Therefore, both C and CP must be non-conserved.

The last requirement must be taken in conjunction with the corollary of the CPT theorem that particle and anti-particle masses have to be the same. Also, in thermal equilibrium, chemical potentials associated with non-conserved quantum numbers vanish. Therefore, the number densities of baryons and anti-baryons are equal and given by the Fermi-Dirac statistics:

$$n_b = n_{\bar{b}} \sim \frac{1}{1 + exp(-\frac{p^2+m^2}{T})} \tag{2}$$

So, even if B, C, and CP are non-conserved we must make sure to be out of thermal equilibrium to satisfy condition (1).

An immediate question that comes to mind is whether a certain theory satisfies Sakharov's requirements. Let us remind ourselves how the SM, together with the standard cosmology, satisfies, at least in principle, all three criteria.

---

[1]For a review see for example [10].

The baryon number is conserved at the classical level but violated by quantum anomalies. In the notation of ref. [3], the baryon current is given by:

$$j_B^\mu = \frac{1}{3} \sum_{i,\alpha} (\bar{u}_\alpha^i \gamma^\mu u_\alpha^i + \bar{d}_\alpha^i \gamma^\mu d_\alpha^i), \tag{3}$$

where i counts fermion families and α - colours. Its divergence is given by:

$$\partial_\mu j_B^\mu = \frac{N_f}{64\pi^2} \epsilon^{\mu\nu\rho\sigma} (g^2 F_{\mu\nu}^a F_{\rho\sigma}^a + g\prime^2 f_{\mu\nu} f_{\rho\sigma}) \tag{4}$$

Here $N_f = 3$ is the number of fermion families, $gF_{\mu\nu}^a$ and $g\prime f_{\mu\nu}$ are the field strengths of $SU(2)$ and $U(1)$ gauge fields, respectively and $\epsilon^{\mu\nu\rho\sigma}$ is the Levi-Civitta tensor.

Baryon number changes by N_f between two successive θ vacua of a non-Abelian gauge theory (see Fig.1). At zero temperature, the only way to go between different vacua is by a quantum mechanical tunnelling process. The tunnelling solutions are called instantons. In 1976 't Hooft [11] calculated the probability of tunnelling and showed that it is negligible:

$$P \sim e^{-4\pi/\alpha_w} \sim 10^{-164} \tag{5}$$

At high temperature there exists another way of going between different θ vacua and that is by classical thermal excitations. Probability for this is given by a Boltzmann factor:

$$P \sim e^{-\frac{M_{sph}}{T}} \tag{6}$$

M_{sph} is the energy of an unstable solution called the sphaleron. In the field configuration space it is the point which is the lowest barrier between two adjacent θ vacua. M_{sph} at zero T is given by:

$$M_{sph} = 4\pi B(\frac{\lambda}{g^2}) \frac{v}{g_w}, \tag{7}$$

with: $B(\frac{\lambda}{g^2}) \in [1.52, 2.70]$, for $\frac{\lambda}{g^2} \in [0, \infty)$, where the λ is Higgs self coupling. Assuming that the same expression is valid at nonzero temperature with v and λ replaced by their

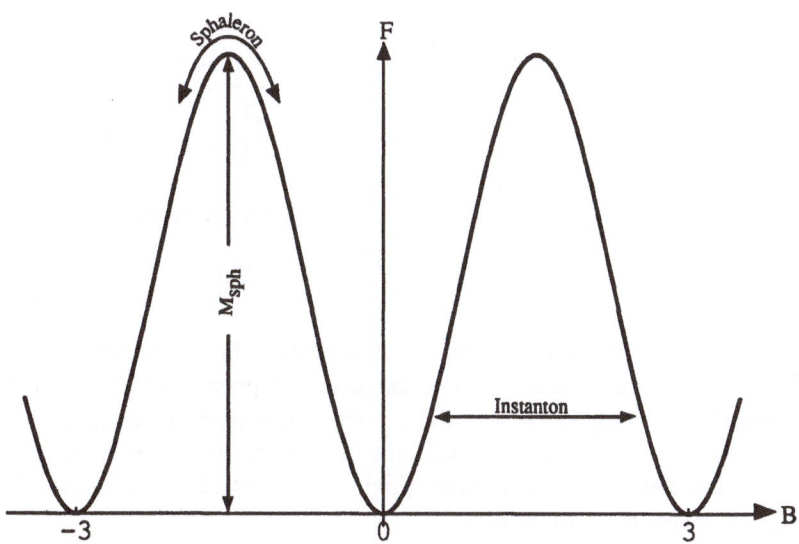

Figure 1. Schematic picture of the ground state of SM

temperature corrected values, from equations (6) and (7) we see that the exponent of the Boltzmann factor is proportional to $v(T)/T$. Today, $v = 246$ GeV and $T \to 0$, therefore probability for B-violating processes via thermal excitations is negligible, but near the EWPT $v \to 0$ and $T \sim 100$ GeV and baryon number is violated very rapidly.

That C and CP are violated is an experimental fact more than a quarter century old. Violation of C invariance was first demonstrated in 1957 in β decay and in π^- and μ^- decays, and violation of CP invariance in 1964 through $K_L^0 \to \pi^+\pi^-$ decays.

Lastly we have to discuss the possibility that out of equilibrium conditions develop. In the standard model of cosmology, the Universe cools as it expands. The rate of expansion is determined by the Hubble constant, H. If the interaction rate for a certain type of process is slower than the expansion rate: $\Gamma < H$, particles do not interact fast enough to follow the change in temperature and non-equilibrium conditions occur. In particular, this may happen during the phase transitions in the early Universe.

Next I will discuss the EWPT in the SM.

ELECTROWEAK PHASE TRANSITION IN THE STANDARD MODEL

The subject of EWPT has been thoroughly discussed in many places.[2] The starting point for our discussion is the scalar potential. In the SM it is just the quartic potential with negative mass squared to ensure the spontaneous symmetry breaking:

$$V_{tree} = -\frac{m^2}{2}\phi^2 + \frac{\lambda^4}{4}\phi^4 \tag{8}$$

To get the one loop effective potential we have to add zero temperature one loop corrections to the previous expression:

$$\Delta V_{1-loop} = \sum_i g_i \frac{m_i^4(\phi)}{64\pi^2} \ln \frac{m_i^2(\phi)}{Q^2} \tag{9}$$

Here i runs over all particles of degeneracy g_i that couple to Higgs and $m_i^2(\phi)$ is their mass as a function of the background Higgs field ϕ. Q is the arbitrary renormalization scale.

When the temperature is nonzero, effective potential gets a contribution [13]:

$$\Delta V_T = \frac{T^4}{2\pi^2} \sum_i g_i I_{\pm} \left[\frac{m_i(\phi)}{T}\right] \tag{10}$$

where $I_-(I_+)$ which are to be used for bosons (fermions) are given by:

$$I_{\pm}(y) = \mp \int_0^\infty x^2 \ln\left(1 \pm e^{-\sqrt{x^2+v^2}}\right) dx \tag{11}$$

This contribution to the effective potential describes the interactions of the Higgs bosons with the thermal bath surrounding them. Expressions (11) are rather difficult to operate with, especially when the masses depend on fields in a complicated way. Fortunately, as shown in ref. [12], one can always use either high temperature (small y) or low temperature expansion (high y), so that the mistake in determining ΔV_T is never bigger than 10 percent.

---

[2] See e.g. ref. [10] or [12].

416

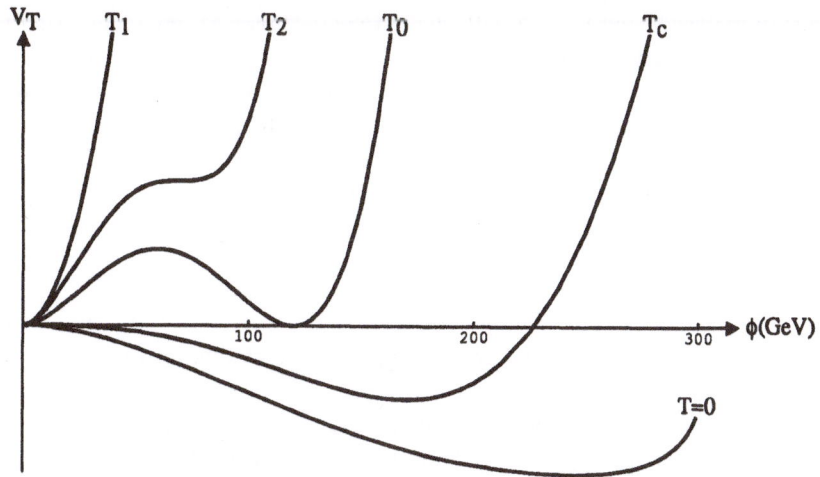

Figure 2. Temperature dependence of V_T

High temperature expansions of (11) are given by:

$$h_-(y) = -\frac{\pi^4}{45} + \frac{\pi^2}{12}y^2 - \frac{\pi}{6}y^3 - \frac{y^4}{32}\ln\left(\frac{y^2}{c_b}\right), h_+(y) = -\frac{7\pi^4}{360} + \frac{\pi^2}{24}y^2 + \frac{y^4}{32}\ln\left(\frac{y^2}{c_f}\right),$$

$$\ln c_b \approx 5.41, \ln c_f \approx 2.64, \tag{12}$$

Whereas the low temperature expansion is:

$$l(y) = -\sqrt{\frac{\pi}{2}}y^{3/2}e^{-y}\left(1 + \frac{15}{8y}\right) \tag{13}$$

Since contributions of different particles to the effective potential enter as positive powers of mass, we can safely neglect all but the heaviest particles. In the SM these are W and Z bosons, t-quark and Higgs itself. By adding their contributions one gets the expression:

$$V^T(\phi) = \gamma(T^2 - T_c^2)\phi^2 - ET\phi^3 + \frac{\lambda_T}{4}\phi^4, \tag{14}$$

where γ, E, and λ_T are determined by masses of particles. For example, λ_T plays the role of temperature corrected Higgs self coupling.[3]

A typical development of V_T as a function of T is shown in Fig.2. For high T, the symmetry is restored. There is only one minimum, $\phi = 0$. For $T = T_2$, a second, asymmetric minimum of higher energy begins to form at $\phi \neq 0$. At $T = T_0$, this minimum becomes of same energy as the symmetric one and for $T_0 > T > T_c$ it becomes the global minimum. The symmetric vacuum, being of higher energy, becomes meta-stable and decays by creation and expansion of bubbles of asymmetric vacuum.[4] Actually [12], this process happens very soon below T_0 and by the time when T has dropped to T_c, and the barrier between two minima disappears, almost all the Universe

---

[3]I say "plays the role" because actually λ_T depends logarithmically on ϕ. However, it will serve to illustrate my purpose. For details please refer to ref. [10].

[4]In principle, as noticed by Gleiser and Kolb, and by Tetradis [14], things may not be so simple. However, for the range of parameters of the SM which is experimentally interesting, the picture described here is quite accurate [15].

is in the asymmetric vacuum. As T falls down to zero, this vacuum expectation value increases until it reaches 246 GeV.

Let me now explain how this gives us limits on the Higgs mass. Assume that sufficient BAU was created at EWPT. Then, we want to freeze it, i.e. we want to make sure that processes which would dilute it happen sufficiently slow: $\Gamma \ll H$. From equation (6) this happens when exponent of the Boltzmann factor is sufficiently large. It was shown [6], [9] that exact criterion for this is :

$$\frac{M_{sph}(T_c)}{T_c} \geq 45 \tag{15}$$

By using equation (7) for the sphaleron energy we convert this into a lower limit on $v(T_c)/T_c$.

For typical values of parameters of SM, T_2, T_0, and T_c are very close. In fact, there is an uncertainty in determining these three temperatures which stems from the inadequacy of using the one-loop effective potential at the origin. I will discuss all the possible sources of error later. For the time being assume just that the phase transition happens near the point $T = T_c$. From equation (14), the minimum is given by:

$$v(T_c) \cong \frac{3ET_c}{\lambda_T} \tag{16}$$

Therefore, the lower limit on $v(T_c)/T_c$ gets converted into the upper limit on λ_T. Since λ_T is temperature corrected λ, this in turn imposes upper limit on λ, and through $m_h^2 = \lambda v^2$, an upper limit on the Higgs mass m_h. Therefore:

$$\Gamma \ll H \Rightarrow m_h < m_h^{crit} \tag{17}$$

When we take higher order corrections in account, and make generous estimates of various uncertainties, [15], the upper limit on the Higgs mass in the SM is given by $m_h \leq 41$ GeV which should be compared to the present lower bound from LEP of 57 GeV.

MINIMAL SUPERSYMMETRIC MODEL

Higgs sector of MSUSY (see for instance ref [16]) contains two complex Higgs doublets with the following $SU(3) \times SU(2) \times U(1)$ quantum numbers:

$$H_1 = \begin{pmatrix} H_1^0 \\ H_1^- \end{pmatrix} \in (1, 2, -1/2) \ , \ H_2 = \begin{pmatrix} H_2^+ \\ H_2^0 \end{pmatrix} \in (1, 2, +1/2) \tag{18}$$

From these eight real fields spontaneous symmetry breaking decouples three unphysical Goldstone bosons and one is left with five physical Higgs bosons, namely: two CP-even scalars, one CP-odd scalar and a pair of charged scalars. The tree level Higgs potential:

$$V = m_1^2|H_1|^2 + m_2^2|H_2|^2 - m_3^2(H_1 H_2 + h.c.) + \frac{g_1^2}{8}(H_1^+ \vec{\sigma} H_1 + H_2^+ \vec{\sigma} H_2)^2$$
$$+ \frac{g_2^2}{8}(|H_1|^2 - |H_2|^2)^2 \tag{19}$$

can be restricted to the real components of the neutral Higgs fields, $\phi_1 = Re\ H_1^0$, $\phi_2 = Re\ H_2^0$:

$$V_{tree} = m_1^2\phi_1^2 + m_2^2\phi_2^2 - m_3^2\phi_1\phi_2 + \frac{g_1^2 + g_2^2}{8}(\phi_1^2 - \phi_2^2)^2 \tag{20}$$

One can always choose such a field basis that m_3^2, v_1, v_2 are real and positive. Constants m_1, m_2 and m_3 have to satisfy certain conditions. Requiring that the potential be bounded from below gives:

$$\frac{m_1^2 + m_2^2}{2} \geq m_3^2 \tag{21}$$

and the spontaneous symmetry breaking condition is:

$$m_3^4 \geq m_1^2 m_2^2 \tag{22}$$

Here v_1 and v_2 are proportional to vacuum expectation values of ϕ_1 and ϕ_2:

$$< \phi_1 > \equiv \frac{v_1}{\sqrt{2}}, < \phi_2 > \equiv \frac{v_2}{\sqrt{2}} \tag{23}$$

and $tan\beta$ is defined to be their ratio: $tan\beta \equiv v2/v1$. Here:

$$\sqrt{v_1^2 + v_2^2} = 246 \; GeV \tag{24}$$

to reproduce the measured values of gauge boson masses:

$$m_w^2 = \frac{g_1^2}{4}(v_1^2 + v_2^2) \; , \; m_z^2 = \frac{g_1^2 + g_2^2}{4}(v_1^2 + v_2^2) \tag{25}$$

Fields ϕ_1 and ϕ_2 couple to down and up type quarks respectively. For example, after the spontaneous symmetry breaking top quark gets the mass:

$$m_t^2 = h_t^2 < \phi_2 >^2 \tag{26}$$

However, if the supersymmetry is softly broken, its scalar superpartner "stop" will have the different mass:

$$m_{\tilde{t}}^2 = h_t^2 < \phi_2 >^2 + \mu^2 \tag{27}$$

where we consider only the case of common soft supersymmetry-breaking mass μ for $\tilde{t}_L$ and $\tilde{t}_R$ and vanishing off-diagonal elements of the 2×2 stop mass matrix. (The investigation of the effects of these terms is in progress.)

Finally, CP-even physical eigenstates h and H, with masses $m_h < m_H$ are obtained by diagonalizing the mass matrix:

$$\mathcal{M} \equiv \frac{1}{2} \left(\frac{\partial^2 V_{tree}}{\partial \phi_i \partial \phi_j} \right)_{min} = \begin{pmatrix} A & C \\ C & B \end{pmatrix} \tag{28}$$

and their masses are given by:

$$m_{h,H}^2 = \frac{1}{2}(A + B \mp \sqrt{(A-B)^2 + 4C^2}) \tag{29}$$

All this was at the tree level, but, as was already mentioned, in order to obtain limits on the Higgs mass it is paramount to include one -loop corrections. These have been extensively studied in literature [17] , [18]. In the effective potential approach used in ref. [17], masses of the Higgs bosons are approximated with the eigenvalues of the matrix of second derivatives of the one-loop effective potential evaluated at its minimum.

The one loop effective potential at zero temperature is given by the expression:

$$V^0 = V_{tree}(Q) + \frac{1}{64\pi^2} Str \left\{ \mathcal{M}^4(\phi) \left[\ln \frac{\mathcal{M}^2(\phi)}{Q^2} - \frac{3}{2} \right] \right\} \tag{30}$$

Here $\mathcal{M}^2(\phi)$ is the field dependent squared mass matrix, Q is the renormalization scale and the supertrace is given by:

$$Str\, f\left(\mathcal{M}^2\right) = \sum_i (-1)^{2J_i} g_i f\left(m_i^2\right) \tag{31}$$

The sum runs over all the physical particles i of spin J_i, field dependent mass eigenvalue m_i and multiplicity g_i that couple to fields ϕ_1 and ϕ_2. In our case the most important ones are W and Z bosons, top and stop quarks and h and H bosons with multiplicities:

$$g_w = 6\ ,\ g_z = 3\ ,\ g_t = g_{\tilde{t}} = 12\ ,\ g_h = g_H = 1 \tag{32}$$

We have neglected the contributions due to other quark-squark flavors. This is justifiable insofar as their masses are small. As was pointed out in ref. [17], also the bottom-sbottom contributions can be non negligible for very large values of $tan\beta$. This case will not be relevant for us.

Next we have to add the nonzero temperature contributions (10) like in the case of the SM. By substituting field dependent values of m_w, m_z, m_t, $m_{\tilde{t}}$, m_h and m_H from equations (25) - (29) into the expressions for the effective potential (30) and (10) one obtains the full zero and nonzero temperature one-loop effective potentials. Similarly to the case of the SM, the critical temperature in this system is close to the point where the temperature dependent effective mass matrix has a zero eigenvalue.

There is a way [19] to get constraints similar to the ones in the SM.[5]

If we take m_t and μ to be our input parameters we have eight unknowns: 3 SUSY parameters m_1, m_2 and m_3; 2 zero-temperature VEV's $<\phi_1>$ and $<\phi_2>$, 2 nonzero-temperature VEV's $<\phi_1>_T$ and $<\phi_2>_T$ and the critical temperature T_c.

We have seven conditions: first the fixed magnitude of the zero T VEV (23) and (24), then 2 zero-T minima:

$$\frac{\partial V^0}{\partial \phi_1} = \frac{\partial V^0}{\partial \phi_2} = 0 \tag{33}$$

next 2 nonzero-T minima:

$$\frac{\partial V^T}{\partial \phi_1} = \frac{\partial V^T}{\partial \phi_2} = 0 \tag{34}$$

the critical temperature condition:

$$\det\left(\frac{\partial^2 V^T}{\partial \phi_i \partial \phi_j}\right)_{\phi_1=\phi_2=0, T_c} \approx 0 \tag{35}$$

and, finally, from (7) and (15), the condition that nonzero-T VEV be sufficiently large:

$$v(T_c) = \sqrt{v_1^2(T_c) + v_2^2(T_c)} \geq v_{crit}(T_c) = \frac{45 g_w T_c}{4\pi B(\lambda/g_w^2)} \tag{36}$$

Altogether, we have seven equations in eight unknowns, therefore we can impose one relation between them. This was done numerically in the form: $m_h = m_h(tan\beta)$ for different values of parameters m_t and μ. Here, m_h is the upper limit on the mass of the lighter CP-even Higgs field.

The argument λ^{eff}/g^2 in the function $B(\lambda^{eff}/g^2)$ was determined in ref. [21] for the general case of a two doublet model:

$$\lambda^{eff} = \lambda_1 \cos^4 \beta_T + \lambda_2 \sin^4 \beta_T + 2h \cos^2 \beta_T \sin^2 \beta_T \tag{37}$$

---

[5] See also ref. [20].

Table 1. Baryogenesis constraints on MSUSY

m_t (GeV)	m_h (GeV)	μ (GeV)
115	51	750
150	55	250
200	63	170

Here β_T is the nonzero-T "mixing angle of VEV's":

$$\tan \beta_T = \frac{v_2(T_c)}{v_1(T_c)} \tag{38}$$

In the MSUSY case:

$$\lambda_1 = \lambda_2 = \frac{g_1^2 + g_2^2}{4}, h = -\frac{g_1^2 + g_2^2}{4} \tag{39}$$

therefore:

$$\lambda^{eff} = \frac{g_1^2 + g_2^2}{4} \cos^2(2\beta_T) \tag{40}$$

Table 1. shows upper limits on the Higgs mass for $m_t = 115, 150$, and 200 GeV respectively. Points are obtained as maxima of curves $m_h(tan\beta)$ for different values of μ. All of the allowed values are found to lie in the region $1.1 < tan\beta < 1.7$. For the considered region of $tan\beta > 1$, there was always a maximal value of μ above which $v(T_c)$ was never big enough to satisfy requirement (36). This gives the upper limit on the soft-SUSY breaking scale of about 750 GeV, 250 GeV and 170 GeV for three top masses considered.

There are several causes of uncertainty in these calculations. First, as already mentioned, due to infrared divergences, we don't know the exact temperature of the phase transition in the SM. A lot of effort is taking place in this direction: [15] and [22]. Second, even if we assume that temperatures T_2, T_0, and T_c are known exactly, we still don't know the exact temperature of the phase transition in MSUSY. (In the SM, as mentioned earlier, this happens just bellow T_0 [12]). Third, we used the fact derived in [9] that the upper limit on the sphaleron mass in MSUSY is the sphaleron mass of the SM. It would be desirable to find the exact expression for the sphaleron mass of the MSUSY. Fourth, substituting temperature corrected parameters in the zero-temperature expression for the sphaleron mass is certainly just an approximation. Efforts to find the real nonzero temperature expression are under way [23].

Finally, as is usual in the study of phase transitions in early universe, we are using effective potential which is a static quantity for a system which not only evolves but evolves out of equilibrium.

CONCLUSION

In the SM, as mentioned before, upper limit on the m_h from baryogenesis is about 41 GeV, whereas the experimental lower limit from LEP is 57 GeV. Therefore EWPT as a source of B-asymmetry is excluded in the SM. In contrast, baryogenesis restricts MSUSY to have upper limit on m_h between 55 and 63 GeV, and the soft SUSY-breaking scale μ to be between 750 and 170 GeV, depending on m_t. This should be compared with experimental lower limits [24], on m_h of 41 GeV and on μ of 150 GeV. One can conclude that there is still an open window for baryogenesis from EWPT in MSUSY.

Even, if experiments exclude this region, there are still parts of parameter space (such as nonzero off-diagonal squark mass matrix elements) which require further study.

I would like to thank organizers of the 1992 Advanced Summer Institute in Quantitative Particle Physics for inviting me to the school and for giving me the opportunity to present this work. I particularly wish to thank professors Raymond Gastmans and Jean-Marc Gérard for their warmth and hospitality. This work was supported in part under NSF contract PHY-9057173, under DOE contracts DE-FG02-91ER40676, DE-AC02-89ER40509 and by funds from the Texas National Research Laboratory Commission under grant RGFY91B6.

References

[1] A. D. Sakharov, JETP Lett. 5:24 (1967)

[2] A. D. Linde, Phys. Lett. B70:306 (1977); S. Dimopoulos and L. Susskind, Phys. Rev. D18:4500 (1978)

[3] F. Klinkhamer and N. Manton, Phys. Rev. D30:2212 (1984)

[4] V. Kuzmin, V. Rubakov, and M. Shaposhnikov, Phys. Lett. B155:36 (1985)

[5] M. Fukugita and T. Yanagida, Phys. Lett. B174:45 (1986)

[6] M. E. Shaposhnikov, JETP Lett. 44:465 (1986), Nucl. Phys. B287:757 (1987), Nucl. Phys. B299:797 (1988)

[7] L. McLerran, Phys. Rev. Lett. 62:1075 (1989); A. I. Bochkarev, S. Yu. Khlebnikov, and M. E. Shaposhnikov, Nucl. Phys. B329:493 (1990); A. Nelson, D. Kaplan, and A. Cohen, Phys. Lett. B245:561 (1990); N. Turok and J. Zadrozny, Phys. Rev. Lett. 65:2331 (1990); M. Dine, P. Huet, R. Singleton, and L. Susskind, Phys. Lett. B263:86 (1991); R. L. Singleton, SLAC Report-380, June 1991; L. McLerran, M. Shaposhnikov, N. Turok, and M. Voloshin, Phys. Lett. B256:451 (1991); N. Turok and J. Zadrozny, Nucl. Phys. B358:471 (1991); A. Nelson, D. Kaplan, and A. Cohen, Phys. Lett. B263:86 (1991), Nucl. Phys. B349:727 (1991), Nucl. Phys. B373:453 (1992); M. Dine, P. Huet, and R. Singleton, Nucl. Phys. B375:625 (1992); N. Turok and J. Zadrozny, Nucl. Phys. B369:729 (1992); M. E. Shaposhnikov, Phys. Lett. B277:324 (1992), Erratum - ibid. B282:483 (1992);

[8] A. Cohen and A. Nelson, Boston University preprint BU-HEP-92-20

[9] A. I Bochkarev and M. E. Shaposhnikov, Mod. Phys. Lett. Vol.2, A6:417 (1987)

[10] E. W. Kolb and M. J. Turner, "The Early Universe", Addison Wesley (1990)

[11] G. 't Hooft, Phys. Rev. Lett. 37:8 (1976)

[12] G. Anderson and L. Hall, Phys. Rev. D45:2685 (1992)

[13] L. Dolan and R. Jackiw, Phys. Rev. D9:3320 (1974)

[14] M. Gleiser and E. W. Kolb, Fermilab preprint FERMILAB-Pub-91/305-A; N. Tetradis, DESY preprint DESY 91-151

[15] M. Dine, R. G. Leigh, P. Huet, A. Linde, and D. Linde, Phys. Lett. B283:319 (1992), Phys. Rev. D46:550 (1992); M. Dine, Santa Cruz preprint SCIPP 92/21

[16] F. Zwirner, "Lectures on phenomenological supersymmetry", ICTP Summer School 1991

[17] J. Ellis, G. Ridolfi, and F. Zwirner, Phys. Lett. B 257:83 (1991), Phys. Lett. B262:477 (1991); F. Zwirner, CERN preprint CERN-TH. 6099/91

[18] Y. Okada, M. Yamaguchi, and T. Yanagida, Prog. Theor. Phys. 85:1 (1991), Phys. Lett. B262:54 (1991); R. Barbieri, M. Frigeni, and F. Caravaglios, Phys. Lett. B258:167 (1991); R. Barbieri and M. Frigeni, Phys. Lett. B 258:395 (1991); H. Haber and R. Hempfling, Phys. Rev. Lett. 66:1815 (1991); A. Yamada, Phys. Lett. B263:233 (1991); P. Chankowski, S. Pokorski, and J. Rosiek, Phys. Lett. B274:191 (1991); A. Brignole, Phys. Lett. B271:123 (1991), Phys. Lett. B277:313 (1992), Phys. Lett. B281:284 (1992); D. Pierce, A. Papadopoulos, and S. Johnson, Phys. Rev. Lett. 68:3678 (1992)

[19] S. Myint, Phys. Lett. B287:325 (1992)

[20] O. Bertolami, Phys. Lett. B176:416 (1986); G. F. Giudice, Phys. Rev. D45:3177 (1992)

[21] A. Bochkarev, S. Kuzmin, and M. Shaposhnikov, Phys. Rev. D43:369 (1991), Phys. Lett. B244:275 (1990)

[22] P. Fendley, Phys. Lett. B196:175 (1987); D. E. Brahm and S. D. Hsu, CALTECH preprints CALT-68-1705,CALT-68-1762; C. G. Boyd, D. E. Brahm, and S. D. Hsu, CALTECH preprint CALT-68-1795; M. E. Carrington, Phys. Rev. D45:2933 (1992); P. Arnold, Univ. of Washington preprint UW/PT-92-06; J. R. Espinosa, M. Quirós, and F. Zwirner, CERN preprint CERN-TH.6451/92; M. Gleiser and E. Kolb, Fermilab preprint FERMILAB-Pub-92/222-A

[23] V. V. Khoze, MIT preprint CTP# 2103

[24] ALEPH Collaboration, CERN preprints PPE/91-111, CERN-PPE/91-149

CARBON 60

T. D. Lee

Columbia University, New York, N.Y. 10027

1. INTRODUCTION

The discovery of C_{60} and the superconductivity of its alloy compounds[1-4] provide a rich field for new theoretical and experimental investigations.[5-8] In this paper, we show that the near degeneracy of the low-lying vector (T_{1u}) and axial-vector (T_{1g}) levels of an isolated C_{60}^- may provide a simple pairing mechanism in C_{60}^{--}, C_{60}^{---} and the K_nC_{60} crystal.

We will begin with a review of the single-electron energy levels of a C_{60} molecule. Above the 60-closed shell lie two triplets possessing the same icosahedral symmetry label T_1 but of opposite parity. We shall derive their wave functions in the tight-binding limit of the molecular orbit approximation, and exhibit these functions in a simple and useful form which prompts us to refer to them as "vector" and "axial vector". Our analysis is then extended to C_{60}^{--}. By taking into account the strong Coulomb energy between electrons, we find that the spectrum of C_{60}^{--} contains two low-lying states invariant under the proper icosahedral symmetry transformations, but again of opposite parity and hence naturally called "scalar" and "pseudoscalar", with an energy degeneracy within 1% of their excitation energy (in our approximate calculation), even closer than the vector axial-vector doublets in C_{60}^-.

Since parity doublets, two states of opposite parity with nearly the same energy, are not a common occurrence in physics, it is worthwhile to examine the mechanism that gives rise to them in C_{60}. For a microscopic system, such as mesons and baryons, composite particles of opposite parities usually have quite different internal structures which make them unlikely to be degenerate. In most macroscopic systems, as in the case of left-handed or right-handed sugar molecules, it is conceptually simple to contstruct coherent mixtures of opposite parities that are nearly degenerate; the difficulty lies in the practical realization of such coherent mixtures.

Quantitative Particle Physics, Edited by
M. Lévy *et al.*, Plenum Press, New York, 1993

2. ICOSAHEDRAL GROUP

In order to have the necessary tools for our subsequent analysis, we give a brief discussion of the molecular orbits of C_{60}. As will be shown below, these can be expressed in terms of the regular representation of the proper

$$\text{icosahedral group } \mathcal{G} = \{g\} \tag{1}$$

where g denotes its sixty group elements. Corresponding to the identity element e, we may take any point $\hat{e}$ on a unit sphere with center 0. The application of the finite rotation that associates with g transforms $\hat{e}$ to $\hat{g}$. For $\hat{e}$ not lying on any of the symmetry axes, the resulting sixty $\hat{g}$ are all different. (A convenient notation is to use the same $\hat{e}$ and $\hat{g}$ to denote the corresponding unit radial vectors of the sphere, as well as their end points on the sphere.) For definiteness, let the unit sphere be the *circumsphere* of a regular icosahedron of vertices $N, A_1, \cdots, A'_5, S$. As in Figure 1, choose the point e (corresponding to the identity element e in $\mathcal{G}$) on the edge $\overline{A_1 N}$, with $\overline{eN} = \frac{1}{3}\overline{A_1 N}$ and $\vec{Oe} \| \hat{e}$. Each unit radial vector $\hat{g}$ intersects a point g on one of the edges of the icosahedron. The location of each point g, thus generated, gives the position of a carbon nucleus in C_{60}, provided the unit scale of the radius is $R \cong 4.03\text{Å}$, so that the circumspherical radius of C_{60} is its actual value $R_0 \cong 3.5\text{Å}$, with the ratio

$$\frac{R}{R_0} = 3\left(5 + \frac{4}{\sqrt{5}}\right)^{-\frac{1}{2}} \cong 1.15. \tag{2}$$

(The same notation g denotes both the group element as well as the position of the carbon nucleus.) Any function $f(g)$ of these sixty positions g can be expressed in terms of the irreducible representations of the icosahedral group $\mathcal{G}$, which consist of a singlet **1**, two triplets **3** and $\tilde{\mathbf{3}}$, a quartet **4** and a quintet **5**. This gives

$$60 = 1^2 + 3^2 + 3^2 + 4^2 + 5^2. \tag{3}$$

In the literature[9,10], these representations are often referred to as A, T_1, T_2, G and H respectively. Our notations follow those used in particle physics, with the dimensionality of the irreducible representation shown explicitly.

The irreducible representations **1**, **3** and **5** can be readily identified with the usual $\ell = 0$ (*s*-wave), $\ell = 1$ (*p*-wave) and $\ell = 2$ (*d*-wave) of the spherical harmonics $Y_{\ell,m}(\theta,\phi)$, where $m = -\ell, -\ell+1, \cdots, \ell$, and θ, ϕ are the polar and azimuthal angles. To derive the remaining two irreducible representations $\tilde{\mathbf{3}}$ and **4**, we designate one of the icosahedral vertices as the north pole N ($\theta = 0$), as in Figure 1. Express any function F of θ, ϕ defined on the twelve vertices of the icosahedron in terms of the sixteen spherical harmonics: $\ell = 0, 1, 2$ and 3 of $Y_{\ell,m}(\theta,\phi)$. (Note that $16 = 1 + 3 + 5 + 7$.) On the other hand, the function F has only

$$12 = 1 + 3 + 5 + 3 \tag{4}$$

values. Hence, we expect four linear combinations of the seven $\ell = 3$ $Y_{\ell,m}(\theta,\phi)$ to be identically zero on these twelve icosahedral vertices, leaving a triplet $\tilde{\mathbf{3}}$. For an explicit

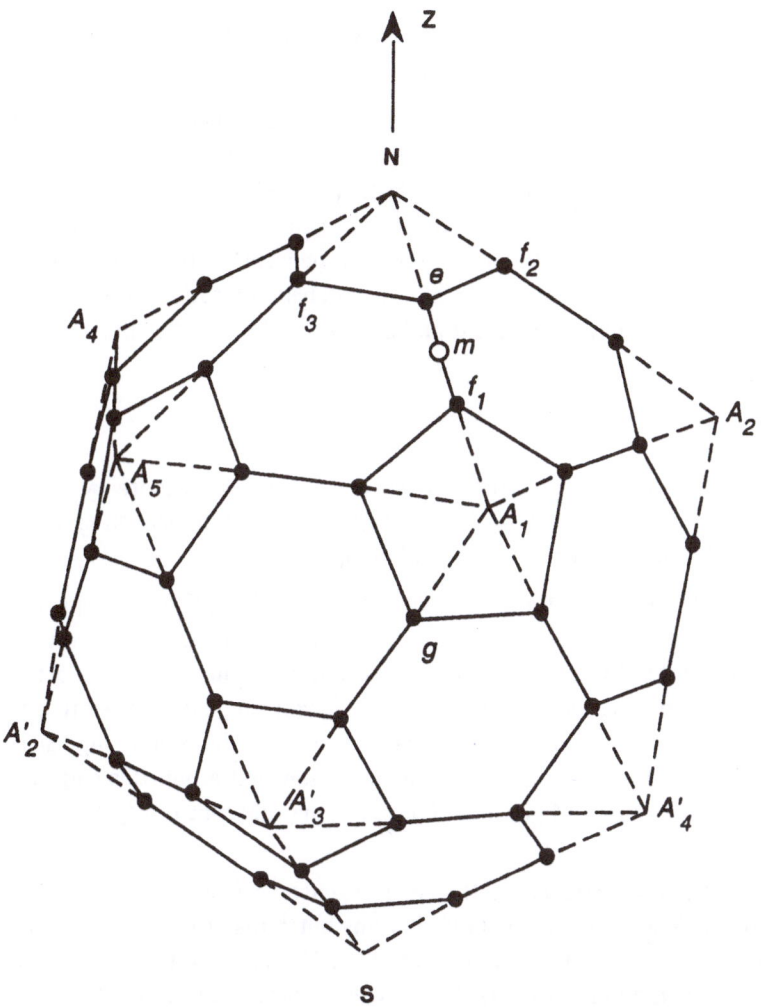

Figure 1. Set on a unit sphere the 12 vertices of an icosahedron: N, S, A_i and A_i' (i = 1, 2, ·, 5) with $\overline{NS}$, $\overline{A_i A_i'}$ the diameters of the sphere whose center is 0. Each of the sixty dots $e, f_1, f_2, f_3, g, \cdots$ denote the positions of C atoms in C_{60}. The three nearest neighbors of e are f_1, f_2 and f_3. The distance $\overline{ef_2} = \overline{ef_3}$ is slightly different from $\overline{ef_1}$.

construction, we observe that the geodesic arc between two nearest neighboring vertices of an icosahedron is $\cos^{-1}(1/\sqrt{5})$; therefore,

$$Y_{3,\pm 1}(\theta,\phi) \propto (5\cos^2\theta - 1)\sin\theta\, e^{\pm i\phi} \tag{5}$$

is zero on all these twelve vertices. Likewise, on these sites

$$Y_{3,\pm 3}(\theta,\phi) \propto Y_{3,\mp 2}(\theta,\phi), \tag{6}$$

from which we can readily form two linear combinations of these $\ell = 3$ spherical harmonics that are also zero. In (6), both sides have the same azimuthal variation because of the five-fold symmetry along the north pole-south pole diameter; hence, (6) holds in the northern hemisphere, since excluding the pole, all the other five icosahedral vertices are at the same latitude. Its validity in the southern hemisphere then follows from inversion symmetry. In this way, we see that under the icosahedral rotations, the seven $Y_{3,m}(\theta,\phi)$ functions decompose into a quartet $\mathbf{4}$ and another triplet $\bar{\mathbf{3}}$, different from $\mathbf{3}$.

3. C_{60}

Returning to the C_{60} structure on the unit sphere, we denote by $|g>$ a vector (*ket*-vector in Dirac's notation) with sixty components. Each element g corresponds to a definite finite rotation of the icosahedral group.

The radial position vector $\hat{g}$ of each C atom on the sphere corresponds to an element g of the proper icosahedral group. Thus the 60-dimensional Hilbert space of the tight-binding limit supports the regular representation of this group. For a typical "hopping" Hamiltonian of icosahedral symmetry, the lowest single-particle state is that having constant wave function, reminiscent of the s-wave (i.e., the orbital angular momentum quantum number $\ell = 0$) on a sphere. In the icosahedral group this singlet representation $\mathbf{1}^+$ is usually labeled A, or more fully A_g (the subscript $g = gerade$ meaning even parity).

The next-lowest state has a threefold degeneracy (excluding spin) which may be described as follows. For each C atom it turns out[11] that there is a certain unit vector $\hat{e}_0(g)$, very close to $\hat{g}$, within an angle of about $1°$; the wave function at site $\hat{g}$ can be chosen to be proportional to $\hat{e}_0(g) \cdot \hat{n}$ where $\hat{n}$ is some fixed vector. Clearly one obtains in this way three independent wave functions of the same energy, corresponding to the three independent choices of $\hat{n}$. If we approximate $\hat{e}_0(g) \cong \hat{g}$ (as we shall do hereafter in this discussion), then these three states correspond to the p-wave triplet ($\ell = 1$) on the sphere. Under the icosahedral group this triplet representation $\mathbf{3}^-$ is usually labeled T_1, or more fully T_{1u} ($u = ungerade$ meaning odd parity, since $\hat{g}$ changes sign for oppositely placed C atoms).

It is natural to call the singlet and triplet states described above "scalar" and "vector" referring to the way they transform when the laboratory coordinate system is rotated or inverted. Thus, the designation "vector" refers to the vector $\hat{n}$ which selects a T_1 state. Since multiplicity and parity play a prominent role in our analysis, we shall write $\mathbf{1}^+$(scalar) for A_g, $\mathbf{3}^-$(vector) for T_{1u}, etc., with $\mathbf{1}^+$ denoting a singlet of even parity, $\mathbf{3}^-$ a triplet of odd parity, etc. (All these multiplicities are doubled by spin.)

In the regular representation, each irreducible representation of dimension d appears d times; thus, the triplet T_1 must appear three times. Our interest will be in the *other* two triplets T_1, whose energy lies far above that of the "p-wave" state described above. Their wave functions can be expressed simply as follows:

Because of the detailed structure of the molecule, each C atom defines not only a radial vector $\hat{g}$ but a local coordinate system, with its three orthonormal basis vectors denoted by

$$\hat{g}, \qquad \hat{e}_-(g) \qquad \text{and} \qquad \hat{e}_+(g), \tag{7}$$

where

$$\hat{e}_+(g) = \hat{g} \times \hat{e}_-(g). \tag{8}$$

Under an inversion, $\hat{g}$ changes sign; consequently, $\hat{e}_-(g)$ and $\hat{e}_+(g)$ are of opposite parities. Let $\hat{e}_-(g)$ be of odd parity, then $\hat{e}_+(g)$ is of even parity. By taking the wave function proportional to $\hat{e}_-(g) \cdot \hat{n}$ (where, as before, $\hat{n}$ is some fixed vector) one obtains a triplet $\mathbf{3}^-$ or T_{1u}, the LUMO (lowest unoccupied molecular orbital) state in neutral C_{60}. By taking $\hat{e}_+(g) \cdot \hat{n}$, one obtains a $\mathbf{3}^+$ or T_{1g}, the second LUMO state in neutral C_{60}. It is natural to call these triplets "vector" and "axial-vector" respectively. Together, (7) determines the set of three triplets T_1 contained in the regular representation.

To transform a C atom to one of its nearest neighbors by an element of the icosahedral group, one must rotate through a large angle ($180°$ for one of the nearest neighbors, from e to f_1 in Figure 1), about an axis rather close to $\hat{g}$. Hence $\hat{g}$ varies slowly between neighbors but, as we shall see, $\hat{e}_-(g)$ and $\hat{e}_+(g)$ vary rapidly and, because of (8), by about the same amount. This is why the $\hat{e}_- \cdot \hat{n}$ and $\hat{e}_+ \cdot \hat{n}$ triplets have energy far above that of the $\hat{g} \cdot \hat{n}$ triplet, but close to each other. (Although the high-lying $\mathbf{3}^-$ and $\mathbf{3}^+$ states have the same formal structure under the *icosahedral group* as the low-lying $\mathbf{3}^-$, if one attempts to fill in a smooth wave function between the C atoms, one will need mostly spherical harmonics of $\ell = 5$ and $\ell = 6$ for $\hat{e}_-$ and $\hat{e}_+$ respectively, instead of $\ell = 1$ as for $\hat{g}$.)

Since $\hat{e}_- \cdot \hat{n}$ and $\hat{e}_+ \cdot \hat{n}$ are the two low-lying levels above the 60-closed shell, in C_{60}^- the extra electron may be in either of these triplets. The nearness of these two levels of opposite parity, combined with their compatibility (both T_1) under the icosahedral group, render C_{60}^- highly polarizable; the polarizability can be readily calculated.

4. PAIRING MECHANISM

In C_{60}^{--}, the Coulomb energy between the two extra electrons plays a dominant role. Without this term, the lowest-energy state would be of the form $T_{1u}T_{1u}$ or $\mathbf{3}^- \times \mathbf{3}^-$. But we find that two other states have considerably lower Coulomb energy. One is obtained by combining T_{1u} and T_{1g} to make a two-particle state $\mathbf{1}^-$:

$T_{1u}T_{1g} + T_{1g}T_{1u}$ with wave function

$$[\hat{e}_-(g) \cdot \hat{e}_+(g') + \hat{e}_+(g) \cdot \hat{e}_-(g')](\uparrow\downarrow' - \downarrow\uparrow'). \tag{9}$$

The other is made by mixing the 1^+ combination of $T_{1u}T_{1u}$ with that of $T_{1g}T_{1g}$:

$T_{1u}T_{1u} + T_{1g}T_{1g}$ with wave function

$$[\hat{e}_-(g) \cdot \hat{e}_-(g') - \hat{e}_+(g) \cdot \hat{e}_+(g')](\uparrow\downarrow' - \downarrow\uparrow'), \qquad (10)$$

where $\hat{g}$, $\uparrow$ or $\downarrow$ and $\hat{g}'$, $\uparrow'$ or $\downarrow'$ are the position and spin variables of the two electrons. Because both wave functions vanish when the two electrons coalesce, $\hat{g} = \hat{g}'$, their mutual Coulomb energy is greatly reduced; this leads to the spin-0 parity doublets in C_{60}^{--}. labeled 1^- and 1^+.

The wave functions (9) and (10) represent singlets in both position and spin; no external vector $\hat{n}$ appears. It is natural to call these paired states pseudoscalar and scalar. It is evident that if the energy difference $\Delta\epsilon$ between 3^- (T_{1u}) and 3^+ (T_{1g}) is small, then it will contribute only to order $(\Delta\epsilon)^2$ to the splitting between scalar and pseudoscalar paired states. This is why the paired states in C_{60}^{--} form a much tighter parity doublet than the component one-particle states in C_{60}^-.

The closeness of these parity doublets makes it natural to isolate their response to strong interactions, separate from other distant levels. This provides a convenient means to derive analytical expressions for many of the important parts of strong interaction effects. Thus, we can calculate the polarization energy of C_{60}^- in a strong electric field E, showing that it changes from the weak field expression $-\frac{1}{2}\alpha E^2$ to one that depends linearly on E. Another example is to derive the final energy in C_{60}^{--}, within the parity doublet approximation, to all powers of the interaction Hamiltonian.

5. K₃C₆₀

We turn our attention to the K_3C_{60} crystal and calculate the Madelung energy. The result shows that all K are ionized, as is commonly accepted.

Above the 60-closed shell, a typical band calculation reveals a low-lying cluster of three overlapping narrow bands, which is the Bloch wave extension of the three components of the vector wave function $\hat{e}_-(g)$ of an isolated C_{60}^-. These overlapping fermion bands are half-filled in the case of K_3C_{60}. On the other hand, the pseudoscalar pairing wave function (9) in C_{60}^{--} suggests a different Bloch wave extension, one that represents the hopping of such a highly correlated two-particle state in the crystal. The orthogonality of these correlated two-particle Bloch wave functions to any product of two one-particle wave functions in the fermion bands follows from the original orthogonality condition in a single C_{60} molecule:

$$\sum_{g=1}^{60} \hat{e}_-(g)_i \, \hat{e}_+(g)_j = 0 \qquad (11)$$

where i and j denote the vector components. We call the Bloch extension of (9) the "boson band" (or the pseudoscalar band).

The two electrons in the fermion bands have an energy advantage over the boson on account of the excitation energy $\Delta\epsilon$ and the lowering in kinetic energy; however, there is a disadvantage to the fermion bands of having a higher Coulomb energy. The important question concerning the role of bosons versus fermions depends on the delicate balance between these two opposing factors. This is examined in Ref. 11. We start with the Coulomb energy ($\sim$ 11eV per lattice cell) between three electrons in the fermion bands, and compare that with the configuration of placing two electrons in the boson band and the remaining one in the fermion bands. We then take into account the kinetic energy difference and the Van der Waals energy difference; the latter is important because of the large polarizability of the C_{60} negative ion. The final energy balance between these two configurations is estimated to be only 0.22eV, slightly in favor of the fermion bands. However, considering the highly approximate nature of our calculations and that the final answer is only about 2% of the initial energy that we begin with, it is not possible to make any definitive statement. Nevertheless, a variety of interesting theoretical possibilities emerges. It seems likely that there exists a narrow boson band, which may lie very close to, or overlapping with, the three fermion bands.

The possibility of a close-by low-lying boson band, in addition to the usual fermion bands, has important consequences for superconductivity. The bosons may undergo Bose-Einstein condensation. The zero momentum nature of the Bose condensate necessarily generates charge fluctuations in the coordinate space; thereby it increases the Coulomb energy. It can be shown that this Coulomb energy increase is compensated for, at near distances, by the monopole-dipole interaction between neighboring C_{60} molecules because of their large polarizability. At large distances there is, in addition, the Debye screening of the Coulomb potential generated by these charge fluctuations.

In Ref. 11, through a simplified but explicit field theoretic model, we are able to examine the effect of the Bloch extension of the correlated scalar wave function (10), the parity doublet partner of the pseudoscalar (9). Except in the somewhat unlikely case that the boson band is lower in energy than two times the bottom energy of the fermion bands, the scalar only provides a resonance to the electrons in the fermion bands. Such a resonance can produce an energy gap, as in the BCS theory of superconductivity. While both members of the bosonic parity doublet can be important to superconductivity, their roles are different. The experimentally observed pressure variation of the critical temperature[6] can be shown to be consistent with the model.

Our thesis is that the 3^+ (T_{1g}) level plays a special role because of its coupling and near degeneracy with the 3^- (T_{1u}) at the Fermi level. Therefore the C_{60}^{--} energy should be calculated by starting with the mixed states of (9) and (10), and treating all levels *outside* the parity doublet perturbatively. This makes it possible to handle strong interactions, and leads to high T_c superconductivity.

The parity doublets provide an essential insight into the nature of the paired wavefunction in C_{60}, as well as the physics associated with polarizability of the C_{60} negative ion. The superconductivity is discussed on the basis of a simplified model Hamiltonian in which there is an effective "attractive and local" four-fermion field interaction term. Its presence is due to the lowering of the strong Coulomb energy in the correlated paired states (9) and (10) in C_{60}^{--} which, in turn, can be justified within the parity doublet approximation. On the other hand, parity doublets are rather special to C_{60}; a natural question is to ask how important are their roles to superconductivity in general. We believe that, for all high T_c superconductors (cupric oxides and C_{60} alloys) with very small coherence length, the essential common feature probably lies in the approximate applicability of an effective "attractive and local" interaction term. For material other than C_{60}, its underlying reason may have nothing to do with parity doublets. The parity doublets of C_{60} simply provide a convenient tool for us to penetrate the maze of strong interactions.

ACKNOWLEDGEMENTS

This research was supported in part by the U.S. Department of Energy.

REFERENCES

[1] H. W. Kroto et al., Nature **318**, 162 (1985); W. Krätschmer et al., Nature **347**, 354 (1990); H. Ajie et al., J.Phys.Chem. **94**, 8630 (1990).

[2] A.F. Hebard et al., Nature **350**, 600 (1991).

[3] M.J. Rosseinsky et al., Phys.Rev.Lett. **66**, 2830 (1991).

[4] K. Holczer et al., Science **252**, 1154 (1991).

[5] K. Holczer et al., Phys.Rev.Lett. **67**, 271 (1991).

[6] G. Sparn et al., Science **252**, 1829 (1991); R.M. Fleming et al., Nature **352**, 787 (1991); O. Zhou et al., Science **255**, 833 (1992).

[7] P.-M. Allemand et al., Science **253**, 301 (1991).

[8] S. Chakravarty, M.P. Gelfand and S. Kivelson, Science **254**, 970 (1991), and UCLA preprint "Electronic Correlation Effects and Superconductivity in Doped Fullerenes".

[9] M. Hamermesh, *Group Theory and Its Application to Physical Problems* (Addison-Wesley Publishing Co., 1962).

[10] F. Albert Cotton, *Chemical Applications of Group Theory*, 2nd edition (Wiley-Interscience, 1963).

[11] R. Friedberg, T. D. Lee and H. C. Ren, Columbia University preprint "Parity Doublets and the Pairing Mechanism in C_{60}", to be published in Physical Review B.

INDEX

Asymmetry
  baryon number-, 414
  CP-, 163, 188, 351
  forward- backward-, 48, 57, 74, 400
  polarization-, 74

Bardeen equation, 119
Bianchi models, 365
Bloch extension, 430
B-mesons
  charmless decays of, 178
  CP-violation in, 187, 296
  exclusive decays of, 61, 287
  factory, 195
  inclusive J/Ψ decays of, 286
  leptonic decays of, 288
  1/2-leptonic decays of, 68, 175, 280
  lifetime of, 59, 60, 64, 176
  mixing in, 57, 182, 291
  production of, 279
B_s-mesons
  search for, 62
Brans-Dicke theory, 362, 364, 378

Cabibbo angle, 174
Callan-Gross relation, 232
Cartan-Maurer one form, 109
Cerenkov light, 307
Coleman-Weinberg potential, 379
Conservation
  of baryon number, 4
  of CPT, 14, 160
  of lepton number, 4
Corrections
  electromagnetic, 74
  electroweak, 82
  oblique, 87, 399
Crab Nebula, 312
Cygnus-X-3, 317

Dark Matter, 349, 386
Deep inelastic scattering, 228

η-mesons
  decays of, 131
  mixing in, 131, 168

Fermi theory, 3
Fermi constant, 73

Friedmann equation, 348, 367

Gell-Mann-Oakes-Renner relation, 114
Gell-Mann-Okubo mass relation, 114, 168
GIM mechanism, 174, 410
Goldstone theorem, 107

Higgs mass, 52, 81, 98, 408, 418, 421
't Hooft consistency conditions, 105
Hubble constant, 341, 347
Hyperon decays, 8

Irreversibility
  macroscopic, 14, 149
  microscopic, 14, 149

Jeans instability, 354
Josephson effect, 72

K-mesons
  CP-violation in, 185
  hadronic decays of, 132, 134
  mixing in, 133
  1/2-leptonic deays of, 125
  rare decays of, 135
KLN theorem, 17
Kobayashi-Maskawa mixing matrix, 111, 158, 173, 184
KSFR relation, 121

Leptoquarks, 240

Michel parameter, 5
MSW oscillations, 207

Nambu-Goldstone realization, 105, 108
Nambu-Jona-Lasinio model, 122
Neutrino
  and astronomy, 320
  and astrophysics, 321
  beam of, 54
  mass of, 304, 345
  number of, 52, 344
  of 17 Kev, 304
  solar, 306
Non-conservation
  of charge conjugation, 12, 150, 154, 414
  of parity, 9, 11, 150 ,154
  of time-reversal, 14, 150, 154, 414
Noether currents, 104, 113

Operator product expansion, 131

Peccei-Quinn symmetry, 363
Penguin modes, 183
Photoproduction, 243
Planck mass, 350

Quark
    confinement, 19
    gluon plasma, 22, 352
    mass, 128, 304

Renormalization, 76
Robertson-Walker metric, 347

Sachs-Wolfe effect, 355, 371, 375
Schwinger-Dyson equation, 408
Small x, 232
SN 1987 A, 321
Strong CP violation, 128, 165
Supersymmetry, 89, 418

Technicolour, 93, 409
Temperature anisotropy, 342

τ-leptons
    lifetime of, 66
    polarization of, 48
θ–τ puzzle, 7
Top quark
    condensate, 408
    mass of, 54, 81, 88, 260, 275, 408, 421

Universality, 66

Vector meson dominance, 94, 121, 243

W-boson
    mass of, 73, 278
Weinberg scattering amplitude, 114, 123
Weinberg theorem, 102, 112, 117
Wess-Zumino-Witten action, 119, 139
Wigner-Weyl realization, 105, 108
Wilson coefficients, 131
WIMPS, 330
Wolfenstein parametrization, 159, 174

Z-boson
    mass of, 47, 73
    width of, 47

434